Benjamin Fine, Anja Moldenhauer, Gerhard Rosenberger, Annika Schürenberg, Leonard Wienke

Topics in Infinite Group Theory

Also of Interest

Abstract Algebra. With Applications to Galois Theory, Algebraic Geometry, Representation Theory and Cryptography
Gerhard Rosenberger, Annika Schürenberg, Leonard Wienke, 2024
ISBN 978-3-11-113951-7, e-ISBN (PDF) 978-3-11-114252-4,
e-ISBN (EPUB) 978-3-11-114284-5

Algebra and Number Theory. A Selection of Highlights
Benjamin Fine, Anja Moldenhauer, Gerhard Rosenberger,
Annika Schürenberg, Leonard Wienke, 2023
ISBN 978-3-11-078998-0, e-ISBN (PDF) 978-3-11-079028-3,
e-ISBN (EPUB) 978-3-11-079039-9

Geometry and Discrete Mathematics. A Selection of Highlights
Benjamin Fine, Anja Moldenhauer, Gerhard Rosenberger,
Annika Schürenberg, Leonard Wienke, 2022
ISBN 978-3-11-074077-6, e-ISBN (PDF) 978-3-11-074078-3,
e-ISBN (EPUB) 978-3-11-074093-6

Elements of Discrete Mathematics. Numbers and Counting, Groups, Graphs, Orders and Lattices
Volker Diekert, Manfred Kufleitner, Gerhard Rosenberger, Ulrich Hertrampf, 2023
ISBN 978-3-11-106069-9, e-ISBN (PDF) 978-3-11-106255-6,
e-ISBN (EPUB) 978-3-11-106288-4

Discrete Algebraic Methods. Arithmetic, Cryptography, Automata and Groups
Volker Diekert, Manfred Kufleitner, Gerhard Rosenberger, Ulrich Hertrampf, 2016
ISBN 978-3-11-041332-8, e-ISBN (PDF) 978-3-11-041333-5,
e-ISBN (EPUB) 978-3-11-041632-9

Benjamin Fine, Anja Moldenhauer,
Gerhard Rosenberger, Annika Schürenberg,
Leonard Wienke

Topics in Infinite Group Theory

Nielsen Methods, Covering Spaces, and Hyperbolic Groups

2nd edition

DE GRUYTER

Mathematics Subject Classification 2020
Primary: 08A50, 20EXX, 20FXX, 20H10, 57MXX; Secondary: 11AXX, 11B37, 11B39, 11D09, 11D25

Authors
Prof. Dr. Benjamin Fine†

Dr. Anja Moldenhauer
20535 Hamburg
Germany

Prof. Dr. Gerhard Rosenberger
University of Hamburg
Department of Mathematics
Bundesstr. 55
20146 Hamburg
Germany

Annika Schürenberg
Grundschule Hoheluft
Wrangelstr. 5
20253 Hamburg
Germany

Dr. Leonard Wienke
University of Bremen
Department of Mathematics
Bibliothekstr. 5
28359 Bremen
Germany

ISBN 978-3-11-133956-6
e-ISBN (PDF) 978-3-11-134004-3
e-ISBN (EPUB) 978-3-11-134018-0

Library of Congress Control Number: 2024942154

Bibliographic information published by the Deutsche Nationalbibliothek
The Deutsche Nationalbibliothek lists this publication in the Deutsche Nationalbibliografie; detailed bibliographic data are available on the Internet at http://dnb.dnb.de.

Cover image: kvkirillov / iStock / Getty Images Plus
Typesetting: VTeX UAB, Lithuania

www.degruyter.com

Preface

The guiding principle of our book is to show how geometric group theory investigates the interaction between algebraic and geometric properties of groups. In this sense we put special emphasis on the possibility to prove theorems with algebraic, topological, and geometric methods. We also place great value on applications, in particular in number theory, and include recent results from research and open problems.

The first chapter deals with free constructions and one-relator groups. We give a thorough introduction to these topics and accompany our exposition with several applications. The broad cancellation method by Nielsen is running throughout our treatment. We briefly discuss the classification of the three-generator fully residually free groups (limit groups), the description of properties of words and subgroups and the solution of equations inside groups. In this second edition, we added Subsection 1.1.4 on universally free groups and Subsection 1.5.3 on arithmetic Fuchsian groups.

In the second chapter, we introduce the fundamental groups of topological spaces, especially CW complexes, and discuss applications of covering spaces in group theory. For instance, we present proofs of Kurosh's theorem for free products and the subgroup separability for certain free products with amalgamation.

The third chapter is dedicated to hyperbolic groups. Methodically, we describe geodesic, quasi isometric and hyperbolic metric spaces. We then proceed with the most important properties of hyperbolic groups. In the following, we deal in detail with more advanced examples of hyperbolic groups and closely link our exposition to recent research. We also discuss Hall's theorem and property S. We devote three sections to the introduction of quasi isometries and quasi geodesics as well as the boundary of a tree and the boundary of a hyperbolic space. We also give a short description of automatic groups. In this second edition, we include a final Section 3.11 on Stallings foldings.

In writing this book, we took the opportunity to especially include those topics that are less covered in the literature. On the other hand, there are certainly several other interesting subjects in combinatorial and geometric group theory that we could not include; however, we think that our selection gives a balanced and reasonable introduction to further, sometimes advanced, topics in infinite group theory.

The material of the book should be accessible to students in their third year. We only assume a basic knowledge of group theory and topology.

We would like to thank the people who were involved in the preparation of the manuscript. Their dedicated participation is gratefully acknowledged. In particular we have to mention Anja Rosenberger as well as the many students who have taken part in the respective courses. Unfortunately, our colleague and friend Ben Fine, who co-authored the first edition of this book, passed away during the preparation of this second edition.

A. Moldenhauer, G. Rosenberger, A. Schürenberg and L. Wienke

https://doi.org/10.1515/9783111340043-201

Contents

1 Nielsen methods

1.1 Free groups and group presentations

Combinatorial group theory is the part of group theory which is presented using generating sets and systems of defining relations. The main principle is the construction of free objects, here called free groups. Their universal property implies that each group is a factor group of a free group.

We start with some notations. Let G be a group. If H is a subgroup of G, then we write $H < G$, and if H is a normal subgroup of G, then we write $H \triangleleft G$. Let $X \subset G$. We denote the subgroup of G generated by X with $\langle X\rangle$ and call X a *generating system* of $\langle X\rangle$. We have

$$\langle X\rangle = \{x_1^{\epsilon_1}x_2^{\epsilon_2}\cdots x_n^{\epsilon_n} \mid x_i \in X, \epsilon_i = \pm 1\},$$

and we call a formal expression

$$w = x_1^{\epsilon_1}x_2^{\epsilon_2}\cdots x_n^{\epsilon_n}, \quad x_i \in X, \epsilon_i = \pm 1,$$

an *X-word*, a word in X or simply a word if X is fixed. A word is just a sequence of elements from $X \cup X^{-1}$ where $X^{-1} = \{x^{-1} \mid x \in X\}$. Each word w represents in G an element g:

$$w =_G g \quad \text{or just} \quad w = g.$$

The neutral element 1 is represented by the trivial or empty word. The inverse word of

$$w = x_1^{\epsilon_1}x_2^{\epsilon_2}\cdots x_n^{\epsilon_n}$$

is

$$x_n^{-\epsilon_n}\cdots x_2^{-\epsilon_2}x_1^{-\epsilon_1}.$$

A word $w = x_1^{\epsilon_1}x_2^{\epsilon_2}\cdots x_n^{\epsilon_n}$ is called *(freely) reduced* if $\epsilon_i + \epsilon_{i+1} \neq 0$ whenever $x_i = x_{i+1}$ (for $i = 1, 2, \ldots, n-1$). We also write w as

$$w = x_1^{\alpha_1}x_2^{\alpha_2}\cdots x_n^{\alpha_n}, \quad x_i \in X,\ x_i \neq x_{i+1},\ \alpha_i \in \mathbb{Z},\ \text{all } \alpha_i \neq 0.$$

Two words v and w are called *freely equivalent* if one can be transformed into the other by inserting and deleting *peaks* $x^{\epsilon}x^{-\epsilon}$, $x \in X$, $\epsilon = \pm 1$. This we also denote as $v \equiv w$. A group G is called an *α-generated group* if G can be generated by a set of cardinality α. In this sense, G is *finitely generated* if G can be generated by a finite set. The least number of elements needed to generate G is often called the *rank of* G and is denoted by $\operatorname{rk}(G)$. If $X \subset G$, then we write $\langle\langle X\rangle\rangle$ or $\langle X\rangle^N$ to denote the *normal closure* of X in G, that is,

https://doi.org/10.1515/9783111340043-001

$$\langle\langle X\rangle\rangle = \langle gxg^{-1} \mid x \in X, g \in G\rangle.$$

If $w \in G$ is a word in X then we often write

$$w = w(x_1, x_2, \ldots, x_n), \quad x_i \in X,$$

to indicate that w is a word in the letters $x_1, x_2, \ldots, x_n$ of X.

Definition 1.1.1. Let $X \neq \emptyset$ be a set, F a group and $i: X \to F$ an injective map. The group F, more concrete the pair (F, i), is called *free on* X, if it satisfies the following universal property: For each group H and each map $f: X \to H$ there exists a uniquely determined homomorphism, $\varphi: F \to H$ with $f = \varphi \circ i$, that is, the following diagram (see Figure 1.1) commutes.

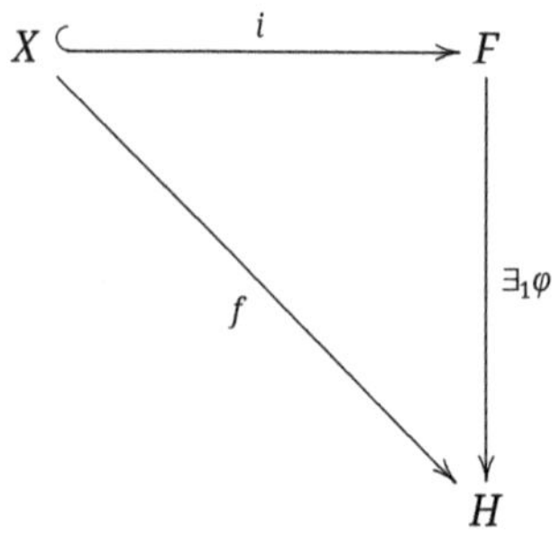

Figure 1.1: Commuting diagram.

In addition, we define the trivial group to be free on the empty set $\emptyset$. Most commonly we consider X as a subset of F with i as the inclusion. In this case we call X a *free generating system or basis* of F, we have $X = \emptyset$ if $F = \{1\}$.

Example 1.1.2. The infinite cyclic group $(\mathbb{Z}, +)$ is free on $\{1\}$.

Theorem 1.1.3. *Let (F_1, i_1) and (F_2, i_2) be free on X. Then there exists an isomorphism $\varphi: F_1 \to F_2$ with $\varphi \circ i_1 = i_2$, that is, F_1 is unique up to isomorphisms.*

Proof. Consider the diagram as in Figure 1.2. By definition we have uniquely determined homomorphisms $\varphi: F_1 \to F_2$ and $\psi: F_2 \to F_1$ such that $\varphi \circ i_1 = i_2$ and $\psi \circ i_2 = i_1$. Also by definition there is only one homomorphism $f: F_1 \to F_1$ with $i_1 = f \circ i_1$, hence $f = \psi \circ \varphi = \mathrm{id}_{F_1}$. Analogously $\varphi \circ \psi = \mathrm{id}_{F_2}$. □

We now discuss the existence of a free group on a given set X, that is, we construct a free group on X. Let $X \neq \emptyset$ be a set. Let $X^{-1} = \{x^{-1} \mid x \in X\}$ be another set with $X \cap X^{-1} = \emptyset$ and a bijection $x \mapsto x^{-1}$ from X to X^{-1}. Let $M(X)$ be the set of all finite sequences $(x_1, x_2, \ldots, x_n)$ with $x_i \in X \cup X^{-1}$, $n \geq 0$, where we have the empty sequence for $n = 0$. We define a multiplication on $M(X)$ in succession notation:

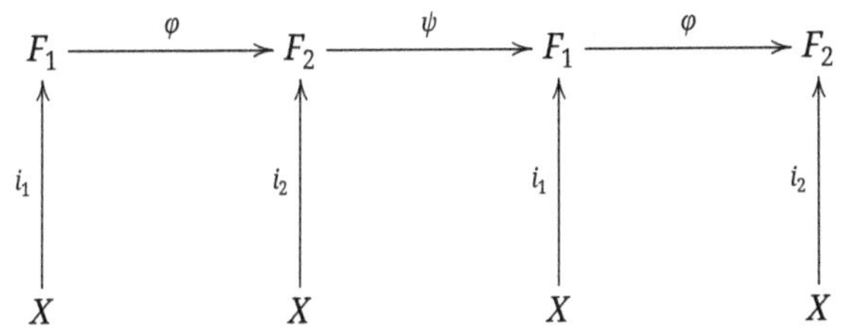

Figure 1.2: Isomorphism diagram.

$$(x_1,x_2,\dots,x_n)\cdot(y_1,y_2,\dots,y_m) := (x_1,x_2,\dots,x_n,y_1,y_2,\dots,y_m).$$

This multiplication is associative with neutral element 1, the empty sequence. The map

$$X\cup X^{-1}\to M(X),\quad x\mapsto (x),\quad x\in X\cup X^{-1},$$

is injective. We identify (x) with x. With this we may write each element $w\in M(X)$ uniquely as

$$w = x_1^{\epsilon_1}x_2^{\epsilon_2}\cdots x_n^{\epsilon_n},\quad \epsilon_i=\pm1,\ x_i\in X,$$

where we identify x_i^{+1} with x_i. The set $M(X)$ is called the *free monoid* on $X\cup X^{-1}$. We call the elements of $X\cup X^{-1}$ *letters.*

As above, the word $w = x_1^{\epsilon_1}x_2^{\epsilon_2}\cdots x_n^{\epsilon_n}$ is called *reduced* if $x_i\neq x_{i+1}$ or $x_i = x_{i+1}$ but $\epsilon_i+\epsilon_{i+1}\neq 0$ for $1\le i<n$. The empty word is also reduced. If a word $w\in M(X)$ is not reduced then we may delete a peak $x^{\epsilon}x^{-\epsilon}$, $\epsilon=\pm1$, to get a word $w'\in M(X)$ and we get w back if we insert $x^{\epsilon}x^{-\epsilon}$ again. This way we get an equivalence relation $\equiv$ on $M(X)$ via the definition: If $w,v\in M(X)$ then $w\equiv v$ if $w=v$ in $M(X)$ or w is *freely equivalent* to v, that is, we get v from w by deleting and inserting peaks. In fact, this is a congruence, that is, if $w\equiv w'$ then $uwv\equiv uw'v$ for all $u,v\in M(X)$, and if $w\equiv w'$, $u\equiv u'$ then $uw\equiv u'w'$. Certainly for each w there exists a reduced w' with $w\equiv w'$.

These facts are straightforward to show. Hence, the multiplication on $M(X)$ implies a multiplication on $F(X) := M(X)/{\equiv}$, the set of equivalence classes of $M(X)$ with respect to $\equiv$, via

$$[u][v] := [uv],$$

where $[w]$ denotes the class of w. This multiplication is associative by transmission to the quotient and has a neutral element 1, the class of the empty word. $F(X)$ is a group because if $w = x_1^{\epsilon_1}x_2^{\epsilon_2}\cdots x_n^{\epsilon_n}$ and $w' = x_n^{-\epsilon_n}x_{n-1}^{-\epsilon_{n-1}}\cdots x_1^{-\epsilon_1}$ then $ww'\equiv 1$. The inclusion $i\colon X\hookrightarrow M(X)$ induces an inclusion $j\colon X\hookrightarrow F(X)$, and $F(X)=\langle j(X)\rangle$. If we consider, as usual, X as a subset of $F(X)$, then $F(X)=\langle X\rangle$.

One also may deduce this from the calculation in [51].

Theorem 1.1.4. *The group $F(X)$ is free on X.*

Proof. Let G be a group and $f: X \to G$ a map. We extend this map via

$$x_1^{\epsilon_1} x_2^{\epsilon_2} \cdots x_n^{\epsilon_n} \mapsto (f(x_1))^{\epsilon_1} (f(x_2))^{\epsilon_2} \cdots (f(x_n))^{\epsilon_n}$$

to a homomorphism $\tilde{f}: M(X) \to G$. If $x_i = x_{i+1}$ and $\epsilon_i = -\epsilon_{i+1}$, then

$$(f(x_1))^{\epsilon_1}(f(x_2))^{\epsilon_2} \cdots (f(x_n))^{\epsilon_n} = (f(x_1))^{\epsilon_1}(f(x_2))^{\epsilon_2} \cdots (f(x_{i-1}))^{\epsilon_{i-1}}(f(x_{i+2}))^{\epsilon_{i+2}} \cdots (f(x_n))^{\epsilon_n}.$$

Hence, if $w \equiv w'$ then w and w' have the same image in G. Hence, $\tilde{f}$ induces a map $\varphi: F(X) \to G$ via $\varphi([w]) = \tilde{f}(w)$, φ is a homomorphism with $\varphi \circ j = f$. Because $F(X) = \langle j(X) \rangle$ there is only one homomorphism from $F(X)$ to G with fixed values on $j(X)$. □

Corollary 1.1.5. *Let G be a group with $G = \langle X \rangle$. Then G is isomorphic to a factor group of $F(X)$.*

Proof. As in the proof of Theorem 1.1.4 we may extend the inclusion $i: X \to G$ to a homomorphism $f: F(X) \to G$. Since $G = \langle X \rangle$, f is surjective. Therefore $G \cong F(X)/\ker(f)$ by the homomorphism theorem for groups. □

Definition 1.1.6. Let G be a group and $G \cong F(X)/N$, $F(X)$ free on X, $N \triangleleft F(X)$.
1. An element $r \in N$ is called a *relator* (relative to X and G).
2. A system R of relators is called a *system of defining relators* if $N = \langle\langle R \rangle\rangle$, that is, every relator is freely equivalent to a word $g_1 r_1^{\epsilon_1} g_1^{-1} g_2 r_2^{\epsilon_2} g_2^{-1} \cdots g_k r_k^{\epsilon_k} g_k^{-1}$ where $r_i \in R$, $\epsilon_i = \pm 1$, $g_i \in F(X)$.
3. A *trivial relator* is freely equivalent to the empty word.

As indicated, the notion of a relator depends on the group G and the system of generators. Without loss of generality we identify G with $F(X)/N$.

Definition 1.1.7. 1. If X is a generating system and R a corresponding system of defining relators for G then we write $G = \langle X \mid R \rangle$ to indicate this and $\langle X \mid R \rangle$ is called a *presentation* for G. In this sense, $F(X) = \langle X \mid \rangle$ indicates that the set of defining relators is the empty word.
2. A group G is called *finitely generated* if it has a finite system of generators, as we already mentioned, and *finitely presented* (or *presentable*) if it has a presentation with a finite number of generators and defining relators.

There are also other forms of the description of G used. If $r \in R$ then we often write $r = 1$ to indicate that the word r is trivial in G. We call an equation $r = 1$ a *relation*. If, for instance, $G = \langle x_1, x_2, \ldots, x_n \mid r_1, r_2, \ldots, r_k \rangle$ then we mostly write

$$G = \langle x_1, x_2, \ldots, x_n \mid r_1 = r_2 = \cdots = r_k = 1 \rangle$$

and call $r_1 = r_2 = \cdots = r_k = 1$ a system of defining relations.

We consider examples of presentations.

Examples 1.1.8. We have
1. $\mathbb{Z} \cong \langle a \mid \rangle$,
2. $\mathbb{Z}_n \cong \langle a \mid a^n = 1\rangle$, $n \geq 2$,
3. $\mathbb{Z} \oplus \mathbb{Z} \cong \langle a, b \mid aba^{-1}b^{-1} = 1\rangle$, and
4. $\mathbb{Z}_n \oplus \mathbb{Z}_m \cong \langle a, b \mid a^n = b^m = aba^{-1}b^{-1} = 1\rangle$, $n, m \geq 2$.

We leave the proofs for these examples as exercises. We discuss presentations of groups in Section 1.3. We now come back to free groups.

Theorem 1.1.9 (Normal form). *There is exactly one reduced word in each equivalence class. In other words: If w is a word in $M(X)$ then there is exactly one reduced word $w' = x_1^{\epsilon_1}x_2^{\epsilon_2}\cdots x_n^{\epsilon_n}$ with $w \equiv w'$.*

The uniquely determined $n \in \mathbb{N}_0$ is called the (free) *length* $|w|$ (or $\ell(w)$) of w.

Proof (van der Waerden 1948). Let S be the set of all reduced words in $M(X)$ and G the group of all permutations of S. If $x \in X$ then we define f_x via $w \mapsto xw$, if w does not start with x^{-1}, and via $w \mapsto u$, if $w = x^{-1}u$. Then f_x is a permutation of S whose inverse maps w to $x^{-1}w$, if w does not start with x, and w to v, if $w = xv$. This is correct because a reduced word does not start with xx^{-1} or $x^{-1}x$. Hence, we have a map $f: X \to G$, $x \mapsto f_x$, which induces a homomorphism $\varphi: F(X) \to G$ with $\varphi \circ j = f$. If $w = x_1^{\epsilon_1}x_2^{\epsilon_2}\cdots x_n^{\epsilon_n}$ is a word, then

$$\varphi([w]) = (f(x_1^{\epsilon_1}))(f(x_2^{\epsilon_2}))\cdots(f(x_n^{\epsilon_n})) = (f_{x_1})^{\epsilon_1}(f_{x_2})^{\epsilon_2}\cdots(f_{x_n})^{\epsilon_n}.$$

Therefore, if $(\,)$ is the empty word, then $\varphi([w])(\,) = x_1^{\epsilon_1}x_2^{\epsilon_2}\cdots x_n^{\epsilon_n}$. Now, if w and w' are reduced with $w \equiv w'$, then $[w] = [w']$ and $\varphi([w]) = \varphi([w'])$. Therefore $\varphi([w])(\,) = \varphi([w'])(\,)$ and $w = w'$. □

Corollary 1.1.10 (Solution of the word problem in $F(X)$). *Two words are (freely) equivalent if and only if they have the same reduced word.*

Corollary 1.1.11 (Solution of the conjugacy problem in $F(X)$). *Two reduced words w_1 and w_2 represent conjugate elements in $F(X)$ if and only if they are of the form $w_1 = hkgh^{-1}$, $w_2 = \ell gk\ell^{-1}$, where kg and gk, respectively, does not end with the inverse of the first letter.*

A reduced word with this last property is called *cyclically reduced*. The cyclically reduced words kg and gk are *cyclically permuted* or come apart by cyclic permutation.

Proof. If w_1 and w_2 have the given form then they are certainly conjugate. Now, let $w_1, w_2 \in F(X)$ be conjugate. We may, possibly after conjugation, assume that w_1, w_2 are reduced and cyclically reduced. Let $w_2 \equiv ww_1w^{-1}$. Since w_2 is cyclically reduced, then w or w^{-1} is completely canceled. Since w_1 is cyclically reduced, a cancellation is possible only at w or w^{-1}. Hence, w_1 and w_2 are cyclically permuted. □

Remarks 1.1.12. 1. If $g \in F(X)$ is cyclically reduced, then also g^n, $n \in \mathbb{Z} \setminus \{0\}$, and $|g^n| = |n||g|$.

2. A free group $F(X)$ is torsion-free.

   Proof. Let $1 \neq g \in F(X)$. We may assume that g is cyclically reduced. Then $g^n \neq 1$ if $n \neq 0$ because $|g^n| = |n||g|$. □

3. Let $g, h \in F(X)$. If $g^k = h^k$, $k \neq 0$, then $g = h$.

   Proof. Let $k > 0$ and without loss of generality g cyclically reduced. Then g^k is cyclically reduced. Let h be freely reduced. If h is not cyclically reduced, then $h^k \neq g^k$ because $h^k = a(h')^k a^{-1}$ if $h = ah'a^{-1}$. If h is cyclically reduced and $h^k = g^k$ then $|g| = |h|$ and therefore $h = g$. □

4. Let $g, h \in F(X)$. If $ghg^{-1}h^{-1} =: [g, h] = 1$, that is, $gh = hg$, then $\langle g, h\rangle$ is cyclic, that is, there is a $u \in F(X)$ with $g = u^r$, $h = u^s$, for some $r, s \in \mathbb{Z}$.

   Proof. If $g = ag'a^{-1}$ and $gh = hg$ then also $g'a^{-1}ha = a^{-1}hag'$. Hence, we may assume that g is cyclically reduced. We also may assume that $g \neq 1 \neq h$. We use induction on $|g|+|h|$. If $|g|+|h| \leq 2$ then there is nothing to show. Now, let $|g|+|h| > 2$. Let first $|gh| = |g| + |h|$. Then also $|hg| = |g| + |h|$. Then, possibly after interchanging g and h, we must have $g = hw$ with $|hw| = |h| + |w|$. This gives $hw = wh$, and the statement follows by induction.
   Now, let $|gh| < |g|+|h|$. Then there must exist a letter x with $g = vx^{\epsilon}$, $|g| = |v|+1$ and $h = x^{-\epsilon}t$, $|h| = |t| + 1$, $\epsilon = \pm 1$. Since g is cyclically reduced we must have $h = g^{-1}w$ or $g = wh^{-1}$, which gives $gw = wg$ or $hw = wh$, and the statement follows by induction. □

This has the following consequence. The group $F(X)$ is commutative transitive, that is, if $[g, h] = [h, k] = 1$, $h \neq 1$, then $[g, k] = 1$. We note this property for future reference:

Definition 1.1.13. A group G is called *commutative transitive*, if $[g, h] = [h, k] = 1$, $h \neq 1$, then $[g, k] = 1$. Here $[g, h] = ghg^{-1}h^{-1}$ is the *commutator* of h and g.

Another way to phrase Definition 1.1.13 is to say that the relation of commutativity is transitive on the non-identity elements of G. We occasionally come up with commutative transitive groups in the following sections. At this stage we just mention that subgroups of $\mathrm{PSL}(2, \mathbb{R})$ are commutative transitive. The reason is that two elements $A, B \in \mathrm{PSL}(2, \mathbb{R})$ commute if and only if they have common fixed points, considered as linear fractional transformations.

Especially Fuchsian groups, that is, discrete subgroups of $\mathrm{PSL}(2, \mathbb{R})$, are commutative transitive.

We remark that the following are equivalent:
1. There exists a bijection $f: X \to Y$ and
2. $|X| = |Y|$ for the cardinals of X and Y.

Theorem 1.1.14 (Solution of the isomorphism problem in the class of free groups).
Let X, Y be sets. Then $F(X) \cong F(Y)$ if and only if there exists a bijection $f: X \to Y$.

Proof. Let $f: X \to Y$ be a bijection. Then f induces a homomorphism $\tilde{f}: F(X) \to F(Y)$, and $g = f^{-1}$ induces a homomorphism $\tilde{g}: F(Y) \to F(X)$. The homomorphism $\tilde{g} \circ \tilde{f}$ and $\mathrm{id}_{F(X)}$ both extend the identity on X. Hence $\tilde{g} \circ \tilde{f} = \mathrm{id}_{F(X)}$. Analogously $\tilde{f} \circ \tilde{g} = \mathrm{id}_{F(Y)}$. Hence, $\tilde{f}$ is an isomorphism.

Now, let $F(X) \cong F(Y)$. Let $N(X) = \langle g^2 | g \in F(X) \rangle < F(X)$. Certainly $N(X) \lhd F(X)$, and $F(X)/N(X)$ may be considered as a $\mathbb{Z}_2$-vector space with basis X and dimension $|X|$, where $|X|$ is the cardinal of X. The analogue statement holds for $F(Y)$. From $F(X) \cong F(Y)$ we get $F(X)/N(X) \cong F(Y)/N(Y)$ and therefore $|X| = |Y|$. □

Remarks 1.1.15.
1. $F(X)$ is countable if and only if X is finite with $|X| \geq 1$ or X is countable. If X is infinite, X and $F(X)$ have the same cardinality.
2. If $F(X) = \langle A \rangle$ then $|A| \geq |X|$. This follows from $|X| \leq |\bar{A}| \leq |A|$ where $\bar{A}$ is the set of equivalence classes of A in $F(X)/N(X)$ (see the proof of Theorem 1.1.14).
3. A group G is called a *free group* if $G \cong F(X)$ for some X. If $f: F(X) \to G$ is an isomorphism then $f(X)$ is called a *basis* or free generating system of G. Then G is free on $f(X)$. If A is a basis of G and $\alpha: G \to G$ an automorphism then also $\alpha(A)$ is a basis of G. If A and B are two bases of G, then $|A| = |B|$, and a bijection $A \to B$ extends to an automorphism of G. The cardinal of a basis of a free group G is called its *rank*, denoted by $\mathrm{rk}(G)$. A free group of rank 0 is trivial; a free group of rank 1 is isomorphic to $\mathbb{Z}$.
4. Let $\varphi: G \twoheadrightarrow F(X)$ be an epimorphism with $\varphi|_S: S \to X$ a bijection for some $S \subset G$. Then $\langle S \rangle = F(S) \cong F(X)$. This can be seen as follows. Let $\psi: X \to S$, $\psi = \varphi|_S^{-1}$; ψ extends to a homomorphism $\tilde{\psi}: X \to S$. Since $\varphi \circ \psi = \varphi \circ \tilde{\psi}|_X = \mathrm{id}_X$ we get $\varphi \circ \tilde{\psi} = \mathrm{id}_{F(X)}$. Hence, $\tilde{\psi}$ is injective and therefore an isomorphism from $F(X)$ to $\langle S \rangle \subset G$.
5. Let $X \subset G$, G a group, with $X \cap X^{-1} = \emptyset$. Then X is a basis for a free subgroup of G if and only if there is no product $w = x_1^{\epsilon_1} x_2^{\epsilon_2} \cdots x_n^{\epsilon_n}$ trivial with $n \geq 1$, all $x_i \in X$, $\epsilon_i = \pm 1$, $\epsilon_i + \epsilon_{i+1} \neq 0$ if $x_i = x_{i+1}$.

   Proof. Assume that there exists such a w with $w = 1$. We have $x_1^{\epsilon_1} x_2^{\epsilon_2} \cdots x_n^{\epsilon_n} \neq 1$ in $F(X)$ which means that there exists no homomorphisms $\langle X \rangle \to F(X)$, and therefore X is no basis of $\langle X \rangle \subset G$.
   Assume now that there is no w with $w = 1$. We have the diagram (see Figure 1.3). Let $u \neq 1$ be reduced in $F(X)$. Then $\varphi(u) \neq 1$. Therefore φ is injective, and $F(X) \cong \langle X \rangle$ and $\langle X \rangle$ are free with basis X. □

6. Let F be free with basis $\{a, b\}$. Let $c_i = a^{-i} b a^i$, $i \in \mathbb{Z}$. Then $G = \langle c_i \mid i \in \mathbb{Z} \rangle$ is free with basis $\{c_i \mid i \in \mathbb{Z}\}$. We remark that G is the kernel of the homomorphisms $\varphi: F \to \mathbb{Z}$, $a \mapsto 1$, $b \mapsto 0$.

   Proof. We have to show that $c_{i_1}^{r_1} c_{i_2}^{r_2} \cdots c_{i_n}^{r_n} \neq 1$ if $n \geq 1$, all $r_i \neq 0$ and $i_k \neq i_{k+1}$, $1 \leq k < n$. But this element is just $a^{-i_1} b^{r_1} a^{i_1 - i_2} \cdots a^{i_{n-1} - i_n} b^{r_n} a^{i_n}$ in F, and this is $\neq 1$. □

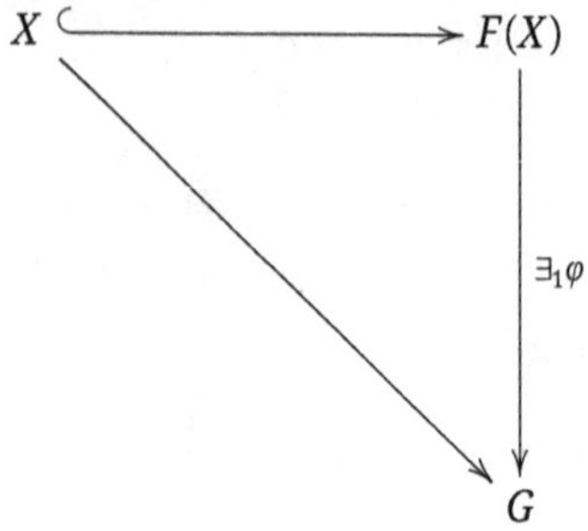

Figure 1.3: Homomorphism diagram.

We get especially: If F_2 is a free group of rank 2 and F a free group of rank $n \geq 1$ or of countable rank, then F is isomorphic to a subgroup of F_2.

1.1.1 Residual finiteness and the Hopfian property

We introduce two important properties for groups.

Definition 1.1.16. A group G is called *residually finite* if for each non-trivial $g \in G$ there exists a normal subgroup $N \triangleleft G$ such that $w \notin N$ and G/N is finite.

Theorem 1.1.17. *Free groups are residually finite, that is, if F is free and $1 \neq w \in F$, then there exists a normal subgroup $N \triangleleft F$ such that $w \notin N$ and F/N is finite.*

Proof. Let $F = F(X)$, $1 \neq w = x_1^{\epsilon_1} x_2^{\epsilon_2} \cdots x_n^{\epsilon_n}$ reduced. We define a homomorphism $\varphi: F \to S_{n+1}$, S_{n+1} the full permutation group on $\{0, 1, \ldots, n\}$ such that $\varphi(w) \neq 1$, more concrete: $\varphi(w)$ maps n to 0. This we obtain through a map $f: X \to S_{n+1}$ such that

$$f(x_r^{\epsilon_r})(r) = r - 1 \quad \text{for all } 1 \leq r \leq n.$$

We need that $f(x)$ maps r to $r-1$ if $x = x_r$, $\epsilon_r = 1$, and $r-1$ to r if $x = x_r$, $\epsilon_r = -1$. If these conditions define an injective map from a subset of $\{0, 1, \ldots, n\}$ to another subset, then we may extend this map to an element of S_{n+1}, and define $f(x)$ finally by this element. This construction is possible because w is reduced. The image of r is defined if $x_r = x$, $\epsilon_r = 1$ or $x_{r+1} = x$, $\epsilon_{r+1} = -1$. Since w is reduced, only one case can occur.

Therefore we have a map from a subset of $\{0, 1, \ldots, n\}$ to a subset $\{0, 1, \ldots, n\}$. Also each s is an image of this map if $x_s = x$, $\epsilon_s = -1$ or $x_{s+1} = x$, $\epsilon_{s+1} = 1$. Again, only one of these cases can occur because w is reduced, and the map is injective. □

There are several different proofs of this statement. Especially it is a consequence of the following result of A. Malcev [187].

Theorem 1.1.18. *Let K be a field and G be a finitely generated subgroup of $\mathrm{GL}(n, K)$, $n \geq 2$. Then G is residually finite.*

Indeed, Theorem 1.1.17 follows from Theorem 1.1.18 for free groups of finite or countable rank if we can show that there is a group $\mathrm{GL}(n, K)$ which contains a free group of rank 2. We show that the free group F of rank 2 has a faithful representation as a subgroup of

$$\mathrm{SL}(2,\mathbb{R}) = \left\{ \begin{pmatrix} a & b \\ c & d \end{pmatrix} \;\middle|\; a,b,c,d \in \mathbb{R}, ad - bc = 1 \right\}.$$

Let $A = \left(\begin{smallmatrix} 0 & 1 \\ -1 & 2 \end{smallmatrix}\right)$, $B = \left(\begin{smallmatrix} 0 & -1 \\ 1 & 2 \end{smallmatrix}\right)$ and $G = \langle A, B\rangle$. Note that A and B both have infinite order. Moreover, $A^n = \left(\begin{smallmatrix} -n+1 & n \\ -n & n+1 \end{smallmatrix}\right)$, $n \geq 0$, $B^n = \left(\begin{smallmatrix} -n+1 & -n \\ n & n+1 \end{smallmatrix}\right)$, $n \geq 0$, $A^{-n} = \left(\begin{smallmatrix} n+1 & -n \\ n & -n+1 \end{smallmatrix}\right)$, $n > 0$, $B^{-n} = \left(\begin{smallmatrix} n+1 & n \\ -n & -n+1 \end{smallmatrix}\right)$, $n > 0$. The group G acts on $\hat{\mathbb{C}} = \mathbb{C} \cup \{\infty\}$ as a group of linear fractional transformations via

$$\begin{pmatrix} a & b \\ c & d \end{pmatrix}(z) = \frac{az+b}{cz+d}, \quad \begin{pmatrix} a & b \\ c & d \end{pmatrix} \in G.$$

We have

$$\begin{aligned} A^n(\mathbb{R}^-) &\subset \mathbb{R}^+ \quad \text{for all } n \neq 0, \\ B^n(\mathbb{R}^+) &\subset \mathbb{R}^- \quad \text{for all } n \neq 0. \end{aligned}$$

Here $\mathbb{R}^+ = \{x \in \mathbb{R} \mid x > 0\}$ and $\mathbb{R}^- = \{x \in \mathbb{R} \mid x < 0\}$. Let $w \in G$, w not a power of A or B. After a suitable conjugation we may assume that

$$w = A^{n_1}B^{m_1}A^{n_2}B^{m_2}\cdots A^{n_k}B^{m_k}A^{n_{k+1}}, \quad \text{all } n_i, m_j \neq 0.$$

Let $z \in \mathbb{R}^-$. Then $w(z) \in \mathbb{R}^+$, and hence $z \neq 1$. Therefore G is a free group of rank 2.

Residually finite groups play a prominent role in combinatorial group theory, and we come back occasionally to this concept. The first direct observation is that residually finite groups have solvable word problem. We now discuss two further properties.

Theorem 1.1.19. *Finitely generated, residually finite groups are Hopfian, that is, every epimorphism $\alpha\colon G \to G$ is already an automorphism.*

Proof. Let N be a normal subgroup with G/N finite. Then $G/\alpha^{-n}(N) \cong G/N$ for all $n \in \mathbb{N}$ because $\alpha^{-1}(N) \lhd G$ and

$$G/\alpha^{-1}(N) \cong \alpha(G)/N = G/N.$$

There exist only finitely many homomorphisms from G onto G/N because each homomorphism is determined by the images of the finite generating system of G and there are only finitely many elements in G/N which come into question as images of the generators. Since $\alpha^{-n}(N)$ is the kernel of a homomorphism for each $n \in \mathbb{N}$, there exist $m, n \in \mathbb{N}$, $m > n$, with $\alpha^{-n}(N) = \alpha^{-m}(N)$. Then $N = \alpha^{-(m-n)}(N)$ and therefore $\ker(\alpha) \subset N$. Since G is residually finite, the intersection of all possible N is $\{1\}$, which means that $\ker(\alpha) \subset \{1\}$, and α is injective. □

Remark 1.1.20. If $|X| = \infty$, then $F(X)$ is not Hopfian. Take $X = \{x_1, x_2, x_3, \ldots\}$ countable. The map $x_1 \mapsto 1$, $x_2 \mapsto x_1$, $x_3 \mapsto x_2, \ldots$ defines an epimorphism $\alpha: F(X) \to F(X)$ with $\ker(\alpha) \neq \{1\}$.

Theorem 1.1.21 ([13]). *Let G be a finitely generated residually finite group. Then* $\mathrm{Aut}(G)$, *the group of automorphisms of G, is residually finite.*

Proof. Let $A = \mathrm{Aut}(G)$, $1 \neq \alpha \in A$. Then there exists a $c \in G$ with $c^* := \alpha(c)c^{-1} \neq 1$. Since G is residually finite there is a normal subgroup $N \triangleleft G$ with $c^* \notin N$ and G/N finite that contains a *characteristic subgroup* K of finite index in G (see the following Theorem 1.1.22), that is, $\beta(K) \subset K$ for all $\beta \in A$.

Characteristic subgroups are normal subgroups because conjugation with an element $g \in G$ defines an automorphism of G. Since K is characteristic, we may define a homomorphism $\psi: A \to \mathrm{Aut}(G/K)$ via $\psi(\beta) := K\beta(g)$, $\beta \in A$. This can be seen as follows. The homomorphism ψ is well defined because $K\beta(g) = K\beta(g')$ if $Kg = Kg'$ and $(\psi(\gamma) \circ \psi(\beta))(Kg) = \psi(\gamma \circ \beta)(Kg)$. Now $\mathrm{Aut}(G/K)$ is finite and $\psi(\alpha) \neq 1$, that is, $\ker(\psi)$ is a normal subgroup of A with finite index and $\alpha \notin \ker(\psi)$. □

1.1.2 Counting processes

Next, we discuss some interesting counting processes.

Theorem 1.1.22 ([121]). *Let G be a finitely generated group. The number of subgroups of index n is finite for each $n \in \mathbb{N}$. If H is a subgroup of finite index then H contains a characteristic subgroup K of G with finite index in G.*

Proof. Let $n \in \mathbb{N}$. For each subgroup H of index n we choose a complete set of representatives (or right transversals) $c_1, c_2, \ldots, c_n$ for the right cosets of H in G with $c_1 = 1$. The group G permutes the cosets Hc_i via multiplication from right. Since this is an action on the set of the right cosets, it induces a homomorphism ψ_H of G in S_n: If $g \in G$ then $\psi_H(g)$ is the permutation which maps i to j if $Hc_ig = Hc_j$. Because $Hc_1 = H$ then $\psi_H(g)$ fixes the number 1 if and only if $g \in H$. Let H and L be two different subgroups of index n in G. Then there exists a $g \in G$ with $g \in H$ but $g \notin L$. This means $\psi_H(g) \neq \psi_L(g)$, and ψ_H and ψ_L are distinct. Since G is finitely generated there are only finitely many homomorphisms from G into S_n. Hence, the number of subgroups of index n is finite.

Now, let $H_1, H_2, \ldots, H_m$ be all subgroups of G with index n. Assume that $H = H_1$ for the given subgroups of index n. Let $K = \bigcap_{i=1}^m H_i$. K has finite index in G because $|G : K| \leq |G : H_1||G : H_2| \cdots |G : H_m|$ by Poincaré's theorem (see [51]). Let $\alpha \in \mathrm{Aut}(G)$. Each $\alpha(H_i)$ is a subgroup of index n. Hence, α permutes the H_i, and we get $\alpha(K) = \bigcap_{i=1}^m \alpha(H_i) = K$, and therefore H is characteristic in G. □

Remark 1.1.23. We use Theorem 1.1.22 to introduce some important growth functions for finitely generated groups. Let G be a finitely generated group. Define

$$\zeta_G(s) = \sum_{n=1}^{\infty} a_n n^{-s},$$

where $a_n = |\{H < G \mid [G : H] = n\}|$, the zeta function of G, and

$$\zeta_G^N(s) = \sum_{n=1}^{\infty} b_n n^{-s},$$

where $b_n = |\{H \lhd G \mid [G : H] = n\}|$. We also define the p-parts for a prime number p:

$$\zeta_G^p(s) = \sum_{n=0}^{\infty} a_{p^n} p^{-ns}$$

with $a_{p^n} = |\{H < G \mid [G : H] = p^n\}|$, and

$$\zeta_G^{N,p}(s) = \sum_{n=0}^{\infty} b_{p^n} p^{-ns}$$

with $b_{p^n} = |\{H \lhd G \mid [G : H] = p^n\}|$.

Theorem 1.1.24 (See [70] and [119]). *Let G be a finitely generated nilpotent group. Then*

$$\zeta_G(s) = \prod_p \zeta_G^p(s) \quad \text{and} \quad \zeta_G^N(s) = \prod_p \zeta_G^{N,p}(s)$$

in the area of convergence, whereby p runs through all prime numbers.

A complete, purely group theoretical proof is given in [70], the case of a finitely generated, torsion-free and nilpotent group is handled in [119]. We remark that a group is called *nilpotent* if there exists an $n \in \mathbb{N}$ with $G_n = \{1\}$ where

$$G = G_1 \rhd G_2 \rhd G_3 \cdots$$

with

$$G_{i+1} = [G_i, G] = \langle [g_i, g] \mid g_i \in G_i, g \in G \rangle$$

the lower central series. For details on nilpotent groups consider the book [218]. Here, we just discuss some examples. For these examples we need the Riemann zeta function

$$\zeta(s) = \sum_{n=1}^{\infty} \frac{1}{n^s}, \quad \text{with } \operatorname{Re}(s) > 1.$$

The Riemann zeta function $\zeta(s)$ has an Euler product

$$\zeta(s) = \prod_p \frac{1}{1-p^{-s}}, \quad \text{for } \operatorname{Re}(s) > 1,$$

where again p runs through all prime numbers.

Examples 1.1.25. 1. Let $G = \mathbb{Z}^r$. Then

$$\zeta_G(s) = \zeta_G^N(s) = \zeta(s)\zeta(s-1)\cdots\zeta(s-r+1),$$

$\zeta_G(s)$ converges for $\operatorname{Re}(s) > r$ (see [119]).

2. Let

$$G = \left\{ \begin{pmatrix} 1 & a & b \\ 0 & 1 & c \\ 0 & 0 & 1 \end{pmatrix} \;\middle|\; a,b,c \in \mathbb{Z} \right\}$$

the discrete Heisenberg group. Then

$$\zeta_G(s) = \frac{\zeta(s)\zeta(s-1)\zeta(2s-2)\zeta(2s-3)}{\zeta(3s-3)},$$
$$\zeta_G^N(s) = \zeta(s)\zeta(s-1)\zeta(3s-2),$$

and $\zeta_G(s), \zeta_G^N(s)$ converge for $\operatorname{Re}(s) > 2$ (see [119]).

3. Let $G = \mathbb{Z} \oplus \mathbb{Z}_q$, q prime number. Then

$$\zeta_G(s) = \zeta_G^N(s) = (1 + q^{-s+1})\zeta(s),$$

$\zeta_G(s)$ converges for $\operatorname{Re}(s) > 1$ (see [70]).

We may introduce additional restrictions like free subgroups and free normal subgroups of finite index (with or without more restrictions). Here we just give an example. Let H_4 be the Hecke group generated by the linear fractional transformation

$$T : z \mapsto -\frac{1}{z} \quad \text{and} \quad U_4 : z \mapsto z + \sqrt{2}.$$

Algebraically, H_4 has a presentation $\langle a, b \mid a^2 = b^4 = 1 \rangle$. A subgroup $H < H_4$ is said to have genus g if the quotient space $\mathfrak{H}/H$ has genus g, where $\mathfrak{H}$ is the upper half plane $\{z \in \mathbb{C} \mid \operatorname{Im}(z) > 0\}$.

Let $b_1(n)$ be the number of normal subgroups of H_4 of index $4n$ and genus 1 (these groups are automatically free). Then $b_1(n) = \frac{1}{4}|\{(x,y) \in \mathbb{Z}^2 \mid x^2 + y^2 = n\}|$, and therefore $b_1(nm) = b_1(n)b_1(m)$ if $\gcd(n,m) = 1$ (see [155]). We remark that $\varphi_4(s) = \sum_{n=1}^{\infty} b_1(n)n^{-s}$ is exactly the zeta function of the field of Gaußian numbers $\mathbb{Q}(i)$ which converges for $\operatorname{Re}(s) > 1$. More concrete we have

$$\varphi_4(s) = \zeta(s)L_{-4}(s),$$

where

$$L_{-4}(s) = \sum_{n=1}^{\infty} \chi_{-4}(n)n^{-s}$$

with

$$\chi_{-4}(s) = \begin{cases} 1 & \text{if } n \equiv 1 \mod 4, \\ -1 & \text{if } n \equiv 3 \mod 4, \\ 0 & \text{if } n \equiv 2 \mod 4. \end{cases}$$

More such examples may be found in the papers [155] and [46].

There is another important growth function. Let G be a finitely generated group with a finite generating system X. We want to count the number $\ell(n)$ of elements whose shortest representation (in terms of X) has the given length n. The number $\ell(n)$ is finite and is described by the Hilbert–Dehn function HD_G of G. Hence we write $\mathrm{HD}_G(n) = \ell(n)$. More and detailed explanations can be found in the book [188] and the paper [164].

Here we just describe the Hilbert–Dehn function for the Hecke group H_4, this time given in the monoid presentation

$$H_4 = \langle a, b, c \mid a^2 = b^4 = bc = cb = 1 \rangle.$$

The first 10 values of the Hilbert–Dehn functions are

$$1,\ 3,\ 5,\ 8,\ 13,\ 21,\ 34,\ 55,\ 89,\ 144.$$

In general we have the initial values

$$\mathrm{HD}_{H_4}(0) = 1, \quad \mathrm{HD}_{H_4}(1) = 3 \quad \text{and} \quad \mathrm{HD}_{H_4}(2) = 5$$

and the recursive relation

$$\mathrm{HD}_{H_4}(n) = \mathrm{HD}_{H_4}(n-1) + HD_{H_4}(n-2)$$

for $n \geq 3$ (with respect to the given presentation). We recognize the Fibonacci numbers f_n given by

$$f_1 = f_2 = 1 \quad \text{and} \quad f_n = f_{n-1} + f_{n-2} \quad \text{for } n \geq 3,$$

and we get

$$\mathrm{HD}_{H_4}(n) = f_{n+3} \quad \text{for } n \geq 1.$$

The recursive relation for HD_{H_4} gives a nice generating function, the Hilbert–Dehn series HDS_{H_4}, of H_4 (with respect to the given presentation):

$$\mathrm{HDS}_{H_4}(z) = \sum_{n=0}^{\infty} \mathrm{HD}_{H_4}(n) z^n = \frac{1 + 2z + 2z^2}{1 - z - z^2},$$

which converges at least for $|z| < \frac{1}{2}$.

We close this subsection with another counting procedure.

Theorem 1.1.26. *Let Δ be a finitely generated group and G a finite group. The number $n_\Delta(G)$ of normal subgroups $K \lhd \Delta$ with $\Delta/K \cong G$ is given by*

$$n_\Delta(G) = \frac{|\operatorname{Epi}(\Delta, G)|}{|\operatorname{Aut}(G)|}.$$

Proof. Let $\Theta, \Theta' : \Delta \to G$ be epimorphisms. Then $\ker(\Theta) = \ker(\Theta')$ if and only if $\Theta' = \alpha \circ \Theta$ for some $\alpha \in \operatorname{Aut}(G)$. To see this, let $G \cong \Delta/\ker(\Theta') = \Delta/\ker(\Theta)$. This implies $\Theta' = \alpha \circ \Theta$ for some $\alpha \in \operatorname{Aut}(G)$.

Now let $\Theta' = \alpha \circ \Theta$ for some $\alpha \in \operatorname{Aut}(G)$. Then $\ker(\Theta) = \ker(\Theta')$. Hence, $\operatorname{Aut}(G)$ acts on $\operatorname{Epi}(\Delta, G)$ by composition. The kernels of the epimorphism $\Theta: \Delta \to G$ correspond to the orbits with respect to this action. Since $\Theta = \alpha \circ \Theta$ if and only if $\alpha = \mathrm{id}$ all orbits have the length $|\operatorname{Aut}(G)|$. □

Example 1.1.27. Let $\Delta = \langle a, b \mid a^5 = b^2 = 1 \rangle$ and $G \cong A_5$. Count the $K \lhd \Delta$ with $\Delta/K \cong A_5$.

The group A_5 has 24 elements of order 5 (the 5-cycle) and 15 elements of order 2 (the double transpositions $(a, b)(c, d)$). This gives $24 \cdot 15 = 360$ pairs $\{x, y\}$ which satisfy the relations of Δ. If $\{x, y\}$ is such a pair and $H = \langle x, y \rangle$ then $H \cong D_{10}$, the dihedral group of order 10, or $H \cong A_5$. There are 6 subgroups $H = \langle x, y \rangle$ with $H \cong D_{10}$, each can be generated by $4 \cdot 5 = 20$ such pairs $\{x, y\}$: If x with $x^5 = \mathrm{id}$, y with $y^2 = \mathrm{id}$ and $\langle x, y \rangle \cong D_{10}$ then H is generated by all pairs $\{x^i, yx^j\}$, $1 \le i \le 4$, $0 \le j \le 4$. Hence, $360 - 120 = 240$ such pairs $\{x, y\}$ generate the whole A_5. Therefore $|\operatorname{Epi}(\Delta, G)| = 240$.

Now $\operatorname{Aut}(A_5) = S_5$ because S_5 acts on A_5 by conjugation. Hence, $|\operatorname{Aut}(A_5)| = 120$ and $n_\Delta(G) = \frac{240}{120} = 2$. Therefore, there are exactly two normal subgroups $K \lhd \Delta$ with $\Delta/K \cong A_5$.

1.1.3 Residual freeness, properties CT and CSA

We conclude Section 1.1 with an interesting observation which goes back to B. Baumslag [10]. Before we start and prove Baumslag's theorem we need some definitions.

Definition 1.1.28. 1. A group G is *residually free* if for each non-trivial $g \in G$ there is a free group F_g and an epimorphism $h_g: G \to F_g$ such that $h_g(g) \neq 1$. Equivalently for each $g \in G$, $g \neq 1$, there is a normal subgroup N_g such that G/N_g is free and $g \notin N_g$.

2. A group G is *n-residually free* for $n \in \mathbb{N}$ provided to every ordered n-tuple $(g_1, g_2, \ldots, g_n) \in (G \setminus \{1\})^n$ of non-identity elements of G there is a free group F and an epimorphism $h: G \to F$ such that $h(g_i) \neq 1$ for all $i = 1, 2, \ldots, n$. G is *fully residually free* provided it is n-residually free for every $n \in \mathbb{N}$.

Remarks 1.1.29. 1. Residually free groups are torsion-free. If G is residually free and $g, h \in G$ then $\langle g, h\rangle$ is either free or Abelian.
2. The following immediate statements are pairwise equivalent for a group G:
   (i) The group G is commutative transitive.
   (ii) The centralizer $C(x)$ of every non-trivial element $x \in G$ is Abelian.
   (iii) Every pair of distinct maximal Abelian subgroups in G has trivial intersection.

Theorem 1.1.30. *Let G be a non-Abelian and residually free group. Then G is fully residually free if and only if G is commutative transitive.*

Proof. Suppose that G is fully residually free and a, b, c are non-trivial elements of G satisfying $[a, b] = 1$ and $[b, c] = 1$. Suppose that $[a, c] \neq 1$. Then there exists a homomorphism $\varphi: G \to F$ with F free such that $\varphi(a) = \bar{a} \neq 1$, $\varphi(b) = \bar{b} \neq 1$, $\varphi(c) = \bar{c} \neq 1$ and $\varphi([a, c]) = [\bar{a}, \bar{c}] \neq 1$. Since φ is a homomorphism $[\bar{a}, \bar{b}] = 1$ and $[\bar{b}, \bar{c}] = 1$. Furthermore, F is a free group so $[\bar{a}, \bar{c}] \neq 1$ giving a contradiction. Therefore $[a, c] = 1$ and G is CT.

For the other direction, we first need the following lemma.

Lemma 1.1.31. *Let A be an Abelian normal subgroup of a non-Abelian residually free group G. Then A is contained in the center of G. In particular if G is CT then A must be trivial.*

Proof. Recall that a non-Abelian free group has no non-trivial normal Abelian subgroup. Suppose that A is an Abelian normal subgroup of G and suppose that A is not contained in the center of G. Then there exist $a \in A$ and $b \in G$ such that $[a, b] \neq 1$. Then from the residual freeness there exists a normal subgroup N of G with G/N free and $[a, b] \neq 1$ modulo N. However, then AN/N is a non-trivial normal Abelian subgroup of the non-Abelian free group G/N, a contradiction. If G is also CT and non-Abelian then it is centerless so from the above it follows that any normal Abelian subgroup must be trivial. □

We now complete the proof of Theorem 1.1.30: Let $g_1, g_2, \ldots, g_n$ be a set of non-trivial elements of G. We want to show that there is a free group F and an epimorphism $h: G \to F$ such that $h(g_i) \neq 1$ for all $i = 1, 2, \ldots, n$. We do this by induction on the size n of the set of elements. Since G is residually free this is true for $n = 1$. Assume that for $n \geq 1$ given non-trivial $g_1, g_2, \ldots, g_n$ in G there exists a non-trivial $g \in G$ such that for all normal subgroups N of G if $g \notin N$ then $g_i \notin N$ for $i = 1, 2, \ldots, n$. By the residual freeness this is true for $n = 1$ (just take $g = g_1$ for $n = 1$). We show that given non-trivial $g_1, g_2, \ldots, g_n, g_{n+1}$ we can find a $g' \neq 1$ such that $g' \notin N$ for any normal subgroup N of G then $g_i \notin N$ for $i = 1, 2, \ldots, n, n+1$. Let g be the assumed element for $g_1, g_2, \ldots, g_n$ and for $x \in G$ let $c(x) = [g, xg_{n+1}x^{-1}]$. If $c(x) = 1$ for all x then each conjugate of g_{n+1} commutes with $g \neq 1$. Then by commutative transitivity the normal closure $\langle\langle g_{n+1}\rangle\rangle$ is Abelian and hence here trivial. But g_{n+1} is in it and non-trivial. Therefore $c(x) \neq 1$ for some $x \in G$. Choose such $c(x)$ as g'. Then if $g' \notin N$ for a normal subgroup N of G it follows that $g_1, g_2, \dots, g_{n+1} \notin N$. It follows from this induction that for each finite set $g_1, g_2, \dots, g_n \in G$ of non-trivial elements there is a $g \in G$ that if $g \notin N$ for some normal subgroup N of G then $g_1, g_2, \dots, g_n$ are also not in N. Since G is residually free it follows that there is such an N and therefore G is fully residually free. □

We remark that the class of fully residually free groups is a proper subclass of the class of residually free groups. The direct square $F \oplus F$, F a non-Abelian free group, is residually free, but not commutative transitive, hence not fully residually free. But we have the following.

Theorem 1.1.32. *Suppose that the group G is 2-residually free. Then G is fully residually free.*

Proof. Suppose that G is 2-residually free. To show that G is fully residually free it suffices to show that G is CT. Suppose that there are elements $a, b, c \in G$ with $b \neq 1$ such that $ab = ba$ and $bc = cb$ but $ac \neq ca$. Then $b \neq 1$ and $[a, c] \neq 1$. There is a free group F and an epimorphism $h: G \to F$, $x \mapsto \bar{x}$, such that $\bar{b} \neq 1$, $[\bar{a}, \bar{c}] \neq 1$. But this is impossible since $\bar{a}\bar{b} = \bar{b}\bar{a}$, $\bar{b}\bar{c} = \bar{c}\bar{b}$ and $\bar{b} \neq 1$ in the free group F. This contradiction shows that G is CT, and, hence, fully residually free as claimed. □

There is another concept which is much related to the described properties. To explain the connections we need some definitions.

Definition 1.1.33. 1. Let G be a group and H a subgroup of G. Then H is *malnormal* in G or conjugately separated in G provided $gHg^{-1} \cap H = \{1\}$ unless $g \in H$.
2. A group G is a *CSA group* or *conjugately separated Abelian group* provided the maximal Abelian subgroups are malnormal.

Lemma 1.1.34. *If G is fully residually free then G is CSA.*

Proof. Let G be a fully residually free group with more than one element. Let $u \in G \setminus \{1\}$ and let $M = C_G(u)$ be its centralizer. Then M is maximal Abelian in G. We claim that M is malnormal in G. If G is Abelian, than $M = G$ and the conclusion follows trivially. Suppose now that G is non-Abelian. Suppose that $w = gzg^{-1} \neq 1$, $z \in M$, lies in $gMg^{-1} \cap M$. If $g \notin M$ then $[g, u] \neq 1$. Thus there is a free group F and an epimorphism $h: G \to F$, $x \mapsto \bar{x}$, such that $\bar{w} \neq 1$ and $[\bar{g}, \bar{u}] \neq 1$. Let $C = C_F(\bar{u})$. Then $\bar{w} \in \bar{g}C\bar{g}^{-1} \cap C$. However, the maximal Abelian subgroups in a free group are certainly malnormal (recall that the centralizer of a non-trivial element in a free group is cyclic). This implies $\bar{g} \in C$, contradicting $[\bar{g}, \bar{u}] \neq 1$. This contradiction shows that $gMg^{-1} \cap M$ implies $g \in M$, and hence the maximal Abelian subgroups are malnormal. □

We now describe the relationship between CSA groups and CT groups.

Lemma 1.1.35. *The class of CSA groups is a proper subclass of the class of CT groups.*

Proof. We first show that every CSA group is commutative transitive. Let G be a group in which maximal Abelian subgroups are malnormal and suppose that M_1 and M_2 are maximal Abelian subgroups in G with $z \neq 1$ lying in $M_1 \cap M_2$. We ask if we could have $M_1 \neq M_2$. Suppose that $w \in M_1 \backslash M_2$. Then $wzw^{-1} = z$ is a non-trivial element of $wM_2w^{-1} \cap M_2$ so that $w \in M_2$. This is impossible and therefore $M_1 \subset M_2$. By maximality we get $M_1 = M_2$. Hence, G is CT whenever all maximal Abelian subgroups are malnormal.

We now show that there do exist CT groups that are not CSA groups. In any non-Abelian CSA group the only Abelian normal subgroup is the trivial subgroup $\{1\}$. To see this suppose that N is any normal Abelian subgroup of the non-Abelian CSA group G. Then N is contained in a maximal Abelian subgroup M. Let $g \notin M$. Then $N = gNg^{-1} \cap N \subset gMg^{-1} \cap M$. The fact $N \neq \{1\}$ would imply that $g \in M$ which is a contradiction.

Now let

$$D_\infty = \langle a, b \mid a^2 = (ab)^2 = 1 \rangle,$$

the infinite dihedral group. Note that D_∞ is CT because the centralizer $C(x)$ of every non-trivial element $x \in D_\infty$ is Abelian, in fact cyclic (we leave this as an exercise). However, the cyclic subgroup $\langle b \rangle$ is a non-trivial Abelian normal subgroup of D_∞. Hence, D_∞ is not a CSA group. □

The papers [79] and [80] provide a very satisfactory description of CT groups which are not CSA groups. We now extend Theorem 1.1.32.

Theorem 1.1.36. *If G is a non-Abelian residually free group then CT is equivalent to CSA.*

Proof. We have just seen that CSA implies CT. Suppose that G is residually free and CT. Then G is fully residually free and, hence, G is CSA. □

There is a classification of fully residually free groups of rank ≤ 3.

Theorem 1.1.37. *Let G be a fully residually free group. Then*
1. *if $\mathrm{rk}(G) = 1$ then G is infinite cyclic,*
2. *if $\mathrm{rk}(G) = 2$ then either G is free of rank 2 or free Abelian of rank 2, and,*
3. *if $\mathrm{rk}(G) = 3$ then G is free of rank 3, free Abelian of rank 3 or G has a one-relator presentation*

$$G = \langle x_1, x_2, x_3 \mid x_3 v x_3^{-1} = v \rangle,$$

where $v = v(x_1, x_2)$ is a non-trivial element of the free group on x_1, x_2 which is not a proper power.

For a proof see [78]. The proof is based on the Nielsen method in HNN groups (see Section 1.6 for HNN groups). We list
1. $\langle a_1, b_1, a_2, b_2, \dots, a_g, b_g \mid \prod_{i=1}^{g} [a_i, b_i] = 1 \rangle$, $g \geq 2$, the orientable surface group of genus $g \geq 2$,

2. $\langle a_1, a_2, \ldots, a_g \mid a_1^2 a_2^2 \cdots a_g^2 = 1\rangle$, $g \geq 4$, the non-orientable surface groups of genus $g \geq 4$, and
3. the Baumslag double

$$\langle a_1, a_2, \ldots, a_n, b_1, b_2, \ldots, b_n \mid w(a_1, a_2, \ldots, a_n) = w(b_1, b_2, \ldots, b_n)\rangle, \quad n \geq 2,$$

$w = w(a_1, a_2, \ldots, a_n)$ a non-trivial element which is neither primitive nor a proper power in $\langle a_1, a_2, \ldots, a_n \mid\rangle$

as further interesting examples of fully residually free groups (see [78]).

We remark that groups $G = \langle a_1, a_2, \ldots, a_p, b_1, b_2, \ldots, b_q \mid u = v\rangle$, $p, q \geq 2$, $u = u(a_1, a_2, \ldots, a_p)$ and $v = v(b_1, b_2, \ldots, b_q)$ non-trivial elements which are neither primitive nor a proper power in $\langle a_1, a_2, \ldots, a_p \mid\rangle$ and $\langle b_1, b_2, \ldots, b_q \mid\rangle$, respectively, are commutative transitive but in general not residually free (see [11]).

We also mention that a finitely generated fully residually free group G has a faithful representation $\varphi: G \to \mathrm{PSL}(2, \mathbb{C})$. A faithful representation can be effectively constructed. Actually G can be embedded in $\mathrm{PSL}(2, \mathbb{R})$; see [78]. We finally remark that the above concepts, especially the equivalence, are extended to more classes of groups and Lie algebras (see [36], [55], and [193]).

1.1.4 Universally free groups

We have observed that being fully residually free implies commutative transitivity. B. Baumslag [10] provided a converse; commutative transitivity together with residually free implies being fully residually free. Gaglione and Spellman [110] and independently Remeslennikov [217] then were able to show that these conditions in the presence of residual freeness are equivalent to universal freeness in the non-Abelian case. Before we state and prove the respective connection we need to explain some first order logic.

We start with a first order language appropriate for free group theory. This language, which we denote by L_0, is the first order language with equality (always interpreted as the identity relation) containing a binary operation symbol $\cdot$ (often suppressed in favor of juxtaposition), unary operation symbol $^{-1}$, and a constant symbol 1.

In particular, this is what it means for L_0 to be appropriate for group theory. A *formula* in this language is a logical expression containing a string of variables $\overline{x} = (x_1, x_2, \ldots, x_n)$, the logical connectives $\vee$, $\wedge$, $\neg$, and the quantifiers $\forall$, $\exists$. Here $\vee$ stands for the conjunction of two propositions, $\wedge$ for the disjunction of two propositions and $\neg$ for the negation.

A variable in a formula is called *bound* (or occurs bound) if it is restricted by a quantifier ($\forall$, $\exists$). Otherwise, the variable is called *free*. A *sentence* is a formula of L_0 in which all variables are bound, or in other words there are no free occurrences of any variable. A *universal sentence* in L_0 is one of the form $\forall\overline{x}(\varphi(\overline{x}))$ where $\overline{x} = (x_1, x_2, \ldots, x_n)$ is a tuple of distinct variables, $\varphi(\overline{x})$ is a formula of L_0 containing no quantifiers and containing at most the variables of $\overline{x}$.

Similarly an existential sentence is one of the form $\exists\overline{x}(\varphi(\overline{x}))$ where $\overline{x}$ and $\varphi(\overline{x})$ are as above.

If G is a group then the *universal theory* of G consists of the set of all universal sentences of L_0 true in G. We denote the *universal theory* of a group G by $Th_\forall(G)$. Since any universal sentence is equivalent to the negation of an existential sentence, it follows that two groups have the same universal theory if and only if they have the same *existential theory*.

We start with the following observation that all non-Abelian free groups have the same universal theory. For the argument we need a little set theory which, for instance, can be found in the book by Halmos [122]. Let α, β be ordinal numbers with $\alpha < \beta$. Let F_α be the free group of rank α, and F_β the free group of rank β. Then there is a natural embedding of F_α into F_β as a free factor (see Section 1.4). Furthermore if w is the first limit ordinal then there is an embedding of F_w into F_2, the free group of rank 2, and hence into F_n for all $2 \leq n < w$ (see point 6 of Remark 1.1.15). However F_w cannot be a free factor for any finite n. It follows that if F and G are any two countable non-Abelian free groups, then each is embeddable in the other. Notice that if $\forall\overline{x}(\varphi(\overline{x}))$ is a universal sentence true in a group then it must also be true in every subgroup. However there may be universal sentences in a subgroup that are not true in the overgroup.

It follows that if H is embedded in G then the universal theory of G is embedded in the universal theory of H, that is

$$Th_\forall(G) \subset Th_\forall(H).$$

Since any two countable non-Abelian free groups are embeddable in each other, this easily shows that any two countable non-Abelian free groups have the same universal theory. By a theorem of Vaught (see [116]) every free group of rank $r \geq w$ is universally equivalent to F_w. Combining these observations we obtain the following Theorem.

Theorem 1.1.38. *Any two non-Abelian free groups have exactly the same universal theory and hence exactly the same existential theory.*

Definition 1.1.39. We call a group *universally free* if it has the same universal theory as the class of non-Abelian free groups.

Lemma 1.1.40. *Any universally free group G is CT.*

Proof. Commutative transitivity is given by the universal sentence

$$\forall x, y, z((y \neq 1) \wedge (xy = yx) \wedge (yz = zy) \to (xz = zx)).$$

This universal sentence holds in any free group and hence in any universally free group. Therefore universally free groups are CT. □

Remarks 1.1.41. We know that two non-Abelian free groups have exactly the same universal theory. Hence, we may formally simplify the definition: A universally free group is a group that has the same universal theory as the class of countable non-Abelian free groups.

Lemma 1.1.42. *Let G be a non-Abelian residually free group. If G is universally free then G is fully residually free.*

Proof. By Lemma 1.1.40 we know that G is CT. Since G is residually free, it follows from Theorem 1.1.30 that it is fully residually free. □

Lemma 1.1.43. *Let G be a non-Abelian fully residually free group. Then G is universally free.*

Proof. Let a, b be non-trivial non-commuting elements of G. Then the subgroup $H = \langle a, b\rangle$ is a rank 2 free subgroup of G. Since all countable non-Abelian free groups have the same universal theory, H is universally free. We claim that H and G are universally equivalent, that is, they have the same universal theory.

We prove this in a bit more general context. For groups A and B we say that B is fully residually A if given any finite set $\{b_1, b_2, \ldots, b_n\}$ of non-trivial elements of B there is an epimorphism $\phi : B \to A$ such that $\phi(b_i) \neq 1$ for all b_i.

Claim. *Suppose that A is a subgroup of B. Then, if B is fully residually A, it follows that A and B are universally equivalent.*

Proof of the claim. Let $Th_\forall(A)$, $Th_\forall(B)$ be respectively the universal theories of A and B. Since $A \subset B$, we have $Th_\forall(B) \subset Th_\forall(A)$.

Now, two groups have the same universal theory if and only if they have the same existential theory, by using negations. Hence it suffices to show that every true existential sentence in the overgroup B is also true in A. From this it is clear that the groups A and B have the same universal theory if and only if the following is true: every finite system

$$\begin{aligned} p_i(x_1, x_2, \ldots, x_n) &= 1, \quad 1 \le i \le I, \\ q_j(x_1, x_2, \ldots, x_n) &\neq 1, \quad 1 \le j \le J \end{aligned}$$

of equations and inequations has a solution in A if and only if it has a solution in B.

Suppose that $(b_1, b_2, \ldots, b_n)$ is a solution for a given system in B. Since B is fully residually A, there is an epimorphism $h : B \to A$ with $h(b_i) \neq 1$ which preserves the equations and inequations. Therefore the image will give a solution in A. It follows that $Th_\forall(A) \subset Th_\forall(B)$ and hence A and B are universally equivalent. This proves the claim. □

Now we show that G is fully residually H completing the proof of Lemma 1.1.43. Now $H = \langle a, b\rangle$ and let $c = [a, b] \neq 1$. Let $g_1, g_2, \ldots, g_n$ be any non-trivial elements of G and let $S = \{g_1, g_2, \dots, g_n, c\}$. Since G is fully residually free, there exists a homomorphism $\phi : G \to F_s$ with F_s a free group and $\phi(g_i) \neq 1$ for $i = 1, 2, \dots, n$ and $\phi(c) \neq 1$.

Since $\phi(c) \neq 1$, it follows that $\phi(a) \neq 1$ and $\phi(b) \neq 1$, and hence F_s is non-Abelian. Using Baumslag's big powers argument [14], we straightforwardly get that therefore F_s is fully residually a rank 2 free group and hence there is a homomorphism $\varphi : F_S \to \langle x, y \rangle$ with $\{x, y\}$ a free group, and such that $\varphi(\phi(g_i)) \neq 1$ for $i = 1, 2, \dots, n$. Now, let α be the isomorphism from $\langle x, y \rangle$ to H given by $x \mapsto a, y \mapsto b$.

Combining all these maps we have an epimorphism

$$\alpha(\varphi(\phi)) : G \to H \quad \text{such that} \quad \alpha(\varphi(\phi))(g_i) \neq 1$$

for $i = 1, 2, \dots, n$.

Therefore G is fully residually H, and H and G are universally equivalent. □

We remark that we will consider the big powers argument later in detail.

If we combine Theorems 1.1.30 and 1.1.36 with Lemmas 1.1.42 and 1.1.43 then we get the Theorem of B. Baumslag, Gaglione-Spellman and Remeslennikov:

Theorem 1.1.44. *Let G be a non-Abelian and residually free group. Then the following are equivalent:*
(1) *G is fully residually free,*
(2) *G is CT,*
(3) *G is CSA,*
(4) *G is universally free.*

There have been various extensions of this theorem. Of special interest is the following result of Remeslennikov [217].

Theorem 1.1.45. *Let G be a finitely generated universally free group. Then G is fully residually free.*

In the final part of this section we want to construct fully residually free groups and universally free groups. Here we use explicitly Baumslag's big powers argument and explain it more fully.

Definition 1.1.46. Let G be a CT group, let $a \in G \setminus \{1\}$ and let $M = C_G(a)$ where $C_G(a)$ is the centralizer of a in G. Suppose A is an Abelian group. Then the group

$$H = \langle G, A \mid rel(G), rel(A), [A, z] = 1, \forall z \in M \rangle$$

is a *centralizer extension* of G by A.

If $A = \langle t \mid \ \rangle$ is infinite cyclic, then

$$H = G(a, t) = \langle G, t \mid rel(G), t^{-1}zt = z, \forall z \in M \rangle$$

and is called the *free rank one extension of the centralizer M of a in G.*

What becomes crucial is the fact that if G is fully residually free and A is a fully residually free Abelian group, then a centralizer extension of G by A is again fully residually free. To prove this result we need to prove it for free rank one extensions of the centralizers and to do this we apply Baumslag's big powers lemma.

Lemma 1.1.47 (Big Powers Lemma). *Let k be any given positive integer and let F be a given free group. Suppose that $u, b_1, b_2, \dots, b_k \in F$.*

Furthermore, suppose that

$$b_1 u^{n_1} b_2 u^{n_2} \cdots b_k u^{n_k} = 1 \tag{1.1}$$

for infinitely many integral values of n_1, infinitely many integral values of $n_2, \dots,$ and infinitely many integral values of n_k. Then there exists an i $(1 \le i \le k)$ such that $b_i u = u b_i$.

Proof. Let F be freely generated by a set X. Then each element $f \in F, f \ne 1$, can be written uniquely in the form

$$f = x_1^{\epsilon_1} x_2^{\epsilon_2} \cdots x_n^{\epsilon_n}, \quad \epsilon_i = \pm 1, \quad x_i \in X, \quad x_i^{\epsilon_i} x_{i+1}^{\epsilon_{i+1}} \ne 1.$$

Then $|f| = n$ is the length of f. If $n = 1$ or $x_1 \ne x_n$ then f is cyclically reduced. A subword $x_1^{e_1} x_2^{e_2} \cdots x_m^{e_m}$, $m \le n$ of f is called an initial segment of f, similarly $x_m^{e_m} x_{m+1}^{e_{m+1}} \cdots x_n^{e_n}$, $m \ge 1$, is called a final segment of f.

If $g \in F$, we say that g reacts with f if $|gf| < |g| + |f|$. If g does not react with f, we shall often write $g \wedge f$. We accomplish the proof of Lemma 1.1.47 by proving two lemmas which concern the way in which elements react in a free group.

Lemma 1.1.48. *Let F be a free group an let $a, b \in F$. Suppose that $a \ne 1 \ne b$ and a is cyclically reduced. Then there exists a positive integer m such that $a^{m+1} b a^{m+1} = a \wedge a^m b a^m \wedge a$, $a^m b a^m \ne 1$.*

Proof. There are four main possibilities that can arise and we shall dispose of them in turn.

(I) $a \wedge b$ and $b \wedge a$.

We may here choose $m = 1$ and the required result follows immediately.

(II) a reacts with b, $b \wedge a$.

This case is best dealt with by splitting $\wedge$ up into four subcases.

Subcase 1: $a = b^{-1}$.

Since a is cyclically reduced, we have $a^2 b a^2 = aabaa = a \wedge a \wedge a$. So we can choose $m = 1$.

Subcase 2: a^{-1} is an initial segment of b, $a \ne b^{-1}$.

We choose i to be the largest positive integer satisfying $b = a^{-i} \wedge b_1, b_1 \in F$. Consider now

$$a^{i+1} b a^{i+1} = a b_1 a^{i+1}.$$

If $a \wedge b_1$ then

$$a^{i+2}ba^{i+2} = a \wedge ab_1a^{i+1} \wedge a$$

since $b_1 \wedge a$. Therefore $ab_1a^{i+1} \neq 1$, and we may take $m = i - 1$. The same reasoning applies if b_1^{-1} is neither a final segment of a nor 1.
If $b_1 = 1$, then the fact that a is cyclically reduced allows us to take $m = i + 1$ once more.
Thus we are left with the possibility that $a = a_1 \wedge b_1^{-1}$, $a_1 \neq 1$. So $a^{i+1}ba^{i+1} = a_1a^{i+1}$.
If a_1^{-1} is not an initial element of a, we may again chose $m = i + 1$.
Alternatively, the supposition that a_1^{-1} is an initial segment of a implies $a_1 = a_1^{-1}$. Since $a_1 \neq 1$, this is impossible, and the proof of subcase 2 is complete.

Subcase 3: b^{-1} is a final segment of a, $a \neq b^{-1}$.

It follows that $a = a_1 \wedge b^{-1}$, $a_1 \neq 1$. Thus $aba = a_1a = a_1a_1 \wedge b^{-1}$. Since $a_1 \neq 1$, it follows that $a_1^2b^{-1} \neq 1$ and

$$a^2ba^2 = a \wedge a_1^2b^{-1} \wedge a.$$

So here we can choose $m = 1$.

Subcase 4: a reacts with b, but none of the subcases 1, 2, 3 occur.

These restrictive assumptions immediately imply that

$$a^2b^2 = a \wedge aba \wedge a \qquad \text{with } aba \neq 1.$$

So again we may choose $m = 1$.

The four subcases exhaust all the possibilities that can occur in case (II).

(III) $a \wedge b$, b reacts with a.
This means that a^{-1} reacts with b^{-1} and $b^{-1} \wedge a^{-1}$. So we can apply (II) to a^{-1} and b^{-1}. The result then follows by taking inverses.

(IV) a reacts with b, b reacts with a.
We choose r as large as possible subject to $b = a^{-r} \wedge \bar{b}$, $\bar{b} \in F$. Next we choose s as large as possible subject to $\bar{b} = b_1 \wedge a^s$. It follows immediately from their definitions that r and s are both non-negative. In addition we have

$$a^{r+s+1}ba^{r+s+1} = a^{s+1}b_1a^{r+1}.$$

By the definition of r and s, it is clear that a^{-1} is neither an initial nor a final segment of b_1. If $b_1 = 1$, it is clear that we may choose $m = r + s + 1$. If $b_1 \neq 1$, then the only way in which further reactions can take place in

$$a^{r+s+1}ba^{r+s+1} = a^{s+1}b_1a^{r+1} \tag{1.2}$$

is by b_1 reacting with the a's on either side, without cancelling away into b_1. Thus, if not all of b_1 react in (1.2), then we may simply take $m = r + s + 1$.
Alternatively, either b_1^{-1} is an initial or a final segment of a. Therefore we can apply the type of argument used in (II), subcase 2. In any event the upshot is that we can take $m = r + s + 2$ in both of these cases.
There is still one further possibility. For we may have

$$b_1 = b_2 \wedge b_3, \quad b_2 \neq 1 \neq b_3,$$

where b_2^{-1} is a final segment and b_3^{-1} is an initial one of a. So $ab_1a = a_1a_2$.
If $a_1a_2 = 1$ then

$$a^{r+s+2}ba^{r+s+2} = a^{s+1} \wedge aa \wedge a^r$$

and so $m = r + s + 2$ will satisfy the requirements of the lemma. There remains now only the possibility $a_1a_2 \neq 1$.
In this case we consider

$$aa_1a_2a.$$

If a_1^{-1} is an initial element of a_2 then we can apply (II); similarly if a_2^{-1} is a final segment of a_1. If neither of these two latter possibilities occur, then

$$a \wedge a_1a_2 \wedge a, \quad a_1a_2 \neq 1.$$

So in all cases the existence of an m is assured.

The possibilities (I), (II), (III) and (IV) are the only ones. This completes the proof of Lemma 1.1.48. □

We now come to the counterpart of Lemma 1.1.48.

Lemma 1.1.49. *Let F be a free group and let $a, b \in F$. Suppose that $a \neq 1$ and a is cyclically reduced. Then there exists a positive integer m such that either*

$$a^{-(m+1)}ba^{m+1} = a^{-1} \wedge a^{-m}ba^m \wedge a, \quad a^{-m}ba^m \neq 1 \quad \text{or} \quad ab = ba.$$

Proof. We do the proof by induction on $|b|$.

If $|b| = 0$ then $b = 1$ and here is nothing to prove.

We suppose now that $|b| = 1$. Then $b = x^e$, $e = \pm 1$, where x is a member of the given basis of F. The only elements of F which commute with b, therefore, are the powers of x. So if $ab \neq ba$, then a is not a power of x. This means $a = x^r \wedge a_1 \wedge x^s$, $a_1 \neq 1$, where a_1 does not begin or end with x or x^{-1} and r and s are (possibly zero) integers. So $a^{-1}ba = x^{-s}a_1^{-1} \wedge b \wedge a_1x^s$, and we can take $m = 1$.

Suppose now that Lemma 1.1.49 has been established for $|b| < n$.

Now, let $|b| = n$ and $n > 1$. Put $b_i = a^{-i}ba^i$, $i = 0, 1, 2, \ldots$. If $|b_i| < |b|$ for some (non-zero) i, we can find inductively $\overline{m} > 0$, such that

$$a^{-(\overline{m}+1)}b_i a^{\overline{m}+1} = a^{-1} \wedge a^{-\overline{m}} b_i a^{\overline{m}} \wedge a, \quad a^{-\overline{m}} b_i \overline{a}^{\overline{m}} \neq 1.$$

Thus, we can choose $m = \overline{m} + 1$.

Let us therefore suppose that $|b_i| \geq |b|$ for all i.

If $|b_i| = |b|$ for all i, then the fact that here are only finitely many elements of given length in a finitely generated free group implies $b_i = b$ for some $i > 0$. So $a^{-i}ba^i = b$; but in a free group such an equation implies $a^{-1}ba = b$.

Alternatively, by suitable relabeling we may suppose that $|a^{-1}ba| > |b|$. Now if neither the whole of a^{-1} nor the whole of a cancels away on forming $a^{-1}ba$, it is clear that we can choose the m of Lemma 1.1.49 to be 1. On the other hand suppose that either a or a^{-1} does cancel away completely on forming $a^{-1}ba$. For definiteness suppose that a^{-1} completely cancels away. Then we can write

$$a = b_1 \wedge a_1, \quad b = b_1 \wedge b_2, \quad a = b_2^{-1} \wedge a_1 \wedge a_2.$$

Hence $a^{-1}ba = a_2$. Clearly $|a_2| \leq |a| - |a_1| = |b_1| \leq |b|$. But $|a_2| = |a^{-1}ba| > |b|$ because $|a^{-1}ba| > |b|$. So we have arrived at a contradiction, and the proof of Lemma 1.1.49 is complete. □

We now may continue with the proof of Lemma 1.1.47, the big powers lemma. We use the notation used in the introduction for Lemma 1.1.47.

We consider first the case $k = 1$. It is clear that the existence of infinitely many integers n_1 satisfies (1.1) implies $b_1 = u = 1$, and therefore $b_1 u = ub_1$, and the big powers lemma holds for $k = 1$.

Let us now assume that $k > 1$. If $u = 1$, there is nothing to prove. Thus we also assume that $u \neq 1$.

In addition, there is no loss of generality in assuming also that u is cyclically reduced. Suppose now that

$$ub_i \neq b_i u, \quad i = 1, 2, \ldots, k.$$

It then follows from Lemma 1.1.48 and Lemma 1.1.49 that for each i, $i = 1, 2, \ldots, k - 1$, there exist positive integers n_i, n_{i+1} such that

$$u^{n_i} b_{i+1} u^{n_{i+1}} = u^{e_i} \wedge c_i \wedge u^{e_{i+1}}, \quad c_i \neq 1;$$

the integers e_i, e_{i+1} can be chosen in such a way, that $|e_i|$, $|e_{i+1}|$ tend to infinity as $|n_i|$, $|n_{i+1}|$ tend to infinity. It follows that there exists an integer s such that for any choice of $n_1, n_2, \ldots, n_k$ satisfying $|n_1|, |n_2|, \ldots, |n_k| > s$,

$$b_1 u^{n_1} b_2 u^{n_2} \cdots b_k u^{n_k} = b_1 \wedge u^{f_1} \wedge d_1 \wedge u^{f_2} \wedge \cdots \wedge d_{k-1} \wedge u^{f_k}$$

and

$$|b_1| < |u^{f_1} d_1 u^{f_2} d_3 \cdots u^{f_k}|,$$

here $f_i = \pm 1$ and $d_i \neq 1$, $i = 1, 2, \ldots, k$.

It follows immediately that there cannot be infinitely many integers n_1, infinitely many integers $n_2, \ldots$, infinitely many integers n_k satisfying (1.1). We have therefore a contradiction which invalidates $ub_i \neq b_i u$, $i = 1, 2, \ldots, k$. This finishes the proof of the big powers Lemma 1.1.47. □

Remarks 1.1.50. 1. A consequence of the big powers argument is that any non-Abelian free group is fully residually a rank 2 free group. In a sense this completes the proof of Lemma 1.1.43 where we used the big powers argument.
2. A rigorous and consequent application of Theorem 1.4.52 also shows that any non-Abelian free group is fully residually a rank 2 free group.

Theorem 1.1.51. *Let G be fully residually free group. Let $a \in G$ with $a \neq 1$ and let $M = C_G(a)$. Then the free rank one extension of the centralizer M of a in G,*

$$G(a, t) = \langle G, t \mid rel(G), t^{-1} z t = z \text{ for all } z \in M \rangle,$$

is also fully residually free.

Proof. Let $g_1, g_2, \ldots, g_k$ be finitely many non-trivial elements of $G(a, t)$. We may write

$$g_j = a_{0,j} t^{m_{1,j}} a_{1,j} t^{m_{2,j}} \cdots a_{N(j)-1,j} t^{m_{N(j),j}} z_j,$$

$j = 1, 2, \ldots, k$, where $N(j) \geq 0$, $a_{i,j} \in G \setminus M$, $m_{i,j} \in \mathbb{Z} \setminus \{0\}$ and $z_j \in M$.

Note that $a_{i,j} \in G \setminus M$ is equivalent to $[a_{i,j}, a] \neq 1$.

Now, since G is fully residually free, there is a free group F and an epimorphism $\phi : G \to F$ such that

$$\phi([a_{i,j}, a]) = [\phi(a_{i,j}), \phi(a)] \neq 1, \quad \text{for all } i, j.$$

Then each $\phi(a_{i,j}) \neq 1$ and $\phi(a) \neq 1$. Let $C_F(\phi(a)) = \langle f \rangle$ be the centralizer of $\phi(a)$ in F. Suppose that $\phi(z_j) = f^{e_j}$, $1 \leq j \leq k$. For each positive integer $n \in \mathbb{N}$ we may define an extension

$$\psi_n : G(a, t) \to F \text{ of } \phi \text{ by } \psi_{n|_G} = \phi, \quad \psi_n(t) = f^n.$$

Now fix a j, $1 \leq j \leq k$. Could we have $\psi_n(g_j) = 1$ for infinitely many $n \in \mathbb{N}$? If so we would have

$$\phi(a_{0,j})f^{m^n_{1,j}}\cdots\phi(a_{N(j)-1,j})f^{m^n_{N(j),j}}f^{e_j}=1$$

for infinitely many values of n. Applying the big powers lemma we conclude that $\phi(a_{i,j})f = f\phi(a_{i,j})$ for some i, $0 \le i \le N(j)-1$. For that $\phi(a_{i,j})$ we have

$$\phi(a_{i,j}) \in C_F(f) = \langle f\rangle = C_F(\phi(a))$$

and so

$$[\phi(a_{i,j}), \phi(a)] = 1$$

which contradicts our choice of ϕ. The contradiction shows that the set $S_j = \{n \in \mathbb{N} \mid \psi_n(g_j) \neq 1\}$ is a cofinite subset of $\mathbb{N}$, that is, its complement $S_j = \mathbb{N} \setminus S$ is finite.

Since this is true for all j, $1 \le j \le k$, we must have

$$S_1 \cap S_2 \cap \cdots \cap S_k \neq \emptyset.$$

To see this, notice that if $S_1 \cap S_2 \cap \cdots \cap S_k = \emptyset$ then the complement of $\emptyset = S_1 \cap S_2 \cap \cdots \cap S_k$ in $\mathbb{N}$ is $\mathbb{N}$, and $\mathbb{N}$ is a finite union of finite sets by the de Morgan's rule which is impossible.

Now choose $n \in S_1 \cap S_2 \cap \cdots \cap S_k$. Then $\psi_n(g_j) \neq 1$ for all j, $1 \le j \le k$. Therefore $G(a,t)$ is fully residually free. □

The proof of the general result depends on the fact that it can be reduced to the case of free rank one extensions of centralizers. Hence we have the following theorem.

Theorem 1.1.52. *Let G be a fully residually free group and A be an Abelian fully residually free group. Then a centralizer extension of G by A is again fully residually free.*

Example 1.1.53 (Examples for fully residually free groups).

1. The orientable surface group

$$S_g = \langle a_1, b_1, a_2, b_2, \ldots, a_g, b_g \mid [a_1,b_1][a_2,b_2]\cdots[a_g,b_g] = 1\rangle$$

   for $g \ge 2$ is fully residually free.
2. The non-orientable surface group

$$T_g = \langle a_1, a_2, \ldots, a_g \mid a_1^2 a_2^2 \cdots a_g^2 = 1\rangle$$

   for $g \ge 4$ is fully residually free.
3. Let F be a finitely generated non-Abelian free group and $u \in F$ be a non-trivial element which is neither primitive nor a proper power in F. Let $\overline{F}$ be an identical copy of F and $\overline{u}$ the corresponding element to u in $\overline{F}$. The group $K = \langle F, \overline{F} \mid u = \overline{u}\rangle$ is fully residually free. K is called a Baumslag double. Baumslag proceeded by embedding K in the free rank one extension of centralizers

$$H = \langle F, t \mid t^{-1}ut = u\rangle \quad \text{by} \quad K = \langle F, t^{-1}Ft\rangle.$$

Remark 1.1.54. In [78] and [95] we have shown the following result.

Theorem 1.1.55. *Let G be a finitely generated fully residually free group. Then G has a faithful representation $\varphi : G \to PSL(2, \mathbb{C})$. Further, a faithful representation $\varphi : G \to PSL(2, \mathbb{C})$ can be effectively constructed.*

Now, finitely generated torsion-free subgroups of $PSL(2, \mathbb{C})$ are CT. Hence, a finitely generated fully residually free group is universally free.

Now we consider the converse. Remeslennikov [217] and Gaglione and Spellmann [109] showed that finitely generated and universally free imply fully residually free.

Theorem 1.1.56. *Let G be a finitely generated universally free group. Then G is fully residually free.*

We consider here only the proof in the finitely presented case.

Proof. Suppose that G is finitely presented universally free. Let

$$G = \langle x_1, x_2, \ldots, x_n \mid R_1 = R_2 = \cdots = R_m = 1\rangle$$

be a finite presentation for G where $R_i = R_i(x_1, x_2, \ldots, x_n)$ are words in $x_1, x_2, \ldots, x_n$. Since a universally free group is CT, it suffices to show that G is residually free. Suppose W is a non-trivial element of G. Then W is given by $W = W(x_1, x_2, \ldots, x_n)$, a word in the given generators. Consider now the existential sentence

$$\exists x_1, x_2, \ldots, x_n \left(\left(\bigwedge_{i=1}^{m} R_i(x_1, x_2, \ldots, x_n) = 1 \right) \wedge (W(x_1, x_2, \ldots, x_n)) \neq 1 \right).$$

This existential sequence is clearly true in G. Since G is universally free it is also existentially free, so this existential sentence must be true in all non-Abelian free groups. Therefore in any non-Abelian free group F there exist elements $a_1, a_2, \ldots, a_n$ such that $R_i(a_1, a_2, \ldots, a_n) = 1$ for $i = 1, 2, \ldots, n$; and $W(a_1, a_2, \ldots, a_n) \neq 1$. Then the map from G into F given by $x_i \mapsto a_i$, for $i = 1, 2, \ldots, n$, defines a homomorphism where the image of W is non-trivial. Now from Theorem 1.1.44 we get that G is fully residually free. □

If we combine this with Theorem 1.1.56 we get finally the following Theorem.

Theorem 1.1.57. *Let G be a finitely generated non-Abelian group. Then G is fully residually free if and only if G is universally free.*

Kharlampovich and Myasnikov [159, 158] and independently Sela [237] described equivalent methods to construct finitely generated residually free groups. These lead to the following Theorem.

Theorem 1.1.58. *Let G be a finitely generated fully residually free group.*
1. *Then G is finitely presented.*
2. *A finitely generated subgroup H of G is also fully residually free.*

Definition 1.1.59. A group G is called *coherent* if finitely generated subgroups are finitely presented.

From Theorem 1.1.58 we get the following Corollary.

Corollary 1.1.60. *A finitely generated fully residually free group is coherent.*

Corollary 1.1.61. *Free groups, surface groups and co-compact Fuchsian groups are coherent.*

Proof. This is clear for free groups. Orientable surface groups of genus $g \geq 2$ and non-orientable surface groups of genus $g \geq 4$ are fully residually free, and hence coherent. We leave it as an exercise to show that orientable surface group of genus 1 are free Abelian groups of rank 2, and non-orientable surface groups of genus ≤ 3 are coherent.

Finitely generated subgroups of co-compact Fuchsian groups are cyclic groups, infinite dihedral groups or isomorphic to co-finite Fuchsian groups, and hence co-compact Fuchsian groups are coherent. □

There are some other prominent examples of coherent groups, for instance, finitely generated fundamental groups of 3-manifolds [234]. We mention some more later in the book.

The standard example of an incoherent group is the direct product of two free groups of rank 2. We will sketch the proof:

In the product $F_2 \oplus F_2$, let the first copy be generated by a, b, and the second copy be generated by x, y. Let $q: F_2 \oplus F_2 \to \mathbb{Z}$ denote the homomorphism that maps the generators a, b, x, y to 1. Then set $B = ba^{-1}, X = xa^{-1}, Y = ya^{-1}$ and use the Reidemeister–Schreier result (see Section 1.3) to obtain the presentation

$$K = \ker(q) = \langle X, Y, B \mid [B, X^i(YX^{-1})X^{-i}] \ (i \in \mathbb{Z})\rangle.$$

We remark that K is finitely generated but not finitely presented. To see this, note that K is an HNN extension of the free group $\langle X, Y\rangle$ with stable letter B where the subgroup generated by the $X^i(YX^{-1})X^{-i}$ for $i \in \mathbb{Z}$, consequently not finitely generated, is centralized by B (see Section 1.6). Now let K' be a group defined by a presentation using a finite subset of the relators in the above presentation. Note that K' is also an HNN extension of $\langle X, Y\rangle$ by B, but the subgroup of $\langle X, Y\rangle$ centralized by B is finitely generated, and so is a proper subgroup of the subgroup centralized by B in K. This shows that $K \neq K'$ and hence we know that K is not finitely presentable.

We would like to further point out that this example can be extended to a whole family of incoherent groups, namely those that contain a copy of $F_2 \oplus F_2$. Recall that we can embed F_2 in $\mathrm{SL}(2, \mathbb{Z})$. To see this, consider the generators $A := \left(\begin{smallmatrix} 1 & 2 \\ 0 & 1 \end{smallmatrix}\right)$ and $B := \left(\begin{smallmatrix} 1 & 0 \\ 2 & 1 \end{smallmatrix}\right)$.

Hence we can embed $F_2 \oplus F_2$ in $\mathrm{SL}(4, \mathbb{Z})$ with generators (A, I_2), (B, I_2), (I_2, A) and (I_2, B). More generally, we can embed $F_2 \oplus F_2$ in $\mathrm{SL}(n, \mathbb{Z})$ with $n \geq 4$ and conclude that these special linear groups are incoherent.

1.2 The Nielsen method in free groups

In this section we gather various notions within the framework of Nielsen's cancellation method in free groups.

1.2.1 Nielsen transformations and subgroups of free groups

We start with some definitions.

Definition 1.2.1. Let G be a group and $U = \{u_1, u_2, \ldots\}$ a subset of G. The following operations are called *elementary Nielsen transformations*:
(N1) replace one u_i by u_i^{-1};
(N2) replace one u_i by $u_i u_j$ where $j \neq i$;
(N3) delete u_i if $u_i = 1$.

Definition 1.2.2. 1. A (finite) product of elementary Nielsen transformations is called a *Nielsen transformation*.
2. A Nielsen transformation is *regular* if it is a finite product of transformations (N1) and (N2) only, otherwise *singular*.
3. The subset U is *Nielsen equivalent* to a subset V of G if there is a regular Nielsen transformation from U to V.

We use this general definition in the following sections where we consider group amalgams. In this section we concentrate on free groups $F = \langle X \mid \rangle$, where $X = \{x_1, x_2, \ldots\}$ is a basis of F.

The regular Nielsen transformations form a group A because each transformation of type (N1) or (N2) has an inverse transformation. This group contains all transpositions which fix all u_i up to finitely many:

$$\begin{aligned}(u_1, u_2) &\mapsto (u_1, u_2^{-1}) \mapsto (u_1 u_2^{-1}, u_2^{-1}) \mapsto (u_2 u_1^{-1}, u_2^{-1}) \\ &\mapsto (u_2 u_1^{-1}, u_1^{-1}) \mapsto (u_2 u_1^{-1}, u_1) \mapsto (u_2, u_1).\end{aligned}$$

Especially, if $U = \{u_1, u_2, \ldots, u_n\}$ then A contains the permutation group S_n, and A is generated by

$$\begin{aligned}\alpha &: (u_1, u_2, \ldots, u_n) \mapsto (u_1^{-1}, u_2, \ldots, u_n),\\ \beta &: (u_1, u_2, \ldots, u_n) \mapsto (u_1 u_2, u_2, \ldots, u_n),\\ \gamma &: (u_1, u_2, \ldots, u_n) \mapsto (u_2, u_1, u_3, \ldots, u_n),\\ \delta &: (u_1, u_2, \ldots, u_n) \mapsto (u_n, u_1, u_2, \ldots, u_{n-1}).\end{aligned}$$

We have $S_n = \langle \gamma, \delta \rangle$. Also we get $\langle U \rangle = \langle V \rangle$ if there is a Nielsen transformation from U to V.

Definition 1.2.3. 1. Let $U = \{u_1, u_2, \ldots\}$ be a subset of reduced words of the free group $F = \langle X \mid \rangle$. We are interested in the subgroup generated by U. Now, for each i there are unique words p_i and q_i such that $|p_i| = |q_i|$ and $u_i = p_i z_i q_i^{-1}$ where $z_i \in X \cup X^{-1}$ if $|u_i|$ is odd and $z_i = 1$ if $|u_i|$ is even. If $|u_i|$ is odd then z_i is the *central letter* of u_i and if $|u_i|$ is even then u_i has two *center letters*, namely the last letter of p_i and the first letter of q_i^{-1}.
2. Let $U = \{u_1, u_2, \ldots\} \subset F$. The set U is called *Nielsen reduced* (or U has the *Nielsen property*) if the following hold:
(NR0) $u_i \neq 1$ for all i;
(NR1) $|u_i^{\epsilon} u_j^{\eta}| \geq |u_i|$ if $i \neq j$ or $i = j$, $\epsilon = \eta$ $(\epsilon, \eta = \pm 1)$;
(NR2) $|u_k^{\epsilon} u_i u_j^{\eta}| > |u_k| - |u_i| + |u_j|$ if $k \neq i \neq j$ or $k = i$, $\epsilon = 1$ or $j = i$, $\eta = 1$ $(\epsilon, \eta = \pm 1)$.

Theorem 1.2.4. *Any finite subset $U = \{u_1, u_2, \ldots, u_n\}$ of reduced words of F can be transformed into a Nielsen reduced set by a finite number of elementary Nielsen transformations.*

Proof. We may assume that the basis X is finite because we only need finitely many $x_i \in X$ to describe U. Let $u_i \neq 1$ for all i (otherwise apply (N3)). Assume that U does not satisfy (NR1). Then there exist $v_i, v_j \in U \cup U^{-1}$ with $|v_i v_j| < |v_i|$ if $i \neq j$ or $i = j$ and $v_i = v_j$. We have $i \neq j$ because $|u^2| \geq |u|$ in F. Now we may apply (N2) and replace v_i by $v_i v_j$ which makes the sum $\sum_{\ell=1}^{n} |u_\ell|$ smaller. This we call a *Nielsen process of the first kind*. By induction we can suppose that, after finitely many steps, U satisfies (NR0) and (NR1).

Now, we assume that U does not satisfy (NR2). Then there exist $x = u_k^{\epsilon}, y = u_i, z = u_j^{\eta}$ with

$$|xy| \geq |x|, |y| \quad \text{and} \quad |yz| \geq |y|, |z| \quad \text{and} \quad |xyz| \leq |x| + |z| - |y|.$$

Then we may write x, y, z as $x \equiv ap^{-1}, y \equiv pbq^{-1}, z \equiv qc$ with $xy \equiv abq^{-1}$ and $yz \equiv pbc$. Since

$$|xyz| = |abc| = |x| + |z| - |y| + 2|b| \leq |x| + |z| - |y|$$

we get $b = 1$. Furthermore, we have $|p| = |q| \leq |a|, |c|$ and $p \neq q$ because $y \neq 1$. If we replace $x \equiv ap^{-1}$ by $xy \equiv aq^{-1}$ or $z \equiv qc$ by $yz \equiv pc$ then $\sum_{\ell=1}^{n} |u_\ell|$ is unchanged.

To make sure which of the two possible replacements is suitable we have to proceed by introducing an order relation $\prec$, somewhat like a lexicographical order. We choose a well-ordering of the set $X \cup X^{-1}$ and consider the induced lexicographic order on the set of the reduced words in F. If $w \in F$ then let $a(w)$ be the initial segment of w with the length $[\frac{|w|+1}{2}]$. Define $\{w_1, w_1^{-1}\} \prec \{w_2, w_2^{-1}\}$ if $|w_1| < |w_2|$ or $|w_1| = |w_2|$ and $\min(a(w_1), a(w_1^{-1})) < \min(a(w_2), a(w_2^{-1}))$ or $|w_1| = |w_2|$ and $\min(a(w_1), a(w_1^{-1})) = \min(a(w_2), a(w_2^{-1}))$ and $\max(a(w_1), a(w_1^{-1})) < \max(a(w_2), a(w_2^{-1}))$.

If $\{w, w^{-1}\}$ is given then there are only finitely many pairs $\{w_1, w_1^{-1}\}$ with $\{w_1, w_1^{-1}\} \prec \{w, w^{-1}\}$. For short we just write $w_1 \prec w$. If in the above situation $p < q$ then $yz \equiv pc \prec qc \equiv z$, and we replace z by yz.

If $q < p$ then $xy \equiv aq^{-1} \prec ap^{-1} \equiv x$, and we replace x by xy. Such a replacement is called *Nielsen process of the second kind*. We may apply the Nielsen process of the second kind only finitely many times in a row. If necessary we start again with the Nielsen process of the first kind and deleting trivial elements, and so on. This gives finally the desired result. □

Remark 1.2.5. In [246] an algorithm, using elementary Nielsen transformations, is presented which, given a finite set S of n words of a free group, returns a set S' of Nielsen reduced words, such that $\langle S\rangle = \langle S'\rangle$; the algorithm runs in $\mathcal{O}(\ell^2 n^2)$, where ℓ is the maximum free length of a word in S.

Corollary 1.2.6. *Let U be a Nielsen reduced subset of F. For each $u \in U \cup U^{-1}$ there exist words $\ell(u)$, $m(u)$ and $r(u)$ with $m(u) \neq 1$ such that*
(a) $u \equiv \ell(u)m(u)r(u)$;
(b) *If $w = u_1u_2\cdots u_t$, $t \geq 0$, $u_i \in U \cup U^{-1}$, all $u_iu_{i+1} \neq 1$, then $m(u_1), m(u_2), \ldots, m(u_t)$ remain uncanceled in the reduced form of w.*
Especially, $|w| = |u_1u_2\cdots \ell(u_i)| + |m(u_i)| + |r(u_i)u_{i+1}\cdots u_t|$ for each i, $1 \leq i \leq t$.

Proof. For each $u \in U \cup U^{-1}$ let $\ell(u)$ and $r(u)$ the longest initial and end segment of u which is canceled in a product $vu \neq 1$ and $uv \neq 1$, $v \in U \cup U^{-1}$, respectively. From (NR2) we have $u = \ell(u)m(u)r(u)$ for some $m(u) \neq 1$.

Now, let w be as above. Let w' be the result after all cancellations between adjacent factors u_i and u_{i+1}. Then $w' = m_1'm_2'\cdots m_t'$ where m_i' is a segment of u_i which contains $m(u_i)$. Then $m_i' \neq 1$ and $w' \equiv m_1'm_2'\cdots m_t'$. □

Corollary 1.2.6 allows us to distinguish a *stable letter* $x_j^{\pm 1} \in m(u)$ for each $u \in U \cup U^{-1}$. We can recognize this letter in each freely reduced word between the elements from U the position where u appears. For u^{-1} we may take as a stable letter the inverse of that from u. We have as a stable letter of u certainly the central letter if $|u|$ is odd and one of the two central letters if $|u|$ is even.

Corollary 1.2.7. *Let U be a Nielsen reduced set and $w = u_1u_2\cdots u_k$, $k \geq 0$, $u_i \in U \cup U^{-1}$, all $u_iu_{i+1} \neq 1$. Then*
(a) $|w| \geq k$;
(b) $|w| \geq \frac{1}{2}|u_1| + \frac{1}{2}|u_k| + k - 2$ *for* $k \geq 2$;
(c) $|w| \geq |u_i|$ *for all* i.

Proof. (a) For each u_i at least one stable letter survives.
(b) Let $u_1 \equiv p_1z_1q_1^{-1}$ and $u_k \equiv p_kz_kq_k^{-1}$ be as above with $|p_1| = |q_1|$, $|p_k| = |q_k|$. At least p_1 and q_k^{-1} survive and at least one stable letter from each u_i, $2 \leq i < k$.
(c) We have $|u_i| \leq |u_iu_{i+1}| \leq \cdots \leq |u_i\cdots u_k| \leq |u_1\cdots u_k| = |w|$. □

Theorem 1.2.8. *A Nielsen reduced set $U \subset F$ freely generates $\langle U\rangle$.*

Proof. Let $w = u_1u_2\cdots u_k$, $k \geq 1$, with $u_iu_{i+1} \neq 1$. Then $w \neq 1$ because $|w| \geq k$. □

Corollary 1.2.9 (Nielsen's subgroup theorem). *Every finitely generated subgroup of F is free.*

Proof. By Theorem 1.2.4 each finitely generated subgroup of F has a Nielsen reduced generating system and, hence, is free by Theorem 1.2.8. □

Corollary 1.2.10. *Let F be free of rank $n \in \mathbb{N}$.*
1. *Then F cannot be generated by less than n elements.*
2. *If $U = \{u_1, u_2, \dots, u_n\}$ generates F then U is a basis of F.*
3. *If $U = \{v_1, v_2, \dots, v_m\}$, $m \geq n$, generates F then there is a Nielsen transformation from $\{v_1, v_2, \dots, v_m\}$ to a system $\{u_1, u_2, \dots, u_n, 1, \dots, 1\}$ and $\{u_1, u_2, \dots, u_n\}$ freely generates F.*

We leave the proof of Corollary 1.2.10 as an exercise.

Corollary 1.2.11. *Let F be free with basis X and let U be a Nielsen reduced subset. Then $(X \cup X^{-1}) \cap \langle U \rangle = (X \cup X^{-1}) \cap (U \cup U^{-1})$. If especially U is a basis of F then $X \cup X^{-1} = U \cup U^{-1}$.*

Proof. Let $x \in X \cap \langle U \rangle$. Then $x = u_1 u_2 \cdots u_k$, $k \geq 0$, $u_i \in U \cup U^{-1}$, all $u_i u_{i+1} \neq 1$. By Corollary 1.2.7 we have $1 = |x| \geq k$, hence, $k = 1$ and $x = u_1 \in U \cup U^{-1}$. Hence $(X \cup X^{-1}) \cap \langle U \rangle \subset (X \cup X^{-1}) \cap (U \cup U^{-1})$. The other inclusion is clear.

If $F = \langle U \rangle$, then

$$X \cup X^{-1} = (X \cup X^{-1}) \cap \langle U \rangle = (X \cup X^{-1}) \cap (U \cup U^{-1}) \subset U \cup U^{-1}.$$

By Theorem 1.2.8, F is free on U. Hence,

$$X \cup X^{-1} = U \cup U^{-1}.$$

□

Corollary 1.2.12. *Let F be a free group of rank n. Then the group* Aut(F) *of the automorphisms of F is generated by the finitely many elementary Nielsen transformations* (N1) *and* (N2).

More explicitely: Each automorphism of F is describable as a Nielsen transformation between two bases of F, and each Nielsen transformation between two bases of F defines an automorphism of F. Especially, Aut(F) *is finitely generated.*

Proof. Let $\varphi \in \operatorname{Aut}(F)$. Then $U = \varphi(X)$ is a basis of F. By Corollary 1.2.11 there is a regular Nielsen transformation which transfers U to X. On the other side, it is clear that each regular Nielsen transformation between two bases defines an automorphism of F. □

Let F be a free group of finite rank n. We now know that the automorphism group $\operatorname{Aut}(F)$ of F is finitely generated. Later in this section we will sketch a proof that $\operatorname{Aut}(F)$ is finitely presentable.

Theorem 1.2.13 (Nielsen–Schreier subgroup theorem). *Each subgroup of a free group is free.*

Proof. Since each non-empty set has a well-ordering, we may introduce the order relation from the proof of Theorem 1.2.4 for any free generating system $X \neq \emptyset$ of the free group $F = F(X)$. Let $X \neq \emptyset$ and G be a subgroup of $F = F(X)$. For each $g \in G$ we define $G_g = \langle h \mid h \in G \text{ and } h \prec g \rangle$. Then $A = \{g \mid g \in G \text{ and } g \notin G_g\}$ is a Nielsen reduced basis for G. This we see as follows.

1. A generates G: Assume that $G \neq \langle A \rangle$. Let $g^{\pm 1}$ be the smallest element (with respect to $\prec$) of $G \setminus \langle A \rangle$. All $h \in G$ with $h \prec g$ are in $\langle A \rangle$, and therefore $g \notin G_g$. But then $g \in A$ by definition which gives a contradiction. Hence, $\langle A \rangle = G$.
2. A is Nielsen reduced: Let x and y be two distinct elements from A with, say, $x \prec y$ and $xy \prec y$. Then $y \in \langle x, xy \rangle$ which contradicts $y \in A$. Hence, A is Nielsen reduced. □

Theorem 1.2.14 (Solution of the generalized word problem). *Given a finitely generated subgroup H of a free group F, there is an effective procedure to determine when an element of F lies in H.*

Proof. By using the method of the proof of Theorem 1.2.4, we may assume that H is given by a Nielsen reduced set of generators. The argument given in Corollary 1.2.7 shows that if a word of length n lies in H then it is a product of at most n occurrences of the generators of H. This can be tested by simply examining all possibilities. □

1.2.2 Special equations in free groups

Definition 1.2.15. Let F be a free group of finite rank and G a free subgroup of finite rank. An element $g \in G$ is called *primitive* element of G if there is a basis U of G with $g \in U$.

Theorem 1.2.16. *Let F be a free group on $a_1, a_2, \ldots, a_n$ and*

$$P(a_1, a_2, \ldots, a_n) = a_1^{\alpha_1} a_2^{\alpha_2} \cdots a_p^{\alpha_p} [a_{p+1}, a_{p+2}] \cdots [a_{n-1}, a_n]$$

with $0 \le p \le n$, $n - p$ even and $\alpha_i \ge 1$ for $i = 1, 2, \ldots, p$. Let $X = \{x_1, x_2, \ldots, x_m\}$ be any finite set in F and let $H = \langle X \rangle$. Suppose that H contains some conjugate of $P(a_1, a_2, \ldots, a_n)$. Then

(a) *$\{x_1, x_2, \ldots, x_m\}$ can be carried by a Nielsen transformation into a free basis for H which contains a conjugate of $P(a_1, a_2, \ldots, a_n)$; or*

(b) *$\{x_1, x_2, \ldots, x_m\}$ can be carried by a Nielsen transformation into a free generating system $\{y_1, y_2, \ldots, y_k\}$ for H with $m \ge k \ge n$, $y_i = z a_i^{\gamma_i} z^{-1}$, $1 \le \gamma_i < \alpha_i$, $\gamma_i \mid \alpha_i$ for $i = 1, 2, \ldots, p$, $y_j = z a_j z^{-1}$ for $j = p + 1, \ldots, n$ and $z \in F$.*

Proof. First of all we may assume that $P(a_1, a_2, \ldots, a_n) \in H = \langle X \rangle$ and that $X = \{x_1, x_2, \ldots, x_m\}$ is Nielsen reduced. In particular, as a freely reduced word in $a_1, a_2, \ldots, a_n$ each x_i^{ϵ}, $\epsilon = \pm 1$, contains an non-cancelable letter a_j^{ν}, $\nu = \pm 1$, which remains unchanged in each freely reduced word in $x_1, x_2, \ldots, x_m$ at that place where x_i^{ϵ} occurs. For each x_i^{ϵ},

$\epsilon = \pm 1$, choose a stable letter a_j^ν, $\nu = \pm 1$, and take the inverse letter for $x_i^{-\epsilon}$. We may choose the stable letter in such a way that $x_i^\epsilon \equiv y a_j^\nu z$ with $|x_i| = |y| + |z| + 1$ and either $|y| = |z|$ or $||y| - |z|| = 1$. It is now more convenient to write x_i^ϵ, $\epsilon = \pm 1$, in the freely reduced form $x_i^\epsilon \equiv u a_j^\beta v$ with $\beta \in \mathbb{Z}$, $\beta \neq 0$, and the reduced form of u does not end and the reduced form of v does not start with a power of a_j. In particular, $|x_i^\epsilon| = |u| + |\beta| + |v|$. For the Nielsen reduced set $\{x_1, x_2, \dots, x_m\}$ we have an equation

$$\prod_{j=1}^{q} x_{\nu_j}^{\epsilon_j} = U(x_1, x_2, \dots, x_m) = P(a_1, a_2, \dots, a_n) \tag{1.3}$$

with $\epsilon_j = \pm 1$, $\epsilon_j = \epsilon_{j+1}$ if $\nu_j = \nu_{j+1}$. Among the equations as in (1.3) there is one for which q is minimal, and let us assume that this is the case in equation (1.3). We may also assume that each x_i occurs in (1.3). If one x_i occurs only once in (1.3) as either x_i or x_i^{-1} case (a) holds, that is, $\{x_1, x_2, \dots, x_m\}$ can be carried by a Nielsen transformation into a free basis for H which also contains $P(a_1, a_2, \dots, a_n)$. Hence we may assume that each x_i either occurs twice in (1.3) with the same exponent $\epsilon = \pm 1$ or occurs in (1.3) exactly once with exponent $+1$ and once with exponent -1. If x_i occurs in (1.3) twice with the same exponent $\epsilon = \pm 1$ and has stable letter a_j^ν, $\nu = \pm 1$, then we must have $1 \le j \le p$ and $\nu = 1$ because of the special form of $P(a_1, a_2, \dots, a_n)$. Therefore, if x_i occurs twice in (1.3) with the same exponent $\epsilon = \pm 1$ we may assume that $\epsilon = +1$ and that x_i^{-1} does not occur in (1.3). If x_i and x_k both occur twice in (1.3) with the same exponent and if both have the same stable letter a_j, then there is no x_h^ϵ, $\epsilon = \pm 1$, in (1.3) between x_i and x_k which has a stable letter different from a_j again because of the special form of $P(a_1, a_2, \dots, a_n)$. In this situation it could be the case that $i \neq k$ and that at least two x_i have the same stable letter a_j, $1 \le j \le p$. Hence, if we again write $x_i^\epsilon \equiv u a_j v$, $\epsilon = \pm 1$, then we must consider subwords in (1.3) of the type

$$u_1 a_j^{\beta_1} v_1 u_i a_j^{\beta_i} v_i \cdots u_1 a_j^{\beta_1} v_1 u_k a_j^{\beta_k} v_k \tag{1.4}$$

and

$$u_i a_j^{\beta_i} v_i u_1 a_j^{\beta_1} v_1 \cdots u_k a_j^{\beta_k} v_k u_1 a_j^{\beta_1} v_1, \tag{1.5}$$

where $1 \le j \le p$, $1 \le i, k \le q$. Recall that each x_h with a stable letter a_t, $1 \le t \le p$, occurs twice in equation (1.3). We first consider the subword (1.4). Then $v_1 = u_i^{-1} = u_k^{-1}$. Let $z_1 \equiv u_i a_j^{\beta_i} v_i$ and $z_2 \equiv u_k a_j^{\beta_k} v_k \equiv u_i a_j^{\beta_k} v_k$. We have $\beta_i, \beta_k \ge 1$. If $z_1 \neq z_2$ then $z_1^{-1} z_2 = v_i^{-1} a_j^{\beta_k - \beta_i} v_k \neq 1$ and $z_2^{-1} z_1 = v_k^{-1} a_j^{\beta_i - \beta_k} v_i \neq 1$ which contradicts the fact that $\{x_1, x_2, \dots, x_m\}$ is Nielsen reduced. Hence we must have $z_1 = z_2$. For a subword (1.5) we get $u_1 = v_i^{-1} = v_k^{-1}$ and analogously $u_i a_j^{\beta_i} v_i \equiv u_k a_j^{\beta_k} v_k$. Altogether we see that a maximal subword $\prod x_{\nu_i}^{\epsilon_i}$ of (1.3) in which each x_{ν_i} has a_j, $1 \le j \le \beta$, as a stable letter (recall that here $\epsilon_i = +1$) must have the reduced form

$$(x_{\nu_{i_0}} \cdots x_{\nu_{i_0+j_0}})^\delta w,$$

where $w = 1$ or $w = x_{v_{i_0}} \cdots x_{v_{i_0+j}}$ and with $\delta \geq 2$, $0 \leq j \leq j_0 - 1$, $x_{v_{i_0+r}} \neq x_{v_{i_0+s}}$ for $r \neq s$ and $0 \leq r, s \leq j_0$. If $j_0 = 0$ then $x_{v_{i_0}}$ is conjugate to a power of a_j. If $w \neq 1$ then we replace $x_{v_{i_0+j_0}}$ by $x_0 = x_{v_{i_0}} \cdots x_{v_{i_0+j_0}}$ which defines a Nielsen transformation if the other x_i are left fixed. In this situation case (a) holds.

Now let $w = 1$. Without loss of generality we may assume that $v_{i_0} < v_{i_0+j_0}$. Then there is a Nielsen transformation from $\{x_1, x_2, \dots, x_m\}$ to the system $\{x_1, x_2, \dots, x_{v_{i_0-1}}, x_{v_{i_0}} \cdots x_{v_{i_0+j_0}}, x_{v_{i_0+1}}, \dots, x_m\}$, and $P(a_1, a_2, \dots, a_n)$ is already contained in the proper subgroup of $H = \langle X \rangle$, generated by

$$x_1, x_2, \dots, x_{v_{i_0-1}}, x_{v_{i_0}} \cdots x_{v_{i_0+j_0}}, x_{v_{i_0+1}}, \dots, x_{v_{i_0+j_0-1}}, x_{v_{i_0+j_0+1}}, \dots, x_m.$$

Hence, there is a Nielsen transformation from $\{x_1, x_2, \dots, x_m\}$ to a system $\{y_1, y_2, \dots, y_m\}$ with $P(a_1, a_2, \dots, a_n) \in \langle y_1, y_2, \dots, y_{m-1} \rangle$. Now we start with the set $\{y_1, y_2, \dots, y_{m-1}\}$ and argue as above. Therefore, without loss of generality, we may assume from the very beginning that there is no Nielsen transformation to a set $\{y_1, y_2, \dots, y_m\}$ with $P(a_1, a_2, \dots, a_n) \in \langle y_1, y_2, \dots, y_{m-1} \rangle$, that is, m is minimal with respect to this property. We also assume now that case (a) does not hold. If we argue as above, with $\{x_1, x_2, \dots, x_m\}$ Nielsen reduced and each x_i occurs in equation (1.3), we get the following. If x_i occurs twice in (1.3) with the same exponent $\epsilon = \pm 1$ then x_i is conjugate to a power of some a_j, $1 \leq j \leq p$, and no x_k with $k \neq i$ is also conjugate to a power of a_j.

On the other hand, for each a_j, $1 \leq j \leq p$, there is an x_i which is conjugate to a power of a_j for otherwise in the abelianized group F^{ab} there is some power a_j^{ν}, $\nu \neq 0$, $1 \leq j \leq p$, in the subgroup of F^{ab} generated by $a_1, a_2, \dots, a_{j-1}, a_{j+1}, \dots, a_n$ (recall that if x_i does not occur twice in (1.3) with the same exponent then it occurs in (1.3) exactly once with exponent +1 and once with exponent −1). Also the centralizer of a non-trivial word in F is cyclic. Therefore, without loss of generality, we may assume that $x_i = a_i^{\gamma_i}$, $1 \leq \gamma_i < \alpha_i$, $\gamma_i \mid \alpha_i$, for $i = 1, 2, \dots, p$, possibly after a suitable Nielsen transformation. Therefore we have reduced the proof to the following situation: For $x_{p+1}, \dots, x_m$ we have an equation

$$\prod_{j=1}^{t} x_{v_j}^{\epsilon_j} = [a_{p+1}, a_{p+2}] \cdots [a_{n-1}, a_n] \tag{1.6}$$

with $\epsilon_j = \pm 1$, $\epsilon_j = \epsilon_{j+1}$ if $v_j = v_{j+1}$, and each x_i, $p + 1 \leq i \leq m$, occurs in (1.6) exactly once with exponent +1 and once with exponent −1. The proof for this situation is quite technical and we refer to Chapter 5 of [281]. If we argue analogously to Chapter 5 we obtain finally $m = n$ and we have a Nielsen transformation from

$$\{a_1^{\gamma_1}, \dots, a_p^{\gamma_p}, x_{p+1}, \dots, x_m\} \quad \text{to} \quad \{a_1^{\gamma_1}, \dots, a_p^{\gamma_p}, a_{p+1}, \dots, a_n\}. \qquad \square$$

Theorem 1.2.16 has two straightforward consequences.

Corollary 1.2.17. *Let $F, P(a_1, a_2, \ldots, a_n)$, $X = \{x_1, x_2, \ldots, x_m\}$ and $H = \langle X\rangle$ be as in Theorem 1.2.16. Suppose that there are $u_1, u_2, \ldots, u_k \in H$, $k \geq 1$, with*

$$u_1^{\beta_1} \cdots u_q^{\beta_q}[u_{q+1}, u_{q+2}] \cdots [u_{k-1}, u_k] = P(a_1, \ldots, a_n),$$

where $0 \leq q \leq k$, $2 \leq \beta_j$ for $j = 1, 2, \ldots, q$ and $\gcd(\beta_1, \beta_2, \ldots, \beta_q) \geq 2$ if $q \geq 1$. Then $\{x_1, x_2, \ldots, x_m\}$ can be carried by a Nielsen transformation to a free generating system $\{y_1, y_2, \ldots, y_\ell\}$ for H with $m \geq \ell \geq n$, $y_i = a_i^{\gamma_i}$, $1 \leq \gamma_i < \alpha_i$, $\gamma_i \mid \alpha_i$ for $i = 1, 2, \ldots, p$, $y_j = a_j$ for $j = p + 1, \ldots, n$. In particular we have $k \geq n$. If in addition α_i, $1 \leq i \leq p$, is a prime number then $H = F$, and if further $m = n$ then X is a free generating system for F.

Proof. We just have to mention that $P(a_1, a_2, \ldots, a_n)$ cannot be a primitive element of H because of $\gcd(\beta_1, \beta_2, \ldots, \beta_q) \geq 2$.

We leave the details as an exercise. □

Corollary 1.2.18. *Let F be the free group on $a_1, a_2, \ldots, a_n$ and $P(a_1, a_2, \ldots, a_n) = a_1^{\alpha_1} a_2^{\alpha_2} \cdots a_p^{\alpha_p}[a_{p+1}, a_{p+2}] \cdots [a_{n-1}, a_n]$ with $0 \leq p \leq n$, $n - p$ even, $\alpha_i \geq 2$ for $i = 1, 2, \ldots, n$ and $\gcd(\alpha_1, \alpha_2, \ldots, \alpha_p) \geq 2$ if $p \geq 1$. If $\varphi: F \to F$ is an endomorphism which fixes $P(a_1, a_2, \ldots, a_n)$, that is, $\varphi(P(a_1, a_2, \ldots, a_n)) = P(a_1, a_2, \ldots, a_n)$ then φ is already an automorphism of F.*

If we make a restriction on m then we can get stronger results, for instance the following. Let F be a free group on $a_1, a_2, \ldots, a_n$ and

$$P(a_1, a_2, \ldots, a_n) = a_1^{\alpha_1} a_2^{\alpha_2} \cdots a_p^{\alpha_p}[a_{p+1}, a_{p+2}] \cdots [a_{n-1}, a_n],$$

with $0 \leq p \leq n$, $n-p$ even and $\alpha_i \geq 1$ for $i = 1, 2, \ldots, p$. Let $X = \{x_1, x_2, \ldots, x_m\}$ with $1 \leq m \leq n$ be a finite set in F and let $H = \langle X\rangle$. Suppose that H contains some $y^{-1}P^{\alpha}(a_1, a_2, \ldots, a_n)y$ with $\alpha \neq 0$ and $y \in F$. Then

(a) The set $\{x_1, x_2, \ldots, x_m\}$ can be carried by a Nielsen transformation into a free basis $\{y_1, y_2, \ldots, y_k\}$ with $y_1 = zP^{\beta}(a_1, a_2, \ldots, a_n)z^{-1}$, $\beta \geq 1$ and $z \in F$; or
(b) We have $m = n$, and $\{x_1, x_2, \ldots, x_n\}$ is Nielsen equivalent to a system $\{y_1, y_2, \ldots, y_n\}$ with $y_i = za_i^{\gamma_i}z^{-1}$, $1 \leq \gamma_i < \alpha_i$, $\gamma_i | \alpha_i$ for $i = 1, 2, \ldots, p$ and $y_j = za_jz^{-1}$ for $j = p + 1, \ldots, n$ and $z \in F$.

Since we referred at the end of the proof of Theorem 1.2.16, we at least want to give the proof for the commutator $[a, b]$ in the free group on $\{a, b\}$.

Theorem 1.2.19. *Let F be a free group of rank 2 on $\{a, b\}$. Let $u, v \in F$. The set $\{u, v\}$ is a free generating pair of F if and only if*

$$[u, v] = w[a, b]^{\pm 1}w^{-1} \quad \textit{for some } w \in F. \tag{1.7}$$

Proof. By Corollary 1.2.12 the automorphism group $\operatorname{Aut}(F)$ is generated by the Nielsen transformation

$$\alpha\colon (a,b) \mapsto (a^{-1},b),$$
$$\beta\colon (a,b) \mapsto (ab,b) \quad \text{and}$$
$$\gamma\colon (a,b) \mapsto (b,a).$$

Then

$$[\alpha(a),\alpha(b)] = [a^{-1},b] = a^{-1}[a,b]^{-1}a,$$
$$[\beta(a),\beta(b)] = [ab,b] = [a,b] \quad \text{and}$$
$$[\gamma(a),\gamma(b)] = [b,a] = [a,b]^{-1}.$$

Hence, if $\{u,v\}$ is a generating pair of F then (1.7) holds.

Now, let (1.7) hold. Let $G = \langle u,v\rangle$. After a suitable Nielsen transformation and considering the above behavior we may assume that

$$[a,b] = [u,v]$$

and $\{u,v\}$ is Nielsen reduced. If $|u| = 1$ then necessarily $u = a$, and we get $v = b$ because $\{u,v\}$ is Nielsen reduced. Analogously we handle the case $|v| = 1$ because then necessarily $v^{-1} = b^{-1}$, that is, $v = b$ and further $u = a$.

Now, let $|u| \geq 2$ and $|v| \geq 2$. We cannot have $|u| \geq 3$ because otherwise

$$4 = |w| \geq \frac{3}{2} + 1 + 4 - 2 = \frac{9}{2}$$

by Corollary 1.2.7. Hence, $|u| = 2$. Analogously $|v| = 2$. Then necessarily $u = ax^{\epsilon}$, $\epsilon = \pm 1$ and $x = a$ or b. The element x^{ϵ} cannot be a stable letter for u because otherwise

$$4 = |w| \geq |u| + 3 = 2 + 3 = 5$$

by Corollary 1.2.7. Hence, a is a stable letter for u. Analogously $v = by^{\eta}$, $\eta = \pm 1$ and $y = a$ or b, and b is a stable letter for v. Then

$$aba^{-1}b^{-1} = ax^{\epsilon}by^{\eta}x^{-\epsilon}a^{-1}y^{-\eta}b^{-1}.$$

But this is impossible because $a^{\pm 1}$ and $b^{\pm 1}$ remain uncanceled on the right hand side. Hence $|u| = 1$ or $v = |1|$, and the result holds. □

Corollary 1.2.20. *Let F be the free group of rank 2 on $\{a,b\}$ and $\varphi\colon F \to F$ be an endomorphism of F with $\varphi([a,b]) = g[a,b]^{\pm 1}g^{-1}$ for some $g \in F$. Then φ is already an automorphism. This especially holds if even $\varphi([a,b]) = [a,b]$.*

Before we continue to consider the properties described in Corollaries 1.2.17 and 1.2.20 we give some additional results which are described in [67]. R. Hidalgo asked whether the following is true: If $\{x,y\}$, $\{u,v\}$ are two sets of generating pairs for a free group F of rank 2 and $[x, y^n] = [u, v^m]$ with $n, m \geq 1$ must $n = m$? Hidalgo's question was motivated by a question on different classes of geodesics in specific Schottky groups. Here we give a proof which is quite unexpected in this context.

We need some preparation. Let $\{S_\nu(t)\}$ be the Chebyshev polynomials (of the second kind). These are defined by

$$\begin{aligned} S_0(t) &= 0, \\ S_1(t) &= 1, \\ S_\nu(t) &= tS_{\nu-1}(t) - S_{\nu-2}(t) \quad \text{for } \nu \geq 2, \quad \text{and} \\ S_\nu(t) &= -S_{-\nu}(t) \quad \text{for } \nu < 0. \end{aligned}$$

Now, let $A, B \in \mathrm{SL}(2, \mathbb{C})$ with

$$A \neq \pm E, \quad B \neq \pm E, \quad E = \begin{pmatrix} 1 & 0 \\ 0 & 1 \end{pmatrix},$$

and $\mathrm{tr}(A) = x$, $\mathrm{tr}(B) = y$ and $\mathrm{tr}(AB) = z$ for the traces.

We here assume $|x|, |y|, |z| \geq 2$. Then A, B and AB have infinite order. Then

(1) $A^n = S_n(x)A - S_{n-1}(x)E$ for $n \in \mathbb{N} \cup \{0\}$,

(2) $\mathrm{tr}([A^n, B^m] - 2) = S_n^2(x)S_m^2(y)(\mathrm{tr}([A, B] - 2))$ for $n, m \in \mathbb{N} \cup \{0\}$, and

(3) $A^n = (A^{-1})^{-n}$ for $n < 0$.

Now, let

(4) $R(A, B) = A^{n_1}B^{m_1}A^{n_2}B^{m_2} \cdots A^{n_k}B^{m_k}$ with $k \geq 1$, $n_i \neq 0 \neq m_i$ for $i = 1, 2, \ldots, k$ a reduced word. Then $\mathrm{tr}(R(A, B)) = a_{2k}z^{2k} + \cdots + a_1 z + a_0$ is a non-constant polynomial in z of degree $2k$ because

$$a_{2k} = S_{n_1}(x)S_{m_1}(y)S_{n_2}(x)S_{m_2}(y) \cdots S_{n_k}(x)S_{m_k}(y) \neq 0.$$

Theorem 1.2.21. *Let F be a free group of rank 2 and suppose that $\{x, y\}$, $\{u, v\}$ are two sets of generating pairs for F. Suppose further that $[x, y^n] = [u, v^m]$ with $n, m \geq 1$. Then $n = m$. Further, if $n = m \geq 2$ then y is conjugate in F to $v^{\pm 1}$.*

Proof. Let $x, y, u, v \in F$ be as above and suppose $m \leq n$. We choose $U, V \in \mathrm{SL}(2, \mathbb{R})$ with $\mathrm{tr}(U) = 3$, $\mathrm{tr}(V) = 2$ and $\mathrm{tr}(UV) = -\pi$. Then $\langle U, V \rangle$ is a discrete, free subgroup of $\mathrm{SL}(2, \mathbb{R})$ with exactly one conjugacy class of maximal parabolic elements (recall that an element $C \in \mathrm{SL}(2, \mathbb{R})$ is called *parabolic* if $C \neq \pm E$ and $|\mathrm{tr}(C)| = 2$), and we have a faithful representation $\rho: F \to \mathrm{SL}(2, \mathbb{R})$ with $\rho(u) = U$, $\rho(v) = V$, $\rho(x) = X$ and $\rho(y) = Y$. This can be easily seen by an inductive consideration of reduced words of the above type (4). Moreover, $|\mathrm{tr}(X)|, |\mathrm{tr}(Y)| \geq 2$. For more details one may look at the paper [92]. We have

$$\mathrm{tr}[U, V^m] - 2 = S_m^2(\mathrm{tr}(V))(\mathrm{tr}([U, V]) - 2) = m^2(\mathrm{tr}([U, V]) - 2)$$

because $\mathrm{tr}(V) = 2$ and

$$\operatorname{tr}([X, Y^n]) - 2 = S_n^2(\operatorname{tr}(Y))(\operatorname{tr}([X, Y]) - 2) = S_n^2(\operatorname{tr}(Y))(\operatorname{tr}[U, V] - 2)$$

by Theorem 1.2.19. Therefore $m^2 = S_n^2(\operatorname{tr}(Y))$.

Now, if $n \geq 2$ and $[\operatorname{tr}(Y)] > 2$ then $n^2 < S_n^2(\operatorname{tr}(Y))$. This gives in any case $n = m$ because $m \leq n$, and $|\operatorname{tr}(Y)| = 2$ if $n = m \geq 2$. Therefore Y is conjugate to a power of V. However, since v and y are primitive in F it follows that Y is conjugate in F to $V^{\pm 1}$, which shows that y is conjugate to $v^{\pm 1}$. □

In a similar fashion we may prove the following.

Theorem 1.2.22. *Let F be a free group of rank 2 and suppose that $\{x, y\}$, $\{u, v\}$ are sets of generating pairs for F. Suppose further that*

$$[u^n, v^m] = [x^p, y^q] \quad \text{with } n, m, p, q \geq 1.$$

Then

(1) *If $n, m, p, q \geq 2$ then*

$$n = p, \quad m = q \quad \text{or} \quad n = q, \quad m = p.$$

(2) *If $n = 1$ and $m, p, q \geq 1$ then $p = 1$ or $q = 1$.*

Proof. Choose, for instance $U, V \in \mathrm{SL}(2, \mathbb{R})$ with $\operatorname{tr}(U) = \operatorname{tr}(V) = 2$, $\operatorname{tr}(UV) = -\pi$ and argue analogously as in the proof of Theorem 1.2.21. □

We consider the respective equations in Theorems 1.2.21 and 1.2.22 for free products of two cyclic groups in Section 1.4 (in a slightly different setting). Boileau, Collins and Zieschang [30] considered such equations with similar methods in high-powered triangle groups to distinguish Heegard decompositions of small Seifert manifolds.

1.2.3 Test elements, almost primitive elements, and generic elements

The statements in Corollaries 1.2.18 and 1.2.20 describe a property which is used for the definition of a test element.

Definition 1.2.23. A *test element* in a group G is an element g with the property that if $f(g) = g$ for an endomorphism $f: G \to G$ then f must be an automorphism of G. If G is a free group then a test element is called a *test word*.

Nielsen in [206] gave the first non-trivial example of a test word by showing that in the free groups on $\{x, y\}$ the commutator $[x, y]$ satisfies the property (see Corollary 1.2.20). Corollary 1.2.18 describes different types of test words. We shortly consider the relationship between test words and two related concepts—almost primitive elements (APEs) and generic elements.

Definition 1.2.24. An *almost primitive element (APE)* is an element of a free group F that is not primitive in F but which is primitive in any proper subgroup of F containing it.

Corollary 1.2.17 describes examples of APEs. Later in this section we consider tame APEs.

Definition 1.2.25. An APE w in a free group F is called a *tame APE* if whenever $w^a \in H$ for a subgroup H of F with $a \geq 1$ minimal, then either w^a is primitive in H or the index $[F : H]$ is a.

Generic elements were introduced by Stallings, see [245].

Definition 1.2.26. Let $\mathcal{U}$ be a variety defined by a set of laws $\mathcal{V}$ (we refer to the book of H. Neumann, [198]). For a group G let $\mathcal{V}(G)$ be the verbal subgroup of G defined by $\mathcal{V}$. An element $g \in G$ is called *$\mathcal{U}$-generic* if $g \in \mathcal{V}(G)$ and whenever H is a group and $f: H \to G$ a homomorphism with $g = f(u)$ for some $u \in \mathcal{V}(H)$ it follows that f is surjective. An element $g \in G$ is called (general) generic if it is $\mathcal{U}$-generic for some variety $\mathcal{U}$.

Equivalently, $g \in G$ is $\mathcal{U}$-generic if $g \in \mathcal{V}(G)$ but $g \notin \mathcal{V}(K)$ for every proper subgroup K of G.

A typical variety is the variety $\mathcal{U}_n$ defined by the set of laws $V_n = \{[x, y], z^n\}$. For $n = 0$ we have $\mathcal{U}_n = \mathcal{A}$ the Abelian variety. If G is Hopfian, which is the case if G is free, then being generic in G certainly implies being a test element. There is no converse. In a free group of rank 3 on $\{x, y, z\}$ the word $w = x^2[y^2, z]$ is a test word but is not generic. Also generic does not imply APE. Suppose F is free of rank two on $\{x, y\}$ and let $w = x^4y^4$. Then w is $\mathcal{U}_4$-generic but w is not an APE since $w \in \langle x^2, y^2\rangle$, and w is not primitive in $\langle x^2, y^2\rangle$.

Further, in general it is not true that being an APE implies being a test word. Again, let F be free on $\{x, y\}$ and let $w = x^2yx^{-1}y^{-1}$. Then w is an APE but not a test word. However, many APEs are indeed generic and therefore test elements. Recall that a variety $\mathcal{U}$ defined by a set of laws $\mathcal{V}$ is a non-trivial variety if it consists of more than just the trivial group. In this case $\mathcal{V}(F) \neq F$ for any free group of rank ≥ 1.

Theorem 1.2.27. *Let F be a free group and $\mathcal{B}$ a non-trivial variety defined by a set of laws $\mathcal{V}$. Let $w \in \mathcal{V}(F)$. If w is an APE then w is $\mathcal{B}$-generic. In particular w is a test word.*

Proof. Let $w \in \mathcal{V}(F)$ be an APE and let $\varphi: H \to F$ be a homomorphism with $\varphi(u) = w$ for some $u \in \mathcal{V}(H)$. As in the statement of the theorem, $\mathcal{V}$ is the set of laws defining the non-trivial variety $\mathcal{B}$. Let K be a proper subgroup of F. If $w \notin K$ then certainly $w \notin \mathcal{V}(K)$. If $w \in K$, then, since w is an APE, w is primitive in K because K is a proper subgroup of F.

Further, since $\mathcal{B}$ is a non-trivial variety and K is free, we have $K \neq \mathcal{V}(K)$. It follows then from the primitivity of w in K that $w \notin \mathcal{V}(K)$. Therefore $w \in \mathcal{V}(F)$ and for any proper subgroup K of F we have $w \notin \mathcal{V}(K)$, and, hence, w is $\mathcal{B}$-generic and further a test word. □

Corollary 1.2.28. *Let $F(n)$ be the subgroup of the free group F generated by all commutators and nth powers, $n \geq 2$ or $n = 0$, that is, $F(n) = \mathcal{V}_n(F)$. Let $w \in F(n)$, $n \geq 2$ or $n = 0$. If w is an APE then w is $\mathcal{U}_n$-generic and a test word.*

Examples 1.2.29. 1. Let F be the free group of rank 2 on $\{x, y\}$. Let $\mathcal{U}_n$ be the variety as above and $\mathcal{L}_n$ be the variety generated by the laws $\mathcal{W}_n = \{[x^n, y^n]\}$, $n \geq 1$. Then
(a) $[x^n, y^n]$ is $\mathcal{L}_n$-generic in F but for $n \geq 2$ it is not $\mathcal{U}_n$-generic in F,
(b) $[x^n, y^m]$, $n, m \geq 1$, is an APE if and only if $n = m = 1$, and
(c) $[x^n, y^m]$ is a test word for any $n, m \geq 1$.
2. Let F be the free group of rank two on $\{x, y\}$. Then
(a) $a^n ba^{-1}b^{-1}$, $n \geq 2$, is an APE,
(b) $a^n ba^{-1}b^{-1}$, $n \geq 3$, is $\mathcal{U}_{n-1}$-generic,
(c) $a^n ba^{-1}b^{-1}$, $n \geq 3$, is a test word in F, and
(d) $a^2 ba^{-1}b^{-1}$ is not a test word in F.
We leave these examples as an exercise. More details and results can be found in [97].

We now want to describe examples of tame *APEs*.

Theorem 1.2.30. *Let F be a free group on $a_1, a_2, \ldots, a_p$, $p \geq 2$, p even, and*

$$P(a_1, a_2, \ldots, a_p) = [a_1, a_2], [a_2, a_3] \cdots [a_{p-1}, a_p] \in F.$$

Let $X = \{x_1, x_2, \ldots, x_m\}$, $m \geq 1$, be a finite set of elements of F and let $H = \langle X \rangle$ be the subgroup of F generated by X. Suppose H contains some conjugate of some non-trivial power of $P(a_1, a_2, \ldots, a_n)$, and let β be the smallest positive integer such that some conjugate of $P^\beta(a_1, a_2, \ldots, a_n)$ lies in H. Then
(a) *There is a Nielsen transformation from X to a free basis for H which contains a conjugate of $P^\beta(a_1, a_2, \ldots, a_p)$; or*
(b) *The index of H in F is β, and*

$$1, P(a_1, a_2, \ldots, a_p), \ldots, P^{\beta-1}(a_1, a_2, \ldots, a_p)$$

form a set of coset representatives for H in F.

Proof. In the following, let α be the smallest positive integer for which

$$y^{-1}P^\alpha(a_1, a_2, \ldots, a_p)y \in H$$

for some $y \in F$. If $\alpha = 1$ the statement holds by Theorem 1.2.16. Hence, from now on let $\alpha \geq 2$. We may assume that $y = 1$ (by replacing the x_i by yx_iy^{-1} if necessary). By means of the Nielsen reduction method we obtain from $\{x_1, x_2, \ldots, x_m\}$ a system $\{y_1, y_2, \ldots, y_k\}$, $1 \leq k \leq m$, which generates H freely and is Nielsen reduced, so in particular each y_i

contains a stable letter a_j^{ϵ}, $\epsilon = \pm 1$, as described before. For each y_i we choose such a stable letter and take the inverse letter for y_i^{-1}. We have an equation

$$\prod_{i=1}^{q} y_{v_i}^{\epsilon_i} = P^{\alpha}(a_1, a_2, \ldots, a_p), \quad \epsilon_i = \pm 1, \epsilon_i = \epsilon_{i+1} \text{ if } v_i = v_{i+1}. \tag{1.8}$$

If a factor $y_{v_t}^{\epsilon_t}$ occurs twice in (1.8) then we have a particular product $\prod_{i=t}^{s} y_{v_i}^{\epsilon_i}$ with $v_t = v_{s+1}$ and no $y_{v_i}^{\epsilon_i}$ occurring twice in $y_{v_t}^{\epsilon_t} \cdots y_{v_s}^{\epsilon_s}$. Then $\prod_{i=t}^{s} y_{v_i}^{\epsilon_i}$ must be conjugate to a power of $P(a_1, a_2, \ldots, a_p)$. Therefore, after a suitable conjugation, we obtain an equation

$$\prod_{i=1}^{r} y_{v_i}^{\epsilon_i} = P^{\beta}(a_1, a_2, \ldots, a_p), \quad 1 \leq \beta \leq \alpha, \epsilon_i = \pm 1, \epsilon_i = \epsilon_{i+1} \text{ if } v_i = v_{i+1}, \tag{1.9}$$

with no factor $y_{v_i}^{\epsilon_i}$ occurring twice (of course we must in fact have $\alpha = \beta$ because α is minimal). If some y_i occurs just once in (1.9) either with exponent +1 or with exponent −1 then case (a) occurs. Now, let in (1.9) each y_i occur exactly once with exponent +1 and once with exponent −1. This implies in particular that $k \geq 2$. We may apply Nielsen transformations for quadratic words as described in Chapter 5 of [281] to get from $\{y_1, y_2, \ldots, y_k\}$ a set $\{z_1, z_2, \ldots, z_k\}$ such that

$$\prod_{i=1}^{r} y_{v_i}^{\epsilon_i} = [z_1, z_2][z_2, z_3] \cdots [z_{\ell-1}, z_\ell] z_{\ell+1} z_{\ell+1}^{-1} \cdots z_k z_k^{-1}, \quad 2 \leq \ell \leq k, \ell \text{ even}. \tag{1.10}$$

Without loss of generality, we may assume that $\ell = k$ and, hence, k is even and $\{z_1, z_2, \ldots, z_k\}$ is a free generating system of H. We remark that $\{z_1, z_2, \ldots, z_k\}$ is not necessarily Nielsen reduced.

We want to show that H has finite index $[F : H] = \beta$. From [25] and also from [281] we may regard F as a cycloidal Fuchsian group of the first kind, that is, of finite co-volume, generated by $\{a_1, a_2, \ldots, a_p, P(a_1, a_2, \ldots, a_p)\}$ such that

(a) $a_1, a_2, \ldots, a_p$ are hyperbolic elements and $\frac{p}{2}$ is the genus of F;
(b) $a = P(a_1, a_2, \ldots, a_p)^{-1}$ is a parabolic element; and
(c) $\{a, a_1, a_2, \ldots, a_p\}$ is a canonical generating system of F.

Recall that a Fuchsian group Γ of the first kind is called *cycloidal* if Γ has exactly one conjugacy class for maximal parabolic generators. Now, H is, as a subgroup of F, also discrete, and no primitive element of H is conjugate in F to a power of $P(a_1, a_2, \ldots, a_n)$. Therefore H is a cycloidal Fuchsian group of the first kind such that

(a) $z_1, z_2, \ldots, z_k$ are hyperbolic and $\frac{k}{2}$ is the genus of H;
(b) $b = a^{\beta} = P(a_1, a_2, \ldots, a_p)^{-1}$ is a parabolic element; and
(c) $\{b, z_1, z_2, \ldots, z_k\}$ is a canonical generating system of H.

As a subgroup of F with finite co-volume, H has finite index $[F : H]$ in F (see [281] or [25]). We now have to show that $[F : H] = \beta$. If $[F : H] = 1$ then $\beta = 1$ and $F = H$.

Now let $[F : H] > 1$, $g \in F \setminus H$, and consider $gP(a_1, a_2, \ldots, a_p)g^{-1}$. There exists a natural number γ such that $gP^{\gamma}(a_1, a_2, \ldots, a_p)g^{-1} \in H$ since $[F : H]$ is finite. Hence, there exists $h \in H$ and an integer $\delta \neq 0$ such that $hgP^{\gamma}(a_1, a_2, \ldots, a_p)g^{-1}h^{-1} = P^{\delta}(a_1, a_2, \ldots, a_p)$, because H is cycloidal. Therefore $\gamma = \delta$ and $hg \in \langle P(a_1, a_2, \ldots, a_p)\rangle$ since F is free. Now $g = dP^{\varphi}(a_1, a_2, \ldots, a_n)$ (whence $Hg = HP^{\varphi}(a_1, a_2, \ldots, a_p)$) for some $d \in H$ and some natural number φ with $1 \leq \varphi < \beta$, since $h \in H$ and $P^{\beta}(a_1, a_2, \ldots, a_p) \in H$. □

Corollary 1.2.31. *Let F be a free group on $a_1, a_2, \ldots, a_p$, $p \geq 1$, p even and*

$$P(a_1, a_2, \ldots, a_p) = [a_1, a_2][a_2, a_3] \cdots [a_{p-1}, a_p] \in F.$$

Let α, $\alpha \geq 2$, and be

$$x_1, y_1, x_2, y_2, \ldots, x_q, y_q \in F, \quad q \geq 1,$$

such that

$$P^{\alpha}(a_1, a_2, \ldots, a_p) = \prod_{i=1}^{q} [x_i, y_i].$$

Suppose that $P^{\alpha}(a_1, a_2, \ldots, a_p)$ is not a power in the subgroup H generated by $x_1, y_1, x_2, y_2, \ldots, x_q, y_q$. Then α is odd and

$$q \geq \frac{1}{2}(\alpha(p-1) + 1).$$

Proof. Note that H has finite index α in F by Theorem 1.2.30. If we regard F as a cycloidal Fuchsian group of the first kind such that $P(a_1, a_2, \ldots, a_p)$ is parabolic then the Riemann–Hurwitz area relation (see [25] or [281]) yields the result that α is odd and $\alpha(p-1) \leq 2q-1$, giving the result. □

Corollary 1.2.32. *Let F be a free group of rank 2 on $\{a, b\}$. Then each $[a, b]^{2^n}$, $n \in \mathbb{N}$, is an $\mathcal{U}_0$-generic element, and, hence a test word.*

Examples 1.2.33. Let F be a free group on a and b.

1. Let $z_1 = aba^{-1}$, $z_2 = b^{-1}aba^{-2}$, $z_3 = b^{-1}ab$ and $z_4 = b^2$, then $H = \langle z_1, z_2, z_3, z_4\rangle$ is free of rank 4 and $[z_1, z_2][z_3, z_4] = [a, b]^3$. H has index 3 in F and $\langle z_1, z_2, z_3, z_4, [a, b]\rangle = F$.
2. Let $x_1 = [a, b]$, $x_2 = a$, $x_3 = a^2$, $x_4 = b$. Then $[x_1, x_2][x_3, x_4] = [a, b]^2$ but $\langle x_1, x_2, x_3, x_4\rangle = F$.

If we regard a free group F on $a_1, a_2, \ldots, a_p$, $p \geq 1$, as a plane discontinuous group $F = \langle a, a_1, a_2, \ldots, a_p \mid aa_1^2a_2^2 \cdots a_p^2 = 1\rangle$ we get analogously the following.

Theorem 1.2.34. *Let F be the free group on $a_1, a_2, \ldots, a_p$, $p \geq 1$, and let*

$$Q(a_1, a_2, \ldots, a_p) = a_1^2 a_2^2 \cdots a_p^2 \in F.$$

Let $\{x_1, x_2, \ldots, x_m\}$, $m \geq 1$, *be any finite set of elements of* F *and* H *be the subgroup of* F *generated by* $x_1, x_2, \ldots, x_m$. *Suppose that* H *contains some conjugate of some non-trivial power of* $Q(a_1, a_2, \ldots, a_p)$, *and let* β *be the smallest positive integer such that some conjugate of* $Q^\beta(a_1, a_2, \ldots, a_p)$ *lies in* H. *Then*

(a) *The set* $\{x_1, x_2, \ldots, x_m\}$ *can be carried by a Nielsen transformation into a free bases for* H *which contains a conjugate of* $Q^\beta(a_1, a_2, \ldots, a_p)$; *or*

(b) *The index of* H *in* F *is* β *and*

$$1, Q(a_1, a_2, \ldots, a_p), \ldots, Q^{\beta-1}(a_1, a_2, \ldots, a_p)$$

for a set of coset representatives for H *in* F.

Corollary 1.2.35. *Let* F *be a free group* $a_1, a_2, \ldots, a_p$ *and* $Q(a_1, a_2, \ldots, a_p) = a_1^2 a_2^2 \cdots a_p^2$. *Let* $\alpha \in \mathbb{N}$, $\alpha \geq 2$, *and* $x_1, x_2, \ldots, x_q \in F$, $q \geq 1$, *be such that*

$$Q^\alpha(a_1, a_2, \ldots, a_p) = x_1^2 x_2^2 \cdots x_q^2.$$

Then, if $Q^\alpha(a_1, a_2, \ldots, a_p)$ *is not a proper power in the subgroup* H *generated by* $x_1, x_2, \ldots,$ x_q, *we have* $q \geq \alpha(p-1) + 1$.

We saw that $a_1^2 a_2^2 \cdots a_p^2$ and $[a_1, a_2][a_2, a_3] \cdots [a_{p-1}, a_p]$ are tame APEs in the free group $\langle a_1, a_2, \ldots, a_p \mid \rangle$, $p \geq 2$, and p even in the second case. It would be of interest to see if there are other tame APEs (up to Nielsen transformations) in a finitely generated non-Abelian free group. We remark that $a_1^{n_1} a_2^{n_2} \cdots a_k^{n_k}$, $2 \leq n_i$ for $i = 1, 2, \ldots, k$ is a tame almost primitive element in the free group $F = \langle a_1, a_2, \ldots, a_k \mid \rangle$ if and only if $n_1 = n_2 = \cdots = n_k = 2$, (see [162]).

Remark 1.2.36. We have already shown that we may extend the concept of test elements and generic elements to arbitrary groups. This is not so clear for primitive and tame primitive elements. There are several possibilities. We suggest the following: Let G be a group, then an element $g \in G$ is called primitive if it is a member of a minimal generating system, that is, a system with $\mathrm{rk}(G)$ elements, where $\mathrm{rk}(G)$ is the rank of G. This then also defines almost primitive elements and tame almost primitive elements in G.

We mention that test elements, generic elements, almost primitive elements and tame almost primitive elements are considered in free products and in surface groups, see [74, 114, 208, 242, 243].

We consider free products and surface groups later in this book.

1.2.4 Vaught problem and powers in free groups

In this connection we mention the following observation.

Remark 1.2.37. We may use the Nielsen method more generally to solve equations $w(x_1, x_2, \ldots, x_n) = g$ in a finitely generated free group, where $g \in F$. We consider the variables $x_1, x_2, \ldots, x_n$ as elements of F and then the subgroup generated by $x_1, x_2, \ldots, x_n$. We then use the Nielsen method with respect to $w(x_1, x_2, \ldots, x_n)$. If, for instance,

$$w(x_1, x_2, \ldots, x_n) = u(x_1, x_2, \ldots, x_n) x_{\nu_1}^{\epsilon_1} x_{\nu_2}^{\epsilon_2} v(x_1, x_2, \ldots, x_n), \quad \nu_1 \neq \nu_2, \epsilon_i = \pm 1,$$

and if $|x_{\nu_1}^{\epsilon_1} x_{\nu_2}^{\epsilon_2}| < |x_{\nu_1}|$, then we define $y = x_{\nu_1}^{\epsilon_1} x_{\nu_2}^{\epsilon_2}$, that is, $x_{\nu_1}^{\epsilon_1} = y x_{\nu_2}^{-\epsilon_2}$ and $x_{\nu_1}^{-\epsilon_1} = x_{\nu_2}^{\epsilon_2} y^{-1}$. Then we replace $x_{\nu_1}^{\pm 1}$ with respect to this description at each place where $x_{\nu_1}^{\pm 1}$ occurs in $w(x_1, x_2, \ldots, x_n)$. Analogously we may perform the replacement with respect to the ordering $\prec$. We finally get, after omitting trivial elements, a set for which we eventually can distinguish stable letters. This procedure is especially convenient for quadratic equations, that are equations for which in $w(x_1, x_2, \ldots, x_n)$ each x_i either does not occur or occurs exactly twice, that means, either exactly twice with exponent +1 or twice with exponent −1 or exactly once with exponent +1 and once with exponent −1.

The described replacements transform a quadratic equation into a quadratic equation. This is essentially the idea in the Whitehead method, which we describe soon in this section.

Example 1.2.38. Let $w = x^2y^2z^2$. Then $w = uy^{-1}uyz^2$ by letting $u = xy$. Then $w = uy^{-1}uvy^{-1}v$ by letting $v = yz$. Now letting $r = uv$ we get

$$w = rv^{-1}y^{-1}ry^{-1}v = [v, v^{-1}rv](v^{-1}ry^{-1}v)^2 =: [a, b]c^2.$$

Using this method we easily get the following.

Theorem 1.2.39. *Let F be a finitely generated free group. Then*
1. $x^2y^2z^2 = 1$ *implies* $xy = yx$;
2. $[x, y] = g^{2k} = (g^k)^2$ *(Vaught problem),* $g \in F$, $k \geq 1$. *Then* $g = 1$ *and* $xy = yx$.

We leave this as an exercise.

This result was at first proved by M.-P. Schützenberger [233]. We remark that Theorem 1.2.39.2 also holds for $[x, y] = g^r$, $r \geq 3$ odd. Here a direct proof is costlier (see for instance [73]). Theorem 1.2.39 also follows from a general theorem of G. Baumslag [15] whose proof is long and very technical.

Theorem 1.2.40. *Let F be a free group of rank n with basis $\{x_1, x_2, \ldots, x_n\}$. Let $w = w(x_1, x_2, \ldots, x_n) \in F$ be not a primitive element and not a proper power in F. Suppose that there are elements $g_1, g_2, \ldots, g_n, g$ in a free group H with $w(g_1, g_2, \ldots, g_n) = g^k$ for some $k \geq 2$. Then $\langle g_1, g_2, \ldots, g_n, g\rangle$ has a rank $\leq n - 1$.*

We come back to such equations in later sections.

1.2.5 The automorphism group of free groups and stabilizers

We now want to briefly discuss the fact that the automorphism group Aut(F) of a finitely generated free group F is finitely presented. We have seen that it was quite easy to show that Aut(F) is finitely generated. Much more difficult to establish is Nielsen's theorem (see [207]).

Theorem 1.2.41. *The automorphism group* Aut(F) *of a finitely generated free group is finitely presented.*

We shall sketch a proof later. A significant method in the theory of automorphisms of free groups was introduced by Whitehead (see [270] and [271]).

Theorem 1.2.42. *There is an algorithm to determine for any two m-tuples* $(u_1, u_2, \ldots, u_m)$ *and* $(v_1, v_2, \ldots, v_m)$ *of elements of a finitely generated free group F whether or not there exists an automorphism of F which carries one m-tuple to the other.*

Corollary 1.2.43. *There is an algorithm to determine for an element u of a finitely generated free group whether or not u is primitive.*

Corollary 1.2.43 is a direct consequence of Theorem 1.2.42. Whitehead's argument uses a representation of a fundamental group of a certain 3-manifold (see for instance [48] or Chapter 3 for a definition of the fundamental group). We follow an algebraic account from [128] and concentrate on the case $m = 1$. We finally give a proof for $m = 1$ at the end of this section when we developed the Whitehead method. It is convenient to work modulo the application of inner automorphisms and thus with conjugacy classes rather than elements of F. We use the following terminology: Let F be a free group with basis $X = \{x_1, x_2, \ldots, x_n\}$.

(1) For any permutation π of $\{1, 2, \ldots, n\}$ and $\epsilon_i = \pm 1$, $i = 1, 2, \ldots, n$, let $\varphi\colon (x_1, x_2, \ldots, x_n) \mapsto (x_{\pi(1)}^{\epsilon_1}, x_{\pi(2)}^{\epsilon_2}, \ldots, x_{\pi(n)}^{\epsilon_n})$; φ is an automorphism of F called a *permutation automorphism.*
(2) An automorphism α of F is called a *Whitehead automorphism* of F when there is a fixed $x \in X \cup X^{-1}$ such that $\alpha(s) \in \{s, sx, x^{-1}s, x^{-1}sx\}$ for any $s \in X \cup X^{-1}$; here $\alpha(x) = x$. There are only finitely many Whitehead automorphisms. We know that Aut(F) is generated by the elementary Nielsen transformation (N1) and (N2). The definition shows that Aut(F) is also generated by the finitely many permutation automorphisms and Whitehead automorphisms.
(3) A *cyclic word* $w \in F$ is a cyclically ordered, reduced string of elements of $X \cup X^{-1}$. It represents a conjugacy class of elements of F. Automorphisms of F act on the set of cyclic words. The *length* $|w|$ of the cyclic word w is the number of letters from $X \cup X^{-1}$ in w.
(4) Two cyclic words u, v are *equivalent* if there is an automorphism $\alpha \in$ Aut(F) such that $v = \alpha(u)$.

(5) A cyclic word w is called *minimal* if it has minimal length among all cyclic words which are equivalent to w.
We remark that the definition of minimality makes extensive use of the set of generators, hence, to be minimal is not a group theoretic property of an element.

Example 1.2.44. Let $X = \{a, b\}$.
1. Any word $a^p b^q$, $p, q \geq 2$, is minimal in $F(X)$.
2. The commutator $[a, b]$ is minimal in $F(X)$.

We leave the proof as an exercise.

The main result is the following.

Theorem 1.2.45 (Whitehead's theorem). *Let u and v be cyclic words of the finitely generated free group F and $\alpha \in \operatorname{Aut}(F)$ be such that $\alpha(u) = v$. Then α can be expressed as a product $\alpha = \rho_n \rho_{n-1} \cdots \rho_2 \rho_1$ of permutation and Whitehead automorphisms such that for some p, q, $1 \leq p \leq q \leq n$, we have*
(a) $|\rho_{i-1}\rho_{i-2}\cdots\rho_1(u)| > |\rho_i\rho_{i-1}\cdots\rho_1(u)|$, $1 \leq i \leq p$.
(b) $|\rho_{j-1}\rho_{j-2}\cdots\rho_1(u)| = |\rho_j\rho_{j-1}\cdots\rho_1(u)|$, $p+1 \leq j \leq q$.
(c) $|\rho_{k-1}\rho_{k-2}\cdots\rho_1(u)| < |\rho_k\rho_{k-1}\cdots\rho_1(u)|$, $q+1 \leq k \leq n$.

The inequalities (a), (b), (c) are illustrated by the following diagram (see Figure 1.4) in which the length is plotted upwards.

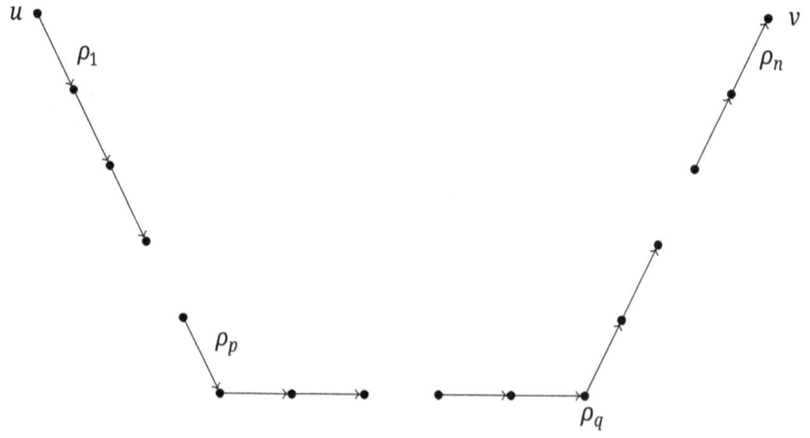

Figure 1.4: Peak reduction.

The main lemma of the Higgins–Lyndon version (see [128]) of the proof of Whitehead's theorem is the following.

Lemma 1.2.46. *Let σ and τ be permutation or Whitehead automorphisms, and let u, w and v be cyclic words such that*
(1) $\sigma(u) = w$, $\tau(w) = v$.
(2) $|u| \leq |w| \geq |v|$.
(3) $|u| < |w|$ *or* $|w| \geq |v|$.

Then $\tau\sigma$ can be expressed as a product $\tau\sigma = \rho_n\rho_{n-1}\cdots\rho_2\rho_1$ of permutation and Whitehead automorphisms such that $|\rho_i\rho_{i-1}\cdots\rho_1(u)| < |w|$ for $1 \leq i \leq n-1$ (see Figure 1.5).

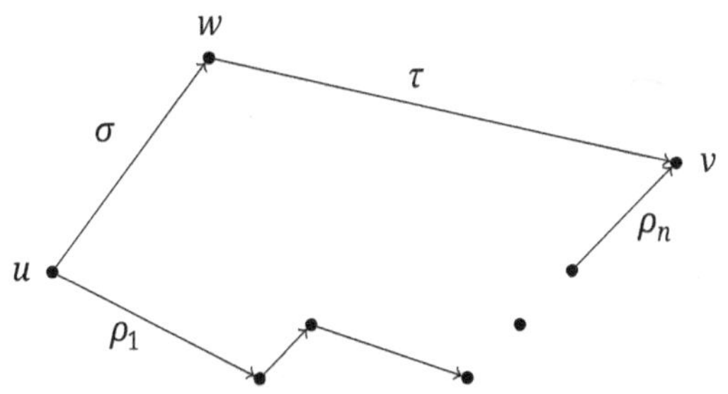

Figure 1.5: Expression of $\tau\sigma$.

The proof is based on an efficient calculation of the number of letters which vanish or have to be added when a Whitehead automorphism is applied to a cyclic word.

Close examination of the detailed argument shows that at most four Whitehead or permutation automorphisms appear in the lower part of the Peak reduction diagram, and this means that a system of defining relations for the Whitehead and permutation automorphisms can be chosen among the relations of length at most six. This, see Theorem 1.2.47, is the basis of McCool's proof for Theorem 1.2.42 (see [189]).

Theorem 1.2.47. *The automorphism group of a free group of finite rank is finitely presented.*

Remark 1.2.48. A striking generalization of Whitehead's argument was introduced in [112] and [113]. Gersten was able to prove the following results.

Theorem 1.2.49. *Let F be a free group of finite rank. Then there is an algorithm to determine of any two subgroups H and K whether or not there is an automorphism of F carrying the conjugacy class of H into the conjugacy class of K.*

The presentation of McCool is not the same as that found by Nielsen. The Peak Reduction Lemma together with McCool's observation enables one to prove the following.

Theorem 1.2.50. *The stabilizer* $\mathrm{Stab}(w) = \{\alpha \in \mathrm{Aut}(F) \mid \alpha(w) = w\}$ *of an element w of a finitely generated free group F is finitely presented. Moreover, there is a procedure to determine a finite presentation of* $\mathrm{Stab}(w)$.

Examples 1.2.51. Let F be a free group with bases $\{x_1, x_2, \ldots, x_n\}$, $n \geq 2$. Then

$$\mathrm{Stab}(x_1 \cdots x_n) = \{\alpha \in \mathrm{Aut}(F) \mid \alpha(x_1 \cdots x_n) = x_1 x_2 \cdots x_n\} = B_n,$$

where B_n is the nth braid group

$$B_n = \langle \sigma_1, \sigma_2, \ldots, \sigma_{n-1} \mid \sigma_i \sigma_j = \sigma_j \sigma_i, 1 \leq i < j - 1 \leq n-2, \sigma_i \sigma_{i+1} \sigma_i = \sigma_{i+1} \sigma_i \sigma_{i+1}, 1 \leq i < n \rangle$$

(see [42]).

Theorem 1.2.52. *Let F be a free group of finite rank. Let $\alpha \in \mathrm{Aut}(F)$. Then the fixed point subgroup*

$$\mathrm{Fix}(\alpha) = \{u \in F \mid \alpha(u) = u\}$$

of α is of finite rank.

We give a proof of Theorem 1.2.52 later as an application of the covering space theory. We remark that Bestvina and Handel [28] showed that $\mathrm{rk}(\mathrm{Fix}(\alpha)) \leq \mathrm{rk}(F)$ for all $\alpha \in \mathrm{Aut}(F)$.

Also, Collins and Zieschang [56, 57] and [60] have extended the Whitehead method to free products of groups (see Section 1.4 for the definition of free products).

Proof. We now finally give a proof of Theorem 1.2.42 for $m = 1$. The algorithm is defined as follows: If the elements u and v are not minimal, then, by the Whitehead theorem, Theorem 1.2.45, a bounded sequence of applications of the finitely many Whitehead automorphisms will replace u and v by minimal words u_0 and v_0. If u_0 and v_0 are equivalent under $\mathrm{Aut}(F)$, then they must have the same length and, again by the Whitehead Theorem 1.2.45, u_0 can be transformed into v_0 by a sequence of applications of permutation and Whitehead automorphisms in such a way that the lengths of the intermediate words obtained are the same as the length of u_0 and v_0. This means that the number of automorphisms in the sequence can be banned in terms of the length of u_0 and the result follows. □

For several purposes, especially for one-relator groups (see Section 1.7 for one-relator groups) the following question is of special interest. Let $F = \langle a_1, a_2, \ldots, a_n \mid \rangle$, $n \geq 1$, be the free groups of rank n with basis $\{a_1, a_2, \ldots, a_n\}$. Let $r \in F$, $r \neq 1$, be freely reduced and suppose that, if $n \geq 2$, there is no Nielsen transformation from $\{a_1, a_2, \ldots, a_n\}$ to a system $\{b_1, b_2, \ldots, b_n\}$ with $r \in \langle b_1, b_2, \ldots, b_{n-1} \rangle$. We call such r *regular*.

We consider generating systems $\{r, x_1, x_2, \ldots, x_n\}$ of F. We say that in a Nielsen transformation from $\{r, x_1, x_2, \ldots, x_n\}$ to a system $\{r, y_1, y_2, \ldots, y_n\}$ the element r is *not replaced* if in all the elementary Nielsen transformations of which φ is composed r remains unchanged or is changed to r^{-1} or is put in a different place of the relevant $(n+1)$-tuples. We refer to such a Nielsen transformation in which r is not replaced and the corresponding Nielsen equivalence classes as *r-stable* (we remark that we have an equivalence relation).

Theorem 1.2.53. *Let F and $r \in F$ be as above. Then there are only finitely many r-stable Nielsen equivalence classes of generating systems $\{r, x_1, x_2, \ldots, x_n\}$ of F.*

Proof. The theorem is certainly true if $n = 1$. Hence, let $n \geq 2$. We assume that in F the free length $|\,|$ and a suitable lexicographic order are introduced relative to the free generators $a_1, a_2, \ldots, a_n$ of F. Let $\{r, x_1, x_2, \ldots, x_n\}$ be a generating system of F. Since r is regular, there is no r-stable Nielsen transformation from $\{r, x_1, x_2, \ldots, x_n\}$ to a system $\{r, y_1, y_2, \ldots, y_{n-1}, 1\}$. Now we perform r-stable Nielsen transformations from $\{r, x_1, x_2, \ldots, x_n\}$ to other systems. Thus we can achieve that

$$|x_i^\eta r^\epsilon| \geq |x_i| \quad \text{and} \quad |x_i^\eta r x_j^\epsilon| > |x_i| - |r| + |x_j|, \quad \eta, \epsilon = \pm 1, i, j = 1, 2, \ldots, n.$$

Since $F = \langle r, x_1, x_2, \ldots, x_n \rangle$ and r is regular we must have either at least once $|x_i^\eta r^\epsilon| < |r|$ or else $|x_j^\eta r^\epsilon| \geq |r|$ but at least once $|r^\eta x_i r^\epsilon| \leq 2|r| - |x_i|$, $\eta, \epsilon = \pm 1$, $i \in \{1, 2, \ldots, n\}$. If at least one $|x_i^\eta r^\epsilon| < |r|$ then $|x_i| < |r|$. If always $|x_i^\eta r^\epsilon| \geq |r|$ but at least once $|r^\eta x_i r^\epsilon| \leq 2|r| - |x_i|$, then all letters in x_i are canceled and $|x_i| \leq |r|$.

Suppose now that $|x_j| \leq |r|$ for $j = 1, 2, \ldots, k$, $1 \leq k < n$. We consider the subgroup U of F generated by $x_1, x_2, \ldots, x_k$ and r. Certainly, $U \neq F$. Suppose that U is free of rank $\leq k$. Then there is a Nielsen transformation from $\{r, x_1, x_2, \ldots, x_k\}$ to a system $\{y_1, y_2, \ldots, y_k, 1\}$; also $U = \langle y_1, y_2, \ldots, y_k \rangle$, and $F = \langle y_1, y_2, \ldots, y_k, x_{k+1}, x_{k+2}, \ldots, x_n \rangle$. In particular, there is a Nielsen transformation from $\{a_1, a_2, \ldots, a_n\}$ to a system $\{b_1, b_2, \ldots, b_n\}$ with $r \in \langle b_1, b_2, \ldots, b_{n-1} \rangle$, which is a contradiction. Hence, U is free of rank $k + 1$ with the basis $\{r, x_1, x_2, \ldots, x_k\}$. There is a Nielsen transformation from $\{r, x_1, x_2, \ldots, x_k\}$ to a system $\{z_1, z_2, \ldots, z_{k+1}\}$ having the Nielsen property. Let us consider this system $\{z_1, z_2, \ldots, z_{k+1}\}$. Since $|x_j| \leq |r|$ for $j = 1, 2, \ldots, k$ we have $|z_i| \leq |r|$ for $i = 1, 2, \ldots, k + 1$. If $g_{k+1}, h_{k+1}, \ldots, g_n$ and h_n are elements of U, then the Nielsen transformation from $\{r, x_1, x_2, \ldots, x_n\}$ to $\{r, x_1, x_2, \ldots, x_k, g_{k+1} x_{k+1} h_{k+1}, \ldots, g_n x_n h_n\}$ is r-stable, that means, we can achieve that

$$|x_i^\eta z_j^\epsilon| \geq |x_i| \quad \text{and} \quad |x_i^\eta z_j x_\ell^\epsilon| > |x_i| - |z_j| + |x_\ell|,$$

$\eta, \epsilon = \pm 1$; $i, \ell \in \{k+1, k+2, \ldots, n\}$; $j \in \{1, 2, \ldots, k+1\}$; furthermore

$$|z_i^\epsilon z_j z_\ell^\eta| > |x_i| - |z_j| + |z_\ell| \quad \text{and} \quad |z_\ell^\eta z_j x_i^\epsilon| > |z_\ell| - |z_j| + |x_i|,$$

$\eta, \epsilon = \pm 1$; $\ell, j \in \{1, 2, \ldots, k+1\}$, $i \in \{k+1, k+2, \ldots, n\}$ and $\eta = 1$ if $\ell = j$. Since $U \neq F = \langle r, x_1, x_2, \ldots, x_k, \ldots, x_n \rangle = \langle z_1, z_2, \ldots, z_{k+1}, x_{k+1}, \ldots, x_n \rangle$ we must have either at least once $|x_i^\eta z_j^\epsilon| < |z_j| \leq |r|$ or else always $|x_i^\eta z_j^\epsilon| \geq |z_j|$, but at least once

$$|z_j^\epsilon x_i z_\ell^\eta| \leq |z_j| - |x_i| + |z_\ell| \leq 2|r| - |x_i|,$$

$\eta, \epsilon = \pm 1$; $j, \ell \in \{1, 2, \ldots, k+1\}$, $i \in \{k+1, k+2, \ldots, n\}$. This means $|x_i| \leq |r|$ for one $i \in \{k+1, k+2, \ldots, n\}$. Now, by induction, we get—possibly after applying r-stable Nielsen transformations—$|x_i| \leq |r|$ for all $i \in \{1, 2, \ldots, n\}$. This completes the proof, for there are only finitely many generating systems $\{r, x_1, x_2, \ldots, x_n\}$ of F for which $|x_i| \leq |r|$ for $i = 1, 2, \ldots, n$. □

Examples 1.2.54. 1. Let $F = \langle a_1, a_2, \ldots, a_n \mid \rangle$, $n \geq 2$ and $r = a_1^{\alpha_1} a_2^{\alpha_2} \cdots a_n^{\alpha_n}$, all $\alpha_i \geq 2$. If there are $x_1, x_2, \ldots, x_n \in F$ with $\langle r, x_1, x_2, \ldots, x_n \rangle = F$ then $\{r, x_1, x_2, \ldots, x_n\}$ is r-stable Nielsen equivalent to a system

$$\{r, a_1, a_2, \ldots, a_{i-1}, a_i^{\gamma_i}, a_{i+1}, \ldots, a_n\},$$

$1 \leq i \leq n$, $1 \leq \gamma_i \leq \frac{1}{2}\alpha_i$, $\gcd(\gamma_i, \alpha_i) = 1$.

2. Let $F = \langle a, b \mid \rangle$ and $r = a^\alpha b a^\beta b^{-1}$, $\alpha, \beta \in \mathbb{Z} \setminus \{0\}$. If there are $x, y \in F$ with $\langle r, x, y \rangle = F$ then $\{r, x, y\}$ is r-stable Nielsen equivalent to a system $\{r, a^\gamma, ba^\delta\}$ with $1 \leq \gamma \leq |\alpha|$, $\gcd(\gamma, \beta) = 1$, $0 \leq \delta < \gamma$ or $1 \leq \gamma \leq |\beta|$, $\gcd(\gamma, \alpha) = 1$, $0 \leq \delta < \gamma$. Especially, if $r = [a, b] = aba^{-1}b^{-1}$, then $\{r, x, y\}$ is r-stable Nielsen equivalent to $\{r, a, b\}$.

1.3 Generators and relators

We already briefly introduced presentations of groups in Section 1.1. Now, we want to discuss these more in detail and formally. Let G be a group, X a set and $\varphi: F(X) \twoheadrightarrow G$ an epimorphism. A set X is called a set of generating symbols for G under φ and $\{\varphi(x) \mid x \in X\}$ a set of generators of G. There is no misunderstanding if we just take X as a set of generators of G. We call $\ker(\varphi)$ the set of relators of G with respect to φ. If $u = x_1^{\epsilon_1} x_2^{\epsilon_2} \cdots x_n^{\epsilon_n}$ and $v = y_1^{\eta_1} y_2^{\eta_2} \cdots y_m^{\eta_m}$ so that $uv^{-1} \in \ker(\varphi)$ and $\varphi(x_i) = a_i$, $\varphi(y_j) = b_j$, then we call $a_1^{\epsilon_1} a_2^{\epsilon_2} \cdots a_n^{\epsilon_n} = b_1^{\eta_1} b_2^{\eta_2} \cdots b_m^{\eta_m}$ a relation in G. If already $u \in \ker(\varphi)$, then $a_1^{\epsilon_1} a_2^{\epsilon_2} \cdots a_n^{\epsilon_n}$ is a relator in G. A set $R \subset F(X)$ is a set of defining relators of G (with respect to φ) if $\ker(\varphi) = \langle\langle R \rangle\rangle$, that is, $\ker(\varphi)$ is the set of consequences of R. In G we have the respective set of defining relations. We also say that a relation $\varphi(u) = \varphi(v)$ is a consequence of a set of defining relators (or defining relations) if $uv^{-1} \in \ker(\varphi)$. A presentation $\langle X \mid R \rangle^\varphi$ of G consists of a set X, an epimorphism $\varphi: F(X) \twoheadrightarrow G$ and a set of defining relators of G (under φ). The group G is called *finitely presented* if X and R both are finite. We write $G = \langle X \mid R \rangle^\varphi$ if $\langle X \mid R \rangle^\varphi$ is a presentation of G. It is convenient to write

$$G = \langle X \mid R \rangle^\varphi \quad \text{or} \quad G = \langle X \mid r = 1, r \in R \rangle$$

for a presentation of G if there is no misunderstanding. We also allow relations like $t = s$ in a presentation of G.

1.3.1 Examples of group presentations

Examples 1.3.1. We have

1. $F(X) = \langle X \mid \rangle$.
2. $\{1\} = \langle x \mid x = 1\rangle$.
3. $\mathbb{Z}_n = \langle x \mid x^n = 1\rangle, n \geq 2$; and
4. $\mathbb{Z}^n = \langle x_1, x_2, \ldots, x_n \mid [x_i, x_j] = 1, 1 \leq i < j \leq n\rangle, n \geq 2$. This presentation has n generators and $\frac{1}{2}n(n-1)$ relations. It is known from linear algebra that n is the minimal number of elements needed to generate $\mathbb{Z}^n$, that is, the rank of $\mathbb{Z}^n$ is $\mathrm{rk}(\mathbb{Z}^n) = n$. It can be shown, using homological methods, that there does not exist a presentation with less than $\frac{1}{2}n(n-1)$ relations (see Chapter 5 in [61]).
5. Consider the dihedral group $D_n, n \geq 2$, the symmetry group of the regular n-gon P_n. It is generated by the rotation x with angle $\frac{2\pi}{n}$ and a reflection y in the line through the center and one of the vertices. Then $x^n = y^2 = 1(= \mathrm{id}), yxy = x^{-1}$. Via these relations every element of D_n can be expressed in the form $x^k y^\ell$, $0 \leq k < n$, $\ell \in \{0, 1\}$. These expressions give different motions and hence

   $$D_n = \langle x, y \mid x^n = y^2 = (xy)^2 = 1\rangle.$$

   Among these is the symmetric group $S_3 = D_3$.
   Similarly, let D_∞ be the infinite dihedral group consisting of the motions $\mathbb{R} \to \mathbb{R}$, $r \mapsto \pm r + k$, $k \in \mathbb{Z}$. It follows $D_\infty = \langle x, y \mid x^2 = (xy)^2 = 1\rangle$. The groups D_n are not Abelian if $n \geq 3$ and have rank 2.
6. Direct calculations, also with any computer algebra program, give

   $$\begin{aligned} A_4 &= \langle x, y \mid x^2 = y^3 = (xy)^3 = 1\rangle, \\ S_4 &= \langle x, y \mid x^2 = y^3 = (xy)^4 = 1\rangle, \\ A_5 &= \langle x, y \mid x^2 = y^3 = (xy)^5 = 1\rangle. \end{aligned}$$

   A group with a presentation

   $$G(p, q, r) = \langle x, y \mid x^p = y^q = (xy)^r = 1\rangle$$

   with $2 \leq p, q, r$, is called a (ordinary) triangle group. $G(p, q, r)$ is finite if and only if $\frac{1}{p} + \frac{1}{q} + \frac{1}{r} > 1$. Hence if $G = G(p, q, r)$ is finite, then $G \cong D_n$, $n \in \mathbb{N}$, $n \geq 2$, $G \cong A_4$, $G \cong S_4$ or $G \cong A_5$.
   The finite triangle groups describe exactly the non-cyclic finite subgroups of the

   $$\mathrm{PSL}(2, \mathbb{C}) = \mathrm{SL}(2, \mathbb{C})/\{\pm E_2\}, \quad E_2 = \begin{pmatrix} 1 & 0 \\ 0 & 1 \end{pmatrix}.$$

   A non-cyclic finite subgroup of the $\mathrm{PSL}(2, \mathbb{C})$ is isomorphic to some D_n, $n \geq 2$, A_4, S_4 or A_5. We remark that $G(p, q, r)$ is infinite and solvable if and only if $\frac{1}{p} + \frac{1}{q} + \frac{1}{r} = 1$, and $G(p, q, r)$ contains a non-Abelian free subgroup if and only if $\frac{1}{p} + \frac{1}{q} + \frac{1}{r} < 1$ (for details and proofs see for instance [66] or [77]).

7. $\langle x, y \mid xy^2 = y^3x, yx^2 = x^3y\rangle = \{1\}$. This can be seen as follows: $xy^4x^{-1} = (xy^2x^{-1})^2 = y^6$, hence $x^2y^4x^{-2} = xy^6x^{-1} = (xy^2x^{-1})^3 = y^9$. This gives

$$x^2y^4x^{-2} = yx^2y^4x^{-2}y^{-1} = yx^2y^{-1}y^4yx^{-2}y^{-1} = x^3y^4x^{-3},$$

hence $xy^4 = y^4x$, but then $y^6 = xy^4x^{-1} = y^4$, that is, $y^2 = 1$. From $xy^2 = y^3x$ we get now $1 = x^{-1}yx$, hence $y = 1$ and $x = 1$ because $yx^2 = x^3y$.

8. Let $\Delta = |ad - bc|$, $a, b, c, d \in \mathbb{N}$. Then

$$\langle x, y \mid x^a y^b = x^c y^d = 1\rangle \cong \mathbb{Z}_\Delta.$$

We leave this as an exercise.

9. We consider the modular group

$$\Gamma = \left\{ z \mapsto \frac{az + b}{cz + d} \;\middle|\; z \in \mathbb{C} \cup \{\infty\}, a, b, c, d \in \mathbb{Z}, ad - bc = 1 \right\}.$$

We have

$$\Gamma \cong \mathrm{SL}(2, \mathbb{Z})/\{\pm E_2\} = \mathrm{PSL}(2, \mathbb{Z}).$$

The group Γ is generated by the linear fractional transformations

$$\overline{U}: z \mapsto z + 1 \quad \text{and} \quad \overline{T}: z \mapsto -\frac{1}{z}.$$

We show that $\mathrm{SL}(2, \mathbb{Z})$ is generated by the matrices

$$U = \begin{pmatrix} 1 & 1 \\ 0 & 1 \end{pmatrix} \quad \text{and} \quad T = \begin{pmatrix} 0 & -1 \\ 1 & 0 \end{pmatrix}.$$

Then Γ is generated by $\overline{U}$ and $\overline{T}$. Let $A = \left(\begin{smallmatrix} a & b \\ c & d \end{smallmatrix}\right) \in \mathrm{SL}(2, \mathbb{Z})$. We first consider the case $c \neq 0$.

(1) Let $|a| < |c|$:
Then $TA = \left(\begin{smallmatrix} 0 & -1 \\ 1 & 0 \end{smallmatrix}\right)\left(\begin{smallmatrix} a & b \\ c & d \end{smallmatrix}\right) = \left(\begin{smallmatrix} -c & -d \\ a & b \end{smallmatrix}\right)$. This is a reduction to the case $|a| \geq |c|$.

(2) Let $|a| > |c|$:
Then there are $q, r \in \mathbb{N}$ with $|a| = q|c| + r$ with $0 < r < |c|$. It follows $a = pc + s$ with $p = \pm q$, $s = \pm r$ and $0 < |s| < |c|$, and

$$U^{-p}A = \begin{pmatrix} 1 & -p \\ 0 & 1 \end{pmatrix}\begin{pmatrix} a & b \\ c & d \end{pmatrix} = \begin{pmatrix} s & b - pd \\ c & d \end{pmatrix}.$$

Together with case (1) we get a reduction to $|a| = |c|$.

(3) Let $|a| = |c|$. We consider subcases.

i. Let $a = c$:
Then $TUTA = \begin{pmatrix} -1 & 0 \\ 1 & -1 \end{pmatrix}\begin{pmatrix} a & b \\ c & d \end{pmatrix} = \begin{pmatrix} -a & -b \\ 0 & b-d \end{pmatrix}$.

ii. Let $a = -c$:
Then $T^{-1}UA = \begin{pmatrix} 0 & 1 \\ -1 & -1 \end{pmatrix}\begin{pmatrix} a & b \\ c & d \end{pmatrix} = \begin{pmatrix} c & d \\ 0 & -b-d \end{pmatrix}$.

Hence we get inductively a $S \in \langle T, U\rangle$ with $SA = \begin{pmatrix} \alpha & \beta \\ 0 & \delta \end{pmatrix} \in \mathrm{SL}(2,\mathbb{Z})$. Since $\alpha\delta = 1$ we have $\alpha = \delta = \pm 1$. Let without loss of generality $\alpha = \delta = 1$ (we have $T^2 = \begin{pmatrix} -1 & 0 \\ 0 & -1 \end{pmatrix}$). Then $SA = U^{\beta}$ and, hence, $A = S^{-1}U^{\beta} \in \langle U, T\rangle$. With $\Gamma = \langle \overline{U}, \overline{T}\rangle$ we also have $\Gamma = \langle \overline{R}, \overline{T}\rangle$ where $\overline{R} = \overline{TU}$. Then

$$\begin{aligned} \overline{R}&: z \mapsto -\frac{1}{z+1}, \\ \overline{R}^2&: z \mapsto -\frac{z+1}{z}, \quad \text{and} \\ \overline{R}^3&: z \mapsto z, \end{aligned}$$

that is, $\overline{R}$ has order 3. Let $\mathbb{R}^- = \{x \in \mathbb{R} \mid x < 0\}$ and $\mathbb{R}^+ = \{x \in \mathbb{R} \mid x > 0\}$. Then $\overline{T}(\mathbb{R}^-) \subset \mathbb{R}^+$ and $\overline{R}^{\alpha}(\mathbb{R}^+) \subset \mathbb{R}^-$ for $\alpha = 1, 2$. The relations $\overline{T}^2 = \overline{R}^3 = 1 = \mathrm{id}$ are defining relations for Γ. This can be seen as follows. Let $\overline{S} \in \Gamma$. Applying $\overline{T}^2 = \overline{R}^3 = 1$ and suitable conjugations we may assume that $\overline{S} = 1$ is a consequence of $\overline{T}^2 = \overline{R}^3 = 1$ or that

$$\overline{S} = \overline{R}^{\alpha_1}\overline{TR}^{\alpha_2}\cdots\overline{TR}^{\alpha_{n+1}}$$

with $1 \le \alpha_i \le 2$, $i = 1, 2, \dots, n+1$, $\alpha_1 = \alpha_{n+1}$ (recall that $\overline{R}$ has order 3).
In the second case we get $\overline{S}(x) \in \mathbb{R}^-$ if $x \in \mathbb{R}^+$, and, hence, $\overline{S} \neq 1$.
Therefore $\Gamma = \langle \overline{T}, \overline{R} \mid \overline{T}^2 = \overline{R}^3 = 1\rangle$. Moreover, the above proof also shows that if $\overline{A} \in \Gamma$ is an element of order 2, then $\overline{A}$ is conjugate to $\overline{T}$.
This has an interesting application.

Theorem 1.3.2 (Two square theorem of Fermat). *Let $n \in \mathbb{N}$.*
(i) *If -1 is a quadratic residue modulo n, that is, $-1 \equiv q^2 \mod n$ for some $q \in \mathbb{Z}$, then n is the sum of two squares over $\mathbb{Z}$, that is, $n = x^2 + y^2$ with $x, y \in \mathbb{Z}$.*
(ii) *If $n = x^2 + y^2$ with $x, y \in \mathbb{Z}$ and $\gcd(x, y) = 1$, then -1 is a quadratic residue modulo n.*

Proof. (i): There exists a $q \in \mathbb{Z}$ with $-1 \equiv q^2 \mod n$ and further a $p \in \mathbb{Z}$ with $-q^2 - pn = 1$. We take the matrix $A = \begin{pmatrix} q & n \\ p & -q \end{pmatrix}$. We have $A \in \mathrm{SL}(2,\mathbb{Z})$ and $A^2 = \begin{pmatrix} -1 & 0 \\ 0 & -1 \end{pmatrix}$.
$\overline{A}: z \mapsto \frac{qz+n}{pz-q}$ has order 2 in Γ.
Hence, there is a $X = \begin{pmatrix} x & y \\ u & v \end{pmatrix} \in \mathrm{SL}(2,\mathbb{Z})$ with $XTX^{-1} = \pm A$. It follows

$$\begin{pmatrix} x & y \\ u & v \end{pmatrix}\begin{pmatrix} 0 & 1 \\ -1 & 0 \end{pmatrix}\begin{pmatrix} v & -y \\ -u & x \end{pmatrix} = \begin{pmatrix} -vy - ux & x^2 + y^2 \\ -v^2 - u^2 & vy + ux \end{pmatrix} = \pm\begin{pmatrix} q & n \\ p & -q \end{pmatrix}.$$

Since $n \in \mathbb{N}$ we have $n = x^2 + y^2$.

(ii): Since $\gcd(x,y) = 1$ there exist $u, v \in \mathbb{Z}$ with $X = \left(\begin{smallmatrix} x & y \\ u & v \end{smallmatrix}\right) \in \mathrm{SL}(2,\mathbb{Z})$. It follows

$$XTX^{-1} = \begin{pmatrix} -vy - ux & x^2 + y^2 \\ -v^2 - u^2 & vy + ux \end{pmatrix} =: \begin{pmatrix} q & x^2 + y^2 \\ p & -q \end{pmatrix} = \begin{pmatrix} q & n \\ p & -q \end{pmatrix}.$$

From $-q^2 - pn = 1$ we get $-1 \equiv q^2 \mod n$. □

Especially: If p is a prime number then p is a sum of squares if and only if $p = 2$ or $p \equiv 1 \mod 4$. We only have to show: If $p = 4q + 1$ for some $q \geq 2$, then -1 is a quadratic residue modulo p. The multiplicative group $((\mathbb{Z}/p\mathbb{Z})^*, \cdot)$ is cyclic of order $p - 1$, that is, there exists a generating element $x \in (\mathbb{Z}/p\mathbb{Z})^*$. Then $1 = x^{p-1} = (x^{2q})^2$, and there exists a $y \in \mathbb{Z}$ with $y^2 \equiv -1 \mod p$.
We know that

$$\Gamma \cong \mathrm{PSL}(2,\mathbb{Z}) = \mathrm{SL}(2,\mathbb{Z})/\{\pm E_2\}$$

has the presentation

$$\Gamma = \langle \overline{T}, \overline{R} \mid \overline{T}^2 = \overline{R}^3 = 1 \rangle.$$

Now, $\mathrm{SL}(2,\mathbb{Z})$ is generated by $T = \left(\begin{smallmatrix} 0 & -1 \\ 1 & 0 \end{smallmatrix}\right)$ and $U = \left(\begin{smallmatrix} 1 & 1 \\ 0 & 1 \end{smallmatrix}\right)$ and, hence, by T and $R = \left(\begin{smallmatrix} 0 & -1 \\ 1 & 1 \end{smallmatrix}\right)$. From $T^2 = \left(\begin{smallmatrix} -1 & 0 \\ 0 & -1 \end{smallmatrix}\right) = R^3$ we get the presentation

$$\mathrm{SL}(2,\mathbb{Z}) = \langle T, R \mid T^2 = R^3, T^4 = 1 \rangle \quad (\text{with } 1 := E_2).$$

We now consider the group $\mathrm{GL}(2,\mathbb{Z})$ of all invertible integer matrices. If we add the matrix $S = \left(\begin{smallmatrix} 0 & 1 \\ 1 & 0 \end{smallmatrix}\right)$ to R and T, then certainly R, T and S generate $\mathrm{GL}(2,\mathbb{Z})$, and the corresponding presentation is

$$\mathrm{Aut}(\mathbb{Z} \oplus \mathbb{Z}) = \mathrm{GL}(2,\mathbb{Z}) = \langle R, T, S \mid T^2 = R^3, T^4 = S^2 = (ST)^2 = (SR)^2 = 1 \rangle.$$

Then

$$\mathrm{PGL}(2,\mathbb{Z}) = \mathrm{GL}(2,\mathbb{Z})/\{\pm E_2\} = \langle R, T, S \mid T^2 = R^3 = S^2 = (ST)^2 = (SR)^2 = 1 \rangle.$$

For later purposes we need a slightly different presentation. The $\mathrm{PGL}(2,\mathbb{Z})$ is also generated by the elements

$$B = R^2 = \pm \begin{pmatrix} -1 & -1 \\ 1 & 0 \end{pmatrix},$$

$$A = SR^2 = \pm \begin{pmatrix} -1 & -1 \\ 0 & 1 \end{pmatrix} \quad \text{and}$$

$$C = ST = \pm \begin{pmatrix} 1 & 0 \\ 0 & -1 \end{pmatrix}.$$

With these generators we rewrite the presentation and get

$$\mathrm{PGL}(2,\mathbb{Z}) = \langle A,B,C \mid A^2 = B^3 = C^2 = (AB)^2 = (CAB)^2 = 1\rangle.$$

10. Let E be the Euclidean plane and F be the regular tessellation of the plane E by squares.

Theorem 1.3.3. *The symmetry group G of F has a presentation*

$$G = \langle \alpha,\beta,\gamma \mid \alpha^2 = \beta^2 = \gamma^2 = (\alpha\gamma)^2 = (\gamma\beta)^4 = (\beta\alpha)^4 = 1\rangle.$$

Proof. We may assume that we have a regular tessellation by unit squares. Let Q be a single square. It is incidental that we get G by the translations of the squares and the symmetry group $G_Q = \mathrm{Sym}(Q) \cong D_4$ of the single square Q; see Figure 1.6.

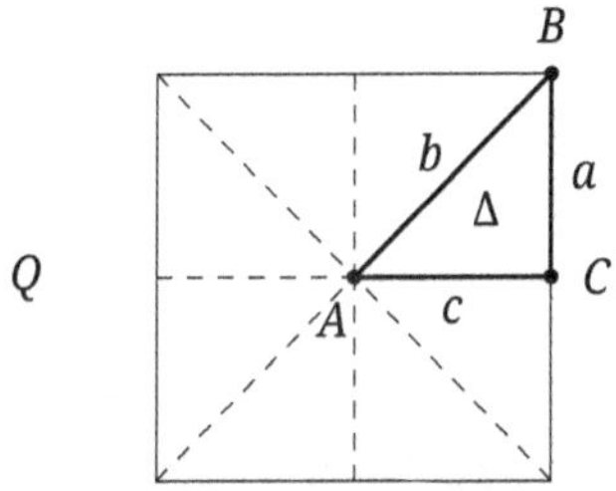

Figure 1.6: Fundamental region for G.

We have

$$\mathbb{R}^2 = \bigcup_{g\in G} g(\Delta)$$

and $\widehat{g(\Delta)} \cap \widehat{h(\Delta)} = \emptyset$ if $g \neq h$, where $\widehat{g(\Delta)} = g(\Delta) \setminus \vartheta(g(\Delta))$ and $\vartheta(g(\Delta)) = g(a) \cup g(b) \cup g(c)$. This in fact means that Δ is a *fundamental region for G*. Let α, β, γ be the reflections at the sides a, b, c of Δ, respectively. Certainly $\alpha,\beta,\gamma \in G$.

Claim. $G = \langle \alpha,\beta,\gamma\rangle$.

Proof. To proof the claim let $H = \langle \alpha,\beta,\gamma\rangle$, H is a subgroup of G. Assume that $H \neq G$. Let

$$U = \bigcup_{h\in H} h(\Delta).$$

By assumption we have $U \neq \mathbb{R}^2$. Thus some side s of some triangle $h(\Delta)$, $h \in H$, must lie on the boundary of U, separating $h(\Delta)$ from some $g(\Delta)$, $g \in G$, $g \notin H$.
Now, s is the image $h(a)$, $h(b)$, $h(c)$ of a side a, b, c of Δ, and the reflection ρ in s is the corresponding element $h \circ \alpha \circ h^{-1}$, $h \circ \beta \circ h^{-1}$ or $h \circ \gamma \circ h^{-1}$. Since $h,\alpha,\beta,\gamma \in H$ we get $\rho \in H$ and $g = \rho \circ h \in H$, which gives a contradiction. Therefore $H = G$ which proves the claim. □

We now seek a set of defining relations among α, β and γ. Since α, β and γ are reflections, we have the relations $\alpha^2 = \beta^2 = \gamma^2 = 1$. Further, $\gamma \circ \beta$ is a rotation at A through an angle twice the interior angle $\frac{2\pi}{8}$ of Δ at A, that is, through $\frac{2\pi}{4}$, therefore $(\gamma \circ \beta)^4 = 1$. Similarly, $(\alpha \circ \gamma)^2 = 1$ and $(\beta \circ \alpha)^4 = 1$.
We show now that these six relations form a full set of defining relations: Let D be the tessellation of $\mathbb{R}^2$ by the regions $g(\Delta)$, $g \in G$. Let O be the center of Δ, that is, the intersection of lines joining vertices to midpoints of opposite sides. We join $g(O)$ and $h(O)$ with a (directed) edge e if and only if $g(\Delta)$ and $h(\Delta)$ have a side s in common. This way we constructed a graph C with vertices all centers $g(O)$ of triangles $g(\Delta)$ and edges as described above; see Figure 1.7. For a general discussion of graphs see Chapter 5 in [77].

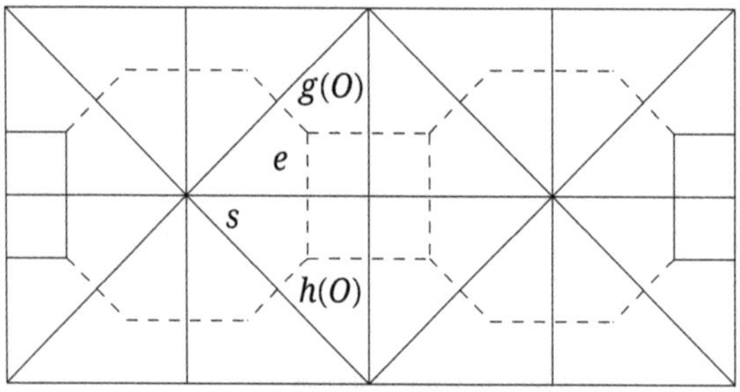

Figure 1.7: Dual graph C.

We assign to e the label $\lambda(e) = \alpha, \beta, \gamma$ according as s is the image $g(a)$, $g(b)$, $g(c)$ of the side a, b, c of Δ. Let $p = e_1 e_2 \cdots e_n$, $n \geq 1$, be a path in C, that is, a sequence of edges such that e_{i+1} begins where e_i ends for $1 \leq i < n$. Let the path p begin at some $g_0(O)$ and have successive vertices $g_0(O), g_1(O), \ldots, g_n(O)$, see Figure 1.8.

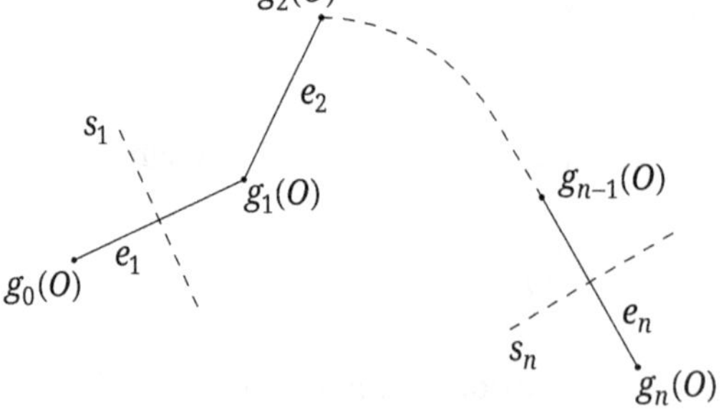

Figure 1.8: Path from $g_0(O)$ to $g_n(O)$.

Let $x_i = \lambda(e_i) = \alpha, \beta, \gamma$, $1 \leq i \leq n$, respectively. The reflection ρ at the edge s_i between $g_i(\Delta)$ and $g_{i+1}(\Delta)$ is $\rho = g_i \circ x_{i+1} \circ g_i^{-1}$, hence $g_{i+1} = \rho \circ g_i = g_i \circ x_{i+1}$. It follows that $g_n = g_0 \circ x_1 \circ x_2 \circ \cdots \circ x_n$. We define $\lambda(p) = x_1 \circ x_2 \circ \cdots \circ x_n$, and hence, $g_n = g_0 \circ \lambda(p)$. Now,

if p is a closed path, that is, $g_0(O) = g_n(O)$, then $g_0 = g_n$, and we have the relation $\lambda(p) = 1$.
On the other hand, if $w = x_1 \circ x_2 \circ \cdots \circ x_n = 1$ in G and $g_0 \in G$ arbitrary, then there exists a unique closed path p at $g_0(O)$ such that $\lambda(p) = 1 = x_1 \circ x_2 \circ \cdots \circ x_n$.
Now we have to show that each relation is a consequence of the described six relations. For this we consider the graph C and the closed paths in C more carefully.
The graph C divides the plane into regions Δ^* which are square or octagonal and defines a tessellation D^* of $\mathbb{R}^2$ (the dual of D). Suppose that a path p runs between points P and Q along an arc on side of the boundary of a region Δ^*. Let p' be the path between P and Q obtained from p by replacing this arc by the other side of Δ^* (see Figure 1.9).

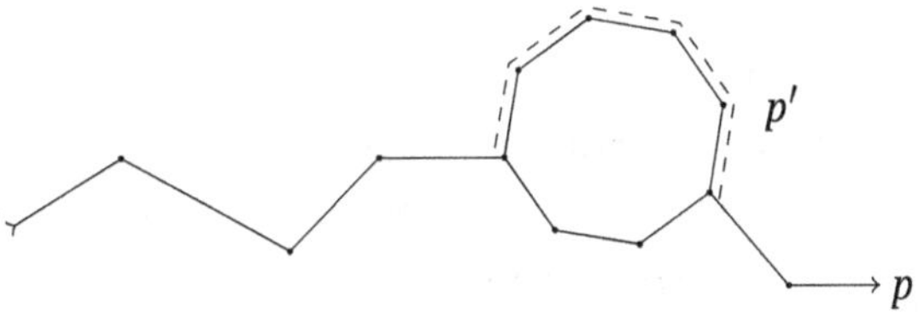

Figure 1.9: Replacing p by p'.

We have $\lambda(p) = \lambda(p')$ because running around the boundary of Δ^* is one of $(\gamma \circ \beta)^{\pm 4}$, $(\alpha \circ \gamma)^{\pm 2}$, $(\beta \circ \alpha)^{\pm 4}$. The relation $\alpha^2 = \beta^2 = \gamma^2 = 1$ corresponds to modifying p by deleting or inserting a spine, that is, an edge e_i followed by an edge e_{i+1}, that is, the same edge except traversed in the opposite direction (see Figure 1.10).

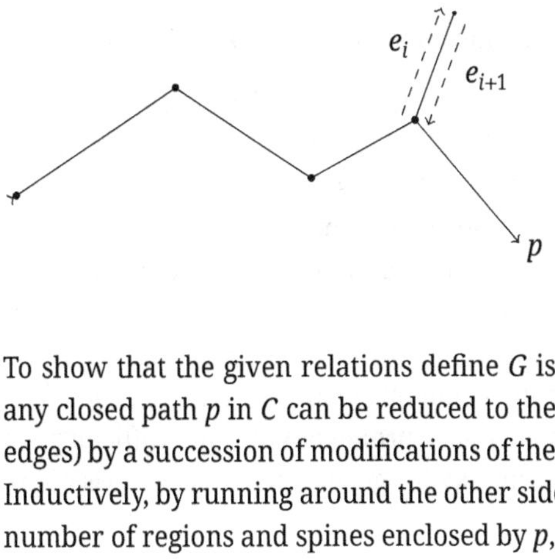

Figure 1.10: Cutting spines.

To show that the given relations define G is therefore equivalent to showing that any closed path p in C can be reduced to the trivial path at some point (with $n = 0$ edges) by a succession of modifications of the two kinds from above. But this is clear. Inductively, by running around the other side of some region Δ^*, we can reduce the number of regions and spines enclosed by p, and finally reduce to the case that p is a simple loop. □

11. Let $G = \langle \{x_n, n \geq 1\} \mid \{x_n = x_{nk}^k, n, k \geq 1\} \rangle$. The mapping $x_n \mapsto \frac{1}{n}$ induces an isomorphism from G to the group $\mathbb{Q}$ of additive rationals. This is an example of a

presentation of a group which is not finitely generated and needs infinitely many defining relations.
We leave this as an exercise. There also exist groups which are finitely generated but not finitely presentable. We give an example in Section 1.6.

1.3.2 Decision problems and Tietze transformations

These examples lead in a canonical manner to the following questions:

Remarks 1.3.4. 1. When do two presentations represent isomorphic groups?
If we formulate this as a decision problem then we get the *isomorphisms problem*: Given two presentations $G_1 = \langle X \mid R \rangle$ and $G_2 = \langle Y \mid S \rangle$. Are G_1 and G_2 isomorphic as groups?
2. The *word problem*:
Given a presentation $G = \langle X \mid R \rangle^{\varphi}$ and a word $w \in F(X)$. Does w represent in G the trivial element? In other words: Do we have $\varphi(w) = 1$? We may also write $w =_G 1$. Obviously, the solution of the word problem is equivalent to the problem to decide if two given words w_1 and w_2 from $F(X)$ are equal in G. This is exactly the case if $w_1 w_2^{-1} =_G 1$.
3. The *conjugacy problem*:
Given a presentation $G = \langle X \mid R \rangle^{\varphi}$ and two words $w_1, w_2 \in F(X)$. Do w_1 and w_2 represent in G conjugate elements, that is, is there a $g \in G$ with $\varphi(w_1) = g^{-1}\varphi(w_2)g$?
4. The *subgroup membership problem* or *generalized word problem*:
Given a group $G = \langle X \mid R \rangle^{\varphi}$, a set of words $W \subset F(X)$ and a word $w \in F(X)$. Is $\varphi(w) \in \langle \varphi(W) \rangle$?
The first three questions were already formulated by Max Dehn in 1911.

In this book these problems and their solutions for special groups keep coming up. In general all these problems are (algorithmically) unsolvable, more concrete: One can show for certain instances that there is no algorithmic solution of these problems.

There are certain relations between these decision problems:

The solution of the conjugacy problem induces the solution of the word problem because $w = 1$ in G if and only if w is in G conjugate to 1. Also, the solution of the membership problem implies the solution of the word problem. Both implications are proper, there are groups with solvable word problem but unsolvable conjugacy problem or unsolvable membership problem. This isomorphism problem is the most difficult problem. Hence, one usually considers this problem only for groups in a certain class of groups, for instance, the class of free groups.

We remark that the isomorphism problem is solvable in the class of finitely generated fully residually free groups. That is, given two finite presentations, that are known to define fully residually free groups, there is an effective algorithm to determine if the defined groups are isomorphic [41].

Certainly one can intensify the problems to search problems.

For the isomorphism problem this is the indication of an isomorphism, for the word problem this is the concrete description of w as the product of elements from $\langle\langle R\rangle\rangle$, for the conjugacy problem this is the indication of a conjugacy factor and for the membership problem this is the description of $\varphi(w)$ as a product in $\varphi(w_1), \varphi(w_2), \ldots, \varphi(w_n)$ with $w_1, w_2, \ldots, w_n \in W \cup W^{-1}$.

We now describe some first general theorems and procedures for group presentations.

Theorem 1.3.5 (Theorem of van Dyck). *Let $G = \langle X \mid R\rangle^{\varphi}$. Let H be a group, $f{:}\,X \to H$ a mapping and $\vartheta{:}\,F(X) \to H$ the corresponding homomorphism which extends f. If $\vartheta(r) = 1$ for all $r \in R$ then there exists a uniquely determined homomorphism $\psi{:}\,G \to H$ with $f(x) = \psi \circ \varphi(x)$ for all $x \in X$.*

Moreover, ψ is an epimorphism iff $f(X)$ is a generating system for H.

Proof. We have $R \subset \ker(\varphi)$. From $\ker(\varphi) = \langle\langle R\rangle\rangle$ we get $\ker(\varphi) \subset \ker(\vartheta)$, and hence Figure 1.11.

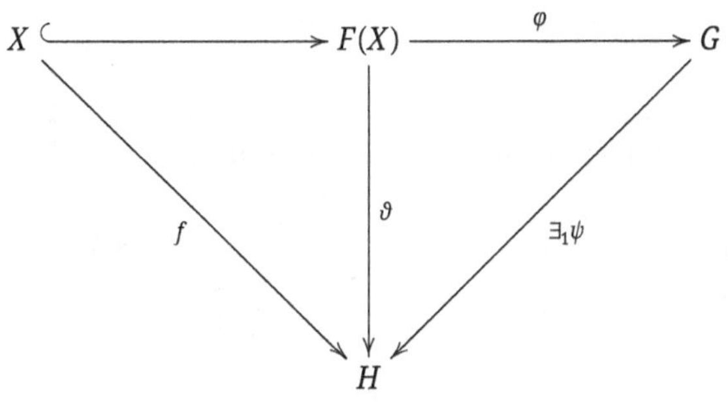

Figure 1.11: Factorization of homomorphisms.

We define $\psi(g) := \vartheta(w)$ for w with $\varphi(w) = g$, ψ is well defined and a homomorphism because $\ker(\varphi) \subset \ker(\vartheta)$ and unique because G is generated by $X(= \varphi(X))$. We have $\psi(G) = \langle f(X)\rangle$, and, hence $\psi(G) = H$ if $H = \langle f(X)\rangle$. □

Remarks 1.3.6. 1. We say that the homomorphism ϑ is factorized via the homomorphism φ. If we consider especially $G = \langle X \mid R\rangle$, $K = \langle X \cup Y \mid R \cup S\rangle$ and the inclusion $i{:}\,X \hookrightarrow X \cup Y$, then this induces a homomorphism $\langle X \mid R\rangle \to \langle X \cup Y \mid R \cup S\rangle$. This justifies also the notation $K = \langle G, Y \mid S\rangle$.

2. With the same argument as in Theorem 1.3.5 we also get:
Let G and H be groups, $R \subset G$ and $\vartheta{:}\,G \to H$ a homomorphism with $\vartheta(R) = \{1\}$. We get $\psi{:}\,G/\langle\langle R\rangle\rangle \to H$ with $\vartheta = \psi \circ \pi$ with $\pi{:}\,G \to G/\langle\langle R\rangle\rangle$ be the canonical map.

Corollary 1.3.7. *Two groups with the same presentation are isomorphic.*

Proof. Let $G_1 = \langle X \mid R\rangle^{\varphi_1}$, $G_2 = \langle X \mid R\rangle^{\varphi_2}$. By Theorem 1.3.5 there exist uniquely defined homomorphisms $\psi_1 : G_1 \to G_2$ and $\psi_2 : G_2 \to G_1$ with $\varphi_2 = \psi_1 \circ \varphi_1$ and $\varphi_1 = \psi_2 \circ \varphi_2$, hence, $\varphi_1 = \psi_2 \circ \psi_1 \circ \varphi_1$. Since φ_1 is surjective we get $\psi_2 \circ \psi_1 = \mathrm{id}_{G_1}$. Analogously $\psi_1 \circ \psi_2 = \mathrm{id}_{G_2}$. Therefore ψ_1 and ψ_2 are isomorphisms. □

Theorem 1.3.8 (Existence theorem). *For each system $(X; R)$ with $R \subset F(X)$ there exists a group $G = \langle X \mid R\rangle$.*

Proof. Let $N = \langle\langle R\rangle\rangle \lhd F(X)$. Then $F(X)/N \cong \langle X \mid R\rangle$. □

It is clear that a group G admits different presentations. We now describe a set of procedures allowing one to move from any presentation of G to any other.

Definition 1.3.9 (Tietze transformations). Let $G = \langle X \mid R\rangle$.

1. Let U be a system of letters disjoint from X and $\{w_u(X) \mid u \in U\}$ a system of words in X. Define $X' = X \cup U$ and $R' = R \cup \{u \cdot (w_u(X))^{-1} \mid u \in U\}$. The procedure $(X; R) \to (X'; R')$ is called *adding (the) new generators* U. The inverse operation is called *deleting (the) generators U.*
2. Let S be a system of relations which are consequences of the defining relations R, possibly including trivial relations. Define $X' = X$ and $R' = R \cup S$. The procedure $(X; R) \to (X'; R')$ is called *adding consequences* and the inverse process is called *deleting redundant relations.*

   These procedures are called *Tietze transformations.* If the sets U and/or S are finite then the respective procedure is called a *finite Tietze transformation.*

Theorem 1.3.10. *Two presentations $\langle X \mid R\rangle$ and $\langle Y \mid S\rangle$ define the same group G if and only if one can be transformed into the other by a sequence of Tietze transformations. If both presentations are finite only finitely many finite Tietze transformations are needed.*

Proof. It is clear that presentations which differ only by a Tietze transformation define isomorphic groups. Now, let $G = \langle X \mid R\rangle = \langle Y \mid S\rangle$. Assume that $w_x(Y)$, $x \in X$, and $w_y(X)$, $y \in Y$ are words such that $w_x(Y)$ and x represent the same element in G and similarly $w_y(X)$ and y. We assume that the formal generator and relator sets are pairwise disjoint. By adding first generators and then consequence relations we obtain

$$\begin{aligned}\langle X \mid R\rangle &\cong \langle X \cup Y \mid R \cup \{y(w_y(X))^{-1} \mid y \in Y\}\rangle \\ &\cong \langle X \cup Y \mid R \cup \{y(w_y(X))^{-1} \mid y \in Y\} \cup \{x(w_x(Y))^{-1} \mid x \in X\} \cup S\rangle.\end{aligned}$$

This last presentation can also be obtained by adding generators and consequence relations starting from $\langle Y \mid S\rangle$. □

Examples 1.3.11. 1. Let $G = \langle a, b, c, d \mid ab = c, bc = d, cd = a, da = b\rangle$. Then

$$G = \langle a, b, c, d \mid ab = c, bc = d, cd = a, da = b\rangle$$

$$
\begin{aligned}
&= \langle a,b,c,d \mid ab = c, bc = d, cd = a, da = b, cbc = a\rangle \\
&= \langle a,b,c,d \mid ab = c, bc = d, da = b, cbc = a\rangle \\
&= \langle a,b,c,d \mid ab = c, bc = d, da = b, cbc = a, ca = 1\rangle \\
&= \langle a,b,c,d \mid ab = c, bc = d, cbc = a, ca = 1\rangle \\
&= \langle a,b,c \mid ab = c, cbc = a, ca = 1\rangle \\
&= \langle a,b,c \mid ab = c, cbc = a, ca = 1, bbab = 1\rangle \\
&= \langle a,b,c \mid ab = c, ca = 1, bbab = 1\rangle \\
&= \langle a,b,c \mid ab = c, bbab = 1, aba = 1\rangle \\
&= \langle a,b \mid bbab = 1, aba = 1\rangle \\
&= \langle a,b \mid a = b^{-3}, aba = 1\rangle \\
&= \langle a,b \mid a = b^{-3}, b^{-5} = 1\rangle \\
&= \langle b \mid b^5 = 1\rangle = \mathbb{Z}_5.
\end{aligned}
$$

2. Let $G = \langle x,y \mid xyx = yxy\rangle = B_3$, the third braid group.

$$
\begin{aligned}
G &= \langle x,y \mid xyx = yxy\rangle \\
&= \langle a,b,x,y \mid xyx = yxy, a = xy, b = yxy\rangle \\
&= \langle a,b,x,y \mid xyx = yxy, a = xy, b = yxy, a^3 = b^2\rangle \\
&= \langle a,b,x,y \mid a = xy, b = yxy, a^3 = b^2\rangle \\
&= \langle a,b,x,y \mid a = xy, ba^{-1} = y, a^3 = b^2\rangle \\
&= \langle a,b,x,y \mid x = ay^{-1}, y = ba^{-1}, a^3 = b^2\rangle \\
&= \langle a,b,y \mid y = ba^{-1}, a^3 = b^2\rangle \\
&= \langle a,b \mid a^3 = b^2\rangle.
\end{aligned}
$$

3. Let $G = \langle a,b,c \mid a^2, aba = b^{-1}, c = ab\rangle$. Then $G = \langle a,c \mid a^2 = c^2 = 1\rangle$. We leave this as an exercise.
4. Let $G = \langle a,b \mid a^2b^3 = a^3b^4 = 1\rangle$. Then $G = \{1\}$. We leave this as an exercise.

Corollary 1.3.12. *If $\langle X \mid R\rangle$ and $\langle Y \mid S\rangle$ are finite presentations of the same group G then the word problem for $\langle X \mid R\rangle$ is solvable if and only if the word problem for $\langle Y \mid S\rangle$ is solvable.*

Proof. This follows from Tietze's theorem since adding generators and consequences of the defining relations—or the inverse operations—preserves solvability of the word problem. Furthermore, since $\langle X \mid R\rangle$ and $\langle Y \mid S\rangle$ present the same group, suitable words $w_x(Y)$ and $w_y(X)$ exist and are, by hypothesis, assumed to be known. □

Corollary 1.3.13. *Let G be finitely generated and $G = \langle Y \mid S\rangle$ be a presentation of G. Then there exists a finite subset of Y which generates G.*

Proof. Let X be finite and $G = \langle X \rangle$. Each $x \in X$ is a product of finitely many $y \in Y$ and their inverses. Hence there exist a finite subset $Y_1 \subset Y$ with $X \subset \langle Y_1 \rangle$. □

In the same manner we get:

Corollary 1.3.14. *Let $G = \langle X \mid R \rangle = \langle Y \mid S \rangle$ with X, R and Y finite. Then there exists a finite subset $S_1 \subset S$ with $G = \langle Y \mid S_1 \rangle$.*

One way to try to deal with the word problem or the isomorphism problem is to map the group $\langle X \mid R \rangle$ to some known group by homomorphism, for instance by abelianizing. Let G be a group and G' be the commutator subgroup of G, that is, $G' = \langle [x, y] \mid x, y \in G \rangle$. A group G is Abelian if and only if $G' = \{1\}$. If $\varphi: G \to H$ is a homomorphism then $\varphi(G') \subset H'$, and thus an automorphism of G maps G' onto itself. The image $\varphi(G)$ is Abelian if and only if $G' < \ker(\varphi)$. The *abelianization* of G is the group $G^{ab} = G/G'$, and the projection is written $ab: G \to G^{ab}$. The abelianization has the following universal property: If $\varphi: G \to A$ is a homomorphism of G to an Abelian group A then there is a uniquely determined homomorphism $\varphi^{ab}: G^{ab} \to A$ such that $\varphi = \varphi^{ab} \circ ab$. Hence, if G and H are isomorphic then G/G' and H/H' are isomorphic, too. Now, we can use the Classification Theorem of finitely generated Abelian groups (see for instance [51]).

Theorem 1.3.15. *Let A be a finitely generated Abelian group.*

1. *Then A has a presentation of the following type:*

$$\begin{aligned} A &= \langle a_1, a_2, \ldots, a_n \mid [a_i, a_j] = 1, 1 \le i < j \le n, a_1^{n_1} = a_2^{n_2} = \cdots = a_r^{n_r} = 1 \rangle \\ &\cong \mathbb{Z}_{n_1} \oplus \mathbb{Z}_{n_2} \oplus \cdots \oplus \mathbb{Z}_{n_r} \oplus \mathbb{Z}^p \end{aligned}$$

   where $0 \le r \le n$, $p = n - r$, $1 < n_i$ for $i = 1, 2, \ldots, r$ and $n_1 \mid n_2, n_2 \mid n_3, \ldots, n_{r-1} \mid n_r$. The numbers $n_1, n_2, \ldots, n_r$ are called the torsion coefficients and p is called the Betti number of A. The rank of A is $\mathrm{rk}(A) = p + r = n$.
2. *Two finitely generated Abelian groups are isomorphic if and only if their Betti numbers and torsion coefficients coincide.*
3. *The elements of finite order of A form the subgroup $\mathrm{Tor}(A)$ generated by $a_1, a_2, \ldots, a_r$, and hence $A = \mathrm{Tor}(A) \oplus \mathbb{Z}^p$. Any subgroup $U < A$ needs at most as many generators as A, that is, $\mathrm{rk}(U) \le \mathrm{rk}(A)$.*

Examples 1.3.16. 1. Let $F_n = \langle x_1, x_2, \ldots, x_n \mid \rangle$. Then $F_n^{ab} \cong \mathbb{Z}^n$. This shows again, that $\mathrm{rk}(F_n) = n$ and free groups of different finite rank are not isomorphic.
2. Let

$$G_g = \left\langle a_1, b_1, a_2, b_2, \ldots, a_g, b_g \,\middle|\, \prod_{i=1}^{g} [a_i, b_i] = 1 \right\rangle$$

and

$$H_g = \langle v_1, v_2, \ldots, v_g \mid v_1^2 v_2^2 \cdots v_g^2 = 1 \rangle$$

with $g \geq 1$. Then $G_g^{ab} \cong \mathbb{Z}^{2g}$ and $H_g^{ab} \cong \mathbb{Z}_2 \oplus \mathbb{Z}^{g-1}$, and this shows that different values of g give non-isomorphic groups and that no G_g can ever be isomorphic to some H_g.

3. Let $D_n = \langle a, b \mid a^2 = b^n = (ab)^2 = 1 \rangle$, $n \geq 2$, the nth dihedral group. Then

$$D_n^{ab} \cong \begin{cases} \mathbb{Z}_2 \oplus \mathbb{Z}_2 & \text{if } 2 \mid n, \\ \mathbb{Z}_2 & \text{otherwise.} \end{cases}$$

4. We have

$$\begin{aligned} \mathrm{PSL}(2,\mathbb{Z})^{ab} &\cong \mathbb{Z}_2 \oplus \mathbb{Z}_3 \cong \mathbb{Z}_6, \\ \mathrm{SL}(2,\mathbb{Z})^{ab} &\cong \mathbb{Z}_{12}, \quad \text{and} \\ \mathrm{GL}(2,\mathbb{Z})^{ab} &\cong \mathbb{Z}_2 \oplus \mathbb{Z}_2. \end{aligned}$$

5. Let $G = \langle a, b \mid a^2 = b^p = (ab)^q = 1 \rangle$ with $3 \leq p < q$ prime numbers. Then $G^{ab} = \{1\}$.
6. Let

$$\begin{aligned} B_n = \langle \sigma_1, \sigma_2, \ldots, \sigma_{n-1} \mid &\{\sigma_i \sigma_j = \sigma_j \sigma_i, 1 \leq i < j-1 \leq n-2\} \\ &\cup \{\sigma_i \sigma_{i+1} \sigma_i = \sigma_{i+1} \sigma_i \sigma_{i+1}; 1 \leq i < n\} \rangle \end{aligned}$$

the nth braid group, $n \geq 3$. Then $B_n^{ab} \cong \mathbb{Z}$.

1.3.3 The Reidemeister–Schreier method

We now describe two methods to get systems of representatives and presentations for subgroups of a group $G = \langle X \mid R \rangle$. We start with the Reidemeister–Schreier method.

Definition 1.3.17 (Schreier property). Let F be a free group and H be a subgroup of F. A *Schreier set of representatives* for H is a complete set of representatives $T \subset F$ of the right cosets Hg of H in F such that
(1) If $t \in T$ then every initial subword of t is in T; and
(2) If $t_1, t_2 \in T$ with $t_1, \neq t_2$ then $Ht_1 \neq Ht_2$.

We say that T has the *Schreier properties* (1) and (2).

Theorem 1.3.18. *Let F be a free group and H a subgroup of F. Then there exists a Schreier set of representatives for H.*

Proof. Let $F = F(X)$. If $T_1 \subset T_2 \subset \cdots$ be an ascending chain of partial sets of representatives which satisfy the Schreier properties (1) and (2). Then also the union $\bigcup_i T_i$ satisfies properties (1) and (2). By Zorn's lemma there exists a maximal partial set T of representatives which satisfies (1) and (2). This T is a complete set of representatives. This can be seen as follows. Assume, that there exists a $w \equiv x_1x_2\cdots x_n \in F$ with $x_i \in X \cup X^{-1}$, without loss of generality of minimal length, such that $Hw \neq Ht$ for all $t \in T$.

We have $|w| \geq 1$, and by virtue of the minimality of w we have $H(x_1x_2\cdots x_{n-1}) = Ht$ for some $t \in T$, and, hence, $Hw = H(tx_n)$. If t ends with x_n^{-1} then we must have $w \in T$ by property (1). If t does not end with x_n^{-1} then we may w add to T without violating the properties (1) and (2), which contradicts the maximality of T. □

Theorem 1.3.19. *Let $F = F(X)$ be a free group with basis X, H a subgroup of F and T a Schreier set of representatives for H. If $w \in F$ then let $\overline{w}$ be the representative of Hw in T and $\gamma(w) = w\overline{w}^{-1}$. Then $Y = \{\gamma(tx) \mid t \in T, x \in X, \gamma(tx) \neq 1\}$ is a basis for H.*

Proof. First we show that Y generates the subgroup H. Certainly $Y \subset H$ because $H = \{w\overline{w}^{-1} \mid w \in F\}$. Assume that there exists a $w \equiv x_1^{\epsilon_1}x_2^{\epsilon_2}\cdots x_n^{\epsilon_n}$, $x_i \in X$, $\epsilon_i = \pm 1$, without loss of generality of minimal length, such that $w\overline{w}^{-1} \notin \langle Y\rangle$. We have $|w| \geq 1$.

Let first $\epsilon_n = 1$, and let $w_1 \equiv x_1^{\epsilon_1}x_2^{\epsilon_2}\cdots x_{n-1}^{\epsilon_{n-1}}$. Then

$$\begin{aligned} w\overline{w}^{-1} &= (w_1x_n)(\overline{w_1x_n})^{-1} \\ &= (w_1\overline{w_1}^{-1})((\overline{w_1}x_n)(\overline{\overline{w_1}x_n})^{-1}) \in \langle Y\rangle \end{aligned}$$

because $Hw_1x_n = H\overline{w_1}x_n = H\overline{w_1x_n}$ and $w_1\overline{w_1}^{-1} \in \langle Y\rangle$ by virtue of the minimality of $|w| = n$ and also $(\overline{w_1}x_n)(\overline{\overline{w_1}x_n})^{-1} \in \langle Y\rangle$ after the construction of Y which contradicts $w\overline{w}^{-1} \notin \langle Y\rangle$. Hence, $w\overline{w}^{-1} \in \langle Y\rangle$.

Now let $\epsilon_n = -1$. Then $wx_n(\overline{wx_n})^{-1} \in \langle Y\rangle$ and $\overline{w}x_n(\overline{\overline{w}x_n})^{-1} \in \langle Y\rangle$, and on the other side

$$wx_n(\overline{wx_n})^{-1} = (w\overline{w}^{-1})(\overline{w}x_n(\overline{\overline{w}x_n})^{-1})$$

which again contradicts $w\overline{w}^{-1} \notin \langle Y\rangle$. Hence, again $w\overline{w}^{-1} \in \langle Y\rangle$. This shows $H = \langle Y\rangle$.

We now show that H is free on Y. We have $\gamma(w) = w\overline{w}^{-1}$. If $t \in T$ then $\gamma(t\overline{t}^{-1}) = 1$. We show that x is a stable letter of $\gamma(tx)$ for all $t \in T$ and $x \in X$ with $\gamma(tx) \in \langle Y\rangle$, that is, we show the following. If $w = \gamma(t_1x_1)^{\epsilon_1}\gamma(t_2x_2)^{\epsilon_2}\cdots\gamma(t_nx_n)^{\epsilon_n}$, with $n \geq 1$, $t_i \in T$, $x_i \in X$, $\epsilon_i = \pm 1$, all $\gamma(t_ix_i) \neq 1$, and if there is no i with $t_i = t_{i+1}$, $x_i = x_{i+1}$ and $\epsilon_i = -\epsilon_{i+1}$, then $x_i^{\epsilon_i}$ remains uncanceled in the ith factor $\gamma(t_ix_i)^{\epsilon_i}$. Especially we then have $w \neq 1$. Let $x_1, x_2 \in X$, $t_1, t_2 \in T$ be with $\gamma(t_1, x_1), \gamma(t_2, x_2) \neq 1$. We look first at a product

$$\gamma(t_1x_1)\gamma(t_2x_2) = t_1x_1(\overline{t_1x_1})^{-1}t_2x_2(\overline{t_2x_2})^{-1}. \tag{1.11}$$

There is no cancellation between t_1 and x_1 because otherwise $t_1x_1 \in T$ and $\gamma(t_1x_1) = 1$. Assume there is a cancellation between x_1 and $\overline{t_1x_1}^{-1}$. Then $\overline{t_1x_1}$ ends with x_1, that is, there is a $t \in T$ with $\overline{t_1x_1} \equiv tx_1 \in T$, which gives $tx_1 = \overline{tx_1} = \overline{t_1x_1}$ and further $t = t_1$. But since

$t_1x_1 \notin T$ we have $t \neq t_1$, which gives a contradiction. Analogously there is no cancellation between t_2 and x_2 and between x_2 and $\overline{t_2x_2}^{-1}$.

Assume now that x_1 or x_2 are canceled somehow different in (1.11). Then $\overline{t_1x_1}x_1^{-1}$ is an initial segment of $t_2x_2\overline{t_2x_2}^{-1}$ or t_2x_2 is an initial segment of $\overline{t_1x_1}x_1^{-1}t_1^{-1}$. We consider the situation that x_1 is canceled.

Now, t_2x_2 is not an initial segment of $\overline{t_1x_1}$ because $t_2x_2 \notin T$. Therefore $\overline{t_1x_1}x_1^{-1}$ is an initial segment of t_2x_2 and, hence, of t_2 because otherwise $t_1x_1 \in T$. This gives $\overline{t_1x_1}x_1^{-1} \in T$ and

$$1 = \overline{t_1x_1}x_1^{-1}\overline{\overline{t_1x_1}x_1^{-1}}^{-1} = \overline{t_1x_1}x_1^{-1}t_1^{-1} = \gamma(t_1x_1)^{-1},$$

which gives a contradiction. We now consider a product

$$\gamma(t_1x_1)\gamma(t_2x_2)^{-1} = t_1x_1\overline{t_1x_1}^{-1} \cdot t_2x_2\overline{t_2x_2}^{-1}. \tag{1.12}$$

If x_1 or x_2 is canceled then $\overline{t_1x_1}x_1^{-1}$ is an initial segment of $\overline{t_2x_2}x_2^{-1}t_2^{-1}$ or $\overline{t_2x_2}x_2^{-1}$ is an initial segment of $\overline{t_1x_1}x_1^{-1}t^{-1}$.

Since, as above, $\overline{t_ix_i}x_i^{-1} \notin T$, we necessarily have $\overline{t_1x_1}x_1^{-1} \equiv \overline{t_2x_2}x_2^{-1}$ and, hence, $x_1 = x_2$, $t_1 = t_2$. □

Remark 1.3.20. Especially we have shown:

If $\gamma(t_1x_1) = \gamma(t_2x_2) \neq 1$, $t_1, t_2 \in T$, $x_1, x_2 \in X$, then $t_1 = t_2$, $x_1 = x_2$.

Theorem 1.3.21. *Let F be free of finite* rank ≥ 2 *and $H < F$ of finite index $[F : H]$, then H has finite rank, and*

$$\text{rk}(H) - 1 = [F : H](\text{rk}(F) - 1).$$

Proof. Let $F = F(X)$, $f = \text{rk}(F)$, $h = \text{rk}(H)$, T a Schreier set of representatives for H in F and $j = [F : H] = |T|$. We count the elements of the basis Y of H in Theorem 1.3.19. Then $h = |T||X| - d$, where d is the number of pairs (t, x) with $t \in T$ and $x \in X$ and $\gamma(tx) = 1$. We have $\gamma(tx) = 1$, if and only if $tx \in T$, which is exactly the case, if t ends with x^{-1} or if t does not end with x^{-1} and $tx \in T$. Hence, d is the number of pairs $(t_1, t_2) \in T^2$ with $t_1 \equiv t_2y$ for some $y \in X \cup X^{-1}$. It follows $d = |T| - 1 = j - 1$ and, hence,

$$h - 1 = |T||X| - d - 1 = jf - j = j(f - 1).$$

□

Remark 1.3.22. Using Theorem 1.3.21 we may calculate the index of $H < F$ if the rank of H is known and if one knows that this rank is finite. This last assumption is necessary because for each $n \in \mathbb{N}$ there exists a subgroup of rank n but with infinite index (see Section 1.1). But, if $\{1\} \neq H \lhd F$ and if H is as a subgroup finitely generated, then H has finite index in F.

In fact, we hence have the following.

Theorem 1.3.23. *Let F be a free group of finite rank and H be a non-trivial normal subgroup of F. Then H has finite index in F if and only if H is finitely generated.*

We give an extension later in this section (Corollary 1.3.34) and another proof in Section 2.3 as an application of the covering space theory.

Theorem 1.3.24. *Let F be free of rank ≥ 2, $H < F$ a finitely generated subgroup and $[F : H] < \infty$. Then H is not properly contained in any subgroup $G < F$ with $\operatorname{rk}(G) > \operatorname{rk}(H)$.*

Proof. Let $f = \operatorname{rk}(F) \geq 2$, $h = \operatorname{rk}(H)$, $H \subset G < F$. Since $[F : H] < \infty$ we have $[F : G]$, $[G : H] < \infty$. Let $\operatorname{rk}(G) = g$. It follows from Theorem 1.3.21 that $[F : G] = \frac{g-1}{f-1} \geq 1$, hence, $g \geq 2$. Furthermore, we have $[G : H] = \frac{h-1}{g-1} \geq 1$, and therefore $g \leq h$ and $g = h \Leftrightarrow [G : H] = 1$, that is, $G = H$. □

We now use the Schreier set of representatives in free groups to get a presentation of a subgroup U of a group G.

Theorem 1.3.25 (Reidemeister–Schreier method). *Let $G = \langle x_i, i \in I \mid R_j, j \in J\rangle$ and U a subgroup of G. Let F be the free group with basis $\{x_i, i \in I\}$ and $\varphi : F \twoheadrightarrow G$, $x_i \mapsto x_i$, the canonical epimorphism. Let $U' = \varphi^{-1}(U)$ and $\{g_k, k \in K\}$ be a Schreier set of representative for U' in F. If $w \in F$ then let again $\overline{w}$ be the representative of $U'w$. Then*
1. *U is generated by the elements $g_k x_i \overline{g_k x_i}^{-1}$.*
2. *If we omit the trivial ones among the generators then $g_\ell R_j g_\ell^{-1}$, written in the non-trivial generators, form a set of defining relations for U.*

Proof. The non-trivial $g_k x_i \overline{g_k x_i}^{-1}$ generate U' freely by Theorem 1.3.19. Let $N = \langle\langle R_j \mid j \in J\rangle\rangle$; we have $N \subset U'$ and $U'/N \cong U$. The subgroup N is generated by wR_jw^{-1}, $w \in F$, and hence, also by the elements $ug_kR_jg_k^{-1}u^{-1}$, $u \in U'$. Hence, N is the normal closure of the $g_kR_jg_k^{-1}$ in U'. If we write each $g_kR_jg_k^{-1}$ as a word S_{kj} in $g_\ell x_i \overline{g_\ell x_i}^{-1} \neq 1$, then we get the desired presentation. □

Corollary 1.3.26. *Let $H < G$ with finite index. If G is finitely generated, then also H. If G is finitely presented, then also H.*

Examples 1.3.27. 1. Let $F = \langle x, y \mid \rangle$, and consider the homomorphism $\varphi : F \to \mathbb{Z}_n$, $\mathbb{Z}_n = \{0, 1, \ldots, n-1\}$, $n \geq 2$, defined by $x \mapsto 1$, $y^n \mapsto 1$, and let U be the kernel of φ. As coset representatives we take $1, y, y^2, \ldots, y^{n-1}$. Certainly, this system fulfills the Schreier condition. The Reidemeister–Schreier generators for U are the non-trivial elements of the set $\{y^i x(\overline{y^{i+1}})^{-1}, y^i y(\overline{y^{i+1}})^{-1} \mid i = 0, 1, \ldots, n-1\}$. We obtain the non-trivial elements y^n, $x_i = y^i x y^{-(i+1)}$ for $i = 0, 1, \ldots n-2$, and $x_{n-1} = y^{n-1}x$. The rank of this subgroup U is $n+1$.

Note that this shows that the free group of rank 2 contains a free group of rank $n+1$ as a subgroup of index n for each $n \in \mathbb{N}$.

2. Let $G = \langle v_1, v_2, \ldots, v_n \mid v_1^2 v_2^2 \cdots v_n^2 = 1\rangle$, $n \geq 2$, and $\alpha : G \to \mathbb{Z}_2$ the homomorphism which maps each v_i onto -1. Let $U = \ker(\alpha)$. As representatives we have 1, v_n, and

the generators are $x_i = 1\cdot v_i(\overline{v_i})^{-1} = v_i v_n^{-1}, i = 1, 2, \dots, n-1, y_i = v_n v_i(\overline{v_n v_i})^{-1} = v_n v_i, i = 1, 2, \dots, n-1$ and $z = v_n^2$. We have two relations $v_1^2 v_2^2 \cdots v_n^2 = 1$ and $v_n v_1^2 v_2^2 \cdots v_n^2 v_n^{-1} = 1$. We may rewrite these as

$$v_1 v_n^{-1} v_n v_1 \cdots v_{n-1} v_n^{-1} v_n v_{n-1} v_n^2 = x_1 y_1 x_2 y_2 \cdots x_{n-1} y_{n-1} z = 1$$

and

$$v_n v_1 v_1 v_n^{-1} v_n v_2 v_2 v_n^{-1} \cdots v_n v_{n-1} v_{n-1} v_n^{-1} v_n^2 = y_1 x_1 y_2 x_2 \cdots y_{n-1} x_{n-1} = 1.$$

Using Tietze transformations we may eliminate z and get

$$\begin{aligned} U &= \langle x_1, x_2, \dots, x_{n-1}, y_1, y_2, \dots y_{n-1} \mid y_1 x_1 y_2 x_2 \cdots y_{n-1} x_{n-1} y_{n-1}^{-1} x_{n-1}^{-1} \cdots y_1^{-1} x_1^{-1} = 1 \rangle \\ &= \left\langle a_1, b_1, a_2, b_2, \dots, a_{n-1}, b_{n-1} \;\middle|\; \prod_{i=1}^{n-1} [a_i, b_i] = 1 \right\rangle \end{aligned}$$

by Nielsen transformations (see Chapter 5 of [281]).

3. Let $G = \langle x, y \mid x^2 = y^3 = 1 \rangle \cong \mathrm{PSL}(2, \mathbb{Z})$. We have $G/G' = \langle a \mid a^6 = 1 \rangle$. We look for a presentation of G'. To do this we write $G = \langle x, z \mid x^2 = (xz)^3 = 1 \rangle$ with $z = x^{-1}y = xy$. A Schreier set of representatives is given by 1, z, z^2, z^3, z^4, z^5. The non-trivial generators are $x_1 = xz^{-3}$, $x_2 = zxz^{-4}$, $x_3 = z^2xz^{-5}$, $x_4 = z^3x$, $x_5 = z^4xz^{-1}$, $x_6 = z^5xz^{-2}$ and $x_7 = z^6$.
We only have to keep in mind that $\overline{x} = z^3$. We have the following relations:

$$\begin{aligned} x^2 &= xz^{-3}z^3x = x_1x_4, \\ zx^2z^{-1} &= x_2x_5, \\ z^2x^2z^{-2} &= x_3x_6, \\ z^3x^2z^{-3} &= x_4x_1, \\ z^4x^2z^{-4} &= x_5x_2, \\ z^5x^2z^{-5} &= x_6x_3, \\ (xz)^3 &= xz^{-3}z^4xz^{-1}z^2xz^{-5}z^6 = x_1x_5x_3x_7, \\ z(xz)^3z^{-1} &= x_2x_6x_4, \\ z^2(xz)^3z^{-2} &= x_3x_7x_1x_5, \\ z^3(xz)^3z^{-3} &= x_4x_2x_6, \\ z^4(xz)^3z^{-4} &= x_5x_3x_7x_1 \quad \text{and} \\ z^5(xz)^3z^{-5} &= x_6x_4x_2. \end{aligned}$$

If we delete redundant relations we are left with the relations

$$x_1x_4,\ x_2x_5,\ x_3x_6,\ x_1x_5x_3x_7 \quad \text{and} \quad x_2x_6x_4.$$

Now we may delete generators with the help of Tietze transformations and get $G' = \langle x_1, x_2, \mid \rangle$, that is, G' is free of rank 2 with basis $\{x_1, x_2\}$. From this presentation we get with the help of Nielsen transformations $G' = \langle [x, y], [x, y^2] \mid \rangle$.

We now describe some interesting number theoretical connections. We know that G, as the group $\mathrm{PSL}(2,\mathbb{Z})$, is generated by $X = \pm\left(\begin{smallmatrix} 0 & 1 \\ -1 & 0 \end{smallmatrix}\right)$ and $Y = \pm\left(\begin{smallmatrix} 0 & -1 \\ 1 & 1 \end{smallmatrix}\right)$. Then

$$A := [X, Y] = XYX^{-1}Y^{-1} = \pm\begin{pmatrix} 2 & 1 \\ 1 & 1 \end{pmatrix} \quad \text{and}$$

$$B := [X, Y^2] = XY^2X^{-1}Y^{-2} = \pm\begin{pmatrix} 1 & 1 \\ 1 & 2 \end{pmatrix}.$$

Further, $AB^{-1} = \pm\left(\begin{smallmatrix} 3 & -1 \\ 1 & 0 \end{smallmatrix}\right)$. Since G' is torsion-free we may G' consider as a subgroup of $\mathrm{SL}(2,\mathbb{Z})$, freely generated by $A = \left(\begin{smallmatrix} 2 & 1 \\ 1 & 1 \end{smallmatrix}\right)$ and $B = \left(\begin{smallmatrix} 1 & 1 \\ 1 & 2 \end{smallmatrix}\right)$. We have $\mathrm{tr}(A) = \mathrm{tr}(B^{-1}) = \mathrm{tr}(AB^{-1}) = 3$ for the traces. The triple $(\mathrm{tr}(A), \mathrm{tr}(B), \mathrm{tr}(AB^{-1}))$ is a solution of the diophantine equation $x^2 + y^2 + z^2 - xyz = 0$. From Theorem 1.2.19 we know the following. The set $\{U, V\} \subset G'$ is a basis of G' if and only if

$$[U, V] = W[A, B]^{\pm 1}W^{-1}$$

for some $W \in G'$, and there is a (regular) Nielsen transformation from $\{A, B\}$ to $\{U, V\}$.

Hence, $\mathrm{tr}[U, V] = \mathrm{tr}[A, B] = -2$. Moreover, $(\mathrm{tr}(U), \mathrm{tr}(V), \mathrm{tr}(UV))$ also is a solution of the diophantine equation $x^2 + y^2 + z^2 - xyz = 0$ because of the general trace formula

$$\mathrm{tr}([U, V]) = (\mathrm{tr}(U))^2 + (\mathrm{tr}(V))^2 + (\mathrm{tr}(UV))^2 - \mathrm{tr}(U) \cdot \mathrm{tr}(V) \cdot \mathrm{tr}(UV) - 2.$$

We define

$$E = \{(U, V) \mid \{U, V\} \text{ is Nielsen equivalent to } \{A, B\}\}$$

and

$$L = \{(\mathrm{tr}(U), \mathrm{tr}(V), \mathrm{tr}(UV)) \mid (U, V) \in E\}.$$

Since $G = \langle U, V \rangle$ and $\mathrm{tr}[U, V] = \mathrm{tr}[A, B]$ for all $\{U, V\} \in E$ the ternary form $F(x, y, z) = x^2 + y^2 + z^2 - xyz$ is invariant under automorphisms of G', that is, under Nielsen transformations from generating pairs of G' to other generating pairs. The automorphism group $\mathrm{Aut}(G')$ is also generated by the Nielsen transformations

$$\alpha\colon (U, V) \mapsto (V, U),$$
$$\beta\colon (U, V) \mapsto (VU, U^{-1}), \quad \text{and}$$

$$\gamma\colon (U,V) \mapsto (VUV, V^{-1}).$$

This follows from direct conversions from the given ones in Section 1.2. These induce birational transformations of L via

$$\begin{aligned} \varphi &\colon (x,y,z) \mapsto (y,x,z), \\ \omega &\colon (x,y,z) \mapsto (z,x,y), \quad \text{and} \\ \psi &\colon (x,y,z) \mapsto (x',y,z) \quad \text{with } x' = yz - x. \end{aligned}$$

Let M be the permutation group of L generated by φ, ω, ψ.

We have (with respect to composition)

$$M = \langle \varphi, \omega, \psi \mid \varphi^2 = \omega^3 = \psi^2 = (\varphi\omega)^2 = (\psi\varphi\omega)^2 = 1 \rangle.$$

This can be seen as follows. All relations are evident by the definition of φ, ω, ψ. We now show the completeness of the relations

$$\varphi^2 = \omega^3 = \psi^2 = (\varphi\omega)^2 = (\psi\varphi\omega)^2 = 1.$$

We define $\psi_0 = \psi$, $\psi_1 = \omega\psi\omega^{-1}$ and $\psi_2 = \omega^{-1}\psi\omega$. Then

$$\begin{aligned} \psi_1(x,y,z) &= (x,y',z) \quad \text{with } y' = xz - y \quad \text{and} \\ \psi_2(x,y,z) &= (x,y,z') \quad \text{with } z' = xy - z. \end{aligned}$$

Let $r = r(\varphi, \omega, \psi)$ a reduced word in which ψ occurs. Using the known relations we may write r as $r = \gamma\psi_{r_m}\psi_{r_{m-1}}\cdots\psi_{r_1}$ with $\gamma \in \langle \varphi, \omega \rangle \cong S_3$, $m \geq 1$, $r_j \in \{0,1,2\}$ for $j = 1,2,\ldots,m$ and $r_i \neq r_{i-1}$, if $m \geq 2$, for $i = 2,3,\ldots,m$ (otherwise there is a cancellation). If $(x,y,z) \in L$ we define the height $h(x,y,z) = h + y + z$.

First of all $2 < x,y,z$ for $(x,y,z) \in L$ because of $x^2 + y^2 + z^2 - xyz = 0$. We choose $(x,y,z) \in L$ such that the components of (x,y,z) are pairwise distinct and that the $(r_1 + 1)$th component of (x,y,z) is not the biggest one. This is always possible after an application of φ, ω, ψ. Then we get:

$h(\psi_{r_1}(x,y,z)) > h(x,y,z)$, the components of $\psi_{r_1}(x,y,z)$ are pairwise distinct, and the (r_1+1)th component of $\psi_{r_1}(x,y,z)$ is now the biggest one. This is clear because if, for instance, $x < y$ then $x' = yz - x > x,y,z$.

Inductively we get

$$h(\gamma\psi_{r_m}\psi_{r_{m-1}}\cdots\psi_{r_1}(x,y,z)) > h(x,y,z),$$

because $m \geq 1$ and, hence, $r \neq 1$. Thus we have proved the following.

Theorem 1.3.28.

$$M = \langle \varphi, \omega, \psi \mid \varphi^2 = \omega^3 = \psi^2 = (\varphi\omega)^2 = (\psi\varphi\omega)^2 = 1 \rangle.$$

If we compare this presentation with the one for PGL(2, ℤ), namely

$$\mathrm{PGL}(2,\mathbb{Z}) = \langle a,b,c \mid a^2 = b^3 = c^2 = (ab)^2 = (cab)^2 = 1\rangle,$$

where $a = \pm\left(\begin{smallmatrix}-1 & -1\\ 0 & 1\end{smallmatrix}\right)$, $b = \pm\left(\begin{smallmatrix}-1 & -1\\ 1 & 0\end{smallmatrix}\right)$ and $c = \pm\left(\begin{smallmatrix}-1 & 0\\ 0 & 1\end{smallmatrix}\right)$, we see that $M \cong \mathrm{PGL}(2,\mathbb{Z})$. If we define M° to be the subgroup of M generated by ω and $\rho = \psi\varphi\omega$, then $M^\circ \cong \mathrm{PSL}(2,\mathbb{Z})$. We remark that

$$\rho(x,y,z) = \psi\varphi\omega(x,y,z) = (yz - x, z, y).$$

Theorem 1.3.29. *The natural numbers x, y, z solve the equation $x^2 + y^2 + z^2 = xyz$ if and only if there are generators U, V of G′ with* $\mathrm{tr}(U) = x$, $\mathrm{tr}(V) = y$ *and* $\mathrm{tr}(UV) = z$.

Proof. Let $x,y,z \in \mathbb{N}$ with $x^2 + y^2 + z^2 = xyz$. We have certainly $2 < x,y,z$. Let $L' = \{(x,y,z) \in \mathbb{N}^3 \mid x^2 + y^2 + z^2 = xyz\}$.

We show that $L = L'$. Starting with $(x,y,z) \in L'$ we may certainly apply the transformations φ, ω and ψ on L', and we get ongoing new triples from L'. We apply φ, ω and ψ in a minimizing manner. Then we have at triple $(x,y,z) \in L'$ with $2 < x \le y \le z$ and $h(x,y,z)$ minimal amongst all triples from L'. Then we have $z \le xy - z$ and altogether $2 < x \le y \le z \le xy - z \le \frac{xy}{2}$. From $x \le y \le z$ we get also $\frac{1}{3}xy \le z$. Hence $\frac{1}{3}xy \le z \le \frac{1}{2}xy$ which gives $\frac{1}{2}xy - z \le \frac{1}{6}xy$. From this we get

$$0 = x^2 + y^2 + z^2 - xyz = x^2 + y^2 + \left(\frac{1}{2}xy - z\right)^2 - \left(\frac{1}{2}xy\right)^2 \le 2y^2\left(1 - \frac{1}{9}x^2\right).$$

Hence, $x = 3$. This gives furthermore

$$y \le z = \frac{3}{2}y - \frac{1}{2}\sqrt{5y^2 - 36},$$

that is, also $y^2 \le 9$ and $y = 3$. Finally from the equation we also get $z = 3$.

Now, for the matrices A, B^{-1} from the above we have $\mathrm{tr}(A) = \mathrm{tr}(B^{-1}) = \mathrm{tr}(AB^{-1}) = 3$, but this gives $L = L'$. □

Corollary 1.3.30. *M acts transitively on the set*

$$L = \{(x,y,z) \in \mathbb{N}^3 \mid x^2 + y^2 + z^2 = xyz\}.$$

This connection leads to some other number theoretical and group theoretical applications (see [221]). We list some of these applications.

1. Let $K = \mathrm{GF}(q)$ be any finite field, $q = p^m$ with p a prime number. We consider the permutation group $G(K)$ from K^3 to K^3, generated by

$$\omega\colon (x,y,z) \mapsto (z,x,y) \quad \text{and}$$
$$\rho\colon (x,y,z) \mapsto (yz - x, z, y).$$

Via $\omega \mapsto \omega$, $\rho \mapsto \rho$ we get an epimorphism $f: M^\circ \to G(K)$. Let $\Gamma^q = \ker(f)$. It is an interesting question whether or not Γ^q is a congruence subgroup of $\Gamma = \mathrm{PSL}(2,\mathbb{Z})$. Recall that a subgroup of Γ is a congruence subgroup if it contains a principal congruence subgroup

$$\Gamma(N) = \left\{ \pm \begin{pmatrix} a & b \\ c & d \end{pmatrix} \in \mathrm{PSL}(2,\mathbb{Z}) \;\middle|\; b \equiv 0 \mod N, c \equiv 0 \mod N \right\}$$

for some $N \in \mathbb{N}$.

At first step to answer this question is to determine the order $O(\gamma)$ of $\gamma = \rho\omega$ in $G(K)$, the so called level of Γ^q. We have

$$\gamma(x,y,z) = (y, yz - x, z).$$

Using properties of the Chebyshev polynomials $S_n(t)$ (see Section 1.2) we get

$$O(\gamma) = \begin{cases} 2(q^2-1) & \text{if } p = 2, q = 2^m, \\ \frac{1}{2}p(q^2-1) & \text{if } p \geq 3, q = p^m. \end{cases}$$

This follows from the observation that

$$\gamma^n(x,y,z) = \big(yS_n(z) - xS_{n-1}(z), yS_{n+1}(z) - xS_n(z), z\big)$$

for $n \geq 1$, hence, $\gamma^n = 1$ if and only if $S_n(z) = 0$ and $S_{n-1}(z) = -1$ for all $z \in K = \mathrm{GF}(q)$. This shows $O(\gamma) \mid |\mathrm{PSL}(2,q)|$ if $q = p$. Using Klein's level concept (see [276]) gives the next theorem.

Theorem 1.3.31. *If $q = p^m$, $p \geq 3$, or $q = 2^m$, m even, then Γ^q is not a congruence subgroup of the modular group.*

This method gives a new method to construct a series of non-congruence subgroups of the modular group.

We remark that we may apply this method also for rings $\mathbb{Z}/n\mathbb{Z}$, $n \in \mathbb{N}$, $n \geq 2$.

2. We call a natural number x a *weak Markoff number* if there are natural numbers y and z such that $x^2 + y^2 + z^2 = 3xyz$. We consider the matrix $\left(\begin{smallmatrix} 0 & 1 \\ 1 & 1 \end{smallmatrix}\right) \in \mathrm{GL}(2,\mathbb{Z})$, It is known that

$$\begin{pmatrix} 0 & 1 \\ 1 & 1 \end{pmatrix}^n = \begin{pmatrix} f_{n-1} & f_n \\ f_n & f_{n+1} \end{pmatrix}, \quad n \geq 1,$$

where the f_n are the Fibonacci numbers $f_0 = 0, f_1 = 1$ and $f_n = f_{n-1} + f_{n-2}$ for $n \geq 2$. Using the connection $\left(\begin{smallmatrix} 0 & 1 \\ 1 & 1 \end{smallmatrix}\right)^2 = \left(\begin{smallmatrix} 1 & 1 \\ 1 & 2 \end{smallmatrix}\right) = B$ we see that f_{2n-1} is a weak Markoff number for all $n \in \mathbb{N}$. One only has to use the observation that $3 \mid x, 3 \mid y, 3 \mid z$ if x, y, z are natural numbers with $x^2 + y^2 + z^2 = xyz$.

Moreover, straightforward calculations lead to the following results:
(1) $(f_{2n-1})^2 + (f_{2n+1})^2 + 1 = 3f_{2n-1}f_{2n+1}$, $n \in \mathbb{N}$.
(2) $(f_{2n+1} + f_{2n-1})^2 + (f_{2m+1} + f_{2m-1})^2 + (f_{2n+1}f_{2m-1} + f_{2m+1}f_{2n-1} + 2f_{2n}f_{2m})^2 - (f_{2n+1} + f_{2n-1})(f_{2m+1} + f_{2m-1})(f_{2n+1}f_{2m-1} + f_{2m+1}f_{2n-1} + 2f_{2n}f_{2m}) + 4f_n^2f_m^2 = 4$, $n, m \in \mathbb{N}$.
(3) For each $n \in \mathbb{N}$ there exist $y, z \in \mathbb{N}$ with $9f_{2n+1}^2 = 4 + x^2 + y^2$.
(4) For each $n \in \mathbb{N}$ there exists a $x \in \mathbb{N}$ with $5f_{2n}^2 + 4 = x^2$.

3. We now consider our last example for the Reidemeister–Schreier method. Let

$$G = \langle x, y, z \mid x^3 = y^5 = z^2 = (xyx^2y^2)^2 = (yz)^2 = (xz)^3 = 1\rangle$$

be a generalized tetrahedron group. One way to consider the structure of G is to find suitable subgroups. We are looking for a subgroup H of index 5 with the representatives $1, y, y^2, y^3$ and y^4. After some rewriting we get as non-trivial generators for such H

$$a = xy, \quad b = yxy^4, \quad c = y^2x, \quad d = y^3xy^2, \quad e = y^4xy^3, \quad u = z, \quad v = yz,$$
$$w = y^2z, \quad r = y^3z \quad \text{and} \quad s = y^4z.$$

This gives the presentation

$$\begin{aligned} H = \langle a, b, c, d, e, u, v, w, r, s \mid aec &= b^3 = d^3 = a^2e^2d^2 = (bca)^2 = dec^2b^2 = u^2 \\ &= v^2 = w^2 = r^2 = s^2 = (as)^3 = (bw)^3 = (cw)^3 \\ &= (dv)^3 = (ev)^3 = 1\rangle. \end{aligned}$$

We may map H onto the infinite dihedral group, hence, H, and also G, is infinite. This is not so easy to see from the presentation for G.
In any case, it is an important problem to decide if a given presentation describes a finite or infinite group.

1.3.4 Hall's theorem

As an application of the Reidemeister–Schreier method we give a first algebraic proof of a theorem of M. Hall. We describe a topological proof in Section 3.6. We remark that we consider free products and free factors in the next section.

Theorem 1.3.32 (M. Hall's theorem). *Let H be a finitely generated subgroup of a free group F. Then there is a subgroup G of finite index in F for which H is a free factor.*

Proof. Let F be free on X and H a finitely generated subgroup of F. Let T_1 be a Schreier system of coset representatives for H in F. If $w \in F$ we again write $\overline{w}$ for the representative of w in T_1, defined by the condition $Hw = H\overline{w}$, $\overline{w} \in T_1$. Then H is free on the set

$$Y_1 = \{\gamma(tx) = tx(\overline{tx})^{-1} \mid t \in T_1, x \in X, \gamma(tx) \neq 1\}.$$

Let T_2 consist of all $t \in T_1$ and $\overline{tx} \in T_1$ for such $y(tx) \in Y_1$. Thus $T_2 \subset T_1$. Let T consist of all initial segments of elements of T_2. Since $T_2 \subset T_1$ and T_1 is a Schreier system, $T \subset T_1$. Since H is finitely generated, Y_1 is finite. Since T_2 is finite we see that T is finite.

For given $x \in X$ let $T(x)$ be the set of all $t \in T$ such that $\overline{tx} \in T$. Define $\varphi(x)\colon T(X) \to T$ by $\varphi(x)(t) = \overline{tx}$. Then $\varphi(x)$ is injective since $\varphi(x)(t_1) = \varphi(x)(t_2)$ implies $\overline{t_1x} = \overline{t_2x}$, $Ht_1x = Ht_2x$, $Ht_1 = Ht_2$ and $t_1 = t_2$ because $t_1, t_2 \in T$. Thus we may extend $\varphi(x)$ to a permutation $\pi(x)$ of T. This defines an action π of F on T. If $t \in T, x \in X$ and $tx \in T$, then $tx \in T_1$, $\overline{tx} = tx$, and $\pi(x)(t) = \overline{tx}$. Interchanging t and tx, this says that $t \in T, x \in X$ and $tx^{-1} \in T$, implies $\pi(x^{-1})(t) = tx^{-1}$. If $t = y_1y_2\cdots y_n, y_i \in X \cup X^{-1}$, we have inductively that

$$\pi(t)(1) = \pi(y_1)\pi(y_2)\cdots\pi(y_n)(1) = y_1y_2\cdots y_n = t.$$

Thus π defines a transitive action of F on T, and T is therefore a complete system of coset representatives for a group G with respect to this action, in fact G is the stabilizer J of 1 under this action. T is closed under taking initial segments, and, hence, T is a Schreier system of coset for G in F. Since T is finite, G has finite index in F. Denote the representative of the coset $Gf, f \in F$, by $\tilde{f}$. Then G is free on

$$Y_2 = \{tx(\tilde{tx})^{-1} \mid t \in T, x \in X, tx(\tilde{tx})^{-1} \neq 1\}.$$

Let $y = y(tx) \neq 1$ belong to the basis Y_1 for H. Then by the definition of T it follows that $t, tx \in T$, and by the definition of π that $\pi(x)(t) = tx$. Consequently $y = tx(\pi(x)(t))^{-1} \neq 1$ belongs to the basis Y_2 for G. Now $Y_1 \subset Y$ implies that H is a free factor in G. □

Wilton, see [273], also extended Hall's theorem for finitely generated fully residually free groups. We may extend the above theorem to the following corollary.

Corollary 1.3.33. *Let F be a free group, A a finite subset of F, and G a finitely generated subgroup of F that is disjoint from A. Then H is a free factor of a group G of finite index in F and disjoint from A.*

We only have to extend the set T_2 and take as the new T_2 all $t \in T_1$ and $\overline{tx} \in T_1$, together with $\overline{a}$ for all $a \in A$, and as the new T all initial segments of elements of T_2. The group G constructed from the new T is disjoint from A. This we see as follows. Let $a \in A$. Since $H \cap A \neq \emptyset$, $a \neq 1$, and $\overline{a} \in \overline{T}, \overline{a} \neq 1$, whence $\overline{a} \notin G$. Since $Ha = H\overline{a}$, $a\overline{a}^{-1} \in H \subset G$ and thus $\overline{a} \notin G$ implies $a \notin G$.

The theorem gives an extension of Theorem 1.3.23.

Corollary 1.3.34. *If a finitely generated subgroup H of a free group F contains a nontrivial normal subgroup of F, then it has finite index in F.*

Proof. Suppose that $N \subset H$, $\{1\} \neq N \lhd F$. Then some $G = H * K$ has finite index in F. Suppose H has infinite index. Then $H \neq G$ whence $K \neq \{1\}$. Let G have a basis $X = X_1 \cup X_2$ with $H = \langle X_1 \rangle, K = \langle X_2 \rangle$. Let $1 \neq u \in N \subset H, 1 \neq k \in K$. Since N is normal, $kuk^{-1} \in N \subset H$. But $k^{-1}uk$ is clearly not a word in the generators X_1 for H. Hence, H cannot have infinite index. □

We close this point with the following technical proposition also based on Hall's theorem. This is useful in the study of one-relator groups.

Proposition 1.3.35. *Let $F = \langle a_1, a_2, \ldots, a_p\rangle$ be a free group of rank p with basis $\{a_1, a_2, \ldots, a_p\}$. Let w_1 be an element of F which is neither a proper power nor primitive in F. Let $w = w_1^k$ with k a natural number and N the normal closure of w in F. Let H be a free subgroup of F with rank p and basis $\{x_1, x_2, \ldots, x_p\}$. Suppose $w_1^\alpha \in \langle x_1, x_2, \ldots, x_n\rangle$ with $1 \le n < p$ for some $\alpha \ge 1$ and further $F = HN$. Then $w_1 \in \langle x_1, x_2, \ldots, x_n\rangle$ and $F/N \cong \langle x_1, x_2, \ldots, x_n \mid w = 1\rangle * K$, where $K = \langle x_{n+1}, x_{n+2}, \ldots, x_p\rangle$.*

Proof. By assumption F/N has rank p. Let $\alpha \ge 1$ be the minimal power such that $w_1^\alpha \in \langle x_1, x_2, \ldots, x_n\rangle$ and assume $\alpha \ge 2$. Since $F/N \cong HN/N \cong H/(N \cap H)$ has rank p, there is no conjugate of a power of w_1 primitive in $\langle x_1, x_2, \ldots, x_n\rangle$. Let $R(x_1, x_2, \ldots, x_n) = w_1^\alpha$ with $R(x_1, x_2, \ldots, x_n)$ not a proper power in $\langle x_1, x_2, \ldots x_n\rangle$. As a subgroup of F, the group G generated by $w_1, x_1, x_2, \ldots, x_n$ is also free. Then by Theorem 1.2.40 G has rank $n - 1$, which contradicts $F = HN$. Therefore, $w_1 \in \langle x_1, x_2, \ldots, x_n\rangle$. Suppose first that H is of finite index in F. Let $v \in N$ with $v \notin H$. Then $v^\beta \in H$ for some $\beta \ge 2$. No conjugate of a power of v is primitive in H because $F/N \cong H/(N \cap H)$. Consider the subgroup M of F generated by $v, x_1, x_2, \ldots, x_p$. Again from Theorem 1.2.40, we see that M has rank $\le p - 1$. This contradicts that

$$F/N \cong H/(N \cap H) \cong M/(N \cap M)$$

has rank p. Therefore, $N \subset H$ if H has finite index and the result holds. Now suppose that H has infinite index in F. Then from Hall's theorem, there is a subgroup H_1 in F such that $H * H_1$ has finite index in F. Let $v \in N$ with $v \notin H * H_1$. As before, $v^\beta \in H * H_1$ for some $\beta \ge 2$. Now consider the subgroup Q generated by v and $H * H_1$. As in the previous case, we get a contradiction unless v^β is contained in the normal closure of H_1 in $H * H_1$. However, in this case, we would have $Q = H * H_2$ for some subgroup H_2, and the index of $H * H_2$ in F is smaller than the index of $H * H_1$ in F. Therefore, by induction we may assume that there is a subgroup P of F such that $H * P$ is a subgroup of F with finite index and $N \subset H * P$. This gives the desired result. □

Proposition 1.3.35 is also a consequence of a recent result by Louder and Wilton [172].

1.3.5 The Todd–Coxeter method

The disadvantage of the Reidemeister–Schreier method is that one has to know a system of coset representatives and must be able to determine the coset representative of an element. This is not possible in general. An alternative construction is the Todd–Coxeter method which leads to a Schreier set of representatives for a finitely generated subgroup of a finitely presented group.

Theorem 1.3.36. *Let $X = \{x_1, x_2, \ldots, x_n\} \in S \subset F(X)$ be a set of finitely many words in X, $G = \langle X \mid R\rangle$ finitely presented, U the subgroup of G generated by S and the index $k = [G : U]$ finite. The following algorithm of Todd–Coxeter determines a complete set T of representatives $g_1, g_2, \ldots, g_k$ of the right cosets Ug for U in G as well as the action of G on the right cosets, that is, for each $g \in G$ we may determine the index i such that $Ug = Ug_i$. Especially, we may decide for each $g \in G$ whether or not $g \in U$.*

Before we prove Theorem 1.3.36 we describe the algorithm of Todd–Coxeter. The above given conditions may hold.

1. Set $T = \emptyset$.
2. For each $s \in S$ initialize a subgroup table

y_1	y_2	$\cdots$	y_t
1		$\cdots$	1

where $s = y_1 y_2 \cdots y_t$ with $y_i \in X \cup X^{-1}$. Here

y_i	
g	h

with $g, h \in G$ means $Ugy_i = Uh$ and $Uhy_i^{-1} = Ug$, respectively.
3. For each $r \in R$ initialize a relation table

y_1	y_2	$\cdots$	y_t
1		$\cdots$	1

where $r = y_1 y_2 \cdots y_t$ with $y_i \in X \cup X^{-1}$. Here

y_i	
$\vdots$	$\vdots$
g	h

means that $Ugy_i = Uh$ and $Uhy_i^{-1} = Ug$, respectively.
4. Initialize the coset table

	x_1	$\cdots$	x_n	x_1^{-1}	$\cdots$	x_n^{-1}
1						

Here

	$\cdots$	y_i	$\cdots$
$\vdots$		$\vdots$	
g	$\cdots$	h	$\cdots$

means that $Ugy_i = Uh$. The appropriate field will be denoted in the following by

$[g, y_i]$ and the listing in this field by $R([g, y_i])$, if such one exist (we remark that here $[g, y_i]$ does not mean the commutator). Go to step 7.

5. If all rows in the coset table are completely filled, then give this table out and stop. The left column forms a complete system of representatives for the right cosets of U in G, the table describes the action of G on the right cosets.
6. Let g be the left listing of the first row of the coset table which is not completely filled. Choose a free field $[g, y]$ in this row and fill in this field the word gy. Add in the coset table below a new row with the listing gy in the left column and the listing g in the field $[gy, y^{-1}]$.
7. Fill the subgroup tables and the relation tables as far as it goes. If it completes a row then cancel this row and go to steps 8. Otherwise go to step 5.
8. If it completes in a subgroup or relation table a row then we get a new information of the type $Ugy = Uh$ or $Uhy^{-1} = Ug$. If the field $[g, y]$ is still free then occupy it with h. If the field $[h, y^{-1}]$ is still free then occupy it with g. If the field $[g, y]$ is already occupied with a $y_1 \neq h$ then add $\{y_1, h\}$ to T. If the field $[h, y^{-1}]$ is already occupied with a $y_2 \neq g$ then add $\{y_2, g\}$ to T.
9. If $T \neq \emptyset$ go to step 7.
10. Remove some $\{g, h\}$ from T. Let g_i be one element from $\{g, h\}$ whose row occurs first in the coset table, and let g_j be the other element. Compare in the coset table column by column the rows of g_i and g_j. If a listing g' stands in the field $[g_j, y]$ and nothing in $[g_i, y]$ then fill in g' in $[g_i, y]$. If a listing g' stands in $[g_j, y]$ and a listing $h' \neq g'$ in $[g_i, y]$ then add $\{g', h'\}$ to T. If the rows of g_i and g_j are processes, then cancel rows of g_j in the coset table and the relation tables, replace in all sets in T the element g_j by g_i and go to step 9.

We now make some remarks about the meaning of the tables. The action of G on the right cosets of U is recoded in the coset table. The representatives of the right cosets are listed in the left column, the representatives of the cosets of the Ugx_i^{ϵ}, $i = 1, 2, \dots, n$, $\epsilon = \pm 1$, are listed in the row of a representative g. If $u \in U$, then $Uu = U$, and $Us = U$ holds especially for each $s \in S$. This is reflected in the subgroup tables. Since $r =_G 1$ for each $r \in R$ we have $Ugr = Ug$ for each right coset Ug and each $r \in R$. And this will be written down in the relation table.

Proof of Theorem 1.3.36. We have two things to show: On the one hand that the algorithm terminates if the index $[G : U]$ is finite, and on the other hand that, in the case of termination, the output $g_1, g_2, \dots, g_k$ actually forms a complete set of representatives for the right cosets Ug of U in G and that the coset tables reflects the action of G on the right cosets correctly. We first show the correctness of the algorithm. For this we assume that the algorithm terminates and gives out a coset table with representatives $g_1, g_2, \dots, g_k$ in which each row is completely filled out. The first row of the coset table will not be canceled hence $g_1 = 1$. If $g \in G$, $g = x_{i_1}^{\epsilon_1} x_{x_2}^{\epsilon_2} \cdots x_{i_\ell}^{\epsilon_\ell}$, then

$$Ug = (\cdots(((U \cdot 1)x_{i_1}^{\epsilon_1})x_{x_2}^{\epsilon_2}) \cdots)x_{i_\ell}^{\epsilon_\ell},$$

hence, g is assigned by the coset table some right coset Ug_i. This can be seen as follows. The coset table will be filled out and modified, respectively, only in steps 6., 8. and 10. Step 6. is correct because G acts on the set of the right cosets. If a row in the relation table of one $r = y_1y_2\cdots y_t \in R$ will be filled, for instance at the place for $Ugy_1y_2\cdots y_t$ with Ug_p, and if $Ugy_1y_2\cdots y_{i+1} = Ug_p$, then we get the equations $Ug_py_{i+1} = Ug_q$ and $Ug_qy_{i+1}^{-1} = Ug_p$. If the fields $[g_p,y_{i+1}]$ and $[g_q,y_{i+1}^{-1}]$, respectively, are still free, then the coset table will be filled correctly in step 8. Otherwise there arises a situation that two distinct representatives g_i, g_j represent the same right coset. Such pairs will be collected in the set T. In step 10. the g_i and g_j represent the same coset, and g_j may be replaced in all tables by g_i. If a new entry g is made in a field $[g_i,y]$ of the coset table, then this is a consequence of $Ug_j = Ug_i$. As well $Ug' = Uh'$ if $\{g', h'\}$ is added to T. Analogously it follows, in the case that a subgroup table is completed, that step 8. results in a correct supplement and manipulation, respectively, of the coset table. Hence the entries in this table are correct, for each $g \in G$ there is a representative g_i with $Ug = Ug_i$, hence we have $[G : U] \leq k$. We now show $k \leq [G : U]$. At the start of step 5. we always have

$$R([g_k,y]) = g_\ell \quad \text{if and only if} \quad R([g_\ell,y^{-1}]) = g_k \tag{1.13}$$

for all $k, \ell \in \{1, 2, \ldots, n\}$ and all $y \in X \cup X^{-1}$, and, hence, also at issue of the representatives. This can be seen as follows. This is clear at the first pass. Step 6. does not change this property. At the end of step 8. we have

$$R([g_k,y]) = g_\ell \quad \text{if and only if} \quad R([g_i,y^{-1}]) = \gamma \tag{1.14}$$

with $\gamma = g_k$ or $\{g_k, \gamma\}$ in T. This also holds at the end of step 10. which we can see as follows. Condition (1.14) is fulfilled before the first pass through step 10. Suppose now that (1.14) holds before the replacement of the g_j. If one $R([g_k,y]) = g_j$ will be replaced by g_i somewhere in the coset table then we had $R([g_j,y^{-1}]) = \gamma$ with $\gamma = g_k$ or $\{\gamma, g_k\}$ in T. That is, we have $R([g_j,y^{-1}]) = \gamma$ with $\gamma = g_k$ or $\{\gamma, g_k\}$ in T after comparing the rows of g_i and g_j. Consider the manipulations which occur when comparing the rows of g_i and g_j. If $R([g_j,y]) = g_j$ and $R([g_j,y^{-1}]) = \gamma$ with $\gamma = g_j$ or $\{g_j, \gamma\}$ in T before the replacement then $R([g_i,y]) = g_i$ and $R([g_i,y^{-1}]) = g_i$ or $R([g_i,y^{-1}]) = \gamma$ with $\{g_i, \gamma\}$ in T at the end of step 10. If $R([g_j,y]) = g' \neq g_j$ and $R([g',y^{-1}]) = \gamma$ with $\gamma = g_j$ or $\{\gamma, g_j\}$ in T before the replacement then $R([g_i,y]) = g'$ and $R([g',y^{-1}]) = \gamma$ with $\gamma = g_i$ or $\{\gamma, g_i\}$ in T at the end of step 10. Since at each run of step 10. we cancel a row of the coset table and since corresponding amounts do not show up anymore in the set T we get eventually $T = \emptyset$. Therefore condition (1.13) holds again if the algorithm goes back to step 5. Condition (1.13) implies that the free group $F(X)$ acts as a group of permutations on the set $K = \{g_1, g_2, \ldots, g_k\}$. Let H be the corresponding permutation group and $\varphi\colon F(X) \to H$ the canonical epimorphism. We have taken care of that $\varphi(r) = 1$ for all $r \in R$ with the help of the relation tables. By Theorem 1.3.5, the theorem of van Dyck, we get an epimorphism $\psi\colon G \to H$ with $\psi(X) = \varphi(X)$ for all $x \in X$, that is, G acts in the same manner as $F(X)$ on K. The subgroup tables ensure that U is contained in the stabilizer $\mathrm{Stab}(1)$ of $1 = g_1$. The nature of the definition of new representatives in step 6. shows that G acts transitively on K, that is, $[G : \mathrm{Stab}(1)] = k$. Therefore $[G : U] \geq k$.

We now show that the algorithm terminates if $[G : U]$ is finite. Since G acts transitively on the cosets of U, there will be sometime introduced in step 6. a representative for each coset.

Since we cancel in the coset table a row only if there is another row which represents the same coset, there exists a g_ℓ such that in the rows up to the row for g_ℓ all cosets are represented at least once. This property remains in the further course of the algorithm because in step 10. at a time only a row is canceled which stands further down. In addition, each row in each table will be completed sometime. As above, G acts transitively on $K = \{g_1, g_2, \ldots\}$ and U is contained in $\mathrm{Stab}(1)$. Hence, $|K| \leq [G : U]$. □

With the help of this algorithm we also may determine the index of a subgroup if we know at least that this index is finite. Also we may solve the membership problem for subgroups of finite index. Unfortunately there is no upper bound for the running time of the Todd–Coxeter algorithm for a given finitely presented group. If there would be one we could decide whether or not a finitely presented group is trivial. This is proved algorithmically undecidable (see [5] and [216]).

Example 1.3.37. Let $G = \langle x_1, x_2, x_3, x_4 \mid x_1^3 = x_2^3 = x_3^2 = x_4^3 = x_1x_2x_3x_4 \rangle$. We want to get a complete set of representatives for the right cosets of the subgroup U of G which is generated by the four elements $x_3^{-1}x_1x_2$, $x_3x_1x_2x_1x_2$, $x_2x_1^{-1}$ and $x_1x_2x_1$. The algorithm starts with $T = \emptyset$, the subgroup tables

x_3^{-1}	x_1	x_2	
1			1

,

x_3	x_1	x_2	x_1	x_2	
1					1

,

x_2	x_1^{-1}	
1		1

,

x_1	x_2	x_1	
1			1

;

the relation tables

x_1	x_1	x_1	
1			1

,

	x_2	x_2	x_2	
1				1

,

	x_3	x_3	x_3	
1				1

,

	x_4	x_4	x_4	
1				1

,

	x_1	x_2	x_3	x_4	
1					1

,

and the coset table

	x_1	x_2	x_3	x_4	x_1^{-1}	x_2^{-1}	x_3^{-1}	x_4^{-1}
1								

.

In no table there is a complete row. Choose x_1 as the first further representation of a right coset. With this we may completely fill the trivial subgroup table, and we get

	x_2	x_1^{-1}	
1		x_1	1

.

This gives the rules

$$U \cdot 1 \cdot x_2 = U \cdot x_1 \quad \text{and}$$
$$U \cdot x_1 \cdot x_1^{-1} = U \cdot 1,$$

which we recode into the coset table. If we arrive again at step 5., after the first runs, then the tables have the form

	x_3^{-1}	x_1	x_2	
1				1

,

	x_3	x_1	x_2	x_1	x_2	
1						1

,

	x_2	x_1^{-1}	
1		x_1	1

.

	x_1	x_2	x_1
1	x_1		1

	x_1	x_1	x_1
1	x_1		1
x_1		1	x_1

	x_2	x_2	x_2
1	x_1		1
x_1		1	x_1

	x_3	x_3	x_3
1			1
x_1			x_1

	x_4	x_4	x_4
1			1
x_1			x_1

	x_1	x_2	x_3	x_4
1	x_1			1,
x_1				x_1

	x_1	x_2	x_3	x_4	x_1^{-1}	x_2^{-1}	x_3^{-1}	x_4^{-1}
1	x_1	x_1						
x_1					1	1		

.

We choose x_3 as new representative and fill the tables as described in the algorithm. We do not get new rules. As the next one we choose x_4, also this does not give new rules. After filling in all tables we have these as follows:

	x_3^{-1}	x_1	x_2
1			1,

	x_3	x_1	x_2	x_1	x_2
1	x_3				1,

	x_2	x_1^{-1}
1	x_1	1.

	x_1	x_2	x_1
1	x_1		1

	x_1	x_1	x_1
1	x_1		1
x_1		1	x_1
x_3			x_3
x_4			x_4

	x_2	x_2	x_2
1	x_1		1
x_1		1	x_1
x_3			x_3
x_4			x_4

	x_3	x_3	x_3
1	x_3		1
x_1			x_1
x_3		1	x_3
x_4			x_4

	x_4	x_4	x_4
1	x_4		1
x_1			x_1
x_3			x_3
x_4		1	x_4

	x_1	x_2	x_3	x_4
1	x_1			1
x_1				x_1,
x_3				x_3
x_4			1	x_4

	x_1	x_2	x_3	x_4	x_1^{-1}	x_2^{-1}	x_3^{-1}	x_4^{-1}
1	x_1	x_1	x_3	x_4				
x_1					1	1		
x_3							1	
x_4								1

.

As the next representative we choose x_1^{-1}. If we fill the tables we get some new rules. Step by step we get

$$\begin{aligned} Ux_1x_2 &= Ux_1^{-1} \quad \text{and} \quad Ux_1^{-1}x_2^{-1} = Ux_1, \\ Ux_1x_2 &= Ux_1^{-1} \quad \text{and} \quad Ux_1^{-1}x_1^{-1} = Ux_1, \\ Ux_1^{-1}x_2 &= U \quad \text{and} \quad Ux_2^{-1} = Ux_1^{-1}, \\ Ux_3^{-1} &= Ux_1 \quad \text{and} \quad Ux_1x_3 = U, \end{aligned}$$

respectively, which leads to the following coset table:

	x_1	x_2	x_3	x_4	x_1^{-1}	x_2^{-1}	x_3^{-1}	x_4^{-1}
1	x_1	x_1	x_3	x_4	x_1^{-1}	x_1^{-1}		
x_1	x_1^{-1}	x_1^{-1}	1		1	1		
x_3							1	
x_4								1
x_1^{-1}		1			x_1	x_1		

with these rules we may complete the second subgroup table

x_3	x_1	x_2	x_1	x_2
1	x_3	1	x_1	x_1^{-1}

which gives the rules

$$Ux_3x_1 = U \quad \text{and} \quad Ux_1^{-1} = Ux_3.$$

Now we have already $R([1, x_1^{-1}]) = x_1^{-1}$ such that we have to add the pair $\{x_3, x_1^{-1}\}$ to T and to apply step 10. If we fill accordingly all tables, we get as another redundancy $Ux_4 = Ux_3$. Finally, at the end of the algorithm we get

	x_1	x_2	x_3	x_4	x_1^{-1}	x_2^{-1}	x_3^{-1}	x_4^{-1}
1	x_1	x_1	x_3	x_3	x_3	x_3	x_1	x_1
x_1	x_3	x_3	1	1	1	1	x_3	x_3
x_3	1	1	x_1	x_1	x_1	x_1	1	1

Hence, $T = \{1, x_1, x_3\}$ and U has index 3 in G.

1.4 Free products

We start with the definition of the free product of groups and then discuss important properties, examples and applications in this section.

Definition 1.4.1. Let $\{G_\alpha\}$ be a family of groups, G a group, and let $i_\alpha : G_\alpha \to G$ be homomorphisms. Then $(G, \{i_\alpha\})$ is called a *free product of the groups* G_α if for each group H and homomorphism $f_\alpha: G_\alpha \to H$ there is a unique homomorphism $f: G \to H$ with $f_\alpha = f \circ i_\alpha$ for all α. We call the groups G_α the *free factors* or just the *factors of the free product*.

Analogously as for free groups we get the following.

Theorem 1.4.2. *Let $(G, \{i_\alpha\})$ and $(H, \{j_\alpha\})$ be both free products of the groups G_α. Then there exists a unique isomorphism $f: G \to H$ such that $j_\alpha = f \circ i_\alpha$ for all α.*

We leave the proof as an exercise.

Hence, we now may speak about *the* free product of the groups G_α.

Theorem 1.4.3. *Let $(G, \{i_\alpha\})$ be the free product of the groups G_α. Then i_α is a monomorphism for all α.*

Proof. We just take $H = G_\alpha$ and $f_\alpha = \mathrm{id}$. □

Theorem 1.4.4. *Each family of groups G_α has a free product.*

Proof. Let $G_\alpha = \langle X_\alpha \mid R_\alpha \rangle^{\varphi_\alpha}$. If necessary, we replace X_α with a bijective set, and, hence, we may assume that $X_\alpha \cap X_\beta = \emptyset$ if $\alpha \neq \beta$. Let $G = F(\bigcup X_\alpha)/\langle\langle \bigcup R_\alpha \rangle\rangle$, that is, $G = \langle \bigcup X_\alpha \mid \bigcup R_\alpha \rangle^{\varphi}$, where φ is the canonical map. By Theorem 1.3.5 the inclusion $X_\alpha \hookrightarrow \bigcup X_\alpha$ induces a homomorphism $i_\alpha: G_\alpha \to G$. Then $(G; \{i_\alpha\})$ is the free product of the G_α. This we may see as follows. Let $f_\alpha: G_\alpha \to H$ be homomorphisms. Each f_α induces a homomorphism $\psi_\alpha: F(X_\alpha) \to H$ with $\psi_\alpha(R_\alpha) = 1$. This gives a homomorphism $\psi: F(\bigcup X_\alpha) \to H$ with $\psi|_{X_\alpha} = \psi_\alpha$. The map ψ induces a homomorphism $f: G \to H$ with $\psi = f \circ \varphi$ because $\psi(\bigcup R_\alpha) = \{1\}$. We have therefore $f \circ i_\alpha(\varphi_\alpha(x_\alpha)) = f \circ \varphi(x_\alpha) = \psi(x_\alpha)$ for each $x_\alpha \in X_\alpha$ by the definition of f and i_α and

$$\psi(x_\alpha) = \psi_\alpha(x_\alpha) = f_\alpha \circ \varphi_\alpha(x_\alpha)$$

by the definition of ψ and ψ_α. Now $f_\alpha = f \circ i_\alpha$ because $G_\alpha = \langle \varphi_\alpha(X_\alpha) \rangle$. Therefore f is unique because $G = \langle \bigcup i_\alpha \circ \varphi_\alpha(X_\alpha) \rangle$. □

Corollary 1.4.5. *Let $(G_A, \{i_\alpha\})$ be the free product of the groups G_α for $\alpha \in A$ and $(G_B, \{i_\beta\})$ be the free product of the groups G_β for $\beta \in B$. Let $B \subset A$. Then $G_B < G_A$, and there is a homomorphism $\varphi: G_A \to G_B$ which is the identity on G_B.*

We now use the notation $*G_\alpha$ for the free product $(G, \{i_\alpha\})$ of the groups G_α. The free product of n groups $G_1, G_2, \ldots, G_n$ is denoted by $G_1 * G_2 * \cdots * G_n$. From Theorem 1.4.4 we get $G_1 * G_2 = G_2 * G_1$ and $G_1 * G_2 * G_3 = (G_1 * G_2) * G_3 = G_1 * (G_2 * G_3)$. Since each i_α is a monomorphism we mostly consider G_α as a subgroup of $*G_\beta$ (where $G_\alpha = G_\beta$ for some β) with the i_α as inclusion.

1.4.1 Examples of free products and the extension problem

Examples 1.4.6. We have

1. $F(X) = *C_x, C_x \cong \mathbb{Z}$ for all $x \in X$,
2. $D_\infty \cong \mathbb{Z}_2 * \mathbb{Z}_2$, and
3. $\mathrm{PSL}(2,\mathbb{Z}) \cong \mathbb{Z}_2 * \mathbb{Z}_3$ (see Section 1.3). We may extend this example. Let

$$A = \pm\begin{pmatrix} 0 & 1 \\ -1 & 2\cos(\frac{\pi}{q}) \end{pmatrix} \quad \text{and} \quad B = \pm\begin{pmatrix} 0 & -1 \\ 1 & 2\cos(\frac{\pi}{p}) \end{pmatrix},$$

$p, q \geq 2$, $(p,q) \neq (2,2)$, be two elements of $\mathrm{PSL}(2,\mathbb{R})$. Then
(1) A has order q and B has order p in $\mathrm{PSL}(2,\mathbb{R})$, and
(2) $\langle A, B\rangle \cong \mathbb{Z}_q * \mathbb{Z}_q$.

Proof. We first consider A as an element of $\mathrm{SL}(2,\mathbb{R})$. We have $A^n = S_n(\mathrm{trace}(A))A - S_{n-1}(\mathrm{trace}(A))E$ for $n \geq 0$ and $E = \begin{pmatrix} 1 & 0 \\ 0 & 1 \end{pmatrix}$, where $S_0(t) = 0$, $S_1(t) = 1$ and $S_n(t) = xS_{n-1}(t) - S_{n-2}(t)$ for $n \geq 2$, the Chebyshev polynomials. Now $\mathrm{trace}(A) = 2\cos(\frac{\pi}{q})$, which gives

$$S_n\left(2\cos\left(\frac{\pi}{q}\right)\right) = \frac{\sin(\frac{n\pi}{q})}{\sin(\frac{\pi}{q})}$$

by induction. We get therefore

$$A^q = S_q\left(2\cos\left(\frac{\pi}{q}\right)\right)A - S_{q-1}\left(2\cos\left(\frac{\pi}{q}\right)\right)E = -E.$$

This gives $A^q = 1$ in $\mathrm{PSL}(2,\mathbb{R})$, and A has at most order q in $\mathrm{PSL}(2,\mathbb{R})$. Let

$$L_A\colon z \mapsto \frac{1}{-z + 2\cos(\frac{\pi}{q})}$$

the linear fractional transformation which corresponds to A. Then $L_A(z) \in \mathbb{R}^+ = \{r \in \mathbb{R} \mid r > 0\}$ if $z \in \mathbb{R}^- = \{r \in \mathbb{R} \mid r < 0\}$. If $1 \leq n < q$ and $z \in \mathbb{R}^-$, then $L_{A^n}(z) \in \mathbb{R}^+$ because $S_n(2\cos(\frac{\pi}{q})) > 0$ and $S_{n-1}(2\cos(\frac{\pi}{q})) \geq 0$ for $1 \leq n < q$. Especially $A^n \neq 1$ in $\mathrm{PSL}(2,\mathbb{R})$ if $1 \leq n < q$. Hence, A has exact order q in $\mathrm{PSL}(2,\mathbb{R})$. Analogously, B has order p because $L_{B^n}(w) \in \mathbb{R}^-$ if $w \in \mathbb{R}^+$ and $1 \leq n < p$. Assume that $\langle A, B\rangle \not\cong \mathbb{Z}_q * \mathbb{Z}_p$. Then there exists a word

$$W = A^{n_1}B^{m_1}A^{n_2}B^{m_2}\cdots A^{n_k}B^{m_k} = 1,$$

$1 \leq n_i < q, 1 \leq m_i < p$. Without loss of generality, let $p \neq 2$. Then there exists a $\beta \neq 0$ such that

$$1 = B^\beta W B^{-\beta} = B^\beta A^{n_1}B^{m_1}\cdots A^{n_k}B^\alpha,$$

$\alpha = m_k - \beta \not\equiv 0 \mod p$.
Let L be the linear fractional transformation corresponding to $B^\beta WB^{-\beta}$. Using the ping-pong argument gives $L(w) \in \mathbb{R}^-$ if $w \in \mathbb{R}^+$, contradicting $1 = W = B^{-\beta}B^\beta$. Hence $\langle A, B\rangle \cong \mathbb{Z}_q * \mathbb{Z}_p$. □

Analogously, let $A = \pm\begin{pmatrix} 0 & 1 \\ -1 & 2\cos(\frac{\pi}{q}) \end{pmatrix}$ and $B = \pm\begin{pmatrix} 0 & -1 \\ 1 & 2 \end{pmatrix}$, $q \geq 2$, be two elements of $\mathrm{PSL}(2,\mathbb{R})$. Then $\langle A, B\rangle \cong \mathbb{Z}_q * \mathbb{Z}$.

We leave this as an exercises.

Corollary 1.4.7. *Let $G = G_1 * G_2 * \cdots * G_n$ with $G_i = \langle a_i \mid a_i^{\alpha_i} = 1\rangle$, $\alpha_i \geq 2$ or $\alpha_i = 0$ (which means $G_i \cong \mathbb{Z}$). Then G can be embedded into* $\mathrm{PSL}(2,\mathbb{R})$.

Proof. Let first $\alpha_i \neq 0$ for all i. Define $\gamma := \mathrm{scm}\{\alpha_1, \alpha_2, \ldots, \alpha_n\}$ and β_i such that $\beta_i \cdot \alpha_i = \gamma$. Let $H = \langle a, b \mid a^\gamma = 1\rangle$, $H_1 = \langle a \mid a^\gamma = 1\rangle$ and $H_2 = \langle b \mid \rangle$. Then $H = H_1 * H_2$, and H is embeddable in $\mathrm{PSL}(2,\mathbb{R})$ via

$$a \mapsto A = \pm\begin{pmatrix} 0 & 1 \\ -1 & 2\cos(\frac{\pi}{\gamma}) \end{pmatrix} \quad \text{and} \quad b \mapsto B = \pm\begin{pmatrix} 0 & -1 \\ 1 & 2 \end{pmatrix}.$$

Define

$$x_1 := a^{\beta_1}, \quad x_2 := ba^{\beta_2}b^{-1}, \quad \ldots, \quad x_n := b^{n-1}a^{\beta_n}b^{1-n}.$$

Let $G' = \langle x_1, x_2, \ldots, x_n\rangle$. Then G' is embeddable in $\mathrm{PSL}(2,\mathbb{R})$ because H is embeddable in $\mathrm{PSL}(2,\mathbb{R})$. Furthermore, $G' = G_1' * G_2' * \cdots * G_n'$ with $G_i' = \langle x_i \mid x_i^{\alpha_i}\rangle \cong G_i$, and, hence, $G' \cong G$. If all $\alpha_i = 0$ then all G_i are free of rank 1 and the result follows (see Section 1.1). Now, let some $\alpha_i = 0$ and the remaining $\alpha_i \geq 2$. Without loss of generality, let $\alpha_i \geq 2$ for $1 \leq i \leq k < n$ and $\alpha_i = 0$ for $k+1 \leq i \leq n$. Let $\gamma := \mathrm{scm}\{\alpha_1, \alpha_2, \ldots, \alpha_k\}$ and $m > 2n$. Let $H = \langle a, b \mid a^\gamma = 1\rangle$ be as above. Define

$$x_1 := a^{\beta_1}, \quad x_2 := ba^{\beta_2}b^{-1}, \quad \ldots, \quad x_k := b^{k-1}a^{\beta_k}b^{1-k}, \quad x_{k+1} := b^m,$$
$$x_{k+2} := b^{m+1}ab^m a^{-1}b^{-m-1}, \quad \ldots, \quad x_{k+j} := b^{m+j-1}ab^m a^{-1}b^{-m-j+1}$$

for $j = n - k$. Let $G' = \langle x_1, x_2, \ldots, x_n\rangle$. Then G' is embeddable in $\mathrm{PSL}(2,\mathbb{R})$.

Further, $G' = G_1' * G_2' * \cdots * G_n'$ with $G_i' = \langle x_i \mid x_i^{\alpha_i} = 1\rangle$ for $i = 1, 2, \ldots, k$ and $G_i' \cong \mathbb{Z}$ for $i \geq k+1$. Therefore $G' \cong G$.

We leave the details as exercises. □

We now use the concept of free products to calculate presentations of extensions of a finitely presented groups.

Theorem 1.4.8. *Let A, B, C be groups and $h: B \to C$ an epimorphism with* $\ker(h) = A$. *Let $A = \langle x_1, x_2, \ldots, x_n \mid S_1(x_1, x_2, \ldots, x_n) = S_2(x_1, x_2, \ldots, x_n) = \cdots = S_\ell(x_1, x_2, \ldots, x_n) = 1\rangle$ and $C = \langle y_1, y_2, \ldots, y_m \mid R_1(y_1, y_2, \ldots, y_m) = R_2(y_1, y_2, \ldots, y_m) = \cdots = R_k(y_1, y_2, \ldots, y_m) = 1\rangle$ be finitely presented. Then B is also finitely presented with generators $x_1, x_2, \ldots, x_n, w_1, w_2, \ldots, w_m$ where $h(w_i) = y_i$, $1 \leq i \leq m$, and defining relations*

RI: $R_i(w_1, w_2, \ldots, w_m) = T_i(x_1, x_2, \ldots, x_n)$, $1 \leq i \leq k$.

RII: $S_i(x_1, x_2, \ldots, x_n) = 1$, $1 \leq i \leq \ell$.

RIII: $w_i^{-1}x_j w_i = T_{ij}(x_1, x_2, \ldots, x_n)$, $1 \leq i \leq m$, $1 \leq j \leq n$, *for certain words $T_i(x_1, x_2, \ldots, x_n)$ and $T_{ij}(x_1, x_2, \ldots, x_n)$.*

We have the following situation

$$\{1\} \to A \overset{i}{\to} B \overset{h}{\to} C \to \{1\},$$

where i is the embedding of A in B.

Proof. Step 1: Generators of B.

Choose for $y_i \in C$ a $w_i \in B$ with $h(w_i) = y_i$, $1 \le i \le m$. Let $w \in B$ and $h(w) = y \in C$. We may write y as $y = U(y_1, y_2, \ldots, y_m)$. Let $w' = U(w_1, w_2, \ldots, w_m) \in B$. Then $h(w') = h(U(w_1, w_2, \ldots, w_m)) = U(y_1, y_2, \ldots, y_m) = y = h(w)$. Hence, $h(w'w^{-1}) = 1_C$, that is, $w'w^{-1} \in \ker(h)$ and therefore $w' = wx$ for some $x \in \ker(h)$. This means that $x_1, x_2, \ldots, x_n, w_1, w_2, \ldots, w_m$ generate the group B.

Step 2: The given relations hold.

Since $R_i(y_1, y_2, \ldots, y_m) = 1$ we have $R_i(w_1, w_2, \ldots, w_m) \in A$, that is, $R_i(w_1, w_2, \ldots, w_m) = T_i(x_1, x_2, \ldots, x_n)$ for some word $T_i(x_1, x_2, \ldots, x_n)$. This is RI.

RII: Certainly $S_i(x_1, x_2, \ldots, x_n) = 1$ because $A \subset B$.

RIII: Now $w_i^{-1} x_j w_i \in \ker(h)$, that is, $w_i^{-1} x_j w_i = T_{ij}(x_1, x_2, \ldots, x_n)$ for some word $T_{ij}(x_1, x_2, \ldots, x_n)$.

Step 3: Each relation is a consequence of RI, RII and RIII.

Let $F_x := \langle x_1, x_2, \ldots, x_n \rangle$, $F_w := \langle w_1, w_2, \ldots, w_m \rangle$ and $F_y := \langle y_1, y_2, \ldots, y_m \rangle$. Consider the following diagram (see Figure 1.12).

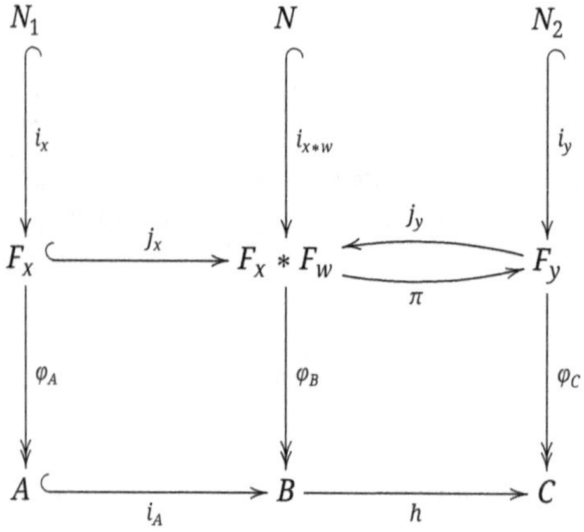

Figure 1.12: Presentations of extensions.

Here j_x, j_y, i_{x*w} and i_A mean inclusions, φ_A, φ_B and φ_C the canonical epimorphisms, π the canonical projection, and h is as above. We have to show that each word from $F_x * F_w$, which is mapped by φ_B to 1, is a consequence from RI, RII, and RIII. For this purpose, let

$$R(x_1, x_2, \ldots, x_n, w_1, w_2, \ldots, w_m) \in F_x * F_w \quad \text{and} \quad \varphi_B(R) = 1.$$

If we repeatedly apply RIII we get from

$$w_i x_j w_i^{-1} = T_{ij}(x_1, x_2, \dots, x_n) \quad \text{if and only if} \quad w_i x_j = T_{ij}(x_1, x_2, \dots, x_n) w_i$$

the result

$$R(x_1, x_2, \dots x_n, w_1, w_2, \dots, w_m) \overset{\text{RIII}}{=:} S(x_1, x_2, \dots, x_n) K(w_1, w_2, \dots, w_m).$$

Therefore

$$\begin{aligned} 1 &= h(R(x_1, x_2, \dots, x_n, w_1, w_2, \dots, w_m)) \\ &= h(S(x_1, x_2, \dots, x_n)) \cdot h(K(w_1, w_2, \dots, w_m)) \\ &= 1 \cdot K(y_1, y_2, \dots, y_m), \end{aligned}$$

that is, $K(y_1, y_2, \dots, y_m)$ is a relation in C. This gives

$$\begin{aligned} K(w_1, w_2, \dots, w_m) &= \prod_i t_i(w_1, w_2, \dots, w_m) R_{v_i}^{\epsilon_i}(w_1, w_2, \dots, w_m) t_i(w_1, w_2, \dots, w_m)^{-1} \\ &\overset{\text{RI}}{=} \prod_i t_i(w_1, w_2, \dots, w_m) T_{v_i}^{\epsilon_i}(x_1, x_2, \dots, x_n) t_i(w_1, w_2, \dots, w_m)^{-1} \\ &\overset{\text{RIII}}{=:} T(x_1, x_2, \dots, x_n), \end{aligned}$$

where $\epsilon_i = \pm 1$. Therefore

$$R(x_1, x_2, \dots, x_n, w_1, w_2, \dots, w_m) = S(x_1, x_2, \dots, x_n) T(x_1, x_2, \dots, x_n)$$

and

$$R(x_1, x_2, \dots, x_n, w_1, w_2, \dots, w_m) = 1.$$

Hence,

$$\begin{aligned} R(x_1, x_2, \dots, x_m, w_1, w_2, \dots, w_m) &= \prod_i p_i(x_1, x_2, \dots, x_n) S_{v_i}^{\epsilon_i}(x_1, x_2, \dots, x_n) p_i(x_1, x_2, \dots, x_n)^{-1} \\ &\overset{\text{RII}}{=} 1, \quad \text{where } \epsilon_i = \pm 1. \end{aligned}$$

Altogether we get

$$R(x_1, x_2, \dots, x_n, w_1, w_2, \dots, w_m) \overset{\text{RI–RIII}}{=} 1. \qquad \square$$

Remark 1.4.9. Given A and C, then there exist in general several possible group presentations B.

Example 1.4.10. Let $G = \langle a, b \mid R_i(a, b) = 1, 1 \leq i \leq m\rangle$ be a finitely generated subgroup of $\mathrm{PSL}(2, \mathbb{R})$ with G non-Abelian and $\mathrm{tr}([a, b]) \neq 2$. Let each automorphism of G be induced by an automorphism of $F_2 = \langle a, b \mid \rangle$, and let each automorphism of F_2 induce an automorphism of G. There are several examples of such groups. We just mention here three of our favorite groups, that is,

1. $G = F_2 = \langle a, b \mid \rangle$,
2. $G = S(n) = \langle a, b \mid [a, b]^n\rangle$, $n \geq 2$, and
3. $G = G(k) = \langle s_1, s_2, s_3, s_4 \mid s_1^2 = s_2^2 = s_3^2 = s_4^{2k+1} = s_1 s_2 s_3 s_4 = 1\rangle$, $k \geq 1$. By substituting $a = s_1 s_2$, $b = s_3 s_1$ we have

$$G(k) = \langle a, b \mid [a, b]^{2k+1} = ([a, b]^k a)^2 = ([a, b]^k ab)^2 = ([a, b]^k aba)^2 = 1\rangle.$$

Now we have the following commutative diagram (see Figure 1.13).

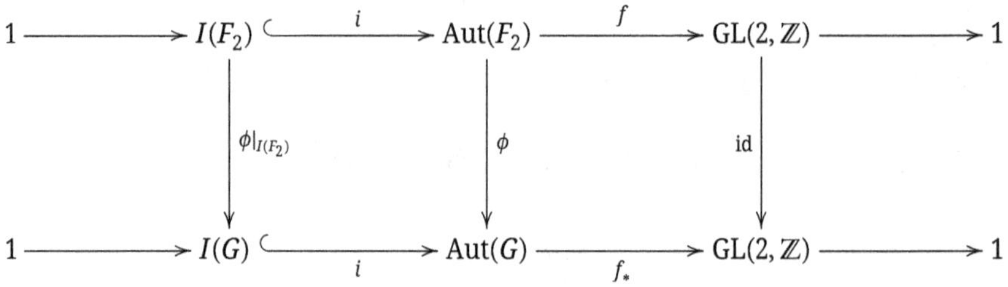

Figure 1.13: Diagram for Example 1.4.10.

Here I denotes the inner automorphisms. It is known that $F_2 \cong \mathrm{GL}(2, \mathbb{Z})/I(F_2)$,

$$\mathrm{Aut}(F_2) = \langle \alpha, \beta, \gamma \mid \alpha^2 = \beta^2 = (\alpha\beta)^4 = (\alpha\beta\alpha\gamma)^2 = (\alpha\beta\gamma)^3 = 1, (\gamma\beta)^2 = (\beta\gamma)^2\rangle,$$

where

$$\alpha\colon a \mapsto b, \quad b \mapsto a,$$
$$\beta\colon a \mapsto a^{-1}, \quad b \mapsto b,$$
$$\gamma\colon a \mapsto ab, \quad b \mapsto b,$$

(see [185], p. 169) and

$$\mathrm{GL}(2, \mathbb{Z}) = \langle \alpha, \beta, \gamma \mid \alpha^2 = \beta^2 = (\alpha\beta)^4 = (\beta\gamma)^2 = (\alpha\beta\alpha\gamma)^2 = (\alpha\beta\gamma)^3 = 1\rangle,$$

where the generators are induced by the generators of $\mathrm{Aut}(F_2)$. We remark that this presentation for $\mathrm{GL}(2, \mathbb{Z})$ is different from that in Section 1.3. We leave it as an exercise to transform one to the other by Tietze transformations.

Anyway: the above presentations give the natural map $f\colon \mathrm{Aut}(F_2) \to \mathrm{GL}(2, \mathbb{Z})$. Now, by the conditions on G we see that G is centerless, that is, especially that $I(G) \cong G$, and that there is a canonical epimorphism $\phi\colon \mathrm{Aut}(F_2) \twoheadrightarrow \mathrm{Aut}(G)$. Also we have $\mathrm{Aut}(G)/I(G) \cong \mathrm{GL}(2,\mathbb{Z})$. This follows easily from the induced trace algorithm described in a special case in Section 1.3 and in full generality in [92]. We use this trace algorithm again in Section 1.7.

The algorithm gives an exact sequence

$$1 \to H \xrightarrow{i} \mathrm{Aut}(G) \xrightarrow{\Delta} \mathrm{PSL}(2,\mathbb{Z}),$$

where H is the normal subgroup of $\mathrm{Aut}(G)$, normally generated by $I(G)$ and the automorphism $a \mapsto a^{-1}$, $b \mapsto b^{-1}$. Hence, we have the induced epimorphism $f_*\colon \mathrm{Aut}(G) \to \mathrm{GL}(2,\mathbb{Z})$ with kernel $I(G)$. We consider the exact sequence

$$1 \to I(G) \xhookrightarrow{i} \mathrm{Aut}(G) \xrightarrow{f_*} \mathrm{GL}(2,\mathbb{Z}) \to 1.$$

Without ambiguity we use α, β, γ for the generators of $\mathrm{Aut}(G)$ via the map $\phi\colon \mathrm{Aut}(F_2) \to \mathrm{Aut}(G)$. Again, as above, we also use α, β, γ for the generators of $\mathrm{GL}(2,\mathbb{Z})$. We want to give a presentation for $\mathrm{Aut}(G)$. We write $A := i(a)$, $B := i(b)$, where $i(g)(x) = g^{-1}xg$. Then $A, B, \alpha, \beta, \gamma$ generate $\mathrm{Aut}(G)$. The relations of the form RI are

$$\alpha^2 = \beta^2 = (\alpha\beta)^2 = (\alpha\beta\alpha\gamma)^2 = (\alpha\beta\gamma)^3 = 1, \quad (\beta\gamma)^2 = B.$$

Since G is centerless, the defining relations of the form RII are

$$R_i(A,B) = 1, \quad 1 \le i \le m.$$

The defining relations of the form RIII are:
(i) $A\alpha = \alpha B$,
(ii) $\alpha A = B\alpha$,
(iii) $(A\beta)^2 = 1$,
(iv) $\beta B = B\beta$,
(v) $\gamma^{-1}A\gamma = AB$ and
(vi) $\gamma^{-1}B\gamma = B$.

We see that (i) and (ii) will reduce to $A = \alpha B\alpha = \alpha(\beta\gamma)^2\alpha$. Hence, by Tietze transformations we can eliminate A and B from the generators. The relation (iv) implies

$$(\alpha\gamma)^2 = \beta(\beta\gamma)^2\beta = (\gamma\beta)^2.$$

Then we easily can show that (iii), (v) and (vi) are redundant. By substituting $A = \alpha(\beta\gamma)^2\alpha$, $B = (\beta\gamma)^2$ into the relations RII and using (i), (ii) and $\alpha^2 = 1$ we get $R_i((\beta\gamma)^2, \alpha(\beta\gamma)^2\alpha)$, $1 \le i \le m$, for the relations RII. Hence we have the presentation

$$\mathrm{Aut}(G) = \langle \alpha,\beta,\gamma \mid \alpha^2 = \beta^2 = (\alpha\beta)^4 = (\alpha\beta\alpha\gamma)^2 = (\alpha\beta\gamma)^3 = 1, (\gamma\beta)^2 = (\beta\gamma)^2, \\ R_i((\beta\gamma)^2, \alpha(\beta\gamma)^2\alpha) = 1, 1 \le i \le m\rangle.$$

1.4.2 Normal form, the Nielsen method in free products, and the theorem of Kurosh

Theorem 1.4.11 (Normal form). *Let (G, i_α) be the free product of the groups G_α. Then:*
(i) *Each i_α is a monomorphism.*
(ii) *If we consider i_α as an inclusion, then each element of G can uniquely be written as $g_1 g_2 \cdots g_n$ with $n \geq 0$, $g_i \in G_{\alpha_i}$ for some α_i, $g_i \neq 1$ and $\alpha_r \neq \alpha_{r+1}$ for $r < n$.*

Proof. We prove (i) and (ii) at the same time (recall that we already proved (i) by Theorem 1.4.3). We write $\overline{g_\alpha}$ for $i_\alpha(g_\alpha)$ and show that $u \in G$ can uniquely be written as $\overline{g}_1 \overline{g}_2 \cdots \overline{g}_n$, $n \geq 0$, $g_i \in G_{\alpha_i}$ for some α_i, $g_i \neq 1$ and $\alpha_r \neq \alpha_{r+1}$ for $r < n$. Since $G = \langle \bigcup i_\alpha(G_\alpha) \rangle$ we certainly may write each $u \in G$ in the form

$$\overline{g}_1 \overline{g}_2 \cdots \overline{g}_n, \quad n \geq 0, g_i \in G_{\alpha_i}, g_i \neq 1. \tag{1.15}$$

If here $\alpha_r = \alpha_{r+1}$ and $g_{r+1} \neq g_r^{-1}$ then we may write u as

$$\overline{g}_1 \overline{g}_2 \cdots \overline{g}_{r-1} \overline{h} \overline{g}_{r+2} \cdots \overline{g}_n \quad \text{with } h = g_r g_{r+1} \neq 1$$

and $h \in G_{\alpha_r}$. If $\alpha_r = \alpha_{r-1}$ and $g_{r+1} = g_r^{-1}$ then we may write u as

$$\overline{g}_1 \overline{g}_2 \cdots \overline{g}_{r-2} \overline{g}_{r+2} \cdots \overline{g}_n.$$

Inductively we now see that we may assume in (1.15) in addition that $\alpha_r \neq \alpha_{r+1}$. We now show the uniqueness and use van der Waerden's method in [260]. Let S be the set of all sequences $(g_n, g_{n-1}, \ldots, g_1)$ with $n \geq 0$, $g_i \in G_{\alpha_i}$, $g_i \neq 1$ and $\alpha_r \neq \alpha_{r-1}$ for $n \geq r > 1$.

Recall that the empty sequence $()$ is in S. Also remark that we indicate now in a different order. Let $g_\alpha \in G_\alpha \setminus \{1\}$. We define a map from S to S by

$$\begin{aligned}
(g_n, g_{n-1}, \ldots, g_1) &\mapsto (g_\alpha, g_n, g_{n-1}, \ldots, g_1) \quad \text{if } \alpha \neq \alpha_n,\\
(g_n, g_{n-1}, \ldots, g_1) &\mapsto (g_\alpha g_n, g_{n-1}, g_{n-2}, \ldots, g_1) \quad \text{if } \alpha = \alpha_n \text{ and } g_\alpha g_n \neq 1,\\
(g_n, g_{n-1}, \ldots, g_1) &\mapsto (g_{n-1}, g_{n-2}, \ldots, g_1) \quad \text{if } \alpha = \alpha_n \text{ and } g_\alpha g_n = 1.
\end{aligned}$$

If we take $\varphi_\alpha(1)$ as the identity, then we get a map $\varphi_\alpha : G_\alpha \to \mathrm{Per}(S)$, the group of permutations of S, because φ_α has the homomorphism property $\varphi_\alpha(g_\alpha h_\alpha) = \varphi_\alpha(g_\alpha) \circ \varphi(h_\alpha)$, and $\varphi_\alpha(g_\alpha)$ has the inverse $\varphi_\alpha(g_\alpha^{-1})$.

Let $\varphi : G \to \mathrm{Per}(S)$ be the homomorphism with $\varphi_\alpha = \varphi \circ i_\alpha$ for all α. Let $u \in G$ given as $u = \overline{h}_m \overline{h}_{m-1} \cdots \overline{h}_1$ with $m \geq 0$, $h_i \neq 1$, $h_i \in G_{\alpha_i}$ and $\alpha_i \neq \alpha_{i-1}$ for $m \geq i > 1$. If we apply $\varphi(u)$ to the empty sequence $()$, then we get the sequence $(h_m, h_{m-1}, \ldots, h_1)$, and the $h_m, h_{m-1}, \ldots, h_1$ are uniquely determined by u. □

This automatically has the consequence: If G_1 and G_2 are finitely generated with solvable word problem, then $G_1 * G_2$ has a solvable word problem. The following three results give in fact stronger statements.

Theorem 1.4.12. *The free product of finitely many finite groups is residually finite.*

Proof. Let $1 \neq u$ be given as in Theorem 1.4.11. We take i_α as the inclusion. Let $S_m \subset S$ be the set of all elements of S with $n \leq m$. We define a map ψ_α from G_α to the set of all mappings from S_m into itself. For this we take φ_α as given in the proof of Theorem 1.4.11 except for the case $n = m$ and $\alpha \neq \alpha_m$. In the latter case ψ_α shall map the sequence $(g_m, g_{m-1}, \ldots, g_1)$ onto itself. We get a homomorphism $\psi: G \to \mathrm{Per}(S_m)$ such that $\psi(u)$ is not the identity (recall that $\psi(u)() \neq ()$ because $1 \neq u$). Now S_m is finite because there are only finitely many G_α which are all finite. □

Corollary 1.4.13. *The free product of finitely many residually finite groups is residually finite.*

Proof. Let $G = G_1 * G_2 * \cdots * G_n$, all G_i residually finite. Let $u \in G \setminus \{1\}$. Then there exist a homomorphism φ of G onto the free product $\overline{G}_1 * \overline{G}_2 * \cdots * \overline{G}_n$ of finite groups $\overline{G}_i$ such that $\varphi(u) \neq 1$. Now apply Theorem 1.4.12. □

In an analogous manner we may show the following (we leave the details as exercises).

Theorem 1.4.14. *The free product of finitely many fully residually free groups is fully residually free. Especially the free product of finitely many fully residually free groups is coherent.*

Theorem 1.4.15. *Let G_α be subgroups of G. Then the following are equivalent:*
(i) *We have $G = *_\alpha G_\alpha$.*
(ii) *Each element of G can be written uniquely as $g_1 g_2 \cdots g_n$ with $n \geq 0$, $g_i \in G_{\alpha_i}$, $g_i \neq 1$ and $\alpha_i \neq \alpha_{i+1}$ for $i < n$.*
(iii) *The group G is generated by the G_α, and 1 cannot be written as a product $g_1 g_2 \cdots g_n$ with $n \geq 1$, $g_i \in G_{\alpha_i}$, $g_i \neq 1$ and $\alpha_i \neq \alpha_{i+1}$ for $1 < n$.*

Proof. If (ii) holds, then certainly also (iii). If (iii) holds then each element from G can be written as in (ii) because G is generated by the G_α, and in fact uniquely because otherwise we get a non-trivial representation of 1 via

$$g_1 g_2 \cdots g_n = h_1 h_2 \cdots h_m \quad \Leftrightarrow \quad g_1 g_2 \cdots g_n h_m^{-1} h_{m-1}^{-1} \cdots h_1^{-1} = 1.$$

Hence (ii) if and only if (iii).

If (i) hold then also (ii) by Theorem 1.4.11. Now, let (iii) hold. We know that $*_\alpha G_\alpha$ exists, and from (iii) we get a homomorphism from $*_\alpha G_\alpha$ to G which is the inclusion for each G_α ($G_\alpha \subset G$). This homomorphism is surjective with trivial kernel. Hence, (i) holds if and only if (ii) holds because (ii) holds if and only if (iii) holds. □

Corollary 1.4.16. 1. *The free product of non-trivial commutative transitive groups is commutative transitive.*
2. *The free product of non-trivial CSA groups with only odd torsions is a CSA group.*

Recall that we should have only odd torsion by Corollary 1.4.16.2 because the infinite dihedral group is not a CSA group.

Definition 1.4.17. Let $u \in *G_\alpha$, $u = g_1g_2,\dots,g_n$, $n \geq 0$, $g_i \in G_{\alpha_i}$, $g_i \neq 1$, $\alpha_i \neq \alpha_{i+1}$.
1. The product $g_1g_2\cdots g_n$ is *reduced* and $|u| = n$ is the *length of* u.
2. If $v = h_1h_2\cdots h_m$ is also reduced uv is said to be *directly reduced* or *reduced as written* if $|uv| = |u| + |v|$, that is, g_n and h_1 are not in the same factor G_α. We write again $w \equiv uv$.
3. If uv is not directly reduced then we talk about *cancellation* if $g_nh_1 = 1$ and *coalescence* if $g_nh_1 \neq 1$. In both cases we have $|uv| < |u| + |v|$, where $|uv| = |u| + |v| - 1$ in the case of a coalescence.
4. Let u be as above, we call u *cyclically reduced* if $u = 1$, that is, $n = 0$, or $\alpha_n \neq \alpha_1$.

Each element $\neq 1$ from $*G_\alpha$ is conjugate to a cyclically reduced element. If v is cyclically reduced and $|v| \geq 2$, then $|v^k| = |k||v|$ for each $k \in \mathbb{Z} \setminus \{1\}$. This has the following consequences.

Corollary 1.4.18. *If $u \in G$, $G = *G_\alpha$, has finite order then u is conjugate to an element of some G_α.*

Remark 1.4.19. Let $u = g_1g_2\cdots g_n$ and $v = h_1h_2\cdots h_m$ be cyclically reduced elements in $G = G_1 * G_2$. If $n \geq 2$ and u and v are conjugate in G, then $m = n$, and v arises from u by a cyclic permutation. If $n \leq 1$ and u and v are conjugate in G then again $m = n$ and u and v are in the same factor G_i, $i = 1$ or 2, and are conjugate in this factor. Hence, if in addition G_1 and G_2 are finitely generated with solvable conjugacy problem, then $G = G_1 * G_2$ has a solvable conjugacy problem. Certainly, if G_1 and G_2 are finitely generated with solvable word problem, then $G = G_1 * G_2$ has a solvable word problem.

The analogous statement holds for $G = G_1 * G_2 * \cdots * G_n$ because $G = G_1 * G_2'$ with $G_2' = G_2 * G_3 * \cdots * G_n$ $(n \geq 3)$.

Definition 1.4.20. Let (as above) $G = *G_\alpha, g \in G, g = a_1a_2\cdots a_n, n \geq 0, a_i \neq 1, a_i \in G_{\alpha_i}$ for $i = 1, 2, \dots, n$, and $\alpha_i \neq \alpha_{i+1}$ $(1 \leq i < n)$. Let $|g| = n$ be the length of g. We call $a_1a_2\cdots a_k$, $n = 2k$ and $n = 2k+1$, respectively, the *left half* $a(g)$ of g and $a(g^{-1})^{-1}$ the *right half* of g.

If $x, y \in G$, $x \neq 1 \neq y$, then we write $x \sim y$ if $x = y^{-1}$ or if both x and y are elements of one conjugate $uG_\alpha u^{-1}$ of some G_α.

The relation is not reflexive because $x \sim x$ if and only if $x \in uG_\alpha u^{-1}$ for some G_α and $u \in G$. If $x \sim y$ and $y \sim z$, then $x \sim z$ if and only if $x \sim x$ (which is equivalent to $y \sim y$).

We now introduce a well-ordering on G. We make this a bit different from the free group case. If $w \equiv pkq$ with $|p| = |q|$, $|k| \leq 1$ in a free group then $k = 1$ or k is just a single letter. This was easily to handle using the overlapping initial segment of length $[\frac{|w|+1}{2}]$. But if $u \equiv pkq$ with $|p| = |q|$, $|k| \leq 1$, in a free product $G = *G_\alpha$ then k is an element of some G_α. We first order the indices α and afterwards the elements of G_α where, for each $x \in G_\alpha$, either $x = x^{-1}$ or x is right in front of x^{-1} or x^{-1} is right in front of x and the elements of G_α stand next to each other. This gives a well-ordering on $\bigcup G_\alpha$. Now G gets a lexicographical order $x <_\ell y$ like the following. Let $x = a_1 a_2 \cdots a_m, y = b_1 b_2 \cdots b_n$ the normal forms. Then $x <_\ell y$ if either $m < n$ or $m = n$ and $a_i = b_i$ for $i < r \leq m$ and a_r is in front if b_r with respect to the order of $\bigcup G_\alpha$. This gives a well-ordering on G, which we denote by $<$, with the following properties:

(1) $x < y$ if $|x| < |y|$.

(2) $x < y$ if $|x| = |y|$ and $a(x) <_\ell a(y)$.

(3) $x < y$ if $|x| = |y|, a(x) = a(y)$ and $a(x^{-1}) <_\ell a(y^{-1})$.

(4) If $u \in G$ and if the normal form of u does not end with an element of G_α then the non-trivial elements of $uG_\alpha u^{-1}$ (all have length $2|u| + 1$ and the same left and right half) stand next to each other.

(5) If $x \in uG_\alpha u^{-1}$, then either $x = x^{-1}$ or x is right in front of x^{-1} or x^{-1} is right in front of x.

Definition 1.4.21. A subset $X \subset G$ is called *irreducible* if

1. $1 \notin X$;
2. if $x \in X$ then $x \leq x^{-1}$; and
3. if $x \in X$ with $x = aub$ and $a, b \in \langle y \mid y \in X, y < x \rangle$, then $x \leq u$.

Theorem 1.4.22. *Let $H < G$. Then H has an irreducible generating system X. If, in addition, H is finitely generated and $r = \operatorname{rk}(H)$, the rank of H, then X may be chosen with $|X| = r$.*

Proof. We argue analogously to the case of free groups. Let $G_h = \langle \{h \mid h \in H, y < h\} \rangle$ and $X = \{h \mid h \in H, h \notin G_h\}$. Then $H = \langle X \rangle$. This we see as follows. Assume that $H \neq \langle X \rangle$. Let $g \in H \setminus \langle X \rangle$ be minimal with respect to $<$. Then $h \in \langle X \rangle$ for all $h \in H$ with $h < g$ because of the minimality of g. Hence, $g \notin G_g$ because otherwise $g \in G_g \subset \langle X \rangle$. But $g \notin G_g$ means $g \in X$ by definition which gives a contradiction. Therefore $H = \langle X \rangle$.

More precisely we even have the following.

Lemma 1.4.23. *If $h \in H$ then $h \in \langle \{x \in X \mid x \leq h\} \rangle =: K_h$.*

Proof. Assume there is a $h \in H$ with $h \notin K_h$. Choose a minimal (with respect to $<$) $h \in H$ with this property. Then we have $k \in \langle \{x \in X \mid x \leq k\} \rangle = K_k \subset G_h$ for all $k \in H$ with $k < h$ and, hence, $\langle \{x \in X \mid x < h\} \rangle = G_h \subset K_h$. It is $h \notin G_h$ because otherwise $h \in K_h$. But then $h \in X$ which gives a contradiction because $h \in X$ means above all $h \in K_h$. □

We now come back to the proof of Theorem 1.4.22. Certainly $1 \notin X$ and $x \notin X$ if $x^{-1} < x$. If $x \in H$ and $x = aub$ with $a, b \in \langle \{y \mid y \in X, y < x\} \rangle$, then $x \notin X$ if $u < x$. Therefore X is irreducible. Now, let $H = \langle h_1, h_2, \dots, h_n \rangle$. If $\{h_1, h_2, \dots, h_n\}$ is reducible then we get another generating system, namely either $\{h_1, h_2, \dots, h_{i-1}, h_{i+1}, h_{i+2}, \dots, h_n\}$ or $\{h_1, h_2, \dots, h_{i-1}, h_i^{-1}, h_{i+1}, h_{i+2}, \dots, h_n\}$ or $\{h_1, h_2, \dots, h_{i-1}, u, h_{i+1}, h_{i+2}, \dots, h_n\}$ depending on whether $h_i = 1, h_i > h_i^{-1}$ or $h_i = aub$ with $u < h_i$ and $a, b \in \langle \{h_j \mid h_j < h_i\} \rangle$. In each case, the new set (considered as a sequence) is smaller than the old one. Since the ordering is a well-ordering, we get an irreducible generating system with at most n elements. Especially we may take $r = n$. □

Remark 1.4.24. As already mentioned, we may define Nielsen transformations in any group, especially in free products. The transformations, described in final parts of the above proof of Theorem 1.4.22, are in fact Nielsen transformations. This means especially, if $H < G$ is finitely generated by a generating system $\{h_1, h_2, \ldots, h_n\}$ then we get from $\{h_1, h_2, \ldots, h_n\}$ in finitely many steps through Nielsen transformations a finite irreducible generating system $\{k_1, k_2, \ldots, k_m\}$ of H with $m \leq n$. At the same time the transition may be chosen such that we can make reductions in each single step. Hence, we also call an irreducible system in G *Nielsen reduced*. In general we do not have the Nielsen property for irreducible systems. For instance, if $x \in G_\alpha$, $\alpha \neq 1$, then

$$|x^3| \leq |x| = |x| - |x| + |x|.$$

From now on, let X be an irreducible generating system of the subgroup H of $G = *G_\alpha$.

Notations: Let N be the union of all subgroups $uG_\alpha u^{-1}$ of G. If $x \in X \cap N$, then let $N(x) = \langle \{y \in X \mid y \sim x\} \rangle$. If $h \in N(x) \setminus \{1\}$ for some $x \in X \cap N$ then $|h| = |x|$, and property (4) of the well-ordering shows that $a > h$ if and only if $a > x$ in the case of $a \not\sim h$ (which is equivalent to $a \not\sim x$).

Especially, if $h \in N(x) \setminus \{1\}$ and $h < a$, $a \not\sim h$, then $h \in \langle \{y \in X \mid y < a\} \rangle$, because if $y \sim x$ then $y \sim h$ and $h < a$ implies $y < a$. Let Y be defined by $y \in Y :\Leftrightarrow \{y \in X \cup X^{-1}$ or $y \in N(x)$ for some $x \in X \cap N\}$. Let $h \in H$. Then $h = z_1 z_2 \cdots z_n$, $z_i \in X \cup X^{-1}$. After possible cancellations and coalescences we may write h as $h = y_1 y_2 \cdots y_m$ where $y_i \in Y$, $i \leq m$, and $y_i \not\sim y_{i+1}$ for $i < m$.

Theorem 1.4.25. *Let $y_1, y_2, \ldots, y_m \in Y$ with $y_i \not\sim y_{i+1}$ for $1 \leq i < m$. Then $|y_i| \leq |y_1 y_2 \cdots y_m|$ for $1 \leq i \leq m$.*

Corollary 1.4.26. *Let $h \in H \cap G_\alpha$. Then $h \in \langle X \cap G_\alpha \rangle$.*

Proof. It is enough to show that $h \in \langle X \cap (\bigcup G_\alpha) \rangle$. Let $h = y_1 y_2 \cdots y_m$ with $y_i \in Y$, $y_i \not\sim y_{i+1}$. By Theorem 1.4.25 we have $|y_i| \leq |h| = 1$ for all i. Now, $y_i \in X \cup X^{-1}$ or $y_i \in \langle \{x' \in X \mid x' \sim x\} \rangle$ for some $x \in X$. Hence, $|x'| = |x| = |y_i|$. Therefore h is generated by elements of length 1 in X. □

We now prove Theorem 1.4.25 by use of several lemmas which we consider first.

Lemma 1.4.27. *If $x, y \in Y$ and $x \not\sim y$, then $|x|, |y| \leq |xy|$.*

Proof. Since $|xy| = |y^{-1}x^{-1}|$ we may assume that $\min(x, x^{-1}) \leq \min(y, y^{-1})$. We have $y \neq x^{-1}$ because $x \not\sim y$. The result is obvious for $y = x$ $(\neq x^{-1})$. Hence, we may assume that $\min(x, x^{-1}) < \min(y, y^{-1})$. Especially, $|x| \leq |y|$. We have to show $|y| \leq |xy|$.

1. Let $y \in X$ or $y \in X^{-1}$.

We cannot have $|xy| < |y|$ for $x \in \langle\{a \in X \mid a < y^{\epsilon}, \epsilon = \pm 1\}\rangle$ because X is irreducible and $y = x^{-1}xy$.
If $x \in X$ then $x \leq x^{-1}$, and, hence, $x < \min(y, y^{-1})$ and $x \in \langle\{a \in X \mid a < y^{\epsilon}, \epsilon = \pm 1\}\rangle$. The analogous statement holds for $x^{-1} \in X$.
Therefore $|xy| \geq |y|$ for $x \in X \cup X^{-1}$. Now, let $x \in N(a)$ for some $a \in X \cap N$. Since $x \sim a \sim x^{-1}$, $\min(x, x^{-1}) < \min(y, y^{-1})$ and $x \nsim y$ we have again

$$x \in \langle\{b \in X \mid b < y^{\epsilon}, \epsilon = \pm 1\}\rangle,$$

that is, again $|xy| \geq |y|$.

2. Let $|x| = |y|$, $y \notin X \cup X^{-1}$, $y \in N(y')$ for some $y' \in X \cap N$. Analogously to above we have $x \in \langle\{a \in X \mid a < y'^{\epsilon}, \epsilon = \pm 1\}\rangle$ because $x \nsim y$ and $x \sim y'$ (property (4) of the order). We have $y' \equiv qgq^{-1}$, $g \in G_\alpha$ for some α.
Assume that $|xy| < |y|$. Then necessarily $x \equiv paq^{-1}$ with $a \in G_\alpha$ because $|x| = |y|$. We have $p < q$ because $x \nsim y'$, $y' = x^{-1}pagq^{-1} = x^{-1}pg'q^{-1}$, $g' \in G_\alpha$, which gives a contradiction. Hence, $|xy| \geq |y|$.
3. Let $|x| < |y|$, $y \notin X \cup X^{-1}$ and $y \in N(y')$ for some $y' \in X \cap N$. Assume that $|xy| < |y|$. We may write $x \equiv paq$, $y \equiv q^{-1}br$ with $|a| = |b| = 1$ and $ab \neq 1$. From $|xy| < |y|$ we get $|p| + 1 \leq |q|$ with $<$ if a and b are not both in the same factor G_α (note that $xy = pabr$).
   (i) If $|r| > |q|$ then y' starts with $q^{-1}b$. If $|r| = |q|$ then y' starts with $q^{-1}b'$ where b and b' belong to the same G_α. In any case $|xy'| < |y'|$.
   (ii) If $|q| > |r|$, then also $|q| > |p| + 1$ because we have $|r| > |p|$ from $|y| > |x|$. Hence, q is more than the right half on x. Let $x \equiv \overline{p}\,\overline{a}\,\overline{q}$ with $|\overline{a}| = 1$ and $\overline{q}$ the right half of x. Then $y \equiv \overline{q}^{-1}\overline{a}^{-1}\overline{r}$. From $|y| > |x|$ we see that $\overline{q}^{-1}$ is less than the left half of y, because otherwise $|y| = |\overline{q}| + 1 + |\overline{r}| \leq 2|\overline{q}| + 1 = |x|$. Hence, $y' \equiv \overline{q}^{-1}\overline{a}^{-1}r'$ and we have $|xy'| < |y'|$. Therefore, in any case $|xy'| < |y'|$. But this is impossible because $x \nsim y \sim y'$, which means also $x \nsim y'$, and $\min(x, x^{-1}) < \min(y', y'^{-1})$ (see part 1). □

Lemma 1.4.28. *If $x, y \in Y$ with $x \nsim y$ and $|xy| = |x|$, then $a(y) \leq_\ell a(y^{-1})$.*

Proof. Let $x \equiv paq^{-1}$, $y \equiv qbr$ with $|a| = |b| = 1$. From $|xy| = |x|$ we see that $|q| = |r| + 1$ or that $|q| = |r|$ and q is the left half of y. By Lemma 1.4.27 we have $|y| \geq |x| = |xy|$ from which we see that $|p| > |r|$ and q^{-1} is at most the right half of x. Therefore $a(x^{-1})$ begins with $a(y)$. If $x \notin X \cup X^{-1}$ and $x \in N(x')$ for some $x' \in X \cap N$, then $x, x' \in uG_\alpha u^{-1}$ with $u = a(x) = a(x') = a(x^{-1}) = a(x'^{-1})$. Since $y \nsim x$ we have $y \notin uG_\alpha u^{-1}$, and it follows that $|x'y| = |x'|$ (we have in any case $|x'y| \leq |x'|$ but $|x'y| < |x'|$ is not possible by Lemma 1.4.27). Therefore we may assume that $x^{\epsilon} \in X$, $\epsilon = \pm 1$. Since $|xy| = |x| \geq |y|$ we see that the left half of x and the right half of y remain in xy. This we see as follows. If $|r| = |q| - 1$ then a and b cannot be in the same G_α, and if $|r| = |q|$ then $|p| \geq |r| = |q|$. Hence, $a(xy) = a(x)$ and $a((xy)^{-1})$ begins with $a(y^{-1})$. Assume that $a(y^{-1}) <_\ell a(y)$. Then $a((xy)^{-1}) <_\ell a(x^{-1})$ because they have the same length and begin with $a(y^{-1})$ and $a(y)$, respectively.

From $|xy| = |x|$ and $a(xy) = a(x)$ we now get $xy < x$ and $(xy)^{-1} < x^{-1}$. From $a(y^{-1}) < a(y)$ we have $y \notin N$, that is, $y^{\epsilon} \in X$, $\epsilon = \pm 1$, and also $y \neq x$ because $y \notin N$ and $|xy| = |x|$. The equations $x = (xy)y^{-1}, y = x^{-1}(xy), y^{-1} = ((xy)^{-1})x, x^{-1} = y((xy)^{-1})$ give now, together with $xy < x$ and $(xy)^{-1} < x^{-1}$, a contradiction to the irreducibility of X, depending on which of $x^{\epsilon}, y^{\eta}, \epsilon, \eta = \pm 1$, is the smaller one. □

Lemma 1.4.29. *Let $x, y, z \in Y$ and $|xy| = |x|$, $|yz| = |z|$. Then $a(y) = a(y^{-1})$, that is, $y \in N$.*

Proof. If $x \sim y$ or $y \sim z$ then $y \in N$ because $xy \neq 1, yz \neq 1$. Otherwise $a(y) \leq_{\ell} a(y^{-1})$ and $a(y^{-1}) \leq_{\ell} a(y)$ from $|xy| = |x|$ and $|yz| = |z|$, which means $|z^{-1}y^{-1}| = |z^{-1}|$, respectively, using Lemma 1.4.28. □

Lemma 1.4.30. *Let $x, y, z \in Y$ and $xy \neq 1 \neq yz$. Then y cannot be written as $y \equiv uv$ such that u is canceled by x and v is canceled by z.*

Proof. Assume we may write y in the form $y \equiv uv$ such that u is canceled by x and v is canceled by z. We do not have $|u| > |v|$ because then $|xy| < |x|$, which contradicts Lemma 1.4.27. Analogously, $|u| < |v|$ is not possible. Hence, $|u| = |v|$ and therefore $|xy| = |x|$ and $|yz| = |z|$. But then $y \notin N$, and we get a contradiction from Lemma 1.4.27 and Lemma 1.4.29. □

Lemma 1.4.31. *Let $x, y, z, w \in Y$ with $|xy| = |x|$, $|yz| = |y| = |z|$ and $|zw| = |w|$. Then $y \sim z$.*

Proof. We have $y \in N$ and $z \in N$ from Lemma 1.4.29. Now we get $y \sim z$ from $y, z \in N$ and $|yz| = |y| = |z|$. □

Lemma 1.4.32. *Let $x, y, z \in Y$ with $x \nsim y, y \nsim z$. Then $|xyz| \geq |x| - |y| + |z|$.*

Proof. We have the following situation $y \equiv ubv$, $x \equiv pau^{-1}$ and $z \equiv v^{-1}cq$ (some of a, b, c, u, v, p, q could be 1), whereby the initial element of b is not canceled by the terminal element of a, and the terminal element of b is not canceled by the initial element of c, but we could have coalescences.

By Lemma 1.4.30 we have $b \neq 1$. If $|b| > 1$, then

$$|xyz| = |p| + |q| + |abc| \geq |p| + |q| + |a| + |b| + |c| - 2 > |x| - |y| + |z|.$$

If $|b| = 1$ and if not $a, b, c \in G_{\alpha}$ for some α such that $abc = 1$, then

$$|xyz| = |p| + |q| + |abc| \geq |p| + |q| + |a| + |c| - 1 = |x| + |z| - |u| - |v| - 1 = |x| + |z| - |y|.$$

In addition, if $|b| = 1$, $a, b, c \in G_{\alpha}$ for some α with $abc \neq 1$, then $|xy| = |x|$, $|yz| = |z|$ and $v = u^{-1}$, that is, $y \in N$ and $|y| \leq |x|, |z|$. This we can see as follows. By Lemma 1.4.27 we have $|u| = |v|$ because if we would have $|u| > |v|$, then $|xy| < |x|$, and if we would have $|v| > |u|$, then $|yz| < |z|$. Hence, $|xy| = |x|$, $|yz| = |z|$, and therefore $v = u^{-1}$ by

Lemma 1.4.29, that is, $y \in N$. Also, if $|b| = 1$ and if not all a, b, c are in some G_α, then $|xyz| > |x| - |y| + |z|$. By Lemma 1.4.27 we have $|y| \le |x|, |z|$.

Now finally let $|b| = 1$, $a, b, c \in G_\alpha$ for some α and $abc = 1$. Then $xyz = pq$, that is, $|xyz| \le |p| + |q| = |x| + |z| - |y| - 1$ and hence, $|xyz| < |x| + |z| - |y|$. We have to show that this case cannot occur: Assume that $|b| = 1$ and $a, b, c \in G_\alpha$ for some α with $abc = 1$ is possible. If $|y| = |x|$ then we cannot have $x \in N$. Because, if $x \in N$, then $x, y \in N$ and $|x| = |y| = |xy|$ which implies $x \sim y$, and this contradicts $x \nsim y$. Analogously $z \notin N$ if $|y| = |z|$. If $|y| < |x|$ and $x \in N(x')$ for some $x' \in X \cap N$, then the equal right halves of x and x' must be at last au^{-1} because $|x| > |y|$, that is, $x' = p'au^{-1}$, and we may replace x by x'. Analogously we may argue for z. Therefore we now may assume that $x \in X$ or $x^{-1} \in X$ and $z \in X$ or $z^{-1} \in X$.

(1) Let $|x| = |y| = |z|$.
Then $|p| = |u| = |q|$. Assume that $u \ge_\ell p$. From $x \nsim y$ we have $p \ne u$. Hence, $u >_\ell p$. Then $x^{-1} > x$, that is, $x \in X$. There exists a $y' \in X$ with $y' = ub'u^{-1}$ (whether $y \in X \cup X^{-1}$ or not). Then $xy' = p(ab')u^{-1}$, in normal form if $ab' \ne 1$, and $|xy'| < |y'|$ if $ab' = 1$. The latter cannot occur by Lemma 1.4.27.
Now, let $ab' \ne 1$. Since $a(y') = u >_\ell a(x) = p$ and $a(xy') = p$ we have $y' > x, xy'$, which contradicts the irreducibility of X. Analogously we get a contradiction if $u \ge_\ell q^{-1}$ (consider for this $z^{-1}y^{-1}$).
Hence, we have $u <_\ell p$ and $u <_\ell q^{-1}$, which means $y < x$ and $y < z$. Therefore $y \in \langle\{x' \in X \mid x' < x, z\}\rangle$ (if $y \in N(y')$ with $y' \in X \cap N$ then use $y \sim y'$ and argue as before). This gives $|xyz| \ge |x| = |z|$ from the irreducibility of x, no matter which of x, x^{-1}, z, z^{-1} is the biggest. For instance, if $x > x^{-1}, z, z^{-1}$ then consider $x^{-1} = (yz)(xyz)^{-1}$. But $|xyz| \ge |x| = |z| = |y|$ contradicts $|xyz| < |x| - |y| + |z|$. Hence, $|z| = |y| = |z|$ does not occur.

(2) Let $|y| = |x| < |z|$.
Then $x^\epsilon < z^\eta, y^\delta < z^\nu, \epsilon, \eta, \delta, \nu = \pm 1$ and $y < x$ as in (1), that is, $y \in \langle\{x' \in X \mid |x'| \le |y|\}\rangle$. It follows $|xyz| \ge |z|$ by the irreducibility of X (write $z = (y^{-1}x^{-1})(xyz)$ and note that $|xy| = |x| < |z|$). This contradicts $|xyz| < |x| - |y| + |z|$. Therefore $|x| = |y| < |z|$ does not occur. Analogously $|y| = |z| < |x|$ does not occur.

(3) Let $|y| < |x|, |z|$.
From the irreducibility of X we get necessarily $xy \ge x$. From $x = pau^{-1}$, $abc = 1$ we get $xy = pabu^{-1} = pc^{-1}u^{-1}$, that is, $|x| = |xy|$ and $a(x) = a(xy)$. Therefore $ua^{-1} \le_\ell uc$ and further $a^{-1} \le_\ell c$. If we compare analogously z and yz, then we get $c \le_\ell a^{-1}$. Hence, $c = a^{-1}$, which is impossible because $abc = 1$, $b \ne 1$. Again, also $|y| < |x|, |z|$ does not occur. This proves the statement. □

Lemma 1.4.33. *Let $y_1, y_2, \ldots, y_m \in Y$ with $y_i \nsim y_{i+1}$ for $1 \le i < m$. Then $|y_1 y_2 \cdots y_m| = \sum_{i=1}^{m} |y_i| - \sum_{i=1}^{m-1} r_i$, where $r_i = |y_i| + |y_{i+1}| - |y_i y_{i+1}|$.*

Proof. We have $y_i \equiv u_i v_i u_{i+1}^{-1}$ (u_i, u_{i+1} eventually equal 1), whereby the initial element of v_i is not canceled by the terminal element of v_{i-1} and the terminal element of v_i is not canceled by the initial element of v_{i+1}; however, we may have coalescence. Especially we have

$$y_1y_2\cdots y_m = u_1v_1\cdots v_m u_{m+1}^{-1}.$$

By Lemma 1.4.30 we have $v_i \neq 1$ for $1 < i < m$. As in the proof of Lemma 1.4.32 we get: If $|v_{i+1}| = 1, 1 \le i < m-1$, and if the terminal element of v_i, the initial element of v_{i+2} and also v_{i+1} belong to one G_α, then

$$|y_iy_{i+1}| = |y_i|, \quad |y_{i+1}y_{i+2}| = |y_{i+2}|,$$

the product $v_iv_{i+1}v_{i+2}$ is non-trivial, and we have $y_{i+1} \in N$. This and Lemma 1.4.31 together with $y_{i+1} \nsim y_{i+2}$ show that the following cannot happen:

There exists an $i < m-2$ such that $|v_{i+1}| = |v_{i+2}| = 1$ and v_{i+1}, v_{i+2}, the terminal element of v_i and the initial element of v_{i+3} are in the same G_α. In $y_1y_2\cdots y_m = u_1v_1\cdots v_m u_{m+1}^{-1}$ we can have further reductions only from combinations of elements from the same G_α, and such a combination cannot produce the 1 by the above arguments. We have $r_i = 2|u_{i+1}| + 1$ or $r_i = 2|u_{i+1}|$ which depends on the terminal element of v_i and the initial element of v_{i+1} are in the same G_α or not. We now get the formula because in $u_iv_1\cdots v_m u_{m+1}^{-1}$ there is no reduction besides the simple coalescences. □

We now give the proof of Theorem 1.4.25.

Proof. We have

$$|y_1y_2\cdots y_m| = \sum_{i=1}^{j-1}(|y_i| - r_i) + |y_j| + \sum_{i=j+1}^{m}(|y_i| - r_i).$$

Now, $|y_i| - r_i = |y_iy_{i+1}| - |y_{i+1}| \ge 0$ by Lemma 1.4.27, and analogously $|y_i| - r_{i-1} \ge 0$. Hence, $|y_1y_2\cdots y_m| \ge |y_j|$. □

Corollary 1.4.34. *Let $x \in X \cap uG_\alpha u^{-1}$. Then $N(x) = H \cap uG_\alpha u^{-1}$, where X is an irreducible generating system of H with $|X| = \mathrm{rk}(H)$, the rank of H, if H is finitely generated.*

Proof. Certainly $N(x) \subset H \cap uG_\alpha u^{-1}$. Let $h \in H \cap uG_\alpha u^{-1}$ and with $h = y_1y_2\cdots y_n, y_i \in Y$, $y_i \nsim y_{i+1}$. If $n = 1$ then $y_i \in N$ and $h = y_1 \in N(x)$. Now, let $n \ge 2$ and $x \nsim y_n$. Then $y_1y_2\cdots y_nx^{-1} = hx^{-1}$ and $x^{-1} \in Y, y_n \nsim x^{-1}$. By Theorem 1.4.25 we have $y_1y_2\cdots y_nx^{-1} \neq 1$ and, hence, $|y_1y_2\cdots y_nx^{-1}| = |hx^{-1}| = |x^{-1}|$. The proof of Theorem 1.4.25 also gives $|y_nx^{-1}| = |x^{-1}|$. By Lemma 1.4.29 we have therefore $y_n \in N$, and then $|y_nx^{-1}| = |x^{-1}|$ contradicts $y_n \nsim x^{-1}$.

Now, let $n \ge 2$ and $y_n \sim x$. Then $y_{n-1} \nsim y_n$ gives also $y_{n-1} \nsim x^{-1}$. Now, the same argument as above, applied to $y_1y_2\cdots y_{n-1}x^{-1}$, gives a contradiction for $n \ge 3$. If $n = 2$, that is, $y_1 = hy_2^{-1}, y_2 \sim x$, then $y_1 \sim y_2$ or $y_1 = 1$, which gives a contradiction. □

Theorem 1.4.35 (Theorem of Kurosh). *Let $H < *G_\alpha$. Then $H = F * (*(H \cap uG_\alpha u^{-1}))$, where F is free and the factors $H \cap uG_\alpha u^{-1}$ are taken over certain conjugates of the G_α.*

Proof. Let X be an irreducible generating system of H. Then H is generated by the subgroups $\langle x \rangle$ for $x \in X \setminus N$ and the subgroups $N(x)$ for $x \in X \cap N$. H is the free product of these, the former infinite cyclic, if each $y_1 y_2 \cdots y_m \neq 1$ for $y_1, y_2, \dots, y_m \in Y$ and $y_i \nsim y_{i+1}$ for $i < m$. But this follows from Theorem 1.4.25. By Corollary 1.4.34 we have $N(x) = H \cap uG_\alpha u^{-1}$ for $x \in X \cap uG_\alpha u^{-1}$. □

Corollary 1.4.36. *The free product of finitely many coherent groups is coherent.*

Corollary 1.4.37. *Let $G = *G_\alpha$ and A be a non-cyclic Abelian subgroup of G. Then there exists an α and an $u \in G$ with $A < uG_\alpha u^{-1}$.*

Proof. This is a direct consequence of the normal form; see Theorem 1.4.11 and Theorem 1.4.35 of Kurosh. □

We now formulate Theorem 1.4.25 in terms of Nielsen transformations (recall that the approaches are equivalent for finite systems in free products).

Theorem 1.4.38. *Let $G = *G_\alpha$. If $\{x_1, x_2, \dots, x_m\}$ is a finite system of elements in G then there is a Nielsen transformation from $\{x_1, x_2, \dots, x_m\}$ to a system $\{y_1, y_2, \dots, y_m\}$ for which one of the following two cases holds:*
(i) *$y_i = 1$ for some $i \in \{1, 2, \dots, m\}$.*
(ii) *Each $w \in \langle y_1, y_2, \dots, y_m \rangle$ can be written as $w = \prod_{i=1}^{q} y_{\nu_i}^{\epsilon_i}$, $\epsilon_i = \pm 1$, $\epsilon_i = \epsilon_{i+1}$ if $\nu_i = \nu_{i+1}$, with $|y_{\nu_i}| \le |w|$ for $i = 1, 2, \dots, q$.*

We remark that we only need finitely many α to describe $\langle x_1, x_2, \dots, x_m \rangle$. If we choose the order in an appropriate manner with respect to these finitely many α, then we need only finitely many elementary Nielsen transformations to go from $\{x_1, x_2, \dots, x_m\}$ to $\{y_1, y_2, \dots, y_m\}$. Also, the Nielsen transformation can be chosen so that $|y_i| \le |x_i|$ for all i and $|x_i| < |y_i|$ fails to hold for at least one i or the lengths of the elements of the system $\{x_1, x_2, \dots, x_m\}$ are preserved.

1.4.3 Special equations in free products

A first application of the above is the following.

Theorem 1.4.39. *Let $G = *G_\alpha$ a free product of the groups G_α. Let $1 \neq a_j \in G_{\alpha_j}$, $\alpha_j \in I$, $j = 1, 2, \dots, n$, $n \ge 1$ and $\alpha_j \neq \alpha_k$ for $j \neq k$. Let p be the number of the a_j which are proper powers in G_{α_j}. Let $\{x_1, x_2, \dots, x_m\} \subset G$, $m \ge 2$, and $H = \langle x_1, x_2, \dots, x_m \rangle$. If $a_1 a_2 \cdots a_n \in H$ then one of the two following cases holds:*
(1) *There is a Nielsen transformation from $\{x_1, x_2, \dots, x_m\}$ to a system $\{y_1, y_2, \dots, y_m\}$ with $y_1 = a_1 a_2 \cdots a_n$, or*

(2) *$m \geq 2n - p$, and there is a Nielsen transformation from $\{x_1, x_2, \ldots, x_m\}$ to a system $\{y_1, y_2, \ldots, y_m\}$ with $y_k \in G_{a_\ell}$, $\ell \in \{1, 2, \ldots, n\}$, for $k = 1, 2, \ldots, 2n - p$.*

Proof. Without loss of generality we may assume that $G = G_1 * G_2 * \cdots * G_n$ with $a_i \in G_i$. We introduce an order in G as above and may assume that $\{x_1, x_2, \ldots, x_m\}$ is Nielsen reduced, that is, irreducible. Note that $\{x_1, x_2, \ldots, x_m\}$ satisfies an equation

$$\prod_{k=1}^{q} x_{\nu_k}^{\epsilon_k} = a_1 a_2 \cdots a_n, \quad \epsilon_k = \pm 1, \epsilon_k = \epsilon_{k+1} \text{ if } \nu_k = \nu_{k+1}. \tag{1.16}$$

We may assume that
(a) $x_i \neq 1$ for all i;
(b) Each x_i occurs in (1.16); and
(c) q is minimal under the numbers for which (1.16) holds.

If one x_i occurs only once in (1.16) then case (1) holds.

Now, suppose that (1) does not hold, and that each x_i occurs at least twice in (1.16). Then the result is certainly correct if $n = 1$ or $q \leq 2$.

Now, let $n \geq 2$ and $q \geq 3$. Because of the special form of $a_1 a_2 \cdots a_n$ we have necessarily

$$|x_{\nu_k}^{\epsilon_k} x_{\nu_{k+1}}^{\epsilon_{k+1}} x_{\nu_{k+2}}^{\epsilon_{k+2}}| = |x_{\nu_k}| - |x_{\nu_{k+1}}| + |x_{\nu_{k+2}}| \quad \text{for all } k \text{ with } 1 \leq k < q - 1.$$

This means that each $x_{\nu_{k+1}}$, $1 \leq k < q-1$, is conjugate to an element of some H_i. The result now follows from the special form of $a_1 a_2 \cdots a_n$. □

In regards to Corollary 1.2.17 we get in general for free products the following Corollary.

Corollary 1.4.40. *Let $G = G_1 * G_2 * \cdots * G_n$, $n \geq 1$, the free product of the groups G_i and $g_i \in G_i$, $i = 1, 2, \ldots, n$.*

Let $\{x_1, x_2, \ldots, x_m\} \subset G$, $m \geq 1$, and X be the subgroup of G generated by $x_1, x_2, \ldots, x_m$. Let there exist $u_1, u_2, \ldots, u_q \in X$ with

$$u_1^{\beta_1} u_2^{\beta_2} \cdots u_p^{\beta_p} [u_{p+1}, u_{p+2}] \cdots [u_{q-1}, u_q] = g_1 g_2 \cdots g_n,$$

$0 \leq p \leq q$, $2 \leq \beta_j$ for $j = 1, 2, \ldots, p$ and $\gcd(\beta_1, \beta_2, \ldots, \beta_p) \geq 2$ and $g_i \neq 1$ for $i = 1, 2, \ldots, n$. Then $m \geq rk(X) \geq n$ and especially $q \geq n$.

More interesting is the following theorem for free products of finite cyclic groups.

Theorem 1.4.41. *Let $G = \langle s_1, s_2, \ldots, s_m \mid s_1^{\alpha_1} = s_2^{\alpha_2} = \cdots = s_m^{\alpha_m} = 1\rangle$, $m \geq 2$, all $\alpha_i \geq 2$, $\{x_1, x_2, \ldots, x_n\} \subset G$ and X be the subgroup generated by $x_1, x_2, \ldots, x_n$, $n \geq 1$. Let $y^{-1}(s_1 s_2 \cdots s_m)^{\alpha} y \in X$ for some $\alpha \neq 0$ and some $y \in G$. Then one of the following cases occurs:*

(1) *There is a Nielsen transformation from* $\{x_1, x_2, \ldots, x_n\}$ *to a system* $\{y_1, y_2, \ldots, y_n\}$ *with* $y_1 = z(s_1 s_2 \cdots s_m)^\beta z^{-1}$, $\beta \geq 1$ *and* $z \in G$.

(2) *There is a Nielsen transformation from* $\{x_1, x_2, \ldots, x_n\}$ *to a system* $\{y_1, y_2, \ldots, y_n\}$ *with* $y_1 = z s_i^{\gamma_i} z^{-1}$ *for some* i $(1 \leq i \leq m)$, $2 \leq \gamma_i < \alpha_i$, $\gcd(\gamma_i, \alpha_i) \geq 2$ *and* $z \in G$.

(3) *The subgroup* X *has finite index* $\beta = [G : X]$, *where* β *is the smallest positive number for which a relation* $y^{-1}(s_1 s_2 \cdots s_m)^\beta y \in X$ *holds for some* $y \in G$, *and a set of coset representatives for* X *in* G *is given by* $\{1, s_1 s_2 \cdots s_m, \ldots, (s_1 s_2 \cdots s_m)^{\beta-1}\}$.

Proof. The statement is certainly correct if $m = 2$ and $\alpha_1 = \alpha_2 = 2$, that is, if G is the infinite dihedral group. Hence, from now on let $\alpha_1 + \alpha_2 \geq 5$ if $m = 2$. Suppose that neither case (1) nor case (2) occurs.

In the following let α be the smallest positive number for which a relation $y^{-1}(s_1 s_2 \cdots s_m)^\alpha y \in X$ holds for some $y \in G$. We may assume that $y = 1$ (replace x_i by $y x_i y^{-1}$ if necessary). In $G = \mathbb{Z}_{\alpha_1} * \mathbb{Z}_{\alpha_2} * \cdots * \mathbb{Z}_{\alpha_m}$ we introduce an order as above. We may assume that

(a) $\{x_1, x_2, \ldots, x_n\}$ is Nielsen reduced.

(b) If $x_i = z s_j^{\gamma_j} z^{-1}$, $z \in G$, $1 \leq \gamma_j < \alpha_j$, then $x_i = z s_j z^{-1}$.

(c) No two of the x_i generate a cyclic group.

The assumptions (b) and (c) are possible. If $x_i = z s_j^{\gamma_j} z^{-1}$, $1 \leq \gamma_j < \alpha_j$, then there exists a δ_j with $\gamma_j \delta_j \equiv 1 \mod \alpha_j$ and we may replace x_i by $x_i^{\delta_j}$ (recall that case (2) does not hold). If $x_i = z s_j^{\gamma_1} z^{-1}$, $x_k = z s_j^{\gamma_2} z^{-1}$, then $\langle x_i, x_k \rangle = \langle z s_j z^{-1} \rangle$, and we may replace $\{x_i, x_k\}$ by $\{z s_j z^{-1}, 1\}$. If $x_i = a$, $x_k = b$, a and b of infinite order, and $\langle a, b \rangle = \langle c \rangle$ then replace $\{x_i, x_k\}$ by $\{c, 1\}$. Repeated applications of these operations and Nielsen transformations lead in finitely many steps to a system with the properties (a)–(c).

Now the system $\{x_1, x_2, \ldots, x_n\}$ satisfies the Nielsen property

$$|x_i^\epsilon x_j x_k^\eta| > |x_i| - |x_j| + |x_k|,$$

$\epsilon, \eta = \pm 1$, if $i \neq j \neq k$ or $i = j \neq k$, $\epsilon = 1$ or $i \neq j = k$, $\eta = 1$. This can be seen as follows. Suppose

$$|x_i^\epsilon x_j x_k^\eta| \leq |x_i| - |x_j| + |x_k|,$$

$\epsilon, \eta = \pm 1$, if $i \neq j \neq k$ or $i = j \neq k$, $\epsilon = 1$ or $i \neq j = k$, $\eta = 1$. Then, for some λ, we have the following situation: $x_j \equiv s_\lambda^\gamma v^{-1}$, $1 \leq \gamma < \alpha_\lambda$, $x_i^\epsilon \equiv u_i s_\lambda^{\delta_i} v^{-1}$, $1 \leq \delta_i < \alpha_\lambda$, $x_k^\eta \equiv v s_\lambda^{\delta_k} u_k$, $1 \leq \delta_k < \alpha_\lambda$. Because case (b) does not occur there exists an δ with $\delta\gamma \equiv 1 \mod \alpha_\lambda$. If $i \neq j$ then $x_i^\epsilon x_j^{-\delta_i \delta} = u_i v^{-1}$ and $|x_i^\epsilon x_j^{-\delta_i \delta}| < |x_i|$. If $j \neq k$ then $x_j^{\delta_k \delta} x_k^\eta = v u_k$ and $|x_j^{-\delta_k \delta} x_k^\eta| < |x_k|$. In both cases we have a contradiction to the fact that $\{x_1, x_2, \ldots, x_n\}$ is Nielsen reduced. If $|x_i^3| \leq |x_i|$, $1 \leq i \leq n$, then x_i is conjugate to a power of some s_λ. We have an equation

$$\prod_{k=1}^{q} x_{v_k}^{\epsilon_k} = (s_1 s_2 \cdots s_m)^{\alpha}, \tag{1.17}$$

where $\epsilon = \pm 1$, $\epsilon_k = \epsilon_{k+1}$ if $v_k = v_{k+1}$. We choose q minimal such that an equation (1.17) holds with this minimal α. Especially, on (1.17) we have no situation $x_{v_k} = x_{v_{k+1}} = x_{v_{k+\ell}}$ and $x_{v_k}^{\ell_1} = 1$ $(1 \le k \le k + \ell \le q)$. Also we may assume that each x_i occurs in (1.17).

Suppose that in (1.17) there are two indices j, i, $1 \le j < i \le q$, with $v_j = v_i = k$, $\epsilon_j = \epsilon_i$ and $v_h \ne k = v_j$ for $j < h < i$. We may assume that $\epsilon_j = \epsilon_i = 1$. It is possible that $v_{j-1} = k$ or $v_{k+1} = k$. If x_k is conjugate to a power for some s_λ then we may assume that $v_{i+1} \ne k$ (change the order in the factor $\mathbb{Z}_{\alpha_\lambda}$, if necessary, and replace x_k by a suitable power). We consider the subword

$$w = x_k \left(\prod_{p=j+1}^{i-1} x_{v_p}^{\epsilon_p} \right) x_k$$

of (1.17). First, let x_k be a conjugate to a power of some s_λ, that is, $x_k \equiv u s_\lambda^{\beta} u^{-1}$, $1 \le \lambda \le n$. Then $v_{i+1} \ne k$ and

$$\begin{aligned} w &= u s_\lambda^{\beta} u^{-1} \left(\prod_{p=j+1}^{i-1} x_{v_p}^{\epsilon_p} \right) u s_\lambda^{\beta} u^{-1} \\ &= u s_\lambda^{\beta'} (s_\lambda \cdots s_m s_1 \cdots s_{\lambda-1})^{\delta} s_\lambda^{\beta''} u^{-1}, \end{aligned}$$

$\delta \ge 0$.

Suppose $\delta > 0$. We get $\beta'' \equiv \beta \equiv 1 \mod \alpha_\lambda$ because $v_{i+1} \ne v_k$, and $\beta' + 1 \equiv \beta \equiv 1 \mod \alpha_\lambda$, that is, $\beta' \equiv 0 \mod \alpha_\lambda$ because $v_h \ne k$ if $j < h < i$. Hence,

$$\left(\prod_{p=j+1}^{i-1} x_{v_p}^{\epsilon_p} \right) x_k = u s_\lambda^{-1} \cdots s_1^{-1} (s_1 s_2 \cdots s_m)^{\delta} s_1 \cdots s_\lambda u^{-1}.$$

If $0 < \delta < \alpha$ then this contradicts the minimality of α, and if $\delta = \alpha$ then this contradicts the minimality of q. Hence, $\delta = 0$ and $j = i - 1$ if x_k is conjugate to a power of some s_λ.

Now, let x_k be not conjugate to a power of some s_λ. Then also $|x_k^3| > |x_k|$, and there is a stable symbol s_λ^{β} of x_k which can be recovered in the normal form of (1.17). We have $x_k \equiv u s_\lambda^{\beta} v$, $1 \le \lambda \le m$, and

$$\begin{aligned} w &= u s_\lambda^{\beta} \left(\prod_{p=j+1}^{i-1} x_{v_p}^{\epsilon_p} \right) u s_\lambda^{\beta} v \\ &= u s_\lambda^{\beta'} (s_\lambda \cdots s_m s_1 \cdots s_{\lambda-1})^{\delta} s_\lambda^{\beta''} v, \end{aligned}$$

$\delta \ge 0$. Suppose $\delta > 0$. Then $\beta' + 1 \equiv \beta \mod \alpha_\lambda$, $\beta'' \equiv 1 \mod \alpha_\lambda$ or $\beta' \equiv 0 \mod \alpha_\lambda$, $\beta'' \equiv \beta \mod \alpha_\lambda$. The case $\beta' \equiv 0 \mod \alpha_\lambda$ can be handled as above.

Now let $\beta' + 1 \equiv \beta \mod \alpha_\lambda$ and $\beta'' \equiv 1 \mod \alpha_\lambda$. Then we obtain

$$\left(\prod_{p=j+1}^{i-1} x_{\nu_p}^{\epsilon_p}\right) x_k = v^{-1} s_\lambda^{-1} \cdots s_1^{-1} (s_1 s_2 \cdots a_m)^\delta s_1 s_2 \cdots s_\lambda v,$$

which also gives a contradiction as above. Hence, $\delta = 0$ and $j = i-1$ if x_k is not conjugate to a power of some s_λ.

The above arguments also lead to the following (replace $x_k = v s_\lambda^\beta v^{-1}$ by a suitable power and regard blocks x_k^δ, if necessary). If some x_k occurs in (1.17) exactly once with exponent +1 and once with exponent −1 then x_k is not conjugate to a power of some s_λ. On the other hand, if some x_k is not conjugate to a power of some s_λ then it occurs in (1.17) exactly once with exponent +1 and once with exponent −1 because case (1) does not occur. As already mentioned, if $x_k = u_k s_\lambda^\beta u_k^{-1}$ then $\langle x_k \rangle = \langle u_k s_\lambda u_k^{-1} \rangle$ because case (2) does not occur. Hence, as in operation (b) explained, we replace $x_k = u_k s_\lambda^\beta u_k^{-1}$ by $y_k = u_k s_\lambda u_k^{-1}$. We apply Nielsen transformations to the so received system, and we get a generating system $\{z_1, z_2, \ldots, z_n\}$ of $\langle x_1, x_2, \ldots, x_m \rangle = X$ such that $X = \langle z_1 \rangle * \langle z_2 \rangle * \cdots * \langle z_n \rangle$ and

$$z_1 z_2 \cdots z_p [z_{p+1}, z_{p+2}] \cdots [z_{\ell-1}, z_\ell] z_{\ell+1} z_{\ell+1}^{-1} \cdots z_n z_n^{-1} = (s_1 s_2 \cdots s_m)^\alpha,$$

$0 \le p \le n$, $p + 1 \le \ell \le n$, $\ell - p$ even, $z_i = v_i s_{+v_i} v_i^{-1}$, $1 \le v_i \le m$ for $i = 1, 2, \ldots, p$ and z_j has infinite order for $j = p+1, \ldots, \ell$ (see [281]). Without loss of generality we may assume that $\ell = n$.

Now we regard G as a cycloidal Fuchsian group of the first kind such that

(i) $s_1, s_2, \ldots, s_m$ are elliptic elements;
(ii) $a := (s_1 s_2 \cdots s_m)^{-1}$ is a parabolic element; and
(iii) $\{s_1, s_2, \ldots, s_m, a\}$ is a canonical generating system of G.

We note that X is, as a subgroup of G, also discrete and that X is not cyclic and not the infinite dihedral group because $\alpha_1 + \alpha_2 \ge 5$ if $m = 2$. No z_j, $p + 1 \le j \le n$, is conjugate in G to a power of $s_1 s_2 \cdots s_m$ because case (1) does not occur. Hence, each z_j, $p+1 \le j \le n$, is a hyperbolic element. Now, $\{a_1, z_1, z_2, \ldots, z_n\}$, $a_1 := a^\alpha = (s_1 s_2 \cdots s_m)^{-\alpha}$ is a canonical generating system for X, and X is a cycloidal Fuchsian group of the first kind.

Especially, X has finite index $[G : X]$ in G (see [281] or [25] for details). If $[G : X] = 1$ then $\alpha = 1$ and $G = X$. Now, let $[G : X] > 1$ and $g \in G \backslash X$. We regard $g(s_1 s_2 \cdots s_m) g^{-1}$. There exists a natural number γ such that $g(s_1 s_2 \cdots s_m)^\gamma g^{-1} \in X$ because $[G : X] < \infty$. Hence, there exists an $h \in X$ and an integer $\delta \ne 0$ such that $hg(s_1 s_2 \cdots s_m)^\gamma g^{-1} h^{-1} = (s_1 s_2 \cdots s_m)^\delta$ because X is cycloidal. Therefore $\gamma = \delta$ and $hg \in \langle s_1 s_2 \cdots s_m \rangle$ because G is a free product of cyclics and $\alpha_1 + \alpha_2 \ge 5$ if $m = 2$. Now, $g = d(s_1 s_2 \cdots s_m)^\varphi$, that is, $Xg = X(s_1 s_2 \cdots s_m)^\varphi$, for some $d \in X$ and an $\varphi \in \mathbb{N}$ with $1 \le \varphi < \alpha$ because $h \in X$ and $(s_1 s_2 \cdots s_m)^\alpha \in X$. □

Remark 1.4.42. If $n < m$ then we may replace case (3) of Theorem 1.4.41 by

(3') m odd, $n = m-1$, all $\alpha_i = 2$, and there is a Nielsen transformation from $\{x_1, x_2, \dots, x_n\}$ to $\{s_1s_2, s_1s_3, \dots, s_1s_m\}$, and $s_1s_2\cdots s_m \notin X$ but $(s_1s_2\cdots s_m)^2 \in X$ (see [211]).

Corollary 1.4.43. *Let $G = \langle s_1, s_2, \dots, s_m \mid s_1^{\alpha_1} = s_2^{\alpha_2} = \cdots = s_m^{\alpha_m} = 1\rangle$, $2 \le n$, all α_i prime numbers, and let $\alpha_1 + \alpha_2 \ge 5$ if $m = 2$. Let $x_1, x_2, \dots, x_m$, $n \ge 1$ be elements in G of finite order and $X = \langle x_1, x_2, \dots, x_n\rangle$. Then X has finite index $[G : X]$ in G if and only if $y(s_1s_2\cdots s_m)^{\alpha}y^{-1} \in X$ for some $\alpha \ne 0$ and $y \in G$.*

Moreover, if X has finite index $[G : X]$ in G and if α is the smallest positive integer with $y(s_1s_2\cdots s_m)^{\alpha}y^{-1} \in X$ for some $y \in G$, then $[G : X] = \alpha$, and a set of coset representatives for X in G is given by $\{1, s_1s_2\cdots s_m, \dots, (s_1s_2\cdots s_m)^{\alpha-1}\}$.

Proof. We have each $x_i \ne 1$ of the form $x_i = z_i s_{v_i}^{\delta_i} z_i^{-1}$, $1 \le \delta_i < \alpha_i$. Since each α_i is a prime number we have $\langle x_i\rangle = \langle z_i s_{v_i} z_i^{-1}\rangle$. Since we may assume that $x_i \ne 1$ for all i we now may assume that $x_i \equiv z_i s_{v_i} z_i^{-1}$ for some v_i. Also we may assume that no two x_i generate a cyclic group. Assume that we have $|x_j| \le |x_i|$ and $|x_i x_j^{\epsilon}| < |x_i|$ or $|x_j^{-\epsilon}x_i| < |x_i|$ for $\epsilon = \pm 1$ and $i \ne j$. Then $|x_j^{-\epsilon}x_i x_j^{\epsilon}| < |x_i|$. This we can see as follows. Suppose that $|x_i x_j| < |x_i|$. Then $z_i^{-1} \equiv r_i^{-1} z_j^{-1}$ with $r_i \ne 1$ because $i \ne j$. We necessarily have $|r_i^{-1} s_{v_j} z_j| < |r_i| + |z_j| + 1$, and hence $|x_j^{-1}x_i x_j| < |x_i|$. Therefore, if $|x_j| \le |x_i|$ and $|x_i x_j^{\epsilon}| < |x_i|$ or $|x_j^{-\epsilon}x_i| < |x_i|$ for $\epsilon = \pm 1$ and $i \ne j$, then $|x_j^{-\epsilon}x_i x_j^{\epsilon}| < |x_i|$, and we may replace x_i by $x_j^{-\epsilon}x_i x_j^{\epsilon}$. Hence, we may assume that

$$|x_i^{\epsilon}x_j^{\eta}| \ge |x_i|, |x_j|, \quad \epsilon, \eta = \pm 1,$$

for $i \ne j$.

Altogether we may assume that $\{x_1, x_2, \dots, x_m\}$ satisfies the conditions (a), (b), (c) in the proof of Theorem 1.4.41. We now may continue as in the proof of Theorem 1.4.41 and get the desired result. □

Remarks 1.4.44. Unfortunately our method in [226] does not give the result without the α_i being prime numbers. Kulkarni [166] showed with the help of a geometric-graphtheoretical method that Corollary 1.4.43 holds as stated for arbitrary $\alpha_i \ge 2$. However, Corollary 1.4.43 has two interesting number-theoretical applications.

(1) We know that the rational modular group

$$\Gamma = \left\{A{:}\, z \mapsto \frac{az+b}{cz+d} \;\middle|\; z \in \mathbb{C} \cup \{\infty\}, a, b, c, d \in \mathbb{Z}, ad - bc = 1\right\} = \mathrm{PSL}(2, \mathbb{Z})$$

is generated by $T{:}\, z \mapsto -\frac{1}{z}$ and $R{:}\, z \mapsto -\frac{1}{z+1}$ with defining relations $T^2 = R^3 = 1$, and, hence, $\Gamma \cong \mathbb{Z}_2 * \mathbb{Z}_3$. We remark that $U(z) = z + 1$ for $U = TR$. The principal congruence group $\Gamma[N]$ of level $N \in \mathbb{N}$ is the subgroup

$$\Gamma[N] = \left\{A{:}\, z \mapsto \frac{az+b}{cz+d} \;\middle|\; A \in \Gamma, \begin{pmatrix} a & b \\ c & d \end{pmatrix} \equiv \pm \begin{pmatrix} 1 & 0 \\ 0 & 1 \end{pmatrix} \mod N\right\}.$$

A subgroup $\Gamma_1 \subset \Gamma$ is called congruence subgroup if $\Gamma[N] \subset \Gamma_1$ for some $N \geq 1$. There exist only finitely many cycloidal congruence subgroups in Γ. The index $\mu = [\Gamma : \Gamma_1]$ of a cycloidal congruence subgroup $\Gamma_1 \subset \Gamma$ is a divisor of $K := 2^4 \cdot 3^2 \cdot 5 \cdot 7 \cdot 11 = 55440$. For each positive divisor μ of K there exist cycloidal congruence subgroups of index μ in Γ. For $\mu = 1, 2, 3, 4, 5, 6, 7, 8, 9, 10, 11, 12, 21$ there exist cycloidal congruence subgroups of index μ in Γ and genus 0; see [212]. Petersson describes in this paper a construction process for cycloidal congruence subgroups of Γ. Unfortunately this process is in general extremely hard to handle.

Now, let $Z_1, Z_2, \ldots, Z_n$ be elements in Γ of finite order, and let $\Gamma_1 = \langle Z_1, Z_2, \ldots, Z_n \rangle$. Furthermore, let $AU^{\alpha}A^{-1} \in \Gamma_1$ for some $\alpha \neq 0$ and $A \in \Gamma$. Then Γ_1 is a cycloidal subgroup of Γ with finite index $[\Gamma : \Gamma_1]$ and genus 0. We just remark that Γ_1 is not a normal subgroup of Γ if $[\Gamma : \Gamma_1] > 6$. Using Petersson's construction process we may determine whether or not Γ_1 is a congruence subgroup of Γ. This way we may classify the cycloidal subgroups Γ_1 of Γ of genus 0.

(2) Let $N \in \mathbb{N}$, and let G_N be the group consisting of all (projective) matrices of type (i) or (ii):

(i) $U = \pm\begin{pmatrix} a & b\sqrt{N} \\ c\sqrt{N} & d \end{pmatrix}$, $a, b, c, d \in \mathbb{Z}$, $ad - Nbc = 1$,

(ii) $U = \pm\begin{pmatrix} a\sqrt{N} & b \\ c & d\sqrt{N} \end{pmatrix}$, $a, b, c, d \in \mathbb{Z}$, $Nad - bc = 1$,

where a matrix is identified with its negative.

We certainly have $T = \pm\begin{pmatrix} 0 & 1 \\ -1 & 0 \end{pmatrix} \in G_N$. Conjugating T by $U \in G_N$ gives a matrix of the form $\begin{pmatrix} * & x^2+Ny^2 \\ * & * \end{pmatrix}$.

The following question arises: How can we use this connection to decide whether or not a natural number n is of the form $n = x^2 + Ny^2$?

This was successfully done for those N with form class number $h(-4N) \leq 2$ where the answer is given by simply congruence conditions (see [76] and [154]).

There is a huge literature in the study of prime numbers of the form $p = x^2 + Ny^2$. An excellent introduction to this topic is given in [65]. There are some N which allow us to provide elementary congruence conditions for when a number $x^2 + Ny^2$ is prime. These N are called convenient numbers. There are 65 convenient numbers N with $1 \leq N \leq 1848$. There is at most one convenient number larger than 1848 (see [257] and [269]). If we look at the described set of convenient numbers, we see that 11 is the first inconvenient number.

Using Corollary 1.4.43 we could show the following.

Theorem 1.4.45. *The group G_{11} has a presentation*

$$G_{11} = \langle t_1, t_2, t_3, t_4, p \mid t_1^2 = t_2^2 = t_3^3 = t_4^2 = t_2 t_3 t_1 t_4 p = 1 \rangle,$$

where $t_1 = \pm\begin{pmatrix} 0 & 1 \\ -1 & 0 \end{pmatrix}$, $t_2 = \pm\begin{pmatrix} -\sqrt{11} & 4 \\ -3 & \sqrt{11} \end{pmatrix}$, $t_3 = \pm\begin{pmatrix} \sqrt{11} & -3 \\ 4 & -\sqrt{11} \end{pmatrix}$, $t_4 = \pm\begin{pmatrix} -\sqrt{11} & -6 \\ 2 & \sqrt{11} \end{pmatrix}$, *and* $p = \pm\begin{pmatrix} -1 & \sqrt{11} \\ 0 & -1 \end{pmatrix}$.

For a proof see [165]. We completely answered the question whether or not a natural number n can be written as $n = x^2 + 11y^2$ by a detailed examination of the conjugacy classes of the t_i, $i = 1, 2, 3, 4$.

1.4.4 The theorem of Grushko and applications for equations in free products

Theorem 1.4.46 (Theorem of Grushko). *Let $G = \mathop{*}\limits_{i=1}^{m} G_i$ and G finitely generated. Then:*
(1) *Each finite system of generators of G can be transferred (in finitely many steps) into a system whose elements all are in one of the factors G_i; and*
(2) $\operatorname{rk}(G) = \sum_{i=1}^{m} \operatorname{rk}(G_i)$, *where $\operatorname{rk}(H)$ denotes the rank of the group H.*

Proof. Without loss of generality, let $m = 2$, $G = G_1 * G_2$. Let $\operatorname{rk}(G_1) = n_1$, $\operatorname{rk}(G_2) = n_2$. We have to show that $n_1 + n_2 = \operatorname{rk}(G) =: n$. Certainly $n \le n_1 + n_2$. Let X be a Nielsen reduced (irreducible) generating system of G with $k \ge n$ elements. By Corollary 1.4.26 we have $G_i = \langle X \cap G_i \rangle$ for $i = 1, 2$. Hence $X \cap G_i$ contains at least n_i elements. Since $1 \notin X$, $(X \cap G_1) \cap (X \cap G_2) = \emptyset$ we have $k \ge n \ge n_1 + n_2$. Hence, $n = n_1 + n_2$.

Assertion (1) now follows from the remarks after Lemma 1.4.23 and because there is none $x \in X$, $x \ne 1$, left with $x \notin X \cap G_i$ for $i = 1$ or 2. If this would be the case we could transfer this x by Nielsen transformation to 1 because $G = \langle X \cap G_1, X \cap G_2 \rangle$, which gives a contradiction to the irreducibility of X. □

Corollary 1.4.47 (General form of the theorem of Grushko). *Let φ be an epimorphism of the free group F of rank n onto a finitely generated free product $*G_\alpha$. Then there exist subgroups $F_\alpha < F$ with $F = *F_\alpha$ and $\varphi(F_\alpha) = G_\alpha$.*

Proof. Let $\{x_1, x_2, \ldots, x_n\}$ be a basis of F such that $\varphi(x_1), \ldots, \varphi(x_n)$ generate $*G_\alpha$, some of the $\varphi(x_i)$ could be 1 in $G = *G_\alpha$. Each reduction in G, starting with $\varphi(x_1), \ldots, \varphi(x_n)$, can be done parallel in F. As a consequence we get a basis $\{y_1, y_2, \ldots, y_n\}$ of F such that $\{\varphi(y_1), \ldots, \varphi(y_n)\}$ is an irreducible generating system of G and $\varphi(y_i) = 1$ for $m + 1 \le i \le n$. Then, $G = *G_\alpha$ is generated by those $\varphi(y_i)$, $1 \le i \le m$, which are in $\bigcup G_\alpha$. By the irreducibility of $\{\varphi(y_1), \ldots, \varphi(y_m)\}$ we have $\varphi(y_i) \in \bigcup G_\alpha$ for all $1 \le i \le m$. Hence there is a basis of F for which all elements are mapped to $\bigcup G_\alpha$. □

Corollary 1.4.48. *Let $G = \langle x, y \mid x^p = y^q = 1 \rangle$, $2 \le p, q$, the free product of the cyclic groups $\langle x \mid x^p = 1 \rangle$ and $\langle y \mid y^q = 1 \rangle$. Then:*
1. *Two elements $u, v \in G$ generate G if and only if*

$$[u, v] = w[x^\alpha, y^\beta]^\epsilon w^{-1}, \quad \epsilon = \pm 1$$

for some $w \in G$ and $1 \le \alpha \le \frac{p}{2}$, $\gcd(\alpha, p) = 1$ and $1 \le \beta \le \frac{q}{2}$, $\gcd(\beta, q) = 1$.
2. *If $G = \langle u, v \rangle$, then $\{u, v\}$ is Nielsen equivalent to exactly one pair $\{x^\alpha, y^\beta\}$ with $1 \le \alpha \le \frac{p}{2}$, $\gcd(\alpha, p) = 1$ and $1 \le \beta \le \frac{q}{2}$, $\gcd(\beta, q) = 1$.*

Proof. The result holds certainly for $(p,q) = (2,2)$. Also let $(p,q) \neq (2,2)$. By Theorem 1.4.46 any generating pair is Nielsen equivalent to a pair $\{x^\alpha, y^\beta\}$ with $1 \leq \alpha \leq \frac{p}{2}$, $\gcd(\alpha,p) = 1$ and $1 \leq \beta \leq \frac{q}{2}$, $\gcd(\beta,q) = 1$. The first assertion follows now in an analogous manner as Theorem 1.2.19 for free groups.

We have to show the uniqueness in the second assertion. For this purpose, it is enough to show that the generating pair $\{x,y\}$ is not Nielsen equivalent to a pair $\{x^\alpha, y^\beta\}$ with $1 \leq \alpha \leq \frac{p}{2}$, $\gcd(\alpha,p) = 1$ and $1 \leq \beta \leq \frac{q}{2}$, $\gcd(\beta,q) = 1$ where $1 < \alpha$ or $1 < \beta$ (replace x and y by powers if necessary). We consider the elements $X = \pm\begin{pmatrix} 0 & -1 \\ 1 & \lambda_p \end{pmatrix} \in \mathrm{PSL}(2,\mathbb{R})$, $\lambda_p = 2\cos(\frac{\pi}{p})$ and $Y = \pm\begin{pmatrix} 0 & 1 \\ -1 & \lambda_q \end{pmatrix} \in \mathrm{PSL}(2,\mathbb{R})$, $\lambda_q = 2\cos(\frac{\pi}{q})$. We saw in example 3 at the beginning of this section that $\langle X, Y\rangle \cong \langle x,y\rangle \cong \mathbb{Z}_p * \mathbb{Z}_q$. Assume that $\{x,y\}$ is Nielsen equivalent to a pair $\{x^\alpha, y^\beta\}$ as above with $1 < \alpha$ or $1 < \beta$. Then

$$\mathrm{tr}[X,Y] - 2 = S_\alpha^2(\lambda_p)S_\beta^2(\lambda_q)(\mathrm{tr}[X,Y] - 2),$$

that is, $S_\alpha^2(\lambda_p)S_\beta^2(\lambda_q) = 1$, where then $S_n(t)$, $n \in \mathbb{N}_0$, are the Chebyshev polynomials as defined already. But this gives a contradiction because if $r \in \mathbb{N}$, $r \geq 3$, then $S_\gamma(2\cos(\frac{\pi}{r})) = \frac{\sin(\frac{\gamma\pi}{r})}{\sin(\frac{\pi}{r})} > 1$ for $1 < \gamma \leq \frac{r}{2}$. This gives the uniqueness, and Corollary 1.4.48 is proved. □

Using Corollary 1.4.7 we easily get the following.

Corollary 1.4.49. *Let $G = \langle x_1, x_2, \ldots, x_n \mid x_1^{k_1} = x_2^{k_2} = \cdots = x_n^{k_n} = 1\rangle$, $n \geq 2$ and $2 \leq k_i$ for $i = 1,2,\ldots,n$. If $G = \langle u_1, u_2, \ldots, u_n\rangle$, then $\{u_1, u_2, \ldots, u_n\}$ is Nielsen equivalent to exactly one system $\{x_1^{\alpha_1}, x_2^{\alpha_2}, \ldots, x_n^{\alpha_n}\}$ with $1 \leq \alpha_i \leq \frac{k_i}{2}$, $\gcd(\alpha_i, k_i)$ for $i = 1,2,\ldots,n$.*

Remark 1.4.50. Our proof is given in the spirit of our approach. We mention that Corollary 1.4.49 is also proved by Lustig [174] with a different, more topological proof. Both proofs do not apply to free products in general. The general unique question is handled by Weidmann [267] using the language of graphs of groups and Stalling's foldings.

Theorem 1.4.51. *Let $G = G_1 * G_2$ with $n_1 = \mathrm{rk}(G_1)$, $n_2 = \mathrm{rk}(G_2)$ and $n = n_1 + n_2 = \mathrm{rk}(G)$. Let $X = \{x_1, x_2, \ldots, x_{n_1}, y_1, y_2, \ldots, y_{n_2}\}$ and $U = \{u_1, u_2, \ldots, u_{n_1}, v_1, v_2, \ldots, v_{n_2}\}$ be two generating systems of G with*

$$x_1, x_2, \ldots, x_{n_1}, u_1, u_2, \ldots, u_{n_1} \in G_1$$

and

$$y_1, y_2, \ldots, y_{n_2}, v_1, v_2, \ldots, v_{n_2} \in G_2.$$

If X and U are Nielsen equivalent then $\{x_1, x_2, \ldots, x_{n_1}\}$ is Nielsen equivalent to $\{u_1, u_2, \ldots, u_{n_1}\}$ and $\{y_1, y_2, \ldots, y_{n_2}\}$ is Nielsen equivalent to $\{v_1, v_2, \ldots, v_{n_2}\}$.

As already mentioned in Section 1.1, analogously as in free groups, we may consider equations $w(x_1, x_2, \ldots, x_n) = 1$ in $*G_\alpha$ with the help of Nielsen transformations relative to $w(x_1, x_2, \dots, x_n)$. We make the description in a more general context to use it later for other groups, too. Let H be a group and F a free group of rank $n \geq 1$ with ordered basis $X = \{x_1, x_2, \dots, x_n\}$. The homomorphism $\phi: F \to H$ are represented uniquely by ordered sets $\phi(X) = U = \{u_1, u_2, \dots, u_n\}$ of elements of H where $\phi(x_i) = u_i$, $i = 1, 2, \dots, n$. If $w = w(x_1, x_2, \dots, x_n) \in F$ is a reduced word then $U \subset H$ is a *solution* of the equation $w = 1$ in H if $\phi(w) = w(u_1, u_2, \dots, u_n) = 1$ in H. If $\alpha: F \to F$ is an automorphism then α is defined by a regular Nielsen transformation on $\{x_1, x_2, \dots, x_n\}$. If $x_i' = \alpha(x_i) = \alpha_i(x_1, x_2, \dots, x_n)$, $1 \leq i \leq n$, then we write $\alpha(U) = U' = \{u_1', u_2', \dots, u_n'\}$ for the Nielsen equivalent system

$$\phi(\alpha(X)) = \{\alpha_1(u_1, u_2, \dots, u_n), \alpha_2(u_1, u_2, \dots, u_n), \dots, \alpha_n(u_1, u_2, \dots, u_n)\}.$$

Suppose U is a solution in H of the equation $w = 1$ and $\alpha : F \to F$ is an automorphism. If $w' = \alpha(w)$ then $U' = \alpha^{-1}(U)$ is a solution of the equation $w' = 1$. In general we call the pairs (w, U) and (w', U') where $w' = \alpha(w)$, $U' = \alpha^{-1}(U)$ with α an automorphism of F, *Nielsen equivalent* (induced by the automorphisms α). Certainly, the study of the solution of $w = 1$ is equivalent to the study of solutions of $w' = 1$.

A reduced word $w \in F$ is called *regular* if there is no automorphism $\alpha: F \to F$ such that $w' = \alpha(w)$, as a word in $x_1, x_2, \dots, x_n$, contains less of the $x_1, x_2, \dots, x_n$ than w itself, that is, w is of minimal rank. Suppose w is not regular.

Then if x_i is in w but not in $w' = \alpha(w)$ the element $u_i' = \phi(\alpha^{-1}(x_i))$ will appear as an arbitrary parameter in each solution of $w' = 1$ and hence in each solution of $w = 1$.

A word $w = w(x_1, x_2, \dots, x_n)$ is *quadratic* if each x_i, which occurs in w, occurs exactly twice, each time as x_i or x_i^{-1}. A quadratic word is *alternating* if each x_i, which occurs in w, occurs exactly once as x_i and exactly once as x_i^{-1}.

Now suppose $G = *G_\alpha$, $G_\alpha \neq \{1\}$ for all α. An ordered set $U = \{u_1, u_2, \dots, u_n\} \subset G$ is *regular* if there is no Nielsen transformation from U to a system $U' = \{u_1', u_2', \dots, u_n'\}$ in which one of the elements is 1. If $U \subset G$ is a solution of an equation $w = 1$ and U is not regular, then $U' = \alpha^{-1}(U) = \{u_1', u_2', \dots, u_n'\}$ with $u_i' = 1$ for some i is a solution of the equation $w' = 1$ with $w' = \alpha(w)$, for some automorphism $\alpha: F \to F$. Then $U_0' = \{u_2', u_3', \dots, u_n'\}$ is essentially a solution of $w_0 = 1$ where $w_0 = w_0(x_2, x_3, \dots, x_n) = w'(1, x_2, x_3, \dots, x_n)$.

Theorem 1.4.52. *Let $G = *G_\alpha$, $G_\alpha \neq \{1\}$ for all i. Let $1 \neq w(x_1, x_2, \dots, x_n)$ be a reduced regular word in F, with F free of rank $n \geq 1$ on the ordered basis $\{x_1, x_2, \dots, x_n\}$. Further, let $\phi: F \to G$ be a homomorphism such that $U = \phi(X)$ is regular and $\phi(w) = 1$. Then the pair (w, U) is Nielsen equivalent to a pair $(w', U') = (\alpha(w), \alpha^{-1}(U))$ with $\alpha: F \to F$ an automorphism such that*

(1) *$w' = r_1 w_1 r_1^{-1} r_2 w_2 r_2^{-1} \cdots r_k w_k r_k^{-1}$ for some $k \geq 1$ where the r_i and w_i are reduced words in F with $w_i \neq 1$ for $1 \leq i \leq k$.*
(2) *For each i, $1 \leq i \leq k$, we have $\phi(\alpha^{-1}(w_i)) = 1$.*
(3) *For each i, $1 \leq i \leq k$, there exists an α_i and a $g_i \in G$ with $\phi(\alpha^{-1}(x_j)) \in g_i G_{\alpha_i} g_i^{-1}$ for each x_j which occurs in w_i.*

Moreover, if ϕ is an epimorphism so that U is a generating set for G, then we may assume that $g_i = 1$ for all i, $1 \le i \le k$.
If in addition the word $w = w(x_1, x_2, \ldots, x_n)$ is quadratic then statement (1) can be replaced by
(1') $w' = r_1 w_1 r_1^{-1} r_2 w_2 r_2^{-1} \cdots r_k w_k r_k^{-1}$ *for some $k \ge 0$, each r_i is reduced and each w_i is a quadratic word in F. Furthermore, if w is alternating then each w_i is also alternating.*

Proof. Without loss of generality we may assume that $U = \phi(X)$ is Nielsen reduced in G. We write $w = w(x_1, x_2, \ldots, x_n) = y_1 y_2 \cdots y_m$, $m \ge 0$, $y_i \in X \cup X^{-1}$, $y_i \ne y_{i+1}^{-1}$. Then $1 = \phi(w) = w(u_1, u_2, \ldots, u_n) = v_1 v_2 \cdots v_m$ with $v_i = \phi(y_i) \in U \cup U^{-1}$ for $i = 1, 2, \ldots, m$. Since $w \ne 1$ and U is regular we have $m \ge 2$. If once $y_j = \cdots = y_{j+k}$ $(1 \le j < j + k \le m)$ and $|v_j^{k+1}| < |v_j|$ then certainly $v_j^{k+1} = 1$, and the statement follows inductively on m. Hence, we may assume that $|v_j^{k+1}| \ge |v_j|$ if $y_j = \cdots = y_{j+k}$ $(1 \le j < j + k \le m)$. Then especially

$$|v_i v_{i+1}| \ge |v_i|, |v_{i+1}|, \quad 1 \le i < m,$$

because U is Nielsen reduced. We consider subwords $y_j y_{j+1} \cdots y_k$ of w with $1 \le j \le k \le m$. The free length of such a subword is $k-j+1$. Because $v_1 v_2 \cdots v_m = 1$ there exists subwords $y_j \cdots y_k$ of w with $|v_j \cdots v_k| < |v_i|$ for some i with $j \le i \le k$. We may choose $k-j+1$ minimal for such a subword $y_j \cdots y_k$. Then $j \le k - 3$ because U is regular and Nielsen reduced. Hence, from Lemma 1.4.32 we must have

$$|v_i v_{i+1} v_{i+2}| \ge |v_i| - |v_{i+1}| + |v_{i+2}|$$

for $j \le i \le k - 2$.

Assume we have once

$$|v_i v_{i+1} v_{i+2}| > |v_i| - |v_{i+1}| + |v_{i+2}|;$$

then we have necessarily a proper subword $y_{j_0} \cdots y_{k_0}$, $j \le j_0 \le k_0 \le k$ of $y_j \cdots y_k$ with $|v_{j_0} \cdots v_{k_0}| < |v_{i_0}|$ for some i_0 with $j_o \le i_0 \le k_0$, and we have decisive reduction in which one of v_i, v_{i+1}, v_{i+2} is not involved. Hence

$$|v_i v_{i+1} v_{i+2}| = |v_i| - |v_{i+1}| + |v_{i+2}|$$

for all i with $j \le i \le k - 2$. From the proof of Lemma 1.4.32 we get $v_i = g r_i g^{-1}$, $j < i < k$, for some $g \in G$ and all $r_i \in G_\alpha \setminus \{1\}$ for some α. Especially $|v_{j+1}| = \cdots = |v_k| =: L \ge 1$.

Furthermore, $|v_j|, |v_k| \ge L$. From the minimality of $y_j \cdots y_k$ we must have $v_j \equiv q_j r_j g^{-1}$ and $v_k \equiv g r_k p_k^{-1}$ with $r_j, r_k \in G_\alpha$. We define $v := v_{j+1} \cdots v_{k-1}$. From the minimality of $y_j \cdots y_k$ we have $v \ne 1$, that is, $|v| = L$.

Assume that $|v_j| > L$. Then especially $v_j \ne v_i^\epsilon$, $\epsilon = \pm 1$, for $j < i < k$. Hence, $|v_k| > L$. From $|v_j v v_k| < |v_i|$ for some i with $j \le i \le k$ and $|v_j|, |v_k| > L$ we get $|v_j v v_k| < |v_j| - |v| + |v_k|$.

But this gives $v = 1$, a contradiction. Therefore also $|v_j| = |v_k| = L$, that is $v_j \equiv q_j r_j g^{-1}$, $v_k \equiv g r_k p_k^{-1}$ with $|q_j| = |g| = |p_k|$. If $v_j = v_i^\epsilon$, $\epsilon = \pm 1$, for some $j < i < k$ then this holds also for v_k because U is Nielsen reduced, and we get $v_i \cdots v_k = 1$ from $|v_j \cdots v_k| < |v_i|$ for some i.

Now let $v_j \neq v_i^\epsilon$, $\epsilon = \pm 1$ for each $j < i < k$. Then necessarily $v_j = v_k^\epsilon$, $\epsilon = \pm 1$, because U is Nielsen reduced. We must have $v_j = v_k$ because of the minimality of $y_j \cdots y_k$. Then $q_j = p_k^{-1} = g$ and $v_j \cdots v_k = 1$. Now, the main statement follows inductively on m. If ϕ is an epimorphism then we may assume that $g_i = 1$ for all i from Grushko's theorem.

The additional statement for quadratic words follows because we choose the Nielsen transformation with respect to w so that the quadratic word w is replaced by a quadratic word. □

Theorem 1.4.52 has some straightforward consequences. We leave the details for the proofs as exercises.

Corollary 1.4.53. *Let F be free of rank n, $n \geq 2$, with ordered basis $\{x_1, x_2, \ldots, x_n\}$. Let $w = w(x_1, x_2, \ldots, x_n)$, and let N be the normal closure of w in F. Then:*
1. *If F/N is free then either $w = 1$ or w is primitive in F.*
2. *If F/N is not free then F/N is decomposable into a non-trivial free product if and only if w is not regular.*

Corollary 1.4.54. *Let $G = G_1 * G_2$, $G_1 \neq \{1\} \neq G_2$. Let $u_1, u_2, u_3, u_4 \in G$ with $[u_1, u_2] \cdot [u_3, u_4] = 1$ in G. Assume that there is no Nielsen form $\{u_1, u_2, u_3, u_4\}$ to a system $\{v_1, v_2, v_3, v_4\}$ with $v_1 = 1$. Then there exists a Nielsen transformation from $\{u_1, u_2, u_3, u_4\}$ to a system $\{v_1, v_2, v_3, v_4\}$ for which one of the following cases hold:*
1. *We have $v_1, v_2, v_3, v_4 \in gG_ig^{-1}$, $i = 1$ or 2, $g \in G$, with $[v_1, v_2][v_3, v_4] = 1$.*
2. *We have $v_1, v_2 \in g_1G_ig_1^{-1}$, $i = 1$ or 2, $g_1 \in G$ with $[v_1, v_2] = 1$, and $v_3, v_4 \in g_2G_jg_2^{-1}$, $j = 1$ or 2, $g_2 \in G$ with $[v_3, v_4] = 1$.*

Proof. We only need to remark that $[x_1, x_2][x_3, x_4]$ is regular in the free group on $\{x_1, x_2, x_3, x_4\}$. □

Corollary 1.4.55. *Let F be a free group, $x_1, y_1, x_2, y_2, \ldots, x_g, y_g \in F$ and $H = \langle x_1, y_1, x_2, y_2, \ldots, x_g, y_g \rangle$. If $[x_1, y_1][x_2, y_2] \cdots [x_g, y_g] = 1$ in F, then H is a free subgroup of rank $\mathrm{rk}(H) \leq g$.*

Proof. Again, we only need to remark that $[x_1, y_1][x_2, y_2] \cdots [x_g, y_g]$ is regular in the free group on $\{x_1, y_1, x_2, y_2, \ldots, x_g, y_g\}$. If there is no Nielsen transformation from $\{x_1, y_1, x_2, y_2, \ldots, x_g, y_g\}$ to a system $\{u_1, u_2, \ldots, u_{2g}\}$ with $u_i = 1$ for some i then the statement follows from Theorem 1.4.52. If such a Nielsen transformation exists, then $\mathrm{rk}(H) \leq 2g - 1$, and the result follows by induction. □

Remark 1.4.56. In an analogous manner we may consider the equation $u_1^2[u_2, u_3] = 1$ or equivalently $u_1^2u_2^2u_3^2 = 1$ (see Section 1.3) in a free product $G = G_1 * G_2$, $G_1 \neq \{1\} \neq G_2$. Let

$u_1, u_2, u_3 \in G$ such that $u_1^2[u_2, u_3] = 1$ and $\langle u_1, u_2, u_3\rangle$ non-cyclic. Assume that $\langle a, b, c\rangle < G_i$, $1 \le i \le 2$, is cyclic whenever $a^2[b, c] = 1$ in G_i. Then G must contain an element of order 2.

We may ask the question. Let $G = G_1 * G_2$, $G_1 \neq \{1\} \neq G_2$. Under what condition on G can a non-trivial commutator in G be a proper power in G? This question is handled completely in [64].

1.5 Free products with amalgamation

We now extend our discussion to free products with amalgamation.

1.5.1 The pushout and first properties of free products with amalgamation

Definition 1.5.1. Let $i_1: G_0 \to G_1$ and $i_2: G_0 \to G_1$ be group homomorphisms. For a diagram (see Figure 1.14)

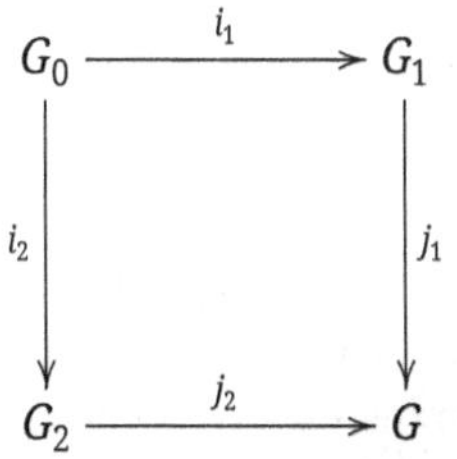

Figure 1.14: Pushout.

of group homomorphisms, a triple (G, j_1, j_2), is called a *pushout* of (i_1, i_2) if
(i) The diagram is commutative, that is, $j_1 \circ i_1 = j_2 \circ i_2$;
(ii) If there exists another commutative diagram of group homomorphisms as in Figure 1.15, then there exists exactly one homomorphism $f: G \to H$ such that $f \circ j_1 = \varphi_1$ and $f \circ j_2 = \varphi_2$ (see Figure 1.16).

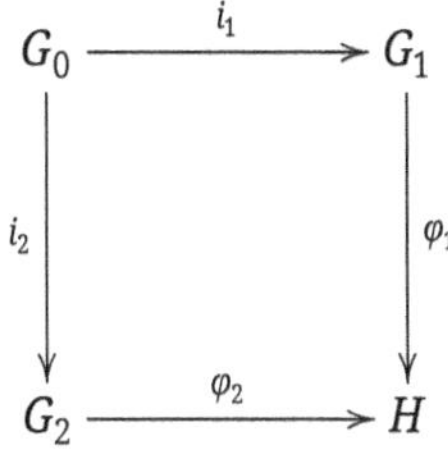

Figure 1.15: Homomorphisms to H.

We also have $\varphi_2 \circ i_2 = f \circ j_2 \circ i_2 = f \circ j_1 \circ i_1 = \varphi_1 \circ i_1$.

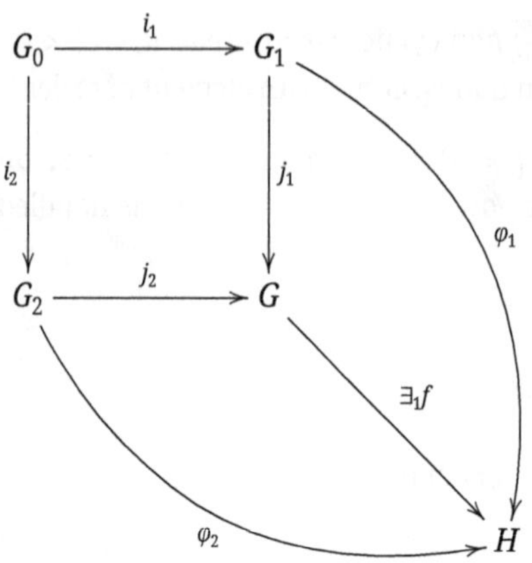

Figure 1.16: Extension of homomorphisms.

If (G, j_1, j_2) is a pushout of (i_1, i_2) then G is uniquely determined up to isomorphy. The proof is analogous to the proof for free groups. Hence, we may talk about *the* pushout of (i_1, i_2). We mostly just write G for the triple (G, j_1, j_2) and call G the pushout of (i_1, i_2).

Theorem 1.5.2. *Each pair (i_1, i_2) has a pushout.*

Proof. Let G_ν, $\nu = 1, 2$, have a presentation $\langle X_\nu \mid R_\nu \rangle^{\vartheta_\nu}$ with $X_1 \cap X_2 = \emptyset$. Let $G_0 = \langle Y \rangle$, and choose for each $y \in Y$ a $w_{y,\nu} \in F(X_\nu)$ with $i_\nu(y) = \vartheta_\nu(w_{y,\nu})$. Let G be the group with the presentation $\langle X_1 \cup X_2 \mid R_1, R_2, \{w_{y,1}^{-1} w_{y,2}\} \rangle$. We have the natural maps $j_\nu : G_\nu \to G$ induced by the inclusions $X_\nu \hookrightarrow X_1 \cup X_2$, and G is generated by $j_1(G_1) \cup j_2(G_2)$. Especially there is at most one homomorphism from G to H with predetermined values on $j_1(G_1) \cup j_2(G_2)$.

Now, let $\varphi_\nu : G_\nu \to H$ with $\varphi_1 \circ i_1 = \varphi_2 \circ i_2$ be given homomorphisms, $\nu = 1, 2$. Then φ_ν defines a homomorphism $\psi_\nu : F(X_\nu) \to H$ with $\psi_\nu(R_\nu) = \{1\}$. By van Dyck's theorem, Theorem 1.3.5, there exists a homomorphism $\psi : F(X_1 \cup X_2) \to H$ with $\psi|_{X_\nu} = \psi_\nu$, which induces a homomorphism $\varphi : G \to H$ such that $\varphi_\nu = \varphi \circ j_\nu$ by construction. □

Remarks 1.5.3. 1. If $G_2 = \{1\}$ then

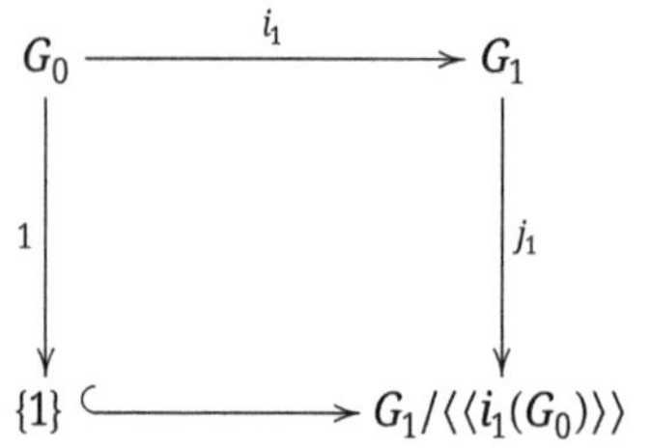

Figure 1.17: Pushout of $(i_1, 1)$.

is the pushout of $(i_1, 1)$ (see Figure 1.17).

2. The maps j_1 and j_2 are not necessarily injective, also if i_1 or i_2 is injective.

Example 1.5.4. Let G_1 be simple, i_1 injective and i_2 surjective but not injective. Take $w \in G_0$, $w \neq 1$ with $i_2(w) = 1$. Then $i_1(w) \neq 1$, but $j_1 \circ i_1(w) = j_2 \circ i_2(w) = 1$. Since G_1 is simple, j_1 is trivial. Since i_2 is surjective and $j_2 \circ i_2 = j_1 \circ i_1$ is trivial, also j_2 is trivial. Hence, the pushout is trivial.

Definition 1.5.5. If both i_1 and i_2 are injective, then the pushout G is called the *free product of G_1 and G_2 with amalgamated subgroup or amalgam G_0*; G is a *free product with amalgamation*. In this case we mostly consider G_0 as a subgroup of G_1 and G_2 and i_1, i_2 as inclusion. In this sense we may write $G_1 *_{G_0} G_2$. If we consider G_0 as a subgroup of G_1 and $H_0 = i_2(G_0)$ as a subgroup of G_2, then we also write $G_1 *_{G_0=H_0} G_2$; i_2 then is an isomorphism between G_0 and H_0.

Examples 1.5.6. 1. The group $G_1 * G_2$ is the free product of G_1 and G_2 with trivial amalgam, that is, $G_0 = \{1\}$.
2. Let $G = SL(2,\mathbb{Z}) = \langle A, B \mid A^2B^{-3} = A^4 = 1\rangle$. Then $G = G_1 *_{G_0} G_2$ with $G_1 = \langle A \mid A^4 = 1\rangle$, $G_2 = \langle B \mid B^6 = 1\rangle$ and $G_0 = \langle A^2\rangle = \langle B^3\rangle \cong \mathbb{Z}_2$.
3. Let $G = \langle a_1, b_1, a_2, b_2, \ldots, a_g, b_g \mid \prod_{i=1}^{g}[a_i, b_i] = 1\rangle, g \geq 2$. Then $G = G_1 *_{G_0} G_2$ with $G_1 = \langle a_1, b_1 \mid \rangle$, $G_2 = \langle a_2, b_2, a_3, b_3, \ldots, a_g, b_g \mid \rangle$ and $A_0 = \langle [a_1, b_1]\rangle = \langle \prod_{i=2}^{g}[a_i, b_i]\rangle \cong \mathbb{Z}$.
4. Let $G = \langle a_1, a_2, \ldots, a_g \mid a_1^2a_2^2 \cdots a_g^2 = 1\rangle, g \geq 2$. Then $G = G_1 *_{G_0} G_2$ with $G_1 = \langle a_1 \mid \rangle$, $G_2 = \langle a_2, a_3, \ldots, a_g \mid \rangle$ and $G_0 = \langle a_1^2\rangle = \langle a_2^2a_3^2 \cdots a_g^2\rangle \cong \mathbb{Z}$.
5. More generally, let $G = \langle a_1, a_2, \ldots, a_p, b_1, b_2, \ldots, b_q \mid WV = 1\rangle$ where $1 \leq p, 1 \leq q$, $1 \neq W = W(a_1, a_2, \ldots, a_p)$ and $1 \neq V(b_1, b_2, \ldots, b_q)$. Then $G = G_1 *_{G_0} G_2$ with $G_1 = \langle a_1, a_2, \ldots, a_p \mid \rangle$, $G_2 = \langle b_1, b_2, \ldots, b_q \mid \rangle$ and $G_0 = \langle W\rangle = \langle V^{-1}\rangle \cong \mathbb{Z}$.
   Then G is called a *cyclically pinched one-relator group*.
   It is clear that G is a free group if and only if U is primitive in $G_1 = \langle a_1, a_2, \ldots, a_p \mid \rangle$ or V is primitive in $G_2 = \langle b_1, b_2, \ldots, b_q \mid \rangle$.
   We remark that if W or V is not a proper power then G has a faithful representation in $PSL(2,\mathbb{R})$ (see Chapter 3 or [83] and [84]).

Remark 1.5.7. We may extend the definition to arbitrary many factors G_α, $\alpha \in I$. If we consider G_0 as a subgroup of all G_α, then we get the free product $G = *_{G_0} G_\alpha$ of the G_α with amalgam G_0. If $I = \{1, 2, \ldots, n\}$, $n \geq 2$, then we may G consider as $G = G_1 *_{G_0} G_2'$ with $G' = G_2 *_{G_0} G_3 *_{G_0} * \cdots *_{G_0} G_n$.

Hence, in the next section we write A instead of G_1, B instead of G_2 and C instead of G_0. Also we write j_A and j_B instead of j_1 and j_2, respectively. We consider also C as a subgroup of A and B.

Hence, let $G = A *_C B$. Let S and T be a complete set of representatives for the left cosets of C in A and B, respectively. Here we represent C by 1.

Theorem 1.5.8 (Normal form). *Let $G = A *_C B$. Then*
1. *j_A and j_B are injective;*
2. *$j_A(A) \cap j_B(B) = j_A(C) = j_B(C)$; and*
3. *If we consider j_A and j_B as inclusions, then we may write each element of G uniquely as $u_1u_2 \cdots u_nc$, where $n \geq 0$, $c \in C$ and the $u_1, u_2, \ldots, u_n$ alternately from $S\setminus\{1\}$ and $T\setminus\{1\}$.*

Remark 1.5.9. By definition $j_A = j_B$ on C. If we, as usual, consider j_A and j_B as inclusion, then we may write 2. as $C = A \cap B$ (recall, we consider i_A and i_B as inclusions and C as a subgroup of A and B).

Proof. We prove 1. and 3. the same time; 2. then follows directly from 3.

As in the case of free products we denote $j_A(a)$ by $\overline{a}$ and $j_B(b)$ by $\overline{b}$. It is enough to show:

If $g \in G = A *_C B$, then there exist uniquely determined $u_1, u_2, \ldots, u_n, c$ with $n \geq 0$, $c \in C$ and $u_1, u_2, \ldots, u_n$ alternately from $S \setminus \{1\}$ and $T \setminus \{1\}$ such that $G = \overline{u}_1\overline{u}_2 \cdots \overline{u}_n\overline{c}$. This we see as follows. The element g can be written as $\overline{g}_1\overline{g}_2 \cdots \overline{g}_k$ where $g_i \in A \cup B$. If g_i and g_{i+1} are both in A or B then $g = \overline{g}_1\overline{g}_2 \cdots \overline{g}_{i-1}\overline{h}\overline{g}_{i+2} \cdots \overline{g}_k$ with $h = g_ig_{i+1}$. Therefore we may assume that $g = \overline{c}$, $c \in C$, or $g = \overline{g}_1\overline{g}_2 \cdots \overline{g}_n$ where the g_i are alternately from $A \setminus C$ or $B \setminus C$. We have $\overline{g}_1 = \overline{u}_1\overline{c}_1$, $c_1 \in C$, $u_1 \in S \cup T$, and $u_1 \neq 1$ because $g_1 \notin C$.

If $n \geq 2$, then further $\overline{g}_1\overline{g}_2 = \overline{u}_1\overline{c}_1\overline{g}_2 = \overline{u}_1\overline{u}_2\overline{c}_2$, $c_2 \in C$, $u_2 \in S \cup T$, we have $u_2 \neq 1$ because $c_1g_2 \notin C$, and u_1 and u_2 are from different factors since g_1 and g_2 are, and so on.

Hence, we may write g as $g = \overline{u}_1\overline{u}_2 \cdots \overline{u}_n\overline{c}$ with $n \geq 0$, $c \in C$, $u_1, u_2, \ldots, u_n$ alternately from $S \setminus \{1\}$ and $T \setminus \{1\}$. The uniqueness we prove analogously as in the case of free products. Again, since we write functional characters from the left, it makes sense to change the representation. We take S^{-1} and T^{-1} as a complete set of representatives for the right cosets of C in A and B, respectively. Let $g^{-1} = \overline{u}_1\overline{u}_2 \cdots \overline{u}_n\overline{c}$ with $n \geq 0$, $c \in C$, $u_1, u_2, \ldots, u_n$ alternately from $S \setminus \{1\}$ and $T \setminus \{1\}$ as above. Then we may write $g = \overline{d}\overline{v}_n\overline{v}_{n-1} \cdots \overline{v}_1$ with $n \geq 0$, $d = c^{-1} \in C$, $v_1 = u_i^{-1}$. Let X be the set of all sequences $(d, v_n, v_{n-1}, \ldots, v_1)$ with $n \geq 0$, $g \in C$ and $v_n, v_{n-1}, \ldots, v_1$ alternately from $S^{-1} \setminus \{1\}$ and $T^{-1} \setminus \{1\}$, X contains certainly the sequence (1). We define a homomorphism φ from G to the group $\operatorname{Per}(X)$ of the permutations of X such that $\varphi(g)(1) = (d, v_n, v_{n-1}, \ldots, v_1)$ if $g = \overline{d}\overline{v}_n\overline{v}_{n-1} \cdots \overline{v}_1$. This proves the uniqueness of the representation of g.

For the construction of φ it is enough (by the definition of free products with amalgamation) to define φ on $A \cup B$ in such a way that the definition for A and B coincides on C. For $a \in A$ we define $\varphi(a)$ by

$$(d, v_n, \ldots, v_1) \mapsto \begin{cases} (f, v, v_n, v_{n-1}, \ldots, v_1) & \text{if } v_n \notin A, ad = fv \text{ with } v \in S^{-1} \setminus \{1\}, f \in C, \\ (f, v_n, v_{n-1}, \ldots, v_1) & \text{if } v_n \notin A, ad = f \in C \text{ (that is, if } a \in C), \\ (f, v, v_{n-1}, \ldots, v_1) & \text{if } v_n \in A, adv_n = fv \text{ with } v \in S^{-1} \setminus \{1\}, f \in C, \\ (f, v_{n-1}, v_{n-2} \ldots, v_1) & \text{if } v_n \in A, adv_n = f \text{ with } f \in C. \end{cases}$$

Then $\varphi(a^{-1})$ is inverse to $\varphi(a)$ and $\varphi(aa') = \varphi(a) \circ \varphi(a')$. Analogously we define $\varphi(b)$ for $b \in B$. This is well defined because the definition coincide for A and B on C. With this we get $\varphi(g)(1) = (d, v_n, v_{n-1}, \ldots, v_1)$ if $g = \overline{d}\overline{v}_n\overline{v}_{n-1} \cdots \overline{v}_1$. □

Corollary 1.5.10. *Let $G = A *_C B$ with j_A, j_B considered as inclusions. Then:*

1. *Each $w \in G \setminus C$ can be written as $g_1g_2 \cdots g_n$ with $n \geq 1$ and the g_i alternately from $A \setminus C$ and $B \setminus C$.*

2. *If we may also write w as $h_1h_2 \cdots h_m$ with $m \geq 0$ and the h_i alternately from $A \setminus C$ and $B \setminus C$, then $m = n$ and $h_1 \in g_1C$, $h_n \in Cg_n$ and $h_i \in Cg_iC$ for all $1 < i < n$.*
3. *If $n \geq 2$ then $w \notin A \cup B$.*
4. *Such a product as in 1. is not contained in C.*

Proof. From the proof of Theorem 1.5.8 we automatically have 1. If w is represented as in 1., then we have for its normal form $u_1u_2 \cdots u_nc$ that $u_1 \in g_1C$, $u_n \in Cg_n$ and $u_i \in Cg_iC$ for all $1 < i < n$ (recall that $g_1g_2 = u_1c_1g_2 = u_1u_3c$ and so on). Now 2. follows from Theorem 1.5.8. The statements 3. and 4. are obvious. □

Remark 1.5.11. There are some straightforward consequences of Corollary 1.5.10. Let $G = A *_C B$ (with j_A, j_B considered as inclusions). Let $w = g_1g_2 \cdots g_n$, $n \geq 1$, the g_i alternately from $A \setminus C$ and $B \setminus C$. We call the representation a *reduced form* of w.

The number n, which depends only on w, is called the *length* $|w|$ of w. We have $|w| = 1$ if and only if $w \in (A \setminus C) \cup (B \setminus C)$. An element $c \in C$ has length 0, we take c as the reduced form for $c \in C$. Any element $g \in G$ is conjugate to a *cyclically reduced element*, that is, an element of length ≤ 1 or an element of the reduced form $g_1g_2 \cdots g_n$, $n \geq 2$, with g_1 and g_n from different factors. If $w \in G$ has finite order, then necessarily $w \in g(A \cup B)g^{-1}$ for some $g \in G$. A finite subgroup of G is conjugate to a subgroup of A or B.

Corollary 1.5.12. *Let $A < G$, $B < G$ and $C = A \cap B$. We have $G = A *_C B$ if and only if each element of $G \setminus C$ can be written as a product $g_1g_2 \cdots g_n$ with the g_i alternately from $A \setminus C$ and $B \setminus C$ and if no such product is equal 1.*

Proof. We note that $A *_C B$ has this property. Now, assume that G has this property. The inclusions $A \hookrightarrow G$, $B \hookrightarrow G$ give a homomorphism $\varphi: A *_C B \to G$, and φ is necessarily a bijection. □

Theorem 1.5.13. *Let $G = A *_C B$, $A_1 < A$, $B_1 < B$ with $C_1 = A_1 \cap C = B_1 \cap C$.*
Then $\langle A_1, B_1 \rangle = A_1 *_{C_1} B_1$ and $\langle A_1, B_1 \rangle \cap A = A_1$, $\langle A_1, B_1 \rangle \cap B = B_1$.

Proof. The first part follows directly from Corollary 1.5.12, applied to $\langle A_1, B_1 \rangle$. The second part follows from Corollary 1.5.10.3, because a product of elements, alternately from $A_1 \setminus C_1$ and $B_1 \setminus C_1$ cannot be in A if its length in G is ≥ 2. □

Theorem 1.5.14. *Let $G = A *_C B$. If G and C are finitely generated then also A and B.*

Proof. We write each element of G as a product of finitely many elements from $A \cup B$. Let A_1 be the subgroup of A, which is generated by C and the finitely many elements from A which occur in the chosen expressions for the generators of G; analogously we define B_1. Then $G = \langle A_1, B_1 \rangle$ by construction. By Theorem 1.5.13 we have $A = A_1$, $B = B_1$, which gives the result. □

Establishing a unique normal form for elements of a group amounts to solving the word problem, provided that normal forms can be effectively calculated. For $G = A *_C B$

the crucial problem in calculating normal forms is to determine when words of A and B define elements of C. We refer to these as the *membership problems* for C in A and B. Hence, we have the following theorem.

Theorem 1.5.15. *Let $G = A *_C B$. Suppose that*
(a) *A and B have solvable word problem; and*
(b) *The membership problems for C in A and B are solvable.*

Then G has a solvable word problem.

The normal form also provides a complete description of conjugacy in free products with amalgamation. Recall that each element of $G = A *_C B$ is conjugate to a cyclically reduced element. We now straightforwardly get the following.

Theorem 1.5.16. *Let $G = A *_C B$ and $u, v \in G$.*
1. *If u and v are conjugate cyclically reduced elements of G of length at least two then some cyclic rearrangement of u and v are conjugate by an element of the amalgamated subgroup.*
2. *If u and v are conjugate elements of length ≤ 1, then either u and v lie in the same factor and are conjugate there or there is a sequence of elements $c_1, c_2, \ldots, c_n$ of C such that, within the factors A and B, the element u is conjugate to c_{i+1} and c_n is conjugate to v.*

Although Theorem 1.5.16 gives an essentially complete specification of when elements are conjugate in G, in practice it is often not easy to apply Theorem 1.5.16 to give an algorithmic solution of the conjugacy problem.

Theorem 1.5.17. *Let $G = A *_C B$. Then $Z(G) = C \cap Z(A) \cap Z(B)$ where $Z(H)$ denotes the center of a group H.*

We leave the proof as an exercise. Very much related to the word problem is the question of residual finiteness. Recall that a residually finite group has a solvable word problem. As an easy consequence of Corollary 1.4.13 we have the following. We just have to modify van der Waerden's method used in the proof of Theorem 1.5.8 analogously to Theorem 1.4.12 and Corollary 1.4.13.

Theorem 1.5.18. *Let $B = A *_C B$ with C finite and A and B residually finite. Then G is residually finite.*

In general it is very complicated to decide whether or not $G = A *_C B$ is residually finite. Even if A and B are residually finite and $C \cong \mathbb{Z}$ then $G = A *_C B$ need not be residually finite (see [129]). However G. Baumslag [13] has shown that $G = A *_C B$ with A and B free and $C \cong \mathbb{Z}$ is residually finite.

There are many results in the literature with different properties for A, B and C, sometimes technically very complicated, under which $G = A *_C B$ is residually finite.

1.5.2 The Nielsen method in free products with amalgamation

We now describe the Nielsen method in free products with amalgamations as developed by H. Zieschang [278] and refined a bit by G. Rosenberger [219] and [220] and R. N. Kalia and G. Rosenberger [141]. A further refinement of the technique was given by D. Collins and H. Zieschang in [59]. Again, we restrict ourself to the free product of two groups with an amalgamated subgroups, although the method works more generally.

For notational reasons we now write $G = H_1 *_A H_2$ to denote the free product of groups H_1 and H_2 with the amalgamated subgroup $A = H_1 \cap H_2$. We always assume that the decomposition is non-trivial, that is, $H_1 \neq A \neq H_2$. If $A = \{1\}$ then G is just the free product $G = H_1 * H_2$. We choose in each H_i, $i = 1, 2$, a system L_i of left coset representatives of A in H_i, normalized by taking 1 to represent A. Each $x \in G$ has a unique representation $X = h_1 \cdots h_n a$ with $a \in A$, $1 \neq h_j \in L_1 \cup L_2$ and $h_{j+1} \notin L_i$ if $h_j \in L_i$. Then length $|x|$ of x is then defined to be n and G is then (partially) ordered by length.

In order to obtain results analogous to those in free groups and free products we need a finer pre-ordering. Take the inverses L_i^{-1} of the left coset representatives as a system of right coset representatives. Then each $x \in G$ has a unique representation

$$x = \ell_1 \ell_2 \cdots \ell_m k_x r_m r_{m-1} \cdots r_1$$

with $m \geq 0$, $k_x \in H_1 \cup H_2$, $1 \neq \ell_j \in L_1 \cup L_2$, $1 \neq r_j \in L_1^{-1} \cup L_2^{-1}$ and $\ell_{j+1} \notin L_i$ if $\ell_j \in L_i$, $r_{j+1} \notin L_i^{-1}$ if $r_j \in L_j^{-1}$.

Furthermore, if $k_x \in A$ then ℓ_m and r_m belong to different H_i (if $m \geq 1$), and if $k_x \in H_i \setminus A$ then $\ell_m \notin H_i$, $r_m \notin H_i$ (if $m \geq 1$). We then have $|x| = 2m$ if $k_x \in A$ and $|x| = 2m + 1$ if $k_x \notin A$. We call $\ell_1 \ell_2 \cdots \ell_m$ the *leading half*, $r_m r_{m-1} \cdots r_1$ the *rear half* and k_x the *kernel* of x. One advantage of this symmetric normal form is that in forming products, cancellations can usually be reduced to free cancellations.

We now introduce an ordering $\leq$ on G. We assume that for each H_i, the system L_i of left coset representatives has a strict total order.

We remark that for our applications we may assume that the groups are countable. This is no restriction if one considers a given finitely generated subgroup of G or a given finite system in G. If G is countable then just enumerate the system L_i and order it correspondingly. Let the elements of L_1 precede those of L_2. Then we order for each m the product $\ell_1 \ell_2 \cdots \ell_m$ of the left coset representatives (where $1 \neq \ell_j \in L_1 \cup L_2$ and $\ell_{j+1} \notin L_i$ if $\ell_j \in L_i$), first by length and second lexicographically. Hence, if $\ell_1 \ell_2 \cdots \ell_m < \ell_1' \ell_2' \cdots \ell_m'$ then for any permitted ℓ_{m+1}, ℓ_{m+1}' we have $\ell_1 \ell_2 \cdots \ell_{m+1} < \ell_1' \ell_2' \cdots \ell_{m+1}'$.

If G is countable then further each product $\ell_1 \ell_2 \cdots \ell_m$ has only finitely many predecessors of the form $\ell_1 \ell_2 \cdots \ell_{m-1} \ell_m'$ (where $\ell_m' \in L_i$ if $\ell_m \in L_i$). Without loss of generality we assume that G is countable if we consider a given finitely generated subgroup of G or a given finite system in G. We define an ordering on the products of right coset representatives in the L_i^{-1} by taking inverses.

We now extend this ordering to the set of pairs $\{g, g^{-1}\}$, $g \in G$, where the notion is chosen so that the leading half of g precedes that of g^{-1} with respect to ordering $<$. Then we set $\{g, g^{-1}\} \prec \{h, h^{-1}\}$ if either $|g| < |h|$ or $|g| = |h|$ and the leading half of g strictly precedes that of h, or $|g| = |h|$, the leading halves coincide, and the leading half of g^{-1} precedes that of h^{-1}. Thus if $\{g, g^{-1}\} \prec \{h, h^{-1}\}$ and $\{h, h^{-1}\} \prec \{g, g^{-1}\}$ then g and h differ only in the kernel. Since this can occur with $g \neq h$ then $<$ is only a pre-order.

For kernels k, k' we could get a differentiation by the situation in A, $H_1 \setminus A$ or $H_2 \setminus A$ and the left coset representatives. But this we do not need for our applications. Basically, the respective distinction in the case of free products was not necessary, it shortened the proofs a bit in the free product case.

We now shorten finite systems with respect to $\prec$ with the help of Nielsen transformations. A finite system $\{g_1, g_2, \dots, g_m\}$ in G is said to be *Nielsen reduced* or *minimal* with respect to $\prec$ if $\{g_1, g_2, \dots, g_m\}$ cannot be carried by a Nielsen transformation into a system $\{h_1, h_2, \dots, h_m\}$ with $h_i = 1$ for some $i \in \{1, 2, \dots, m\}$ or there is no system Nielsen equivalent to $\{g_1, g_2, \dots, g_m\}$ which is shorter. If G is countable then every finite system can be carried by a Nielsen transformation into a minimal system.

In general, as already mentioned, for a given finite system, a suitable order can always be chosen such that this finite system can be carried by a Nielsen transformation into a minimal system.

We now make the following conventions on notation. We write $u_1 u_2 \cdots u_q \equiv v_1 v_2 \cdots v_n$ to stand for the equality together with the fact that $|v_1 v_2 \cdots v_n| = |v_1| + |v_2| + \cdots + |v_n|$. If x is given in its symmetric normal form

$$x = \ell_1 \ell_2 \cdots \ell_m k_x r_m r_{m-1} \cdots r_1$$

then we write $x \equiv p_x k_x r_x$ where $p_x = \ell_1 \ell_2 \cdots \ell_m$ is the leading half and $q_x = r_m r_{m-1} \cdots r_1$ is the rear half. Furthermore, $x = p_x k_x r_x$ refers to the symmetric normal form as above. The Nielsen reduction method in G now refers to shorter systems and the resulting investigation of minimal systems.

This produces the following result.

Theorem 1.5.19. *Let $G = H_1 *_A H_2$, $H_1 \neq A \neq H_2$. If $\{x_1, x_2, \dots, x_m\}$ is a finite system of elements in G then there is a Nielsen transformation from $\{x_1, x_2, \dots, x_m\}$ to a system $\{y_1, y_2, \dots, y_m\}$ for which one of the following cases holds:*

(i) *$y_i = 1$ for some $i \in \{1, 2, \dots, m\}$.*

(ii) *Each $w \in \langle y_1, y_2, \dots, y_m \rangle$ can be written as $w = \prod_{i=1}^{q} y_{v_i}^{\epsilon_i}$, $\epsilon_i = \pm 1$, $\epsilon_i = \epsilon_{i+1}$ if $v_i = v_{i+1}$ with $|y_{v_i}| \leq |w|$ for $i = 1, 2, \dots, q$.*

(iii) *There is a product $a = \prod_{i=1}^{q} y_{v_i}^{\epsilon_i}$, $a \neq 1$ with $y_{v_i} \in A$ $(i = 1, 2, \dots, q)$ and in one of the factors H_j there is an element $x \notin A$ with $x^{-1} a x \in A$.*

(iv) *There is a $g \in G$ such that for some $i \in \{1, 2, \dots, m\}$ we have $y_i \notin gAg^{-1}$, but for a suitable $k \in \mathbb{N}$ we have $y_1^k \in gAg^{-1}$.*

(v) *Of the y_i there are $p \geq 1$ contained in a subgroup of G conjugate to H_1 or H_2 and a certain product of them is conjugate to a non-trivial element of A.*

The Nielsen transformation can be chosen so that $\{y_1, y_2, \ldots, y_m\}$ is shorter than $\{x_1, x_2, \ldots, x_m\}$ or the lengths of the elements of $\{x_1, x_2, \ldots, x_m\}$ are preserved.

Furthermore, if $\{x_1, x_2, \ldots, x_m\}$ is a generating system of G then in case (v) we find $p \geq 2$ for in this case conjugations determine a Nielsen transformation. If we assume that $\{y_1, y_2, \ldots, y_m\}$ is Nielsen reduced then we easily may refine case (iii) by considering parts of reduced words to the following: There is a product $y_v^\epsilon(\prod_{i=1}^q y_{v_i}^{\epsilon_i})y_v^{-\epsilon}$, $\epsilon, \epsilon_i = \pm 1$, $\epsilon_i = \epsilon_{i+1}$ if $v_i = v_{i+1}$, $a := \prod_{i=1}^q y_{v_i}^{\epsilon_i} \neq 1$, $y_{v_i} \in A$ $(i = 1, 2, \ldots, q)$, $y_v \notin A$, $y_v^\epsilon = \ell_1\ell_2 \cdots \ell_n k r_n r_{n-1} \cdots r_1$ in symmetric normal form, such that $r_1 a r_1^{-1} \in A$ if $n \geq 1$ and $kak^{-1} \in A$ if $n = 0$.

If we are interested in the combinatorial structure and description of $\langle x_1, x_2, \ldots, x_m\rangle$ in terms of generators and relations then we may refine case (v), possibly after a suitable conjugation, to the following:

Of the y_i there are $p \geq 2$ contained in H_1 or H_2, at least one of them is not in A, and a certain product of them is a non-trivial element of A. If $G = H_1 *_A H_2$ with A malnormal in G then case (iii) and (iv) do not occur.

This gives the straightforward consequence.

Corollary 1.5.20. *Let $G = H_1 *_A H_2$, $H_1 \neq \{A\} \neq H_2$ and A be malnormal in G. Let $H = \langle x, y\rangle$ a two-generator subgroup. Then H is contained in a conjugate of H_1 or H_2, or H is a free product of cyclics.*

The proof starts with a Nielsen reduced set $\{x_1, x_2, \ldots, x_m\}$ and follows the lines described in the case of a free product. The main tools in the proof are the following two lemmas which handle the obstruction we have in general free products with amalgamation. The rest is straightforward, analogously to the case of free products.

Lemma 1.5.21. *Suppose $x, y, z \in G$, $1 \neq x$, $1 \neq y$, $1 \neq z$ with $|xy| \geq |x|, |y|$ and $|yz| \geq |y|, |z|$. Suppose further that neither xy precedes x nor yz precedes z. Then the following holds.*
1. *If $|xyz| \leq |x| - |y| + |z|$ then y is conjugate to an element of H_1 or H_2.*
2. *If $|x|, |z| > |y|$ and $|xyz| < |x| - |y| + |z|$ then y is conjugate to an element of A.*
3. *If $X \subset G$ is a finite Nielsen reduced system, $x, y, z \in X \cup X^{-1}$ and $|xyz| < |x| - |y| + |z|$, then y is conjugate to an element of A or $x = y = z$.*

Proof. Let $x \equiv p_x k_x q_x$, $y \equiv p_y k_y q_y$, $z \equiv p_z k_z q_z$ be the symmetric normal forms of x, y, z.
1. Let $|xyz| \leq |x| - |y| + |z|$. If $y \in H_1$ or H_2 then there is nothing to prove. Hence suppose $p_y \neq 1$, $q_y \neq 1$. Since $|xyz| \leq |x| - |y| + |z|$, $|xy| \geq |x|, |y|$ and $|yz| \geq |y|, |z|$ we must have $q_x \equiv r_x p_y^{-1}$ and $p_z \equiv q_y^{-1}\ell_y$. Since neither xy precedes x nor yz precedes z we must have $p_y^{-1} \leq q_y$ and $q_y^{-1} \leq p_y$, that is, $p_y = q_y^{-1}$.
2. Let $|x|, |z| > |y|$ and $|xyz| < |x| - |y| + |z|$. If $y \in A$ there is nothing to prove. Hence suppose $|y| \geq 1$. We know from part 1. that y is conjugate to an element of H_1 or H_2 and therefore $k_y \notin A$.

   Now we must have $q_x \equiv r_x r p_y^{-1}$, $r \in L_i^{-1}$, $r \neq 1$, and $p_z \equiv q_y^{-1}\ell\ell_y$, $\ell \in L_i$, $\ell \neq 1$, $p_y = q_y^{-1}$ and $\mathrm{r}k_y\ell \in A$ $(i = 1$ or $2)$. Let r' be the right coset representative of $\mathrm{r}k_y$ and ℓ' the left coset representative of $k_y\ell$. Since xy does not precede x we must have $r' \geq r$.

Analogously $\ell' \geq \ell$. Hence $r' = \ell^{-1}$, $\ell' = r^{-1}$ and $r^{-1} = \ell$ since $\mathrm{rk}_y \ell \in A$. Therefore k_y is conjugate to an element of A and hence also y.

3. Let $x, y, z \in X \cup X^{-1}$ with $X \subset G$ a finite Nielsen reduced system and suppose $|xyz| < |x| - |y| + |z|$. If $y \in A$ then there is nothing to prove. Hence, suppose that $y \notin A$. From part 1. we have $k_y \notin A$ and $q_x \equiv r_x p_y^{-1}$, $p_z \equiv q_y^{-1}\ell_y$. If $r_x \neq 1$, $\ell_y \neq 1$ then the statement follows from part 2. Now suppose $r_x = 1$ or $\ell_y = 1$. Then $r_x = \ell_y = 1$ since X is Nielsen reduced. Since $k_y \notin A$ we have $|xyz| < |x|, |y|, |z|$. This implies $x = y = z$ since x is Nielsen reduced. □

We note again that if $|g^n| < |g|$ for $g \in G$, $n \in \mathbb{N}$, then g is conjugate to an element of H_1 or H_2 and g^n is conjugate to an element of A.

Lemma 1.5.22. *Let $x, y, z \in G$, $|x| \geq 1$, $|z| \geq 1$, $1 \neq y \in A$ with $|xy| \geq |x|$ and $|yz| \geq |z|$. Suppose further that neither xy precedes x nor yz precedes z. If $|xyz| < |x| - |y| + |z| - 1 = |x| + |z| - 1$ then there exists an $a \in H_1 \cup H_2$, $a \notin A$ with $aya^{-1} \in A$.*

Proof. We use the notations as in Lemma 1.5.21 and its proof. Now $y = k_y \in A$ and $\mathrm{rk}_y \ell = rk_y r^{-1} \in A$. □

We have more straightforward consequences.

Corollary 1.5.23 (See [198]). *Let $G = H_1 *_A H_2$, $H_1 \neq A \neq H_2$, and $H < G$ with $gHg^{-1} \cap A = \{1\}$ for all $g \in G$. Then $H = F * (*_{j \in J} H_j)$, where F is a free group and each H_j is conjugate to a subgroup of H_1 or H_2.*

We call a group G *n-free* if each subgroup $H < G$ with $\mathrm{rk}(H) \leq n$ is free.

Corollary 1.5.24. *Suppose that $G = H_1 *_A H_2$ with $H_1 \neq A \neq H_2$. Suppose further that A is malnormal in both H_1 and H_2 and that both H_1 and H_2 are 3-free. Then G is 3-free. Especially, if $x, y, z \in G$ with $x^2y^2z^2 = 1$ then $\langle x, y, z\rangle$ is cyclic.*

Proof. Let $x_1, x_2, x_3 \in G$ and $H = \langle x_1, x_2, x_3\rangle$. Then without loss of generality we may assume that H is free of rank ≤ 3 or $x_1, x_2 \in H_1$ after a suitable conjugation if necessary. We show that $\langle H_1, x_3\rangle$ is 3-free and hence H is free.

Consider $K = \langle H_1, x_3\rangle$. The element $x_3 \in G$ has a unique representation $x_3 = h_1 h_2 \cdots h_m b$ with $m \geq 0$, $b \in A$, and the h_i are left coset representatives and lie alternately in distinct factors of G and are non-trivial. If $x_3 \in H_1$ then $H \subset H_1$ and H is free since H_1 is 3-free. Suppose then that $x_3 \notin H_1$. Then in particular $m \geq 1$. Since $H_1 \subset K$ and $x_3 \notin H_1$ we may assume that $b = 1$, $h_1 \in H_2 \setminus A$ and $h_m \in H_2 \setminus A$. But then $|r| \geq |x_3| \geq 1$ for every freely reduced word $r \in K$ in which x_3 occurs. It follows that K is 3-free and hence H is free. Therefore G is 3-free.

The additional statement is clear. The equation $x^2y^2z^2 = 1$ is equivalent to the equation $x^2[y, z] = 1$, and in a non-Abelian free group an equation $x^2[y, z] = 1$ only holds in cyclic subgroups. □

The malnormality is essential in G. For instance if $G = \langle x, y, z \mid x^2y^2z^2 = 1\rangle$ then the equation holds trivially and $\langle x, y, z\rangle$ is non-Abelian. We have $G = H_1 *_A H_2$ with $H_1 = \langle x, y \mid \rangle$, $H_2 = \langle z \mid \rangle$ and $A = \langle x^2y^2\rangle = \langle z^{-2}\rangle$. But A is not malnormal in H_2. Especially we have the following corollary.

Corollary 1.5.25. *Let $G = \langle a_1, a_2, \ldots, a_n, b_1, b_2, \ldots, b_m \mid UV = 1\rangle$, $n \geq 2$, $m \geq 2$, $U = U(a_1, a_2, \ldots, a_n) \neq 1$ cyclically reduced and not a proper power in $\langle a_1, a_2, \ldots, a_n \mid \rangle$, $V = V(b_1, b_2, \ldots, b_m) \neq 1$ cyclically reduced and not a proper power in $\langle b_1, b_2, \ldots, b_m \mid \rangle$. Then each subgroup H of G with rank* $\mathrm{rk}(H) \leq 3$ *is free of rank* $\mathrm{rk}(H)$.

Proof. Observe that $\langle U\rangle$ is malnormal in $\langle a_1, a_2, \ldots, a_n \mid \rangle$ because U is not a proper power in $\langle a_1, a_2, \ldots, a_n \mid \rangle$. Analogously $\langle V\rangle$ is malnormal in $\langle b_1, b_2, \ldots, b_m\rangle$. □

Remarks 1.5.26. 1. Let $G = H_1 *_A H_2$ with $H_1 \neq A \neq H_2$, and let $H < G$. Then it is possible that H is not decomposable as a non-trivial free product with amalgamation.

Example 1.5.27. Let

$$\begin{aligned} G &= \langle s_1, s_2, s_3, s_4 \mid s_1^2 = s_2^2 = s_3^2 = s_4^2 = s_1s_2s_3s_4 = 1\rangle \\ &= H_1 *_A H_2 \end{aligned}$$

with $H_1 = \langle s_1, s_2 \mid s_1^2 = s_2^2 = 1\rangle$, $H_2 = \langle s_3, s_4 \mid s_3^2 = s_4^2 = 1\rangle$ and $A = \langle s_1s_2\rangle = \langle s_4s_3\rangle$. Let $x = s_1s_2$, $y = s_3s_1$ and $H = \langle s_1s_2, s_3s_1\rangle$. Then H is free Abelian of rank 2 and, hence, not decomposable into a non-trivial free product with amalgamation. A first general result was given by Karrass and Solidar [151].

Let $G = H_1 *_A H_2$, $H_1 \neq A \neq H_2$, and $H < G$. Then H is obtained by two constructions from the intersection of H and certain conjugates of H_1, H_2 and A. The constructions are those of a free product, a special kind of a graph of groups, and of an HNN group. We consider HNN groups and graphs of groups later in this book. There are several refinements, and extensions for special cases, in the literature.
The above example also shows that non-cyclic Abelian subgroups are not necessarily in a subgroup conjugate to H_1 or H_2.

2. Let $G = H_1 *_A H_2$, $H_1 \neq A \neq H_2$. From Grushko's theorem for free products we easily get the following theorem.

Theorem 1.5.28. *Let $A \triangleleft G$ be a normal subgroup of G. If G is finitely generated and X a finite generating system of G then there exists a Nielsen transformation from X to a system $Y \subset A \cup B$.*

This does not hold in general.

Example 1.5.29. Let

$$G = \langle s_1, s_2, s_3, s_4 \mid s_1^2 = s_2^2 = s_3^2 = s_4^3 = s_1s_2s_3s_4\rangle = H_1 *_A H_2$$

with $H_1 = \langle s_1, s_2 \mid s_1^2 = s_2^2 = 1\rangle$, $H_2 = \langle s_3, s_4 \mid s_3^2 = s_4^3 = 1\rangle$ and $A = \langle s_1s_2\rangle = \langle s_3s_4\rangle$.

Let $x = s_1s_2, y = s_3s_1$ and $H = \langle x,y\rangle$. We have $[x,y] = (s_1s_2s_3)^2 = s_4^{-2} = s_4 \in H$. $s_4xy = s_4s_1s_2s_3s_1 = s_1 \in H$, $s_1x = s_2 \in H$, $ys_1 = s_3 \in H$, and hence, $H = G$. However, G cannot be generated by two elements from $H_1 \cup H_2$. This example also is a counter example against the obvious conjecture

$$\mathrm{rk}(G) \geq \mathrm{rk}(H_1) + \mathrm{rk}(H_2) - \mathrm{rk}(A).$$

In this example we have $\mathrm{rk}(G) = 2$, $\mathrm{rk}(H_1) = \mathrm{rk}(H_2) = 2$ and $\mathrm{rk}(A) = 1$, hence, $\mathrm{rk}(H_1) + \mathrm{rk}(H_2) - \mathrm{rk}(A) = 3$. Also, further general conjectures like $\mathrm{rk}(G) \geq \mathrm{rk}(H_1) + \mathrm{rk}(H_2) - \mathrm{rk}(A) - k$ for one from $G = H_1 *_A H_2$ independent constant k do not hold.

For each $m \in \mathbb{N}$ there exists a $G = H_1 *_A H_2$ with $\mathrm{rk}(H_1) \geq m$, $\mathrm{rk}(H_2) \geq m$ and $\mathrm{rk}(A) = \mathrm{rk}(G) = 2$ (see [262]).

Hence, in general there is no non-trivial rank formula for free products with amalgamation. There are many activities to determine the rank for interesting special cases. We here just mention some results. Let $G = H_1 *_A H_2$ with $H_1 \neq A \neq H_2, A \neq \{1\}$.

(1) If A is malnormal in G then $\mathrm{rk}(G) \geq 3$ (see [263]). The lower bound is realized for instance for G with $H_1 = \langle s_1 s_2 \mid s_1^3 = s_2^3 = 1\rangle$, $H_2 = \langle s_3, s_4 \mid s_3^3 = s_4^3 = 1\rangle$ and $A = \langle s_1s_2\rangle = \langle s_3s_4\rangle$.
(2) If A is malnormal in G then $\mathrm{rk}(G) \geq \frac{1}{3}(\mathrm{rk}(H_1) + \mathrm{rk}(H_2) - 2\,\mathrm{rk}(A) + 5)$ (see [264]). In this paper Weidmann constructs for each $n \in \mathbb{N}$ an amalgamated product $G = H_1 *_A H_2$ with A malnormal in G and $A \cong \mathbb{Z}$ such that $\mathrm{rk}(G) \leq n + 2$, $\mathrm{rk}(H_1) \geq n + 1$ and $\mathrm{rk}(H_2) \geq n + 1$.
Bumagin [40] gave an example which realizes the above lower bound.
(3) If A is finite then $\mathrm{rk}(G) \geq \frac{1}{2^{\ell(A)+1}}(\mathrm{rk}(H_1) + \mathrm{rk}(H_2))$ where $\ell(A)$ is the length of the longest strictly ascending chain for non-trivial subgroups (see [266]).
(4) Let

$$G = \langle s_1, s_2, \ldots, s_m, a_1, a_2, \ldots, a_p \mid s_1^{\gamma_1} = s_2^{\gamma_2} = \cdots = s_m^{\gamma_m} = s_1s_2\cdots s_m P(a_1, a_2, \ldots, a_p) = 1\rangle$$

with $m \geq 0$, $p \geq 0$, all $\gamma_i \geq 2$ and

$$P(a_1, a_2, \ldots, a_p) = a_1^{\alpha_1}a_2^{\alpha_2}\cdots a_n^{\alpha_n}[a_{n+1}, a_{n+2}]\cdots[a_{p-1}, a_p],$$

$0 \leq n \leq p$, $p - n$ even, all $\alpha_i \geq 2$ such that
(i) $p \geq 2$ or
(ii) $p = 1$ and $m \geq 2$ or
(iii) $p = 0$ and $m \geq 4$ or
(iv) $p = 0$, $m = 3$ and $\frac{1}{\gamma_1} + \frac{1}{\gamma_2} + \frac{1}{\gamma_3} \leq 1$.
If $n = 0$ and $p - 2 + \sum_{i=1}^{m}(1 - \frac{1}{\alpha_i}) > 0$ then G is a co-compact Fuchsian group. With the help of Theorem 1.2.16, Theorem 1.4.41 and the results and methods in [211] we get the following.

Theorem 1.5.30. *Let G be as above. The rank* $\mathrm{rk}(G)$ *of G is*

(a) p *if* $m = 0$;
(b) $m - 2$ *if* $p = 0$, *m is even, and all* γ_i *equal to* 2 *except for one, which is odd; and*
(c) $p + m - 1$ *in all other cases.*

In fact, if $p = 0$ and $m = 3$ then G is a triangle group which certainly has rank 2. If $n = 0, p = 2$ and $m = 1$ then $G \cong \langle a, b \mid [a,b]^{\alpha} = 1\rangle$, $\alpha \geq 1$, and G has rank 2. In all the other cases (i), (ii), (iii), where $m \geq 2$ if $n = 0, p = 2$, then G has a natural non-trivial decomposition as a free product with amalgamation, and we may apply the Nielsen method in free products with amalgamation. For instance, if $p \geq 2$ and $m \geq 3$ then $G = H_1 *_A H_2$ with $H_1 = \langle s_1, s_2, \ldots, s_m \mid s_1^{\alpha_1} = s_2^{\alpha_2} = \cdots = s_m^{\alpha_m} = 1\rangle$, $H_2 = \langle a_1, a_2, \ldots, a_p \mid \rangle$ and $A = \langle s_1 s_2 \cdots s_m\rangle = \langle P(a_1, a_2, \ldots, a_p)\rangle$. In this case, $\operatorname{rk}(G) \leq p + m - 1$.
Certainly, $\overline{H} = \langle a_1, a_2, \ldots, a_p \mid P(a_1, a_2, \ldots, a_m) = 1\rangle$ has rank p.
If the group $\overline{H}_1 = \langle s_1, s_2, \ldots, s_m \mid s_1^{\alpha_1} = s_2^{\alpha_2} = \cdots = s_m^{\alpha_m} = s_1 s_2 \cdots s_m = 1\rangle$ has rank $m - 1$ then $\operatorname{rk}(G) \geq p + m - 1$ by Grushko's theorem.
The combinatorially long part is to calculate the rank of $\langle s_1, s_2, \ldots, s_m \mid s_1^{\alpha_1} = s_2^{\alpha_2} = \cdots = s_m^{\alpha_m} = s_1 s_2 \cdots s_m = 1\rangle$. This is done in [211]. Here the case $\tilde{G} = \langle s_1, s_2, \ldots, s_m \mid s_1^2 = s_2^2 = \cdots = s_{m-1}^2 = s_m^{2k+1} = s_1 s_2 \cdots s_m = 1\rangle$, $m \geq 4$, m even, $k \geq 1$ gives an exceptional situation. $\tilde{G}$ has rank $m - 2$ and is generated by $\{s_1 s_2, s_1 s_3, \ldots, s_1 s_{m-1}\}$ (see [211]).
Much more complicated than the rank problem is the problem to describe the generating systems of finitely generated free products G with amalgamation, even the minimal generating systems, that is, the generating systems $\{x_1, x_2, \ldots, x_m\}$ with $m = \operatorname{rk}(G)$. We can do this in part for the groups above in (4).

Theorem 1.5.31. *Let G be as above.*

1. *Let* $m \leq 1$ *and* $\{x_1, x_2, \ldots, x_p\}$ *be a (minimal) generating system for G.*
   (a) *If* $m = 1$ *then* $\{x_1, x_2, \ldots, x_p\}$ *is Nielsen equivalent to* $\{a_1, a_2, \ldots, a_p\}$.
   (b) *If* $m = 0$ *and* $p \geq 3$ *then* $\{x_1, x_2, \ldots, x_p\}$ *is Nielsen equivalent to a system*
   $$\{a_1, a_2, \ldots, a_{i-1}, a_i^{\beta_i}, a_{i+1}, a_{i+2}, \ldots, a_n, a_{n+1}, a_{n+2}, \ldots, a_p\}$$
   for some i $(1 \leq i \leq n)$ *with* $\gcd(\beta_i, \alpha_i) = 1$ *and* $1 \leq \beta_i \leq \frac{1}{2}\alpha_i$.
   (c) *If* $m = n = 0$ *and* $p = 2$ *then* $\{x_1, x_2\}$ *is Nielsen equivalent to* $\{a_1, a_2\}$.
   (d) *If* $m = 0$ *and* $n = p = 2$ *then* $\{x_1, x_2\}$ *is Nielsen equivalent to a system* $\{a_1^{\beta_1}, a_2^{\beta_2}\}$ *with* $\gcd(\beta_1, \beta_2) = \gcd(\beta_1, \alpha_1) = \gcd(\beta_2, \alpha_2) = 1$ *and* $1 \leq \beta_1 \leq \frac{1}{2}\beta_2\alpha_1$, $1 \leq \beta_2 \leq \frac{1}{2}\beta_1\alpha_2$.
2. *Let* $p \geq 2$, $m \geq 2$ *and* $\{x_1, x_2, \ldots, x_{p+m-1}\}$ *be a (minimal) generating system for G. Then* $\{x_1, x_2, \ldots, x_{p+m-1}\}$ *is Nielsen equivalent to a system*
   $$\{s_{\nu_1}^{\beta_1}, s_{\nu_2}^{\beta_2}, \ldots, s_{\nu_{m-1}}^{\beta_{m-1}}, a_1, a_2, \ldots, a_p\}$$
   with $\nu_i \in \{1, 2, \ldots, m\}$, $\nu_1 < \nu_2 < \cdots < \nu_{m-1}$ *(if* $m \geq 3$*) and* $1 \leq \beta_i < \frac{1}{2}\gamma_{\nu_i}$, $\gcd(\beta_i, \gamma_{\nu_i}) = 1$.
3. *Let* $p = 1$, $m \geq 2$ *and* $\{x_1, x_2, \ldots, x_m\}$ *be a (minimal) generating system for G. Suppose that one of the following hold:*

(a) *At least two* γ_i *are greater than* 2.
(b) *All* γ_i *are equal to* 2 *except for one which is even and greater than* 2.
(c) m *is odd, all* γ_i *are equal to* 2 *except for one which is odd and* α_1 *is even.*
(d) m *is even and all* γ_i *are equal to* 2.
Then $\{x_1, x_2, \ldots, x_m\}$ *is Nielsen equivalent to a system*

$$\{s_{v_1}^{\beta_1}, s_{v_2}^{\beta_2}, \ldots, s_{v_{m-1}}^{\beta_{m-1}}, a_1\}$$

with $v_i \in \{1, 2, \ldots, m\}$, $v_1 < v_2 < \cdots < v_{m-1}$ *(if* $m \geq 3$*) and* $1 \leq \beta_i \leq \frac{1}{2}\gamma_{v_i}$ *and* $\gcd(\beta_i, \gamma_{v_i})$.

4. *Let* $p = 0$, $m \geq 4$ *and* $\mathrm{rk}(G) = m - 1$. *Let* $\{x_1, x_2, \ldots, x_{m-1}\}$ *be a (minimal) generating system.*
Suppose that one of the following holds:
(a) *At least three* γ_i *are greater than* 2;
(b) *All* γ_i *are equal to* 2 *except for two which are both even and greater than* 2; *and*
(c) m *is even and all* γ_i *are equal to* 2 *except for two which are both odd.*
Then $\{x_1, x_2, \ldots, x_{m-1}\}$ *is Nielsen equivalent to a system*

$$\{s_{v_1}^{\beta_1}, s_{v_2}^{\beta_2}, \ldots, s_{v_{m-1}}^{\beta_{m-1}}\}$$

with $v_i \in \{1, 2, \ldots, m\}$, $v_1 < v_2 < \cdots < v_{m-1}$ *and* $1 \leq \beta_i \leq \frac{1}{2}\gamma_i$, $\gcd(\beta_i, \alpha_{v_i}) = 1$.

We remark that Lustig and Moriah [175] gave a complete classification of minimal generating systems in a very general class of Fuchsian groups. This class includes, for example, any G which has at least seven non-conjugate cyclic subgroups of order $\gamma_i \geq 3$.

For the proof we just use Theorem 1.2.16, Theorem 1.4.41 and a detailed use of the Nielsen cancellation method in free products with amalgamation. Details may be found in [227], [229] and [279].

We give the proof here for the case of orientable surface groups of genius $g \geq 2$.

Theorem 1.5.32. *Let* $G = \langle a_1, b_1, a_2, b_2, \ldots, a_g, b_g \mid \prod_{i=1}^{g}[a_i, b_i] = 1\rangle$, $g \geq 2$. *If* $\{x_1, x_2, \ldots, x_{2g}\}$ *is a (minimal) generating system, then* $\{x_1, x_2, \ldots, x_{2g}\}$ *is Nielsen equivalent to* $\{a_1, b_1, a_2, b_2, \ldots, a_g, b_g\}$.

A first proof for $g \neq 3$ was given by Zieschang in [278]. The technical assumption can be handled by changing the decomposition. Certainly $\mathrm{rk}(G) = 2g$ because $G/G' = \mathbb{Z}^{2g}$.

Proof. We write $G = H_1 *_A H_2$ with $H_1 = \langle a_1, b_1, a_2, b_2, \ldots, a_i, b_i \mid \rangle$, $H_2 = \langle a_{i+1}, b_{i+1}, a_{i+2}, b_{i+2}, \ldots, a_g, b_g \mid \rangle$ and $A = \langle \prod_{j=1}^{i}[a_j, b_j]\rangle = \langle \prod_{j=i+1}^{g}[a_j, b_j]\rangle$ with $1 \leq i < g$.

We remark that A is malnormal in G and that no x_k can be conjugate to a power of $\prod_{j=1}^{i}[a_j, b_j]$ because the group $G/\langle\langle A\rangle\rangle$ has also rank $2g$. Hence, using Theorem 1.2.16 and Theorem 1.5.19, we may assume without loss of generality $\{x_1, x_2, \ldots, x_{2i}\} = \{a_1, b_1, a_2, b_2, \ldots, a_i, b_i\}$. Now we have several possibilities to change the decompositions as a free product with amalgamation. We take a new decomposition and use the Nielsen cancellation method for this one starting with $\{a_1, b_1, a_2, b_2, \ldots, a_i, b_i, x_{i+1}, x_{i+2}, \ldots, x_{2g}\}$. We finally, without loss of generality, come to a system $\{a_1, b_1, a_2, b_2, a_{g-1}, b_{g-1}, x_{2g-1}, x_{2g}\}$. But now, possibly after a further Nielsen transformation, we may assume that $\langle a_g, b_g\rangle = \langle r, x_{2g-1}, x_{2g}\rangle$ with $r = [a_g, b_g]$, the system $\{r, x_{2g-1}, x_{2g}\}$ is r-stable Nielsen equivalent to $\{r, a_g, b_g\}$ (see Section 1.2). Hence $\{x_1, x_2, \ldots, x_{2g}\}$ altogether is Nielsen equivalent to $\{a_1, b_2, a_2, b_2, \ldots, a_g, b_g\}$. □

Louder [171] proved, using graph theoretic methods, the following remarkable result.

Theorem 1.5.33. *Let $G = \langle a_1, b_1, a_2, b_2, \ldots, a_g, b_g \mid \prod_{i=1}^{g}[a_i, b_i] = 1\rangle$, $g \geq 2$, and let $\{x_1, x_2, \ldots, x_n\}$ be a generating system of G. Then there is a Nielsen transformation from $\{x_1, x_2, \ldots, x_n\}$ to $\{a_1, b_1, a_2, b_2, \ldots, a_g, b_g, 1, \ldots, 1\}$.*

The exceptional case (b) in Theorem 1.5.30 may be handled in a similar manner using the addendum to Theorem 1.4.45. Without loss of generality we may assume here m even, $m \geq 4$, $\gamma_1 = \gamma_2 = \cdots = \gamma_{m-1} = 2$ and $\gamma_m = 2k + 1$, $k \geq 1$.

Theorem 1.5.34. *Let $G = \langle a_1, s_2, \ldots, s_m \mid s_1^2 = s_2^2 = \cdots = s_{m-1}^2 = s_m^{2k+1} = 1\rangle$ with $m \geq 4$, m even and $k \geq 1$. Let $\{x_1, x_2, \ldots, x_{m-2}\}$ be a (minimal) generating system of G. Then $\{x_1, x_2, \ldots, x_{m-2}\}$ is Nielsen equivalent to $\{s_1s_2, s_1s_3, \ldots, s_1s_{m-1}\}$.*

We should remark that in this exceptional case the geometric rank of G is $m - 1$.

Let

$$G = \left\langle s_1, s_2, \ldots, s_m, a_1b_1, a_2, b_2, \ldots, a_g, b_g \;\middle|\; s_1^{\gamma_1} = s_2^{\gamma_2} = \cdots = s_m^{\gamma_m} = s_1s_2\cdots s_m \prod_{i=1} g[a_i, b_i] = 1\right\rangle$$

with $m \geq 0$, $g \geq 0$, all $\gamma_i \geq 2$ and $2g - 2 + \sum_{i=1}^{m}(1 - \frac{1}{a_i}) > 0$. Again, G is a co-compact Fuchsian group.

A connected subcomplex of the upper half plane containing exactly one face from each equivalence class with respect to the action of G on the upper half plane, together with their boundaries, is a fundamental domain for G. G has a fundamental domain which is simply connected.

Now, let F be a fundamental domain for G. Let r_F denote the number of pairs $\{x, x^{-1}\} \subset G$ such that $F \cap xF$ is 1-dimensional. The *geometric rank* of G is the minimum of the r_F where F varies over all fundamental domains.

Note that G has geometric rank $2g$ if $m = 0$ and $2g + m - 1$ if $m \geq 1$. For the proof see [211].

Remarks 1.5.35. 1. Let G be as in Theorem 1.5.31.4. For the complete classification of minimal generating systems for G we only have to regard the cases which are left out in Theorem 1.5.32.

If $n = 0, p = 2$ and $m = 1$ then $G = \langle a, b \mid [a, b]^{\alpha} = 1\rangle$, $\alpha \geq 1$, then any generating pair $\{x, y\}$ is Nielsen equivalent to $\{a, b\}$. This is clear for $\alpha = 1$ because then G is free Abelian of rank 2. A proof for $\alpha \geq 2$ can be found in [92].
If $p = 0$, $m = 3$ and $\frac{1}{\gamma_1} + \frac{1}{\gamma_2} + \frac{1}{\gamma_3} = 1$ then, up to permutations, $(\gamma_1, \gamma_2, \gamma_3) = (3, 3, 3), (2, 4, 4)$ or $(2, 3, 6)$, and G is an Euclidean plane discontinuous group and, hence, the generating pair for G are well known. If $\frac{1}{\gamma_1} + \frac{1}{\gamma_2} + \frac{1}{\gamma_3} < 1$ one may find a complete description of the generating pairs in [92].
We here just state the result.
Let $G = \langle a, b \mid a^p = b^q = (ab)^r = 1\rangle$ with $2 \leq p \leq q \leq r$ and $\frac{1}{p} + \frac{1}{q} + \frac{1}{r} < 1$ a hyperbolic Fuchsian (p, q, r) triangle group. Let $\{u, v\}$ be a generating pair of G. If u has finite order n then we call a transformation $\{u, v\} \mapsto \{u^m, v\}$ with $1 \leq m < n$ and $\gcd(m, n) = 1$ an *elementary extended Nielsen transformation.* An extended Nielsen transformation is a finite sequence of Nielsen transformations and elementary extended Nielsen transformations. We have the following result.

Theorem 1.5.36. *Let G, as above, be a hyperbolic (p, q, r) triangle group. Then one and only one of the following cases holds:*
(a) *There is an extended Nielsen transformation from $\{u, v\}$ to $\{a, b\}$,*
(b) *G is a $(2, 3, r)$ triangle group with $\gcd(r, 6) = 1$ and $\{u, v\}$ is Nielsen equivalent to $\{abab^2, b^2aba\}$,*
(c) *G is a $(2, 4, r)$ triangle group with $\gcd(r, 2) = 1$ and $\{u, v\}$ is Nielsen equivalent to $\{ab^2, b^3ab^3\}$,*
(d) *G is a $(3, 3, r)$ triangle group with $\gcd(r, 3) = 1$ and $\{u, v\}$ is Nielsen equivalent to $\{ab^2, b^2a\}$, or*
(e) *G is a $(2, 3, 7)$ triangle group and $\{u, v\}$ is Nielsen equivalent to $\{ab^2abab^2ab^2ab, b^2abab^2ababa\}$.*

Remark 1.5.37. It is interesting to consider the hyperbolic (p, q, r) triangle groups as subgroups of $PSL(2, \mathbb{R})$ and give a description of Theorem 1.5.36 in terms of traces. Let P denote the natural map $P : SL(2, \mathbb{R}) \to PSL(2, \mathbb{R})$. We then use $tr(A)$ in an appropriate manner. Since the description is obvious for case (a) we just consider the cases (b)–(e).

Corollary 1.5.38. *Let G be a hyperbolic (p, q, r) triangle group, given as a discrete subgroup of $PSL(2, \mathbb{R})$ with $tr(a) = 2\cos(\frac{\pi}{p})$, $tr(b) = 2\cos(\frac{\pi}{q})$ and $tr(ab) = -2\cos(\frac{\pi}{r})$ (we may assume this without loss of generality), and let $\{u, v\}$ with $0 \leq tr(u), tr(v)$ be a generating pair of G. Then, relative to the cases (b)–(e):*
(b) *If G is a $(2, 3, r)$ triangle group with $r \geq 7$ and $\gcd(r, 6) = 1$ then $tr([u, v]) = -2\cos(\frac{6\pi}{r})$ and $\{u, v\}$ is Nielsen equivalent to a pair $\{x, y\}$ which satisfies $tr(x) = tr(y) = tr(xy)$.*
(c) *If G is a $(2, 4, r)$ triangle group with $r \geq 5$ and $\gcd(r, 2) = 1$ then $tr([u, v]) = -2\cos(\frac{4\pi}{r})$ and $\{u, v\}$ is Nielsen equivalent to a pair $\{x, y\}$ which satisfies $tr(x) = tr(y)$ and $tr(xy) = \frac{1}{2}(tr(x))^2$.*

(d) *If G is a $(3,3,r)$ triangle group with $r \geq 4$ and $\gcd(r,3) = 1$ then $tr([u,v]) = -2\cos(\frac{3\pi}{r})$ and $\{u,v\}$ is Nielsen equivalent to a pair $\{x,y\}$ which satisfies $tr(x) = tr(y) = tr(xy)$.*

(e) *If G is a $(2,3,7)$ triangle group then $tr([u,v]) = -2\cos(\frac{4\pi}{7})$ and $\{u,v\}$ is Nielsen equivalent to a pair $\{x,y\}$ which satisfies $tr(x) = tr(xy) = tr(y) + 1$.*

The geometric and number theoretical details are given in [92].

We are left with certain cases for which $p = 1, m \geq 2$ or $p = 0, m \geq 4$. Of course, Theorem 1.5.32 is not the best possible, but on the other side it does not hold analogously in general (see [227]).

For instance, let $G = \langle s_1, s_2, a \mid s_1^2 = s_2^3 = s_1 s_2 a^2 = 1\rangle$. Let $x = a^2, y = s_1 a$ and $X = \langle x,y\rangle$. Then $X = G$, and $\{x,y\}$ is not Nielsen equivalent to $\{s_1, a\}$ or $\{s_2, a\}$. We leave the details as exercises.

2. Let G be as in Theorem 1.5.31.4. We remark
   (a) If $m = 0$ and $p \geq 3$ then G has a one-relator presentation for the minimal generating system $\{a_1, a_2, \ldots, a_{i-1}, a_i^{\beta_i}, a_{i+1}, \ldots, a_n, a_{n-1}, \ldots, a_p\}, 1 \leq i \leq n, 1 \leq \beta_i \leq \frac{1}{2}\alpha_i$, $\gcd(\beta_i, \alpha_i) = 1$. This follows from direct rewriting.
   (b) If $m = 0$ and $n = p = 2$ then G has a one-relator presentation for $\{a_1^{\beta_1}, a_2^{\beta_2}\}$ with $\gcd(\beta_1, \beta_2) = \gcd(\beta_1, \alpha_1) = \gcd(\beta_2, \alpha_2) = 1, 1 \leq \beta_1 \leq \frac{1}{2}\alpha_1, 1 \leq \beta_2 \leq \frac{1}{2}\alpha_2$ if and only if $\beta_1 = 1$ or $\beta_2 = 1$ (see [62]).

In a finite system $\{x_1, x_2, \ldots, x_m\}$, $m \geq 1$, it can happen that some of the x_i are conjugate to elements of H_1 or H_2. In applying Nielsen reduction to results like the Kurosh theorem it is important to preserve this property if one tries to replace $\{x_1, x_2, \ldots, x_m\}$ by a shorter system.

1.5.3 Arithmetic Fuchsian groups

In Sections 1.4.4 and 1.5.2 we considered, among others, minimal generating systems of co-finite Fuchsian groups. Recall that a co-finite Fuchsian group has a presentation

$$G = \left\langle s_1, \ldots, s_m, p_1, \ldots, p_n, a_1, b_1, \ldots, a_g, b_g \mid s_1^{\gamma_1} = s_2^{\gamma_2} = \cdots = s_m^{\gamma_m} \right.$$
$$\left. = s_1 \cdots s_m \cdot p_1 \cdots p_n \prod_{i=1}^{g} [a_i, b_i] = 1 \right\rangle,$$

with $m \geq 0, n \geq 0, g \geq 0$, all $\gamma_i \geq 2$ and

$$2g - 2 + n + \sum_{i=1}^{m}\left(1 - \frac{1}{\gamma_i}\right) > 0.$$

G is co-compact if $n = 0$. The p_j are parabolic elements which have exactly one fixed point. To describe G we write $(g; n; \gamma_1, \ldots, \gamma_m)$ and call this the *signature of G*.

In this section we want to consider discreteness in connection with arithmeticity.

Let G be a finitely generated subgroup of $PSL(2,\mathbb{R})$. The subgroup $G^{(2)} = \langle g^2 \mid g \in G\rangle$ is a normal subgroup of finite index with quotient group being a finite Abelian 2-group.

We call

$$k_1 = \mathbb{Q}(tr(g) \mid g \in G)$$

the *trace field* and

$$k_2 = \mathbb{Q}(tr(g^2) \mid g \in G)$$

the *invariant trace field* of G. We use, with slight ambiguity, the term $tr(g)$ for the trace of $g \in PSL(2,\mathbb{R})$. We always assume that G is *non-elementary*. Recall that a subgroup of $PSL(2,\mathbb{R})$ is called elementary if the commutator of any two elements of infinite order has trace 2.

For any finite index subgroup G_1 of a non-elementary group G we have $k_2 \subset \mathbb{Q}(tr(g_1) \mid g_1 \in G_1)$ and k_2 is an invariant of the commensurability class of G. Here, two subgroups Γ_1, Γ_2 of a group Γ are *commensurable (in the wide sense)* if there exists an element $\gamma \in \Gamma$ such that $\Gamma_1 \cap \gamma\Gamma_2\gamma^{-1}$ is of finite index in both Γ_1 and $\gamma\Gamma_2\gamma^{-1}$.

Theorem 1.5.39 ([111]). *Let G be a finitely generated subgroup of $PSL(2,\mathbb{R})$ such that*
(1) *$G^{(2)}$ contains elements g_1 and g_2 which have no common fixed point.*
(2) *$tr(G) = \{tr(g) \mid g \in G\}$ consists of algebraic integers.*
(3) *For each embedding $\sigma : k_2 \to \mathbb{R}$ such that $\sigma \neq id$, the set $\{\sigma(tr(f)) \mid f \in G^{(2)}\}$ is bounded.*

Then G is discrete, that is, G does not contain any convergent sequence of pairwise distinct elements.

The condition of Theorem 1.5.39 reflects the definition of an arithmetic Fuchsian group. For such groups there is a standard construction via a quaternion algebra over invariant trace field.

Let k be a subfield of $\mathbb{R}$. A *quaternion algebra A over k* is a 4-dimensional central simple algebra with a multiplicative identity 1.

Here, central means that the center of A is just k, and simple means that A has no proper two-sided ideal. A has a basis of the form $\{1, i, j, ij\}$ with $i^2 = a$, $j^2 = b$ where $a, b \in k^*$. A is denoted by a *Hilbert symbol* $(\frac{a,b}{k})$. A does not uniquely determine the Hilbert symbol. In particular,

$$\left(\frac{a,b}{k}\right) \cong \left(\frac{ax^2, by^2}{k}\right) \quad \text{for } x, y \in k^*.$$

For example, over $\mathbb{R}$ there are just two isomorphism classes of quaternion algebras represented by

$$\left(\frac{1,1}{\mathbb{R}}\right) \cong \left(\frac{1,-1}{\mathbb{R}}\right) \cong M(2,\mathbb{R})$$

by taking $i = \left(\begin{smallmatrix} 1 & 0 \\ 0 & -1 \end{smallmatrix}\right)$ and $j = \left(\begin{smallmatrix} 0 & 1 \\ 1 & 0 \end{smallmatrix}\right)$ or $j = \left(\begin{smallmatrix} 0 & 1 \\ -1 & 0 \end{smallmatrix}\right)$, respectively, and $(\frac{-1,-1}{\mathbb{R}}) \cong \mathbb{H}$, Hamilton's quaternions.

In general, an element $\alpha = a_0 + a_1 i + a_2 j + a_3 ij$ can be send to the matrix

$$\begin{pmatrix} a_0 + \sqrt{a}a_1 & \sqrt{b}(a_2 + \sqrt{a}a_3) \\ \sqrt{b}(a_2 - \sqrt{a}a_3) & a_0 - \sqrt{a}a_1 \end{pmatrix},$$

where $\sqrt{a}$ and $\sqrt{b}$ are given in $\mathbb{C}$ as usual. Now, let k be an algebraic number field, that is a finite extension of $\mathbb{Q}$, and let R_k denote the ring of integers in k. An order $\mathcal{O}$ in A is a subring with 1 which is a finitely generated R_k-module such that $\mathcal{O} \otimes_{R_k} k \cong A$. We denote by $\mathcal{O}^1$ the set of elements of norm 1 in $\mathcal{O}$. Here, if $\alpha = a_0 + a_1 i + a_2 j + a_3 ij \in A$ with $i^2 = a$, $j^2 = b$, then the norm of α is

$$n(\alpha) = a_0^2 - aa_1^2 - ba_2^2 + aba_3^2.$$

The norm is the determinant on matrix algebras. If there is an embedding φ of A into $M(2,\mathbb{R})$, then $P\rho(\mathcal{O}^1) \subset PSL(2,\mathbb{R})$ need not be discrete in $PSL(2,\mathbb{R})$, where P denotes the projection from $SL(2,\mathbb{R})$ to $PSL(2,\mathbb{R})$.

To generate discreteness we can use arguments by Borel and Harish-Chandra (see [31] and [32]) on discrete arithmetic groups in algebraic groups.

Theorem 1.5.40. *Let k be a real algebraic number field. Let $(\frac{a,b}{k})$ be a quaternion algebra over k such that $(\frac{a,b}{\mathbb{R}}) \cong M(2,\mathbb{R})$, that is, such that there is an embedding*

$$\rho : \left(\frac{a,b}{k}\right) \to M(2,\mathbb{R}).$$

Let $\mathcal{O}$ be an order in $(\frac{a,b}{k})$. Then $P\rho(\mathcal{O}^1)$ is discrete of finite co-volume in $PSL(2,\mathbb{R})$ if and only if k is totally real, that is $\sigma(k) \subset \mathbb{R}$ for each $\mathbb{Q}$-embedding $\sigma : k \to \mathbb{C}$, and $(\frac{\sigma(a),\sigma(b)}{\mathbb{R}}) \cong \mathbb{H}$ for each σ with $\sigma \neq id$.

This leads to the following definition.

Definition 1.5.41. Let $A = (\frac{a,b}{k})$ be a quaternion algebra over a totally real number field k such that there is an embedding

$$\rho : \left(\frac{a,b}{k}\right) \to M(2,\mathbb{R}) \quad \text{with} \left(\frac{\sigma(a),\sigma(b)}{k}\right) \cong H$$

for every $\mathbb{Q}$-embedding $\sigma : k \to \mathbb{R}$, $\sigma \neq id$. Then any subgroup of $PSL(2,\mathbb{R})$ which is commensurable with some $P\rho(\mathcal{O}^1)$ is an *arithmetic Fuchsian group*.

A detailed analysis of this definition, using results from [111] and [153] in connection with simultanous conjugations in quaternion algebras [213] gives the following Theorem.

Theorem 1.5.42. *Let G be a Fuchsian group of finite co-volume. Then G is arithmetic if and only if the following conditions are satisfied:*
(1) $k_2 = \mathbb{Q}(tr(g^2) \mid g \in G)$ *is an algebraic number field.*
(2) $tr(G) = \{tr(g) \mid g \in G\}$ *consists of algebraic integers.*
(3) $\sigma(k_2) \subset \mathbb{R}$ *for every* $\mathbb{Q}$*-embedding* $\sigma : k_2 \to \mathbb{C}$ *with* $\sigma \neq id$*, and there are* $g_1, g_2 \in G^{(2)}$ *such that* $|\sigma(tr(g_1))| < 2$ *and* $\sigma(tr([g_1, g_2])) < 2$ *for every such* σ *with* $\sigma \neq id$*.*

Corollary 1.5.43. *Let G be an arithmetic Fuchsian group which is not co-compact. Then G is commensurable (in the wide sense) with* $PSL(2, \mathbb{Z})$*.*

Proof. G contains a parabolic element g_0. Let $g_1 = g_0^2 \in G^{(2)}$ and $k_2 = \mathbb{Q}(tr(g^2) \mid g \in G)$.
We may assume that $g_0 = \pm\left(\begin{smallmatrix} 1 & 1 \\ 0 & 1 \end{smallmatrix}\right)$, then $g_1 = \pm\left(\begin{smallmatrix} 1 & 2 \\ 0 & 1 \end{smallmatrix}\right)$.
Let $g_2 \in G^{(2)}$, $g_2 = \pm\left(\begin{smallmatrix} a & b \\ c & d \end{smallmatrix}\right)$. Then $tr([g_1, g_2]) = 2 + 4c^2 \geq 2$.
Hence $\sigma = id$ is the only $\mathbb{Q}$-embedding $\sigma : k_2 \to \mathbb{C}$. Therefore $k_2 = \mathbb{Q}$, and the statement holds because G is not co-compact. □

Remarks 1.5.44. 1. A more detailed description of the Fuchsian groups which are commensurable with the modular group $PSL(2, \mathbb{Z})$ is given by Helling [123] and [125].
2. A nice statement for a more general result is given by Theorem 8.2.7 in [179].
3. Takeuchi [255] classified the subgroups of the modular group with signature $(0; n, \gamma_1, \ldots, \gamma_m)$ and $1 \leq n \leq 4$, $0 \leq m \leq 3$, $n + m = 4$.
Nakanishi, Näätänen and Rosenberger [197] gave the complete list of arithmetic Fuchsian groups with signature $(0; n, \gamma_1, \ldots, \gamma_m)$ and $1 \leq n \leq 4$, $0 \leq m \leq 3$, $n + m = 4$.

From now on let G be a co-compact Fuchsian group, that is, G has a presentation

$$G = \left\langle s_1, \ldots, s_m, a_1, b_1, \ldots, a_g, b_g \mid s_1^{\gamma_1} = s_2^{\gamma_2} = \cdots = s_m^{\gamma_m} = s_1 \cdots s_m \prod_{i=1}^{g} [a_i, b_i] = 1 \right\rangle,$$

with $m \geq 0$, $g \geq 0$, all $\gamma_i \geq 2$ and

$$2g - 2 + \sum_{i=1}^{m} \left(1 - \frac{1}{\gamma_i}\right) > 0.$$

We write $(g; \gamma_1, \gamma_2, \ldots, \gamma_m)$ for the signature of G.

Up to conjugacy in $PGL(2, \mathbb{Z})$, for a fixed $n \in \mathbb{N} \setminus \{1\}$ there are only finitely many arithmetic Fuchsian groups which can be generated by $\leq n$ elements [180].

This, especially means that for each signature $(g; \gamma_1, \ldots, \gamma_m)$ there are only finitely many points in its related Teichmüller space which belongs to arithmetic Fuchsian groups. The problem now is to classify the arithmetic Fuchsian groups of a given signature $(0;\gamma_1,\dots,\gamma_m)$. Two-generator co-compact Fuchsian groups have a signature $(0;\gamma_1,\gamma_2,\gamma_3)$, $(0;2,2,q)$, $q \geq 3$ and odd or $(1;\gamma_1)$. The two generator co-compact arithmetic Fuchsian groups are completely classified (see [251, 253, 254, 180, 181] and [4]).

We should remark that Maclachlan and Rosenberger [182] determined when two two-generator Fuchsian groups of finite co-volume are commensurable and, in addition, the relationship between two such groups, by obtaining part of the lattice of the commensurability class which contains one representative from each conjugacy class, in $PGL(2,\mathbb{R})$, of two-generator groups.

The next interesting case is the class of arithmetic surface groups with signature $(2;--)$, that is of genus 2. Fuchsian groups whose signature appear on the list below are those that contain a Fuchsian group of genus 2 as a subgroup of finite index (see [2]):

$$(0;2,3,7),(0;2,3,8),(0;2,3,9),(0;2,3,10),(0;2,3,12),$$
$$(0;2,4,5),(0;2,4,6),(0;2,4,8),(0;2,4,12),(0;2,5,5),$$
$$(0;2,5,6,),(0;2,5,10),(0;2,6,6),(0;2,8,8),(0;3,3,4),$$
$$(0;3,3,5),(0;3,3,6),(0;3,3,9),(0;3,4,4),(0;3,6,6),$$
$$(0;4,4,4),(0;5,5,5),(0;2,2,2,3),(0;2,2,2,4),(0;2,2,2,6),$$
$$(0;2,2,3,3),(0;2,2,4,4),(0;3,3,3,3),(0;2,2,2,2,2),(0;2,2,2,2,2,2),$$
$$(1;2),(1;3),(1;2,2),(2;--).$$

We should remark that covering space arguments in combination with the use of GAP low index calculations also determines the list (see [136]). This list allows to determine the arithmetic data of all arithmetic groups of signature $(2;--)$ and to describe all commensurability classes of arithmetic surface groups of genus two.

Finally we want to mention the so-called *Vierecksgruppen, for short, VE groups*, that are the co-compact Fuchsian groups of signature $(0;\gamma_1,\gamma_2,\gamma_3,\gamma_4)$. Some of them already occurred in the previous parts.

A big part was done by Baer [24]. She classified the arithmetic Fuchsian groups with signature $(0;\gamma_1,\gamma_2,\gamma_3,\gamma_4)$ and at most one of $\gamma_1,\gamma_2,\gamma_3,\gamma_4$ is even.

Any VE group is either a subgroup of a triangle group or a subgroup of a VE group (see [241]). We first consider the case that the VE group is a subgroup of a triangle group. Neumann-Brosig and Rosenberger [203] gave a complete list of VE subgroups of maximal arithmetic triangle groups. Especially, all VE groups which occur in the list are arithmetic. We first gave bounds for the possible indices n of VE subgroups of maximal arithmetic triangle groups. To find these maximal triangle groups we used Takeuchi's result in [252]. Then we used GAP to determine the signature of the VE subgroups and the number of conjugacy classes in the respective maximal arithmetic triangle groups.

Then we got the following lists:

Group: [2, 4, 6]		
Index	Signature	Number of conjugacy classes
2	[2, 2, 2, 3]	1
3	[2, 2, 2, 4]	1
4	[2, 2, 2, 6]	1
4	[2, 2, 3, 3]	2
5	[2, 2, 3, 4]	1
6	[2, 2, 3, 6]	1
6	[2, 2, 4, 4]	4
7	[2, 2, 4, 6]	4
8	[2, 2, 6, 6]	4
8	[2, 3, 4, 4]	1
8	[3, 3, 3, 3]	1
9	[2, 3, 4, 6]	4
10	[2, 4, 4, 6]	3
10	[3, 3, 4, 4]	1
12	[2, 6, 6, 6]	1
12	[3, 3, 6, 6]	1
12	[4, 4, 4, 4]	2
14	[4, 4, 6, 6]	5
16	[6, 6, 6, 6]	2

Group: [2, 6, 8]		
Index	Signature	Number of conjugacy classes
2	[2, 2, 3, 4]	1
3	[2, 2, 4, 8]	1
4	[2, 3, 6, 6]	1
4	[3, 3, 4, 4]	1
6	[4, 4, 8, 8]	1

Group: [2, 3, 8]		
Index	Signature	Number of conjugacy classes
4	[2, 2, 2, 3]	1
6	[2, 2, 2, 4]	1
8	[2, 2, 3, 3]	5
9	[2, 2, 2, 8]	2
10	[2, 2, 3, 4]	3
12	[2, 2, 4, 4]	2
12	[2, 3, 3, 3]	1
13	[2, 2, 3, 8]	4
14	[2, 3, 3, 4]	2
15	[2, 2, 4, 8]	2
16	[3, 3, 3, 3]	3
17	[2, 3, 3, 8]	9
18	[2, 2, 8, 8]	8
18	[3, 3, 3, 4]	1
19	[2, 3, 4, 8]	10
20	[3, 3, 4, 4]	4
24	[2, 4, 8, 8]	1
24	[4, 4, 4, 4]	1
26	[3, 3, 8, 8]	5
27	[2, 8, 8, 8]	2
28	[3, 4, 8, 8]	2
30	[4, 4, 8, 8]	3
36	[8, 8, 8, 8]	4

Group: [2, 3, 12]		
Index	Signature	Number of conjugacy classes
3	[2, 2, 2, 4]	1
4	[2, 2, 3, 3]	1
6	[2, 2, 3, 6]	1
6	[2, 2, 4, 4]	1
6	[2, 3, 3, 3]	1
7	[2, 2, 3, 12]	2
7	[2, 3, 3, 4]	1
8	[2, 3, 3, 6]	1
8	[3, 3, 3, 3]	1
9	[2, 2, 6, 12]	1
9	[2, 3, 4, 6]	1
12	[2, 4, 6, 12]	1
12	[3, 3, 6, 6]	1
12	[4, 4, 4, 4]	1
14	[3, 3, 12, 12]	4
15	[2, 12, 12, 12]	2
16	[3, 6, 12, 12]	3
18	[4, 12, 12, 12]	1
18	[6, 6, 12, 12]	1

Group: [2, 4, 12]		
Index	Signature	Number of conjugacy classes
2	[2, 2, 2, 6]	1
3	[2, 2, 4, 4]	1
4	[2, 4, 4, 12]	1
4	[2, 2, 6, 6]	2
4	[2, 3, 4, 4]	1
5	[2, 3, 4, 12]	2
5	[2, 4, 4, 6]	1
6	[2, 4, 6, 12]	1
6	[4, 4, 4, 4]	1
8	[3, 6, 12, 12]	1
8	[4, 4, 12, 12]	1
8	[6, 6, 6, 6]	1

Group: [2, 4, 5]		
Index	Signature	Number of conjugacy classes
5	[2, 2, 2, 4]	2
6	[2, 2, 2, 5]	2
10	[2, 2, 4, 4]	7
11	[2, 2, 4, 5]	5
12	[2, 2, 5, 5]	9
15	[2, 4, 4, 4]	1
16	[2, 4, 4, 5]	3
17	[2, 4, 5, 5]	8
20	[4, 4, 4, 4]	3
22	[4, 4, 5, 5]	5
24	[5, 5, 5, 5]	5

Group: [2, 4, 10]		
Index	Signature	Number of conjugacy classes
2	[2, 2, 2, 5]	1
4	[2, 2, 5, 5]	2
6	[2, 4, 4, 10]	1
8	[5, 5, 5, 5]	1

Group: [2, 5, 6]		
Index	Signature	Number of conjugacy classes
5	[2, 2, 6, 6]	1
5	[2, 3, 3, 6]	1
6	[2, 3, 5, 6]	1
7	[2, 5, 5, 6]	3
10	[6, 6, 6, 6]	1

Group: [2, 3, 10]		
Index	Signature	Number of conjugacy classes
5	[2, 2, 3, 3]	1
6	[2, 2, 2, 10]	1
7	[2, 2, 3, 5]	1
10	[3, 3, 3, 3]	1
11	[2, 3, 3, 10]	6
12	[2, 2, 10, 10]	5
12	[3, 3, 3, 5]	1
13	[2, 3, 5, 10]	6
14	[3, 3, 5, 5]	1
15	[2, 5, 5, 10]	1
18	[2, 10, 10, 10]	1
24	[10, 10, 10, 10]	3

Group: [3, 4, 6]		
Index	Signature	Number of conjugacy classes
2	[2, 3, 3, 3]	1
3	[2, 3, 4, 6]	1
3	[2, 4, 4, 4]	1
4	[3, 3, 6, 6]	1

Group: [2, 4, 7]		
Index	Signature	Number of conjugacy classes
7	[2, 4, 4, 4]	1
8	[2, 4, 4, 7]	2
9	[2, 4, 7, 7]	4

Group: [2, 3, 7]		
Index	Signature	Number of conjugacy classes
7	[2, 2, 2, 3]	2
14	[2, 2, 3, 3]	9
15	[2, 2, 2, 7]	3
21	[2, 3, 3, 3]	4
22	[2, 2, 3, 7]	13
28	[3, 3, 3, 3]	5
29	[2, 3, 3, 7]	14
30	[2, 2, 7, 7]	12
36	[3, 3, 3, 7]	4
37	[2, 3, 7, 7]	15
44	[3, 3, 7, 7]	16
45	[2, 7, 7, 7]	2
52	[3, 7, 7, 7]	1
60	[7, 7, 7, 7]	6

Group: [2, 3, 14]		
Index	Signature	Number of conjugacy classes
8	[2, 3, 3, 14]	1
9	[2, 2, 14, 14]	1
10	[2, 3, 7, 14]	1
18	[14, 14, 14, 14]	1

Group: [2, 3, 9]		
Index	Signature	Number of conjugacy classes
3	[2, 2, 2, 3]	1
6	[2, 2, 3, 3]	1
9	[2, 3, 3, 3]	2
10	[2, 2, 3, 9]	6
12	[3, 3, 3, 3]	2
13	[2, 3, 3, 9]	4
16	[3, 3, 3, 9]	1
20	[3, 3, 9, 9]	9
21	[2, 9, 9, 9]	2

Group: [2, 3, 18]		
Index	Signature	Number of conjugacy classes
3	[2, 2, 2, 6]	1
6	[2, 2, 6, 6]	1
6	[3, 3, 3, 3]	1
7	[2, 3, 3, 18]	2
8	[3, 3, 3, 9]	1
9	[2, 3, 9, 18]	1
12	[2, 18, 18, 18]	1
12	[3, 6, 9, 18]	1
12	[6, 6, 6, 6]	1

Group: [2, 4, 18]		
Index	Signature	Number of conjugacy classes
2	[2, 2, 2, 9]	1
3	[2, 2, 4, 6]	1
4	[2, 2, 6, 18]	1
4	[2, 2, 9, 9]	2
5	[2, 4, 6, 9]	1
6	[2, 6, 9, 18]	1
6	[4, 4, 6, 6]	1
8	[6, 6, 18, 18]	1
8	[9, 9, 9, 9]	1

Group: [2, 3, 16]		
Index	Signature	Number of conjugacy classes
4	[2, 2, 3, 4]	1
6	[2, 2, 4, 8]	1
8	[3, 3, 4, 4]	1
10	[2, 3, 16, 16]	1
12	[2, 8, 16, 16]	1
12	[4, 4, 8, 8]	1

Group: [2, 5, 20]		
Index	Signature	Number of conjugacy classes
5	[2, 10, 10, 20]	1
6	[5, 5, 20, 20]	1

Group: [2, 3, 30]		
Index	Signature	Number of conjugacy classes
3	[2, 2, 2, 10]	1
5	[2, 3, 3, 6]	1
6	[2, 2, 6, 30]	1
6	[2, 2, 10, 10]	1
6	[3, 3, 3, 5]	1
7	[2, 3, 5, 30]	2
7	[2, 3, 6, 15]	1
8	[3, 3, 5, 15]	1
9	[2, 5, 15, 30]	1
9	[2, 6, 10, 30]	1
10	[3, 6, 10, 15]	1
12	[5, 10, 15, 30]	1
12	[6, 6, 30, 30]	1
12	[10, 10, 10, 10]	1

Group: [2, 5, 8]		
Index	Signature	Number of conjugacy classes
5	[2, 4, 4, 8]	1
6	[2, 5, 8, 8]	1

Group: [2, 3, 11]		
Index	Signature	Number of conjugacy classes
12	[3, 3, 3, 11]	2
13	[2, 3, 11, 11]	6

We remark that a VE group may be contained in several triangle groups (see [182] and [252]).

So, we are left with the VE groups $(0; \gamma_1, \gamma_2, \gamma_3, \gamma_4)$ where at least two of the γ_i are even and which are not subgroups of triangle groups.

All arithmetic VE groups of signature of the form $(0; N, N, N, N)$, $(0; 2, 2, N, N)$, $(0; 2, 2, 2, N)$ have been determined (see [181] and [254]). Note also that any VE group of signature $(0; N_1, N_1, N_2, N_2)$ is of index 2 in a VE group of signature $(0; 2, 2, N_1, N_2)$.

Now, let G be a (co-compact) VE group of signature $(0; \gamma_1, \gamma_2, \gamma_3, \gamma_4)$ with at least two of the γ_i are even and which are not subgroups of arithmetic triangle groups.

Again, let $k_2 = \mathbb{Q}(tr(g^2) \mid g \in G)$ be the invariant trace field of G. Sunaga [249] and [250] determined the arithmetic groups G with signature $(0; \gamma_1, \gamma_2, \gamma_3, \gamma_4)$ and invariant trace field $k_2 = \mathbb{Q}$ and gave the set of generators of them. Maclachlan and Rosenberger [183] gave statements about the possible torsion in maximal arithmetic VE groups and the signatures and conjugacy classes of maximal arithmetic VE groups. Using these results we may handle the arithmetic VE groups with $deg(k_2) \geq 4$. The cases $2 \leq deg(k_2) \leq 3$ are essentially open.

1.5.4 E-transformations

The above discussion leads to the following ideas developed by Kalia and Rosenberger [141].

Definition 1.5.45. A finite set $X = \{x_1, x_2, \dots, x_m\}$, $m \geq 1$, in $G = H_1 *_A H_2$, $H_1 \neq A \neq H_2$, is called an *E-set* if it is composed as follows:
(a) $X = X_1 \cup X_2$, $X_1 \cap X_2 \neq \emptyset$.
(b) Each $x_j \in X_1$ is conjugate to an element of H_1 or H_2.
(c) Each $x_j \in X_2$ is not conjugate to an element of H_1 or H_2.

Definition 1.5.46. On an E-set $X = \{x_1, x_2, \dots, x_m\}$ with partition $X = X_1 \cup X_2$ as above we define the following types of transformations which produce E-set Y with partition $Y = Y_1 \cup Y_2$.
(E1) Replace some $x_j \in X_1$ by $x_j' = x_k^{\epsilon} x_j x_k^{-\epsilon}$, $k \neq j$, $\epsilon = \pm 1$, and leave the remaining x_i, $i \neq j$ fixed.
(E2) Replace some $x_j \in X_2$ by $x_j' = x_k^{\epsilon} x_j$ or $x_j' = x_j x_k^{\epsilon}$, $k \neq j$, $\epsilon = \pm 1$, and leave the remaining x_i, $i \neq j$, fixed.
(E3) Replace some $x_j \in X$ by $x_j' = x_j^{-1}$ and leave the remaining x_i, $i \neq j$, fixed.
(E4) Permute in X_1 and leave X_2 fixed.
(E5) Permute in X_2 and leave X_1 fixed.
(E6) Delete some $x_j \in X$ where $x_j = 1$, $1 \leq j \leq m$.

We call these the *elementary E-transformations*. An *E-transformation* is a finite product of elementary E-transformations. An E-set Y is *derivable* from an E-set X if there is an E-transformation from X to Y. An E-set $X = \{x_1, x_2, \dots, x_m\} \subset G$ is *E-reduced* or *E-minimal* if there is no E-set $Y = \{y_1, y_2, \dots, y_m\}$ derivable from X, such that one of the following holds:
(a) $y_1 = 1$ for some $i \in \{1, 2, \dots, m\}$.
(b) $y_1 \neq 1$ for all $i \in \{1, 2, \dots, m\}$ and Y is shorter than X.

We note that using the elementary E-transformations (E2) we can go from an E-set X to an E-set Y which has more elements conjugate to an element of H_1 or H_2, hence, in the corresponding partitions $X = X_1 \cup X_2$, $Y = Y_1 \cup Y_2$ the set Y_1 contains more elements than X_1. Also we note that in $X = X_1 \cup X_2$ we could have $X_1 = \emptyset$ or $X_2 = \emptyset$.

Furthermore, we have the following three direct observations:
(a) If the E-set Y is derivable from E-set X then $\langle X \rangle = \langle Y \rangle$.
(b) Suppose $X = \langle x_1, x_2, \dots, x_m \rangle \subset G$, $m \geq 1$ is an E-set and suppose G is countable. Then there exists an E-reduced E-set Y which is derivable from X.
(c) The regular E-transformations (E1)–(E5) form a finitely generated group.

As already mentioned if we consider a given finite system in G we may assume that G is countable. Analogously to Theorem 1.5.19 we obtain the following.

Theorem 1.5.47. *Let $G = H_1 *_A H_2$, $H_1 \neq A \neq H_2$. If $\{x_1, x_2, \dots, x_m\}$, $m \geq 1$, is a finite E-set of elements in G then there is an E-transformation from $\{x_1, x_2, \dots, x_m\}$ to an E-set $\{y_1, y_2, \dots, y_m\}$ for which one of the following cases holds:*

(i) $y_i = 1$ *for some* $i \in \{1, 2, \ldots, m\}$.
(ii) *Each* $w \in \langle y_1, y_2, \ldots, y_m\rangle$ *can be written as* $w = \prod_{i=1}^{q} y_{v_i}^{\epsilon_i}$, $\epsilon_i = \pm 1$, $\epsilon_i = \epsilon_{i+1}$ *if* $v_i = v_{i+1}$ *with* $|y_{v_i}| \leq |w|$ *for* $i = 1, 2, \ldots, q$.
(iii) *Of the* y_i *there are* p, $p \geq 1$, *contained in a subgroup of* G *conjugate to* H_1 *or* H_2 *and a certain product of them is conjugate to a non-trivial element of* A.

As stated this theorem looks weaker than Theorem 1.5.19. The advantage is that we have a reasonable control over the elements which are conjugate to elements of H_1 or H_2. However, we note that it can be refined in a manner similar to the earlier theorem, especially if A is malnormal in G (see [141]). The main tools for the proof of Theorem 1.5.47 are the following lemmas.

Lemma 1.5.48. *Suppose* $x, y \in G$ *with* $|x| \leq |y|$ *and* $y \equiv p_y k_y q_y$, $x \equiv p_x k_x p_x^{-1}$, $k_x \in H_i \setminus A$, $i = 1$ *or* 2, *in symmetric form. If* $|xy^\epsilon| < |x|$ *or* $|y^{-1}x| < |x|$, $\epsilon = \pm 1$, *then one of the following cases occurs:*
(a) $|y^{-\epsilon}xy^\epsilon| < |x|$.
(b) $q_y = p_y^{-1} = p_x^{-1}$ *and* $k_x k_y^\epsilon$ *or* $k_y^{-\epsilon}k_x$ *is contained in* A.
(c) $q_y \neq p_y^{-1}$, $k_y \in H_i \setminus A$, $|x| = |y|$, $|y^{-\epsilon}xy^\epsilon| = |x|$ *and* $k_x k_y^\epsilon$ *or* $k_y^{-\epsilon}k_x$ *is contained in* A, *especially* $|xy^\epsilon| < |y|$ *or* $|y^{-\epsilon}x| < |y|$.

Proof. Suppose $|xy| < |x|$. Since $k_x \in H_i \setminus A$ we have $|x| = 2|p_x| + 1$. From $|y| \leq |x|$ and $|xy| < |x|$ we get $p_x^{-1} \equiv r_x^{-1}p_y^{-1}$.

If $r_x \neq 1$ then necessarily $|r_x^{-1}k_y q_y| < |r_x| + |q_r| + 1$ and case (a) holds.

Suppose now that $p_x^{-1} = p_y^{-1}$. If $k_y \in A$ then necessarily $|k_x k_y q_y| < |q_y| + 1$ and $|y^{-1}xy| = |q_y^{-1}k_y^{-1}k_x k_y q_y| < |x|$ and case (a) holds. Now suppose $k_y \notin A$. Then $|x| = |y|$, $xy = p_x k_x k_y q_y$ and $k_x k_y \in A$ since $|xy| < |x|$. If $q_y = p_x^{-1}$ then case (b) holds while if $q_y \neq p_x^{-1}$ case (c) holds. □

Lemma 1.5.49. *Suppose* $x, y \in G$ *with* $|y| \leq |x|$ *and* $x \equiv p_x k_x p_x^{-1}$, $k_x \in H_i \setminus A$, $i = 1$ *or* 2, *in symmetric normal form. If* $|xy^\epsilon| = |x|$ *or* $|y^{-\epsilon}x| = |x|$, $\epsilon = \pm 1$, *then* $|y^{-\epsilon}xy^\epsilon| \leq |x|$.

Proof. Let $|xy| = |x|$ and $y \equiv p_y k_y q_y$ in symmetric normal form. If $p_x = 1$ then also $p_y = 1$ and $x, y \in H_i$ since $|y| \leq |x| = |xy|$, and $|y^{-1}xy| \leq |x|$.

Suppose $p_x \neq 1$. We must have $p_x^{-1} \equiv r_x^{-1}p_y^{-1}$ because $|xy| = |x|$. Then $xy = p_x k_x r_x^{-1} k_y q_y$. If $r_x \neq 1$ then $|r_x^{-1}k_y q_y| = |p_x|$ and hence $|y^{-1}xy| = |x|$ since $|y| \leq |x|$.

Now suppose $r_x = 1$, that is, $xy = p_x k_x k_y q_y$. Then $|k_x k_y| = 1$ since $|xy| = |x|$ and $|y^{-1}xy| = |q_y^{-1}k_y^{-1}k_x k_y q_y| \leq |x|$ since $|y| \leq |x|$. □

1.5.5 Equations in free products with amalgamation

We also may modify Theorem 1.4.52 for free products to a theorem on equations in free products with amalgamations. Let $G = H_1 *_A H_2$, $H_1 \neq A \neq H_2$. We now call an ordered set $U = \{u_1, u_2, \ldots, u_n\} \subset G$ *regular* if there is no Nielsen transformation from U to a system $U' = \{u'_1, u'_2, \ldots, u'_n\}$ in which one of the elements is conjugate to an element of A.

We cannot get the analogous theorem in full generality because an element $aga'g^{-1}$ with $a, a' \in A$ is in general not in A. This holds if $A \lhd G$. In the proof we have to replace the equation $v_1v_2 \cdots v_m = 1$ by the equation $v_1v_2 \cdots v_m = a \in A$. But this we may handle using Lemmas 1.5.21 and 1.5.22.

We leave the details as an exercise (see also [222]).

Theorem 1.5.50. *Suppose $G = H_1 *_A H_2$, $H_1 \neq A \neq H_2$. Let $1 \neq w(x_1, x_2, \ldots, x_n)$ be a reduced regular word in F, with F free of rank $n \geq 1$ on the ordered basis $X = \{x_1, x_2, \ldots, x_n\}$. Furthermore, let $\phi: F \to G$ be a homomorphism such that $U = \phi(X)$ is a regular, Nielsen reduced set and $\phi(w) = hah^{-1}$, $a \in A$, $h \in G$.*

Then there exists a cyclic permutation $w' \in F$ of w, induced by an automorphism $\alpha: F \to F$, such that

(a) *$w' = \alpha(w) = w_1w_2$ with $w_1 \neq 1$;*
(b) *$\phi(\alpha^{-1}(w_1))$ is conjugate in G to an element of A; and*
(c) *there exists an $i \in \{1, 2\}$ and a $g \in G$ with $\phi(\alpha(x_j)) \in gH_ig^{-1}$ for each x_j which occurs in w_1.*

We remark that we do not need the regularity of G because if U is not regular we get anyway some conjugate of an element of A. But if A is malnormal in G then the regularity of U shows that at least two x_j occur in w_1.

If A is normal in G then we get the complete vision of Theorem 1.4.52 for G because $aga'g^{-1} \in A$, is also in A.

Theorem 1.5.51. *Suppose $G = H_1 *_A H_2$, $H_1 \neq A \neq H_2$, with $A \lhd G$. Let $1 \neq w(x_1, x_2, \ldots, x_n)$ be a reduced regular word in F, with F free of rank $n \geq 1$ on the ordered basis $X = \{x_1, x_2, \ldots, x_n\}$, and further let $\phi: F \to G$ be a homomorphism such that $U = \phi(X)$ is regular and $\varphi(w) = a \in A$. Then the pair (w, U) is Nielsen equivalent to a pair $(w', U') = (\alpha(w), \alpha^{-1}(U))$ with $\alpha: F \to F$ an automorphism such that:*

(a) *$w' = r_1w_1r_1^{-1} \cdots r_kw_kr_k^{-1}$ for some $k \geq 0$ where, for $1 \leq i \leq k$, r_i and w_i are reduced words in F.*
(b) *For each i, $1 \leq i \leq k$, we have $\phi\alpha^{-1}(w_i) \in A$.*
(c) *For each i, $1 \leq i \leq k$, there exists a $v_i = 1$ or 2 and a $g_i \in G$ with $\phi\alpha^{-1}(x_j) \in g_iH_{v_i}g_i^{-1}$ for each x_j which occurs in w_i.*

If in addition $w = w(x_1, x_2, \ldots, x_n)$ is a quadratic word then statement

(a) *can be replaced by*
(a') *$w' = w_1w_2 \cdots w_k$ for some $k \geq 0$ and where each w_i, $1 \leq i \leq k$, is a quadratic word in F. Furthermore, if w is alternating then each w_i is also alternating.*
Moreover, if ϕ is an epimorphism, so that U is a generating system for G, then we may assume that $g_i = 1$ for all i in (c).

We note that the theorem does not hold without the assumption that A is normal in G, even if w is a quadratic word. This can be seen from the following example. Let $F = \langle x_1, x_2, x_3 \mid \rangle$ and $w = x_1^2 x_2 x_3 x_2^{-1} x_3^{-1}$. Then w is a regular quadratic word. Now let G be the group presentation $G = \langle t, s_1, s_2, s_3, s_4 \mid s_1^2 = s_2^2 = s_3^2 = s_4^2 = t^2 s_1 s_2 s_3 s_4 = 1\rangle$. Then $G = H_1 *_A H_2$ with $H_1 = \langle t \mid \rangle$, $H_2 = \langle s_1, s_2, s_3, s_4 \mid s_1^2 = s_2^2 = s_3^2 = s_4^2 = 1\rangle$ and $A = \langle t^2 \rangle = \langle s_1 s_2 s_3 s_4 \rangle$. Note that A is not normal in H_2 and therefore A is not normal in G. Let the homomorphism $\phi: F \to G$ be defined by $\phi(x_1) = u_1 = t$, $\phi(x_2) = u_2 = s_1 s_2$, $\phi(x_3) = u_3 = s_3 s_1$. Then $\phi(X) = U = \{t, s_1 s_2, s_3 s_1\}$. The set U is regular but Theorem 1.5.51 does not hold (see Theorem 1.4.41 and the remark after Theorem 1.4.41).

If the amalgamated subgroup A is malnormal in both factors then we can obtain the following somewhat weaker result.

Theorem 1.5.52. *Suppose $G = H_1 *_A H_2$, $H_1 \neq A \neq H_2$, with A malnormal in G. Let $1 \neq w = (x_1, x_2, \dots, x_n)$ be a reduced, regular quadratic word in F, with F free of rank n, $1 \leq n \leq 4$, on the ordered basis $\{x_1, x_2, \dots, x_n\}$. Furthermore, let $\phi: F \to G$ be a homomorphism such that $U = \phi(X)$ is regular and $\phi(w) = 1$. Then the pair (w, U) is Nielsen equivalent to a pair $(w', U') = (\alpha(w), \alpha^{-1}(U))$ with $\alpha: F \to F$ and automorphism such that:*
(a) *$w' = w_1 w_2$ where w_1 and w_2 are also quadratic in F.*
(b) *For $i = 1, 2$, $\phi\alpha^{-1}(w_i)$ is conjugate to an element of A.*
(c) *For $i = 1, 2$ there is a $v_i = 1$ or 2 and a $g_i \in G$ with $\phi(\alpha^{-1}(x_j)) \in g_i H_{v_i} g_i^{-1}$ for each x_j which occurs in w_i.*

1.5.6 Lyndon property LZ, restricted Gromov groups and Baumslag doubles

We now give some applications. For convenience we give a short definition:

Definition 1.5.53. A group G has the *Lyndon property LZ* if whenever $x^2 y^2 z^2 = 1$ for $x, y, z \in G$ the subgroup $\langle x, y, z\rangle$ is cyclic. Recall that the equation $x^2 y^2 z^2 = 1$ in G is equivalent to the equation $x^2 [y, t] = 1$.

Examples 1.5.54. We consider examples of groups that have the Lyndon property LZ.
1. Free groups.
2. 3-free groups.
3. $G = H_1 * H_2$ where H_1 and H_2 do have the property LZ and where G has no element of order 2 (see [64]).

Theorem 1.5.55. *Suppose that H_1 and H_2 are groups with no element of order 2 and that G is the amalgamated product $G = H_1 *_A H_2$, $H_1 \neq A \neq H_2$, with A malnormal in G. Then, if both H_1 and H_2 have property LZ then G also has property LZ.*

Proof. Certainly in a free group F of rank 3 with basis $X = \{x_1, x_2, x_3\}$ the word $x_1^2 x_2^2 x_3^2$ is regular. Let $\phi: F \to G$ with $x_1 \to x$, $x_2 \to y$ and $x_3 \to z$ be a homomorphism from F into G. Let $U = \{x, y, z\}$ with $x^2 y^2 z^2 = 1$.

Let $H = \langle x, y, z\rangle$. If $\mathrm{rk}(H) = 1$ then we are done. Assume that $\mathrm{rk}(H) > 1$. H cannot be a free product $H = B * C$ with $B \neq \{1\} \neq C$ because G has no elements of order 2 (see [64]).

Since A is malnormal in G we may assume in any case that $x, y \in H_1$. It does not matter if U is regular or not. Consider $K = \langle H_1, z\rangle$. The element $z \in G$ has a unique representation $z = h_1 h_2 \cdots h_m b$ with $m \geq 0$, $b \in A$ and the h_i are left coset representative, and lie alternately in distinct factors of G and are non-trivial. If $z \in H_1$ then $H \subset H_1$ and the result holds because H_1 has property LZ.

Suppose then $z \notin H_1$. Because $H_1 \subset K$ we may assume that $b = 1$, $m \geq 1$, $h_1 \in H_2 \setminus A$ and $h_m \in H_2 \setminus A$. But then $|r| \geq |z| \geq 1$ for every freely reduced word $r \in K$ which involves z. It follows that $K = H_1 * \langle z\rangle$ which gives a contradiction because G has no element of order 2. Therefore G has property LZ. □

We remark that more related properties are considered in [88]. Next we define RG groups, compare [91].

Definition 1.5.56. A group G is a *restricted Gromov group* or *RG group*, if G satisfies the following property: If g and h are elements in G then either the subgroup $\langle g, h\rangle$ is cyclic or there exists a positive integer t with $g^t \neq 1 \neq h^t$ and $\langle g^t, h^t\rangle = \langle g^t\rangle * \langle h^t\rangle$.

An RG group G has the property that every Abelian subgroup is locally cyclic. Hence, it is often convenient to assume that every element of G is contained in a maximal cyclic subgroup of G. This last property is certainly satisfied in hyperbolic groups (see Subsection 3.4.2).

Theorem 1.5.57. *An RG group G cannot contain an infinite dihedral subgroup. Furthermore, G can contain an element of finite even order only if it is locally finite cyclic.*

Proof. Consider the infinite dihedral group $\mathbb{Z}_2 * \mathbb{Z}_2$ with the presentation $\langle a, b \mid a^2 = b^2 = 1\rangle$. Let $u = a$, $v = ab$. Then $\langle u, v\rangle = \langle a, b\rangle$ non-cyclic. Furthermore, for any positive integer k we have $(uv^k)^2 = 1$. Therefore there cannot exist a positive integer t with $u^t \neq 1 \neq v^t$ and $\langle u^t, v^t\rangle = \langle u^t\rangle * \langle v^t\rangle$. Therefore $\mathbb{Z}_2 * \mathbb{Z}_2$ is not an RG group. Since subgroups of RG groups are RG groups it follows that no RG groups can contain an infinite dihedral group.

Now suppose G is an RG group and $g \in G$ is an element of order 2. Suppose there is an element h in G with $\langle g, h\rangle$ non-cyclic. From the RG property there is an integer t such that $\langle g^t, h^t\rangle = \langle g^t\rangle * \langle h^t\rangle = \langle g\rangle * \langle h^t\rangle \cong \mathbb{Z}_2 * \mathbb{Z}$ or $\mathbb{Z}_2 * \mathbb{Z}_k$ for some integer k.

In either cases the group contains a subgroup isomorphic to $\mathbb{Z}_2 * \mathbb{Z}_2$ which is impossible. Therefore $\langle g, h\rangle$ must be cyclic for all h in G. Furthermore, G can contain no element of infinite order. Otherwise if x had infinite order then $\langle g, x\rangle$ would be infinite cyclic, which is impossible since g has order 2. □

So far we have the following examples of RG groups:
(1) Free groups.

(2) Let G_1, G_2 be RG groups with only odd torsions. Then $G_1 * G_2$ in an RG group. This is a direct consequence. Especially,

$$G = \langle a_1, a_2, \dots, a_n \mid a_1^{\alpha_1} = a_2^{\alpha_2} = \dots = a_n^{\alpha_n} = 1\rangle$$

with $e_i = 0$, that is, a_i has infinite order; or $e_i \geq 2$ and e_i odd. Then G is an RG group.

(3) 2-free groups.

Theorem 1.5.58. *If G is an RG group then G is commutative transitive.*

Proof. Certainly, every Abelian subgroup of G is locally cyclic. Let $x, y, z \in G$ with $y \neq 1$ and $[x, y] = 1 = [y, z]$. Then we have $x = g^{\alpha}, y = g^{\beta} = h^{\gamma}$ and $z = h^{\delta}$, $\alpha, \beta, \gamma, \delta \in \mathbb{Z}$. This implies $[x, z] = 1$. □

Theorem 1.5.59. *Let H_1, H_2 be RG groups with only odd torsion and $G = H_1 *_A H_2$, $H_1 \neq A \neq H_2$, with A malnormal in G. Then G is an RG group.*

Proof. Let $a, b \in G$ with $K = \langle a, b\rangle$ non-cyclic. Then K is either the free product of two cyclic groups or is conjugate to a subgroup of G_1 or of G_2. This implies that G is an RG group. □

Corollary 1.5.60. *Let H_1, H_2 be RG groups and $G = H_1 *_A H_2$ with $A \neq \{1\}$. Let A be malnormal in G. Then G is an RG group.*

Proof. If $A = H_1$ or $A = H_2$ then G clearly is an RG group. If either H_1 or H_2 is Abelian then $A = H_1$ or $A = H_2$ since A is malnormal in G and non-trivial. Therefore G is an RG group in this case. Now assume that both H_1 and H_2 are non-Abelian. Then, by Theorem 1.5.57, the factors H_1 and H_2 have only odd torsion. Now the result follows analogously as for Theorem 1.5.59. □

Theorem 1.5.61. *Let $G = \langle a_1, a_2, \dots, a_p, b_1, b_2, \dots, b_q \mid wv = 1\rangle$ where $1 \leq p$, $1 \leq q$, $1 \neq w = w(a_1, a_2, \dots, a_p)$ not a proper power in the free group on $a_1, a_2, \dots, a_p$ and $1 \neq v = v(b_1, b_2, \dots, b_q)$ not a proper power in the free group on $b_1, b_2, \dots, b_q$. Then:*

1. *G is 3-free.*
2. *Let H be a four generator subgroup. Then one of the following cases occurs:*
   (a) *H is free of rank ≤ 4.*
   (b) *If $\{x_1, x_2, x_3, x_4\}$ is a generating system of H then there is a Nielsen transformation from $\{x_1, x_2, x_3, x_4\}$ to a system $\{y_1, y_2, y_3, y_4\}$ with $y_1, y_2 \in z\langle a_1, a_2, \dots, a_p\rangle z^{-1}$ and $y_3, y_4 \in z\langle b_1, b_2, \dots, b_q\rangle z^{-1}$ for a suitable $z \in G$. Furthermore, for the generating system $\{x_1, x_2, x_3, x_4\}$ of H there is a presentation of H with one defining relation.*

A long combinatorial proof can be found in [225]. We should remark that we can shorten some arguments because of later refinements of the Nielsen method or by using results we were not aware of at that time, for instance Baumslag's deep Theorem 1.2.40

which makes section 2 in [225] almost superfluous. Still, the proof is a long combinatorial one.

Theorem 1.5.61 is in parts reproved in a different manner by Baumslag and Shalen [23]. We give a sketch of the proof which covers the main steps.

Sketch of the proof of Theorem 1.5.61. If G is a free group then is nothing to show. G is a free group if and only if w is primitive in $\langle a_1, a_2, \dots, a_p\rangle$ or v is primitive in $\langle b_1, b_2, \dots, b_q\rangle$. Hence assume that w is not primitive in $\langle a_1, a_2, \dots, a_p\rangle$ and v is not primitive in $\langle b_1, b_2, \dots, b_q\rangle$. We may write G as a free product $G = H_1 *_A H_2$ with amalgamation where $H_1 = \langle a_1, a_2, \dots, a_p \mid\rangle$, $H_2 = \langle b_1, b_2, \dots, b_q \mid\rangle$ and $A = \langle w\rangle = \langle v\rangle$. We note that A is malnormal in G.

Statement 1. is already given in Corollary 1.5.25. Hence, let $\mathrm{rk}(H) \leq 4$ and $H = \langle x_1, x_2, x_3, x_4\rangle$. We apply the Nielsen method for $\{x_1, x_2, x_3, x_4\}$. If $\mathrm{rk}(H) \leq 3$ then H is free.

Now assume that $\mathrm{rk}(H) = 4$. The main part is to show the following. If there is a Nielsen transformation from $\{x_1, x_2, x_3, x_4\}$ to a system $\{y_1, y_2, y_3, y_4\}$ with $y_1 = gv^\alpha g^{-1}$, $\alpha \neq 0$, then H is free. Here we use a combination of Nielsen's method in free products with amalgamation and in HNN groups. We consider HNN groups in the next section.

Now, assume that there is no Nielsen transformation to a system $\{y_1, y_2, y_3, y_4\}$ where y_1 is conjugate to a power of v. Then, especially, the system $\{x_1, x_2, x_3, x_4\}$ is regular. If there is a Nielsen transformation to a system $\{y_1, y_2, y_3, y_4\}$ with $y_1, y_2, y_3 \in gH_ig^{-1}$ for $i = 1$ or 2, $g \in G$, then H is free from the proof of Corollary 1.5.25. Now, let H be not free. Then there is a regular word $U(X_1, X_2, X_3, X_4)$ in the free group on $\{X_1, X_2, X_3, X_4\}$ with $u(x_1, x_2, x_3, x_4) = 1$ in G because G is 3-free and H is not free. Now we may consider the equation $u(x_1, x_2, x_3, x_4) = 1$.

We may apply Theorem 1.5.51 in combination with E-transformations. We finally may assume without loss of generality that $x_1, x_2 \in H_1$, $w \in \langle x_1, x_2\rangle$ and $x_3, x_4 \in H_2$, $\langle v, x_3, x_4\rangle$ free of rank 2. Relative to $\{v, x_3, x_4\}$ there is a presentation of $\langle v, x_3, x_4\rangle$ with a single defining relation $r_0(v, x_3, x_4) = 1$. We obtain a presentation of $\langle x_1, x_2, x_3, x_4\rangle$ with a single defining relation by expressing $v = w^{-1}$ as a word in x_1 and x_2. □

We give an application of Theorem 1.5.61. Recall that a *Baumslag double* is an amalgamated product of the form $G = F *_A \overline{F}$, $A = \langle w\rangle = \langle\overline{w}\rangle$, where F is a finitely generated free group, $\overline{F}$ is an isomorphic copy, w a reduced non-trivial word in F and $\overline{w}$ is its copy in $\overline{F}$.

We call G a *hyperbolic Baumslag double* if w is not a proper power in F. We will later see that a hyperbolic Baumslag double is a hyperbolic group in the sense of Gromov. We consider hyperbolic groups later in the book.

A question of Gromov states, in this situation, that a hyperbolic Baumslag double contains a subgroup that is isomorphic to a fundamental group of a closed hyperbolic surface. There are many activities concerning Gromov's question (see [161] and the references given there).

Theorem 1.5.62. *Let $G = \langle F, \bar{F} \mid w = \bar{w} \rangle$ be a hyperbolic Baumslag double. Then*
1. *G contains an orientable surface group of genus 2 if and only if w is a commutator $w = [U, V]$ for some $U, V \in F$ with $[U, V] \neq 1$, and*
2. *G contains a non-orientable surface group of genus 4 if and only if $w = X^2Y^2$ for some $X, Y \in F \setminus \{1\}$.*

Proof. We consider the orientable case (the non-orientable case is analogous).

Suppose that $w = [U, V] \neq 1$ in F. Then $w(\overline{w})^{-1} = [U, V][\overline{V}, \overline{U}] = 1$, and the subgroup $H = \langle U, V, \overline{U}, \overline{V} \rangle$ is certainly an orientable surface group of genus 2 by the proof of Theorem 1.5.61. Recall that if $F = \langle a, b \mid \rangle$ is generated by $\{r, s, [a, b]\}$ then there is a $[a, b]$-stable Nielsen transformation from $\{r, s, [a, b]\}$ to $\{a, b, [a, b]\}$. Conversely, let H be a subgroups of G that is an orientable surface group of genus 2. Hence, H has a presentation $H = \langle x_1, x_2, x_3, x_4 \mid [x_1, x_2][x_3, x_4] = 1 \rangle$. We consider the system $\{x_1, x_2, x_3, x_4\}$ in G and apply Nielsen cancellations within G with respect to the quadratic word $v = [x_1, x_2][x_3, x_4]$. From Theorem 1.5.61 we see, without loss of generality, that there are elements $U, V \in F$ with $[U, V] = w^n$ for some non-zero integer n. Since in a free group a commutator is not a proper power we have $n = \pm 1$. This proves the theorem. □

A more detailed description is given in [94].

Corollary 1.5.63. *Let $G = \langle F, \bar{F} \mid w = \bar{w} \rangle$ be a hyperbolic Baumslag double. Then G contains orientable surface groups of all finite genus $g \geq 2$ if and only if w is a commutator in F.*

Proof. An orientable surface group of genus 2 contains an orientable surface group of any finite genus $g \geq 2$ as a subgroup. □

We finally mention the useful lifting theorem whose proof is clear.

Theorem 1.5.64. *Let $f: G \to H$ be a group epimorphism and $H = A *_K B$ a free product with amalgam K. Then $G = f^{-1}(A) *_{f^{-1}(K)} f^{-1}(B)$.*

Example 1.5.65. Let $G = \langle a, b \mid [a, b]^n = 1 \rangle$, $n = 2^r p$ with $p \geq 3$, p odd. Write $p = 2k + 1$, $k \geq 1$. Let $H = \langle s_1, s_2, s_3, s_4 \mid s_1^2 = s_2^2 = s_3^2 = s_4^p = s_1s_2s_3s_4 = 1 \rangle = \langle a, b \mid [a, b]^p = ([a, b]^k a)^2 = ([a, b]^k ab)^2 = ([a, b]^k aba)^2 = 1 \rangle$ with $a = s_1s_2$, $b = s_3s_1$. Then there is an obvious epimorphism from G to H, and hence G is a free product with amalgamation.

We come back to the decomposition problem in the next sections.

1.6 HNN extensions

This construction is named after G. Higman, B. H. Neumann and H. Neumann, who considered this construction first.

1.6.1 Definition and first properties of HNN extensions

Definition 1.6.1. Let H and A be groups and $i_0, i_1 : A \hookrightarrow H$ monomorphisms. Let $T = \langle t \rangle \cong \mathbb{Z}$ and $N = \langle\langle\{t^{-1}(i_0(a))t(i_1(a))^{-1} \mid a \in A\}\rangle\rangle^{H*T}$, the normal closure, where a goes through A completely.

It is enough to assume that a just goes through to a generating system of A.

A group G is called *HNN extension* or *HNN group* of the base (group) H with *stable letter t* and *associated subgroups* $i_0(A)$ and $i_1(A)$.

We again consider i_0 as an embedding and $i_0(A) = A$ as subgroup of H. Then A and $i_1(A)$ are isomorphic via $i := i_1 : A \to i_1(A) =: B$ (B is a subgroup of H isomorphic to A). We write $G = \langle H, t \mid t^{-1}At = B\rangle$. If we will emphasize the isomorphism i explicitly, then we also write $G = \langle H, t \mid t^{-1}At = i(A)\rangle$ or $G = \langle H, t \mid t^{-1}At \overset{i}{=} B\rangle$. If $H = \langle X \mid R\rangle$ and Y a generating system of A then we have the presentation $G = \langle X, t \mid R, t^{-1}yt = i(y), y \in Y\rangle$. If $g_0, g_1 \in H$ with $j_\nu(a) := g_\nu^{-1} i_\nu(a) g_\nu$, $\nu = 0, 1$ then $\langle H, t \mid t^{-1}(i_0(a))t = i_1(a)\rangle$ and $\langle H, t' \mid t'^{-1}(j_0(a))t' = j_1(a)\rangle$ are isomorphic via $h \mapsto h$, $h \in H$, and $t \mapsto g_0 t' g_1^{-1}$.

The construction has an obvious generalization to the case of a family A_α, $\alpha \in I$, of subgroups and monomorphisms $i_{0\alpha}, i_{1\alpha} : A_\alpha \to H$.

Let $T = \bigoplus_{\alpha \in I} \langle t_\alpha \rangle$ with $\langle t_\alpha \rangle \cong \mathbb{Z}$ for all $\alpha \in I$. Let

$$N = \langle\langle\{t_\alpha^{-1}(i_{0\alpha}(a_\alpha))t_\alpha(i_{1\alpha}(a_\alpha))^{-1} \mid \alpha \in I, a_\alpha \in A_\alpha\}\rangle\rangle^{H*T}.$$

Then $G = H * T/N$ is called the HNN extension of the base H with stable letters t_α and associated pairs of subgroups $i_{0\alpha}(A_\alpha)$, $i_{1\alpha}(A_\alpha)$.

However, we consider here essentially the case $G = \langle H, t \mid t^{-1}At = B\rangle$. The two constructions of free products with amalgamation and HNN extensions are closely interwoven. These connections mean that results about amalgamated free products can sometimes be translated into results about HNN extensions and vice versa.

However we decided to consider these constructions in different sections because of many different properties and examples.

Examples 1.6.2. 1. The group $\langle a, t \mid t^{-1}a^m t = a^n\rangle$, the *Baumslag–Solitar group* $\mathrm{BS}(m, n)$, is an HNN extension of $\langle a\rangle$ with stable letter t and associated subgroups $\langle a^m\rangle$ and $\langle a^n\rangle$.

2. Let $G = \langle a_1, a_2, \ldots, a_p, t \mid tUt^{-1} = V\rangle$, $p \geq 1$, where $U, V \in \langle a_1, a_2, \ldots, a_p\rangle$ and $U \neq 1 \neq V$. Then G is called a *conjugacy pinched one-relator group*.
   The Baumslag–Solitar group $\mathrm{BS}(m, n)$ is a special case.
   We remark that G is a free group if and only if one of the following holds:
   (a) We have $\langle a_1, a_2, \ldots, a_p \mid \rangle$ has a basis $\{U, x_1, x_2, \ldots, x_{p-1}\}$ such that V is conjugate in $\langle a_1, a_2, \ldots, a_p\rangle$ to some $V_1 \in \langle x_1, x_2, \ldots, x_{p-1}\rangle$.
   (b) We have $\langle a_1, a_2, \ldots, a_p \mid \rangle$ has a basis $\{V, x_1, x_2, \ldots, x_{p-1}\}$ such that U is conjugate in $\langle a_1, a_2, \ldots, a_p\rangle$ to some $U_1 \in \langle x_1, x_2, \ldots, x_{p-1}\rangle$.

Conjugacy pinched one-relator groups are the HNN analogs of cyclically pinched one-relator groups and are motivated by the structure of hyperbolic orientable surface groups $S_g = \langle a_1, b_1, a_2, b_2, \ldots, a_g, b_g \mid [a_1, b_1][a_2, b_2] \cdots [a_g, b_g] = 1\rangle$, $g \geq 2$. Then S_g is an HNN group of the form $S_g = \langle a_1, b_1, a_2, b_2, \ldots, a_g, t, tUt^{-1} = V\rangle$ where $U = a_g$ and $V = [a_1, b_1][a_2, b_2] \cdots [a_{g-1}, b_{g-1}]a_g$.

Theorem 1.6.3 (Universal property). *Let $G = \langle H, t \mid t^{-1}At \stackrel{i}{=} B\rangle$ and $j: H \to G$ be as above. Let $H \to K$ be a homomorphism. Let K contain an element k with $k^{-1}\varphi(a)k = \varphi(i(a))$ for all $h \in H$. Then there exists a unique homomorphism $\psi: G \to K$ with $\varphi = \psi \circ j$ and $k = \psi(t)$.*

Especially, there exists a homomorphism $G \to \langle t\rangle$ with $t \mapsto t$, $h \mapsto 1$ for all $h \in H$. The image of t in $G = H * \langle t\rangle/N$ (as above) generates necessarily in G an infinite cyclic group. Hence, we may consider $\langle t\rangle$ as a subgroup of G.

From now on we write mostly $A_1 = A$ and $A_{-1} = B$, hence, $G = \langle H, t \mid t^{-1}A_1t \stackrel{i}{=} A_{-1}\rangle$. Let $j: H \to G$ as above. We choose a complete set of representatives R_1 and R_{-1}, respectively, for the left cosets of A_1 and A_{-1}, respectively, in H. Again we assume always that A_1 and A_{-1}, respectively, is represented by 1.

Theorem 1.6.4 (Normal form). 1. *The map j is a monomorphism.*

2. *If we consider H as a subgroup of G, then each $g \in G$ can be written uniquely as $g = u_1t^{\epsilon_1}u_2t^{\epsilon_2} \cdots u_nt^{\epsilon_n}h$ with $n \geq 0$, $h \in H$, $\epsilon = \pm 1$, $u_i \in R_{\epsilon_i}$ and $\epsilon_i = \epsilon_{i+1}$ if $u_{i+1} = 1$ ($n = 0$ means $g = h$).*

If we consider H as a subgroup of G and if we take again the inverses R_1^{-1}, R_{-1}^{-1} as coset representatives for the right cosets, then each $g \in G$ can be written uniquely as $g = g_0t^{\epsilon_1}v_1 \cdots t^{\epsilon_n}v_n$ with $n \geq 0$, $v_i \in R_{-\epsilon_i}^{-1}$, $\epsilon_i = \pm 1$, $g_0 \in H$ and $\epsilon_i = \epsilon_{i+1}$ if $v_i = 1$.

Proof. In fact, 1. follows from a modification of the writing as in the case of amalgamated products. Hence, let $H < G$. We work with right cosets, apart from that the proof is analogous as for free products with amalgamation. It is clear that each $g \in G$ has such a representation (if $H < G$). Let S be the set of normal form sequences $(g_0, t^{\epsilon_1}, v_1, \ldots, t^{\epsilon_n}, v_n)$ with $n \geq 0$, $v_i \in R_{-\epsilon_i}^{-1}$, $\epsilon = \pm 1$, $g_0 \in H$, and $\epsilon_i = \epsilon_{i+1}$ if $v_i = 1$.

We define a homomorphism $\varphi: G \to \mathrm{Per}(S)$. It is enough to define φ for H and t and to show the homomorphism property. For $h \in H$ we define

$$\varphi(h)(g_0, t^{\epsilon_1}, v_1, \ldots, t^{\epsilon_n}, v_n) = (hg_0, t^{\epsilon_1}, v_1, \ldots, t^{\epsilon_n}, v_n).$$

If $\epsilon_1 = -1$ and $g_0 \in A_{-1}$ then we define

$$\varphi(t)(g_0, t^{-1}, v_1, \ldots, t^{\epsilon_n}, v_n) = (i^{-1}(g_0)v_1, t^{\epsilon_2}, \ldots, t^{\epsilon_n}, v_n),$$

(remark that $tg_0t^{-1} = i^{-1}(g_0)$ if $g_0 \in A_{-1}$), otherwise

$$\varphi(t)(g_0, t^{\epsilon_1}, v_1, \ldots, t^{\epsilon_n}, v_n) = (i^{-1}(k), t, v_0, t^{\epsilon_1}, \ldots, t^{\epsilon_n}, v_n),$$

where $g_0 = kv_0$ with $k \in A_{-1}$, $v_0 \in R_{-1}^{-1}$. The map φ defines a homomorphism from G to $\mathrm{Per}(S)$, and we have

$$\varphi(g_0 t^{\epsilon_1} v_1 \cdots t^{\epsilon_n} v_n)(1) = (g_0, t^{\epsilon_1}, v_1, \ldots, t^{\epsilon_n}, v_n). \qquad \square$$

For a detailed proof with van der Waerden's method, see [177, pp. 182–184]. In the following we consider H as a subgroup of G (and j as an embedding). A word $g = h_1 t^{\epsilon_1} h_2 t^{\epsilon_2} \cdots h_n t^{\epsilon_n} h$ with $n \geq 0$, $h, h_1, h_2, \ldots, h_n \in H$, $\epsilon_i = \pm 1$ is called *reduced* if never $\epsilon_{i-1} = -1$, $h_i \in A_1$, $\epsilon_i = 1$ or $\epsilon_{i-1} = 1$, $h_i \in A_{-1}$, $\epsilon_i = -1$ ($i \in \{2, 3, \ldots, n\}$).

As a direct consequence from Theorem 1.6.4 we get

Corollary 1.6.5 (Britton's lemma). *If $g = h_1 t^{\epsilon_1} h_2 t^{\epsilon_2} \cdots h_n t^{\epsilon_n} h$, $n \geq 1$, $h, h_1, h_2, \ldots, h_n \in H$, $\epsilon_i = \pm 1$, is reduced then $g \neq 1$ in G.*

Remarks 1.6.6. 1. The normal form theorem follows reversely also from Britton's lemma, together with the embedding $j: H \to G$ (the actual result of Higman, Neumann and Neumann is indeed that $j: H \to G$ is an embedding).
This can be seen as follows. Let

$$g = u_1 t^{\epsilon_1} u_2 t^{\epsilon_2} \cdots u_n t^{\epsilon_n} h = v_1 t^{\delta_1} v_2 t^{\delta_2} \cdots v_m t^{\delta_m} h_0,$$

both representations as in the normal form theorem (normal forms). If $m = n = 0$ then $h = h_0$ because j is an embedding.
Now, let $m \geq 1$. Then

$$1 = h_0^{-1} t^{-\delta_m} \cdots t^{-\delta_1} v_1^{-1} u_1 t^{\epsilon_1} \cdots u_n t^{\epsilon_n} h.$$

By Britton's lemma we have necessary $\delta_1 = \epsilon_1$, $v_1^{-1} u_1 \in A_{\epsilon_1}$, hence $u_1 = v_1$ because the u_i and v_i are left coset representatives.
It follows

$$1 = h_0^{-1} t^{-\delta_n} \cdots t^{-\delta_2} v_2^{-1} u_2 t^{\epsilon_2} \cdots u_n t^{\epsilon_n} h,$$

and hence $m = n$, $u_i = v_i$ for all i and $h_0 = h$ inductively.

2. Let $g = h_1 t^{\epsilon_1} h_2 t^{\epsilon_2} \cdots h_n t^{\epsilon_n} h \in G$, not necessarily reduced. We get a reduced form by *t-reductions*:
   (i) Replace a subword $t^{-1} a t$, $a \in A_i$, by $i(a) \in A_{-1}$.
   (ii) Replace tat^{-1}, $a \in A_{-1}$, by $i^{-1}(a) \in A_1$.
   After finally many steps we get $g = h_1' t^{\delta_1} t^{\delta_2} \cdots t^{\delta_k} h'$ reduced. If $k \geq 1$ then $g \neq 1$ in G; if $k = 0$ then $g = 1$ in G if and only if $g = 1$ in H (j is an embedding). Therefore we get

Corollary 1.6.7. *Let $G = \langle H, t \mid t^{-1} A_1 t \stackrel{i}{=} A_{-1} \rangle$. Suppose that*
(a) *H has a solvable word problem.*

(b) *The membership problems for* A_1 *and* A_{-1} *in* H *are solvable.*
(c) *The isomorphism* i *is effectively computable.*
Then G *has a solvable word problem.*

We remark that the isomorphism i between A_1 and A_{-1} is effectively computable if A_1 and A_{-1} are finitely generated.

For many purposes it is enough to work with reduced words and Britton's lemma (together with j an embedding).

Lemma 1.6.8. *Let* $g = g_1 t^{\epsilon_1} g_2 t^{\epsilon_2} \cdots g_n t^{\epsilon_n} g_0$ *and* $h = h_1 t^{\delta_1} h_2 t^{\delta_2} \cdots h_m t^{\delta_m} h_0$ *reduced. If* $g = h$ *then* $m = n$ *and* $\epsilon_i = \delta_i$ *for* $i = 1, 2, \ldots, n$.

Proof. From $h = g$ we get

$$1 = h_0^{-1} t^{-\delta_m} \cdots t^{\delta_1} h_1^{-1} g_1 t^{\epsilon_1} \cdots t^{\epsilon_n} g_0.$$

Since g and h are reduced, this can only hold if $\delta_1 = \epsilon_1$ and $h_1^{-1} g_1 \in A_{\epsilon_1}$. Now t-reductions give the result. □

Remark 1.6.9. Let $g = g_1 t^{\epsilon_1} g_2 t^{\epsilon_2} \cdots g_n t^{\epsilon_n} g_0$, $n \geq 0$, $\epsilon_i = \pm 1$, $g_0, g_1, g_2, \ldots, g_n \in H$, reduced. The $n = |g|$ is called the *length* of g, and n is the number of occurrences of t^{ϵ}, $\epsilon = \pm 1$, in g. If $g \in H$, then $|g| = 0$. The length $|g|$ is well defined by Lemma 1.6.8.

An element g is called *cyclically reduced* if all cyclic permutations of g are reduced, especially $t^{\epsilon_n} g_0 g_1 t^{\epsilon_1} \cdots t^{\epsilon_{n-1}} g_n$. Each element of G is conjugate to a cyclically reduced element.

Theorem 1.6.10 (Torsion theorem). *Let* $G = \langle H, t \mid t^{-1} A_1 t = A_{-1} \rangle$. *Each element of finite order in* G *is conjugate to an element of finite order in the base* H. *Hence,* G *has elements of finite order* n *only if* H *has any.*

A finite subgroup of G *is conjugate to a finite subgroup of* H.

Proof. Let $g \in G$ and $g' = g_1 t^{\epsilon_1} g_2 t^{\epsilon_2} \cdots g_n t^{\epsilon_n} g_0$ a cyclically reduced conjugate of g.

If $n \geq 1$ then

$$(g')^k = g_1 t^{\epsilon_1} \cdots t^{\epsilon_n} g_0 g_1 t^{\epsilon_1} \cdots t^{\epsilon_n} g_0 \neq 1,$$

$k \geq 2$, because there is no t-reduction at $t^{\epsilon_n} g_0 g_1 t^{\epsilon_1}$ possible.

We take again the inverse R_1^{-1}, R_{-1}^{-1} as sets of representatives for the right cosets. Then we see that each $g \in G$ has a reduced representation

$$g = u_1 t^{\epsilon_1} \cdots u_m t^{\epsilon_m} k t^{\eta_m} v_m \cdots t^{\eta_1} v_1$$

with $m \geq 0$, $u_i \in R_{\epsilon_i}$, $v_i \in R_{-\eta_k}^{\epsilon_k}$ and $k = h_1 t^{\epsilon} h_2$, $h_1, h_2 \in H$, if $|g|$ is odd and $k \in H$ if $|g|$ is even, the symmetric form for g.

Now let $K < G$, $|K| < \infty$, $g, h \in K$, $g, h \neq 1$. Without any loss of generality let $g \in H$ and

$$g = u_1 t^{\epsilon_1} \cdots u_m t^{\epsilon_m} k t^{\eta_m} v_m \cdots t^{\eta_1} v_1$$

in symmetric form. Since g has finite order we necessarily have $k \in H$, $\eta_i = -\epsilon_i$ and $v_i = u_i^{-1}$ for $i = 1, 2, \ldots, m$. If $u_1^{-1} h u_1 \in A_{\epsilon_1}$ then we can make a reduction for g by replacing h with $t^{-\epsilon_i} u_1^{-1} h u_1 t^{\epsilon_1} \in H$ and g with

$$u_2 t^{\epsilon_2} \cdots u_m t^{\epsilon_m} k t^{-\epsilon_m} u_m^{-1} \cdots t^{\epsilon_2} u_2^{-1}$$

and so on.

Now, let $u_1^{-1} h u_1 \notin A_{\epsilon_1}$ if $m \geq 1$. If $m \geq 1$ then

$$hg = h u_1 t^{\epsilon_1} \cdots u_m t^{-\epsilon_m} k t^{-\epsilon_m} u_m^{-1} \cdots t^{-\epsilon_1} u_1^{-1}$$

is cyclically reduced and has infinite order by Britton's lemma. Hence, here $m = 0$, that is, $g = k \in H$ and therefore $K < H$. □

Theorem 1.6.11 (Lemma of Collins on conjugacy in HNN extensions). *Let $G = \langle H, t \mid t^{-1} A_1 t = A_{-1} \rangle$. Let $g = g_1 t^{\epsilon_1} g_2 t^{\epsilon_2} \cdots g_n t^{\epsilon_n}$, $n \geq 1$, and h be conjugate cyclically reduced elements of G. Then $|u| = |v|$, and g can be obtained from h by taking a suitable cyclic permutation of h, which ends in t^{ϵ_n}, and then conjugating by an element z, where $z \in A_{-\epsilon_n}$.*

The proof of Theorem 1.6.11 is an easy calculation, parallel to Theorem 1.5.16. A proof with all simple details can be found in [177, p. 186]. Analogously to the case of free products and amalgamated free products we get the following theorem.

Theorem 1.6.12. 1. *If H is finite then $G = \langle H, t \mid t^{-1} A_1 t = A_{-1} \rangle$ is residually finite.*
2. *If H is residually finite and A_1 finite then $G = \langle H, t \mid t^{-1} A_1 t = A_{-1} \rangle$ is residually finite.*

Proof. For 1. consider $G/\langle\langle t^n \rangle\rangle$ for n sufficiently large. For 2. apply that for each $h \in H \setminus \{1\}$ there is a homomorphism $\varphi\colon H \to \overline{H}$, $\overline{H}$ finite, with $\varphi(h) \neq 1$. □

Remark 1.6.13. We should mention that Theorem 1.6.12 cannot be really generalized. We take the Baumslag–Solitar group $G = \mathrm{BS}(2, 3) = \langle a, t \mid t^{-1} a^2 t = a^3 \rangle$ and claim that it is non-Hopfian. This we see as follows:

Let $\varphi\colon G \to G$, $\varphi(t) = t$, $\varphi(a) = a^2$. The map φ is well defined and an endomorphism because

$$\varphi(t^{-1} a^2 t) = t^{-1} a^4 t = a^6 = \varphi(a^3).$$

The map φ is surjective because $a = a^{-2} t^{-1} a^2 t$. Let $g = t^{-1} a t a t^{-1} a^{-1} t a^{-1}$. We have $g \neq 1$ by the normal form theorem. However

$$\varphi(g) = t^{-1} a^2 t a^2 t^{-1} a^{-2} t a^{-2} = a^3 a^2 a^{-3} a^{-2} = 1.$$

Therefore, G is non-Hopfian, and also not residually finite by Theorem 1.1.21. From Meskin [191] and Collins and Levin [63] we have the following.

Theorem 1.6.14. *Let* $\mathrm{BS}(m,n) = \langle a,t \mid t^{-1}a^m t = a^n\rangle$. *Then*
1. $\mathrm{BS}(m,n)$ *is residually finite if and only if* $|m| = 1$ *or* $|n| = 1$ *or* $|m| = |n|$; *and*
2. $\mathrm{BS}(m,n)$ *is Hopfian if and only if* m *and* n *have the same prime divisors.*

Hence, the Baumslag–Solitar groups are divided into three classes: those that are residually finite, those that are Hopfian and those that are non-Hopfian. They play a surprisingly useful role in combinatorial and geometric group theory. In a number of situations they have provided examples and counterexamples which mark boundaries between different classes of groups and they often provide a testbed for theories and techniques. Especially $\mathrm{BS}(m,n)$ can only have a linear representation if $|m| = 1$ or $|n| = 1$ or $|m| = |n|$ by Malzév's Theorem 1.1.18.

However, if $|m| = |n|$ or $|m| = 1$ or $|n| = 1$ then $\mathrm{BS}(m,n)$ is linear (see [138] and [126]). For example, $\mathrm{BS}(1,-1)$ is the fundamental group of the Klein bottle, and $\mathrm{BS}(1,1)$ is the fundamental group of the torus. We discuss fundamental groups and covering spaces later in this book.

It is nice to mention that $\mathrm{BS}(1,n)$, $n > 1$, has a faithful representation into $\mathrm{PSL}(2,\mathbb{R})$ via

$$t \mapsto \pm\begin{pmatrix} \frac{1}{\sqrt{n}} & 0 \\ 0 & \sqrt{n} \end{pmatrix}, \quad a \mapsto \pm\begin{pmatrix} 1 & 1 \\ 0 & 1 \end{pmatrix}.$$

We remark that the image group is not discrete for $n > 1$. We may ask about a possible connection between linearity and residually finiteness. There seems to be no general connection. Drutu and Sapir [72] showed that the group $\langle a,t \mid t^2at^{-2} = a^2\rangle$ is residually finite but not linear.

1.6.2 Embeddings of groups using HNN extensions

In this subsection we discuss the following theorem and its consequences.

Theorem 1.6.15 (Higman–Neumann–Neumann). *Each countable group* C *can be embedded in a group* G *which is generated by two elements of infinite order. Then* G *has an element of finite order* n *if and only if* C *has one. If* C *is finitely presented then also* G.

Proof. Let $C = \langle c_1, c_2, \ldots \mid s_1, s_2, \ldots\rangle$ with a countable set of generators.

Let $F = C * \langle a,b \mid \rangle$. The set $\{a, b^{-1}ab, b^{-2}ab^2, \ldots, b^{-n}ab^n, \ldots\}$ generates freely a subgroup of $\langle a,b \mid \rangle$, $n \in \mathbb{N}$ (see Section 1.1). Then the set $\{b, c_1a^{-1}ba, \ldots, c_na^{-n}ba^n, \ldots\}$ generates freely a free subgroup of F, $n \in \mathbb{N}$.

This we can see as follows. Let π be the projection from F onto $\langle a,b \mid \rangle$ defined by $a \mapsto a$, $b \mapsto b$, $c_i \mapsto 1$ for all i. The images $b, a^{-i}ba^i$, $i \geq 1$, generate freely a free subgroup of $\langle a, b \mid \rangle$, and hence $\{b, c_1, a^{-1}ba, \dots, c_n a^{-n} b a^n, \dots\}$ generates freely a free subgroup of F. Therefore the group

$$G = \langle F, t \mid t^{-1}at = b, t^{-1}b^{-i}ab^i t = c_i a^{-i} b a^i, \text{ for all } i \geq 1 \rangle$$

is an HNN extension of F. Hence, C can be embedded in G. G is generated by t and a which follows from the defining relations. By the torsion theorem, Theorem 1.6.10, we see that G has an element of finite order if and only if C has one.

Now, let $C = \langle c_1, c_2, \dots, c_m \mid s_1, s_2, \dots, s_k \rangle$ be finitely presented. The relations $t^{-1}at = b$, $t^{-1}b^{-i}ab^i t = c_i a^{-i} b a^i$ may be eliminated by Tietze transformations because in each relation a generating element b or c_i occurs exactly once. □

Corollary 1.6.16. *There are exactly $2^{\aleph_0}$ non-isomorphic two-generator groups.*

Proof. Let S be some set of prime numbers. Let $T_S = \oplus_{p \in S} \mathbb{Z}_p$. Embed T_S into a two-generator group G_s as in Theorem 1.6.15. T_S, and therefore G_S, has an element of order p if and only if $p \in S$. Since there are $2^{\aleph_0}$ distinct sets of prime numbers, we have at least $2^{\aleph_0}$ non-isomorphic two-generator groups. Now each two-generator group is a quotient of $F_2 = \langle a, b \mid \rangle$. F_2 has $2^{\aleph_0}$ subsets, hence there are at most $2^{\aleph_0}$ non-isomorphic two-generator groups. □

Theorem 1.6.17. *Every countable group C can be embedded into a countable group G in which all elements with the same order are conjugate.*

Proof. Let $\{(a_i, b_i) \mid i \in I\}$ be the set of all ordered pairs of elements from C which have the same order. In

$$C^* := \langle C, t_i, i \in I \mid t_i^{-1} a_i t_i = b_i, i \in I \rangle$$

any two elements of C, which have the same order in C, are conjugate. Define $G_0 = C$. We argue inductively. Let G_i be defined. We embed G_i in a group G_{i+1} in which any two elements of G_i with the same order are conjugate. Then $G^* = \bigcup_{i=0}^{\infty} G_i$ is the desired group. □

We call a group G *divisible* if for each $g \in G$ and $n \in \mathbb{N}$ there exists a $y \in G$ with $g = y^n$.

Theorem 1.6.18. *Each countable group C can be embedded in a countable simple and divisible group.*

Proof. First of all we embed C in a countable group K which contains elements of all orders (take for instance $C \oplus \mathbb{Z} \oplus_{n \in \mathbb{N}} \mathbb{Z}_n$). Now we embed $K * \langle x \rangle$ in a two-generator group U in which both generators have infinite order (see Theorem 1.6.15). Now embed U in a countable group G in which all elements with the same order are conjugate. We have the following embeddings

$$C \hookrightarrow K \hookrightarrow K * \langle x \rangle \hookrightarrow U \hookrightarrow G.$$

Now, G is simple and divisible. This we see as follows:

Let $\{1\} \neq N \triangleleft G$. Since K has elements of all orders and all elements of the same order are conjugate, N must contain some element $z \in K \setminus \{1\}$. Now $x^{-1}z^{-1}x \in N$, and therefore, $x^{-1}z^{-1}xz \in N$ has infinite order. Hence, N contains the generators of U by the conjugation property of G. Therefore $K \subset U \subset N$. Since K contains elements of all orders, and since all elements of the same order are conjugate, we get $N = G$, that is, G is simple.

Now, let $g \in G$, $n \in \mathbb{N}$. Suppose that g has order $m \in \mathbb{N} \cup \{\infty\}$. Since G contains elements of all orders, G contains an element z of order mn. Now, z^n has order m. Therefore, we have an element $v \in G$ with $g = v^{-1}z^n v = (v^{-1}zv)^n$, that is, g is divisible. □

There exist $2^{\aleph_0}$ non-isomorphic countable simple groups by Corollary 1.6.16. There are many embedding theorems in the literature. We just mention two more:

(1) Each countable group C can be embedded in a simple group with six generators (see [192]).
(2) There exists a finitely presented group G in which every finite group can be embedded (see [20]).

At this point we would like to discuss some SQ-universality.

Definition 1.6.19. A group G is called *SQ-universal* if every countable group is embeddable in some factor group.

From Theorem 1.6.15 we know that the free group of rank 2, and hence each free group of finite rank ≥ 2, is SQ-universal. Besides some single examples the first deep result is in [199].

Theorem 1.6.20. *If H is a subgroup of finite index in a group G, then G is SQ-universal if and only if H is SQ-universal.*

We show that a subgroup of finite index in a SQ-universal group is SQ-universal and use the following observation: If G is an SQ-universal group and N is a normal subgroup of G then either N is SQ-universal or the factor group G/N is SQ-universal.

Proof. Suppose that N is not SQ-universal. Then there exist a countable group K that cannot be embedded in a factor group of N. Let H be any countable group. Then $H \oplus K$ is also countable and hence can be embedded in a countable simple group S. S can be embedded in a factor group G/M of G. We have $MN/M \cong N/(M \cap N)$ and $MN/M \triangleleft G/M$.

Since S is a simple subgroup of G/M then $MN/M \cap S = \{1\}$ or $S \subset MN/M \cong N/(M \cap N)$. The latter implies

$$K \subset H \oplus K \subset S \subset N/(M \cap N)$$

contrary to the choice of K. It follows that S can be embedded in

$$(G/M)/(MN/N) \cong G/MN \cong (G/N)/(MN/N)$$

by the isomorphism theorems. Thus S can be embedded in a factor group of G/N. Since $H \subset S$ was an arbitrary countable group, it follows that G/N is SQ-universal. □

Now, by Theorem 1.1.22 each subgroup H of finite index in a group G contains a normal subgroup N of finite index. Hence, if G is SQ-universal then every subgroup of finite index is SQ-universal.

Corollary 1.6.21. *Let $G = \langle s_1, s_2, \ldots, s_n \mid s_1^{e_1} = s_2^{e_2} = \cdots = s_n^{e_n} = 1\rangle$, $n \geq 2$, $e_i = 0$ or $e_i \geq 2$ for all i and $e_1 + e_2 \geq 5$ if $n = 2$ and $e_1 \neq 0 \neq e_2$. Then G is SQ-universal.*

Proof. We just have to mention that G contains a non-Abelian free subgroup of finite index which follows directly from the abelianization of G. □

We should mention that Levin [167] has shown that $\mathbb{Z}_m * \mathbb{Z}_n$ with $m \geq 2$ and $n \geq 3$ is SQ-universal. His result easily goes through if $G = H_1 * H_2$ with H_1 and H_2 both finite and $|H_1| \geq 2$, $|H_2| \geq 3$. In this case the kernel N of the epimorphism of $H_1 * H_2$ onto $H_1 \oplus H_2$ is free of rank $(H_1 - 1)(H_2 - 1)$. P. Neumann observed that in general the free product $H_1 * H_2$ with $|H_1| \geq 2$ and $|H_2| \geq 3$ is SQ-universal. He also observed that the amalgamated product $G = H_1 *_A H_2$ with A finite and $|H_1 : A| \geq 2$, $|H_2 : A| \geq 3$ is SQ-universal. A complete proof is given in [170].

The next result to mention is by Baumslag and Pride [12].

Theorem 1.6.22. *Let G be a group with a finite presentation of g generators and r relators, and suppose that $g - r \geq 2$. Then G has a subgroup H of finite index which may be mapped homomorphically onto the free group F_2 of rank 2. Especially G is SQ-universal.*

Proof. By using the structure theorem for finitely generated Abelian groups we may assume that G has a presentation with generators $t, a_1, a_2, \ldots, a_{g-1}$ and relators $u_1, u_2, \ldots, u_r$ where the u_k have zero exponents sum in t. For each integer n let H_n denote the normal closure of $t^n, a_1, a_2, \ldots, a_{g-1}$. Note that H_n depends on n.

If j is an integer, let $a_{i,j} = t^{-j}a_it^j$, $1 \leq i \leq g-1$ and $u_{k,j} = t^{-j}u_kt^j$, $1 \leq k \leq r$. By conjugating the u_k by suitable powers of t if necessary, we may assume that there is a positive integer m such that the u_k can be expressed as words in the $a_{i,j}$, $0 \leq j \leq m$. Note that m is independent of n, so we may assume that $n > m$. Let $s = t^n$. Since $1, t, t^2, \ldots, t^{n-1}$ is a Schreier transversal for H_n in G it follows that H_n is generated by s and the $a_{i,j}$ with defining relators $v_{k,j}$ where $v_{k,j}$ is obtained from $u_{k,j}$ using the Reidemeister–Schreier rewriting process. Now, if $u_k = u_k(a_{i,j})$, $0 \leq j \leq m$, then for each integer r we get $u_{k,j} = u_k(a_{i,(j+r)})$, $0 \leq j \leq m$. The relator $v_{k,j}$ is obtained from $u_{k,j}$ by replacing symbols $a_{i,(j+n)}$, $0 \leq j \leq m-1$, by $sa_{i,j}s^{-1}$.

To obtain a suitable homomorphic image $\overline{H}_n$ of H_n we add the relations $a_{i,j} = 1$ for $0 \leq j \leq m-1$. Applying Tietze transformations to remove the superfluous generators

$a_{i,j}$, $0 \le j \le m-1$, we find that $\overline{H}_n$ is generated by $(g-1)(n-m)+1$ elements, namely s and the $a_{i,j}$, $m \le j \le n-1$, and that on these generators $\overline{H}_n$ has nr defining relators, none of which involves s. Thus $\overline{H}_n = \langle s \rangle * K$, where K has $(g-1)(n-m)$ generators and nr defining relators. Thus, for the group K,

$$\text{number of generators} - \text{number of relators} = (g-r-1)n + \text{constant},$$

and since $(g-r-1) \ge 1$, this difference can be made greater than 2 by choosing n sufficiently large. Hence K has an infinite cyclic group as homomorphic image, and so $\overline{H}_n$ can be mapped onto F_2. Since H_n is of finite index in G the theorem follows. □

Corollary 1.6.23. *One-relator groups with at least three generators are SQ-universal (see also [232]).*

Corollary 1.6.24. 1. *Let $G = \langle a_1, a_2, \dots, a_p, b_1, b_2, \dots, b_q \mid UV = 1\rangle$ where $1 \ne U \in \langle a_1, a_2, \dots, a_p\rangle$ and $1 \ne V \in \langle b_1, b_2, \dots, b_q\rangle$.*
If $p+q \ge 3$ then G is SQ-universal. Especially the surface groups

$$S_g = \left\langle a_1, b_1, a_2, b_2, \dots, a_g, b_g \;\middle|\; \prod_{i=1}^{g} [a_i, b_i] = 1 \right\rangle$$

with $g \ge 2$ and $N_g = \langle a_1, a_2, \dots, a_g \mid a_1^2 a_2^2 \cdots a_g^2 = 1\rangle$ with $g \ge 3$ are SQ-universal.
2. *Let $G = \langle a_1, a_2, \dots, a_p, t \mid tUt^{-1} = V\rangle$ where $U, V \in \langle a_1, a_2, \dots, a_p\rangle$, $U \ne 1 \ne V$. If $p \ge 2$ then G is SQ-universal.*
3. *Let $G = \langle a, b \mid a^p = b^q\rangle$, $p, q \ge 2$. If $p+q \ge 5$ then G is SQ-universal.*

In 3. just consider the quotient $\overline{G} = \langle a, b \mid a^p = b^q = 1\rangle$.

Corollary 1.6.25. *Let*

$$G = \left\langle s_1, s_2, \dots, s_n, a_1, b_1, a_2, b_2, \dots, a_g, b_g \;\middle|\; s_1^{\alpha_1} = s_2^{\alpha_2} = \dots = s_n^{\alpha_n} = s_1 s_2 \cdots s_n \prod_{i=1}^{g} [a_i, b_i] = 1 \right\rangle,$$

$n \ge 0$, $2 \le \alpha_i$ for $i = 1, 2, \dots, n$, $g \ge 0$ and $2g - g + \sum_{i=1}^{n}(1 - \frac{1}{\alpha_i}) > 0$ be a co-compact Fuchsian group. Then G is SQ-universal.

Proof. Note that G contains a surface group S_g, $g \ge 2$, of finite index. Then the result follows from Theorem 1.6.20 and Corollary 1.6.24. □

Stöhr [248] extended the result by Baumslag and Pride as follows.

Theorem 1.6.26. *Let $G = \langle a_1, a_2, \dots, a_n \mid R_1^{\alpha}, R_2, \dots, R_{n-1}\rangle$ where $\alpha, n \ge 2$. Then G has a subgroup of finite index which can be mapped homomorphically onto the free group of rank 2. Especially, G is SQ-universal.*

Corollary 1.6.27. *Let $G = \langle a, b \mid r^m = 1\rangle$, $m \geq 2$, be a two-generator one-relator group with torsion. Then G is SQ-universal.*

We mention a very interesting conjecture by P. Neumann:

A non-cyclic one-relator group is SQ-universal or it is a Baumslag–Solitar group $BS(1, n)$. There is a partial solution by Button and Kropholler [44] which makes the conjecture even more interesting: A non-cyclic two-generator one-relator group is either SQ-universal or residually finite.

We now describe an application of the result by Baumslag and Pride.

Let G be a group with a presentation

$$G = \langle a_1, a_2, \ldots, a_n \mid a_1^{e_1} = a_2^{e_2} = \cdots = a_n^{e_n} = R_1^{m_1} = R_2^{m_2} = \cdots = R_k^{m_k} = 1\rangle, \tag{1.18}$$

where $n \geq 2$, $e_i = 0$ or $e_i \geq 2$ for $i = 1, 2, \ldots, n$, $m_j \geq 2$ for $j = 1, 2, \ldots, k$ and each R_i is a cyclically reduced word in the free product of the cyclic groups $\langle a_1\rangle, \langle a_2\rangle, \ldots, \langle a_n\rangle$ of syllable length at least two.

A representation $\rho\colon G \to H$, H a linear group over a field of characteristic 0, is called an *essential representation* if for each $i = 1, 2, \ldots, n$ the image $\rho(a_i)$ has infinite order if $e_i = 0$ or exact order if $e_i \geq 2$ and for each $j = 1, 2, \ldots, k$ the image $\rho(R_j)$ has order m_j.

Theorem 1.6.28. *Let G be a group with a presentation* (1.18). *Suppose that the extended deficiency $d^* = n - k \geq 3$ and that G admits an essential representation. Then G contains a subgroup of finite index which maps homomorphically onto a non-Abelian free group. In particular G is SQ-universal. The result is still true if $d^* = 2$ but not all $m_j = 2$ and not all $e_i = 2$.*

Proof. Let first $d^* \geq 3$. We may assume, without loss of generality, that $e_i \geq 2$ for $i = 1, 2, \ldots, n$, passing to a quotient if necessary. Let $\rho\colon G \to H$, H a linear group over a field of characteristic 0, be an essential representation. From Selberg's theorem about finitely generated subgroups of linear groups [238] there exists a normal torsion-free subgroup H of finite index in $\rho(G)$. Thus the composition of maps (where π is the canonical map)

$$G \xrightarrow{\rho} \rho(G) \xrightarrow{\pi} \rho(G)/H$$

gives a representation φ of G onto a finite group such that $\varphi(a_i)$ has order e_i, $i = 1, 2, \ldots, n$, and $\rho(R_j)$ has order m_j, $j = 1, 2, \ldots, k$. Let $X = \langle a_1, a_2, \ldots, a_n \mid a_1^{e_1} = a_2^{e_2} = \cdots = a_n^{e_n} = 1\rangle$ the free product of the cyclic group $\langle a_1\rangle, \langle a_2\rangle, \ldots, \langle a_n\rangle$. There is a canonical epimorphism β from X onto G. We therefore have the sequence

$$X \xrightarrow{\beta} G \xrightarrow{\varphi} G/H.$$

Let $Y = \ker(\varphi \circ \beta)$. Then Y is a normal subgroup of finite index in X and Y is torsion-free. Since X is the free product of cyclics and Y is torsion-free it follows that Y is a free group of finite rank r. Suppose $|X : Y| = j$. Regard X as a Fuchsian group with finite co-area

$\mu(X)$. From the Riemann–Hurwitz formula we see that $j \cdot \mu(X) = \mu(Y)$ where $\mu(Y) = r-1$ and $\mu(X) = (n-1) - (\frac{1}{e_1} + \frac{1}{e_2} + \cdots + \frac{1}{e_n})$. Equating these expressions we obtain

$$r = 1 - j\left(\frac{1}{e_1} + \frac{1}{e_2} + \cdots + \frac{1}{e_n} - (n-1)\right).$$

The group G is obtained from X by adjoining the relations $R_1^{m_1}, R_2^{m_2}, \ldots, R_k^{m_k}$, and so $G = X/K$ where K is the normal closure of $R_1^{m_1}, R_2^{m_2}, \ldots, R_k^{m_k}$. Since K is contained in Y the quotient Y/K can be considered as a subgroup of finite index in G. Applying the Reidemeister–Schreier process Y/K can be defined on r generators subject to $(\frac{j}{m_1} + \frac{j}{m_2} + \cdots + \frac{j}{m_k})$ relations. The deficiency d of this presentation for Y/K is then

$$d = r - \left(\frac{j}{m_1} + \frac{j}{m_2} + \cdots + \frac{j}{m_k}\right).$$

Since each $e_i \geq 2$ and each $m_j \geq 2$ we have the inequality

$$d \geq 1 + j\left(\frac{n-k}{2} - 1\right).$$

By assumption the extended deficiency $n - k \geq 3$ and hence the deficiency d of the above presentation is at least two. Now the main parts of the theorem follows from Theorem 1.6.22 of Baumslag and Pride.

Now let the extended deficiency $d^* = 2$ and not all $m_j = 2$ and not all $e_i = 2$. If $e_i = 0$ for some i then we may pass to a quotient with the respective $e_j \geq 3$ sufficiently large. Hence, we may assume that $e_i \geq 3$ for some i or $m_j \geq 3$ for some j. Then the above inequality becomes proper, that is,

$$d > 1 + \left(\frac{n-k}{2} - 1\right) = 1.$$

Hence $d \geq 2$, and the proof goes through as before. □

Remark 1.6.29. A detailed analysis of the proof gives results for $0 \leq d^* \leq 1$, $d^* = n - k$. We take 0 for $\frac{1}{e_i}$ if $e_i = 0$.

Corollary 1.6.30. *Let G be a group with a presentation* (1.18) *with* $0 \leq d^* \leq 1$. *Suppose that G admits an essential representation.*

If $n > 1 + \frac{1}{e_1} + \frac{1}{e_2} + \cdots + \frac{1}{e_n} + \frac{1}{m_1} + \frac{1}{m_2} + \cdots + \frac{1}{m_k}$ then G contains a subgroup of finite index which maps homomorphically onto a non-Abelian free group. In particular G is SQ-universal.

We leave the details as an exercise. Now, we present examples of groups which have an essential representation into $\mathrm{PSL}(2, \mathbb{C})$:

(1) *One-relator products of cyclics:*
$G = \langle a_1, a_2, \ldots, a_n \mid a_1^{e_1} = a_2^{e_2} = \cdots = a_n^{e_n} = R^m(a_1, a_2, \ldots, a_n) = 1\rangle$ with $n \geq 2$, $m \geq 2$, $e_i = 0$ or $e_i \geq 2$ for $i = 1, 2, \ldots, n$ and $R(a_1, a_2, \ldots, a_n)$ a cyclically reduced word in the free product on $a_1, a_2, \ldots, a_n$ which involves all the generators.
In this special case $n = 2$ the group G is called a *generalized triangle group.*
(2) *Generalized tetrahedron groups:*
$T = \langle a, b, c \mid a^p = b^q = c^r = (R_1(a, b))^m = (R_2(a, c))^n = (R_3(b, c))^t = 1\rangle$ with $p, q, r, m, n, t \geq 2$ and $R_1(a, c)$, $R_2(a, c)$, $R_3(b, c)$ are cyclically reduced words in the respective free products involving both generators $\{a, b\}$, $\{a, c\}$ and $\{b, c\}$, respectively.
(3) *Generalized modular groups:*
$M = \langle a, b, c \mid a^p = b^q = c^r = R_1(a, b)^m = (R_2(a, c))^n = (abc)^t = 1\rangle$ with $2 \leq p, q, r, m, n, t$ and $R_1(a, b)$, $R_2(a, c)$ are cyclically reduced words involving both generators $\{a, b\}$ and $\{a, c\}$, respectively.
(4) *Groups of special NEC-type:*
$G = \langle a_1, a_2, \ldots, a_n, b_1, b_2, \ldots, b_k \mid a_1^{e_1} = a_2^{e_2} = \cdots = a_n^{e_n} = b_1^{f_1} = b_2^{f_2} = \cdots = b_k^{f_k} = R_1^{m_1} = R_2^{m_2} = \cdots = R_k^{m_k} = 1\rangle$ where $n \geq 1$, $k \geq 2$, $e_j = 0$ or $e_j \geq 2$ for $j = 1, 2, \ldots, n$, $f_i = 0$ or $f_i \geq 2$ for $i = 1, 2, \ldots, k$, $m_i \geq 2$ for $i = 1, 2, \ldots, k$ and for each $i = 1, 2, \ldots, k$, $R_i = R_i(a_1, a_2, \ldots, a_n, b_i)$ is a cyclically reduced word in the free product on $a_1, a_2, \ldots, a_n, b_i$, which involves b_i and at least one of the a_j's and which is not a proper power.
(5) *Groups of F-type:*
$G = \langle a_1, a_2, \ldots, a_n \mid a_1^{e_1} = a_2^{e_2} = \cdots = a_n^{e_n} = U(a_1, a_2, \ldots, a_p)V(a_{p+1}, a_{p+2}, \ldots, a_n) = 1\rangle$ where $n \geq 2$, $e_1 = 0$ or $e_i \geq 2$, $1 \leq p \leq n - 1$, $U = U(a_1, a_2, \ldots, a_p)$ is a cyclically reduced word in the free product on $a_1, a_2, \ldots, a_p$ which is of finite order and $V = V(a_{p+1}, a_{p+2}, \ldots, a_n)$ is a cyclically reduced word in the free product on $a_{p+1}, a_{p+2}, \ldots, a_n$ which is of finite order.

Proofs for the examples can be found in [93] and [195]. Occasionally we come back to these examples. Groups of F-type are in general not of the form as given in Theorem 1.6.26. We consider them in detail later in the book.

Remark 1.6.31. There are many activities concerning the question under what conditions an HNN group $G = \langle H, t \mid t^{-1}At \overset{i}{=} B\rangle$ is SQ-universal.

Unfortunately there is no general result as in the case of free products with amalgamation but many partial results (see [103, 8, 232, 43, 44]).

1.6.3 The Nielsen method and equations in HNN extensions

We now discuss the Nielsen method in HNN groups. Let $G = \langle H, t \mid t^{-1}A_1t = A_{-1}\rangle$. Choose left transversals R_1 of A_1 and R_{-1} of A_{-1} on H where A_1 and A_{-1} are represented by 1. As right transversal of A_1 and A_{-1} we take the inverses R_1^{-1}, R_{-1}^{-1}. Then each $x \in G$ has the reduced representation

$$x = \ell_1 t^{\epsilon_1} \ell_2 t^{\epsilon_2} \cdots \ell_m t^{\epsilon_m} k_x t^{\nu_m} r_m \cdots t^{\nu_1} r_1$$

with $m \geq 0$, $\epsilon_i, \nu_i = \pm 1$, $\ell_i \in R_1$, $r_i \in R_{-\nu_i}^{-1}$ and $k_x = h_1 t^{\epsilon} h_2$, $h_1, h_2 \in H$, $\epsilon = \pm 1$ if $|x|$ is odd or $k_x \in H$ if $|x|$ is even. In this representation $\ell_1 t^{\epsilon_1} \ell_2 t^{\epsilon_2} \cdots \ell_m t^{\epsilon_m}$ is called the leading half, $t^{\nu_m} r_m \cdots t^{\nu_1} r_1$ the rear half and k_x the kernel of x. Again, we call the above representation a symmetric form of x.

We now introduce, as in the free product with amalgamation, an ordering of G. We assume that the involved groups are countable. This is no restriction if we consider a given finite system in G, for a given finite system a suitable order can always be chosen so that this system can be carried by a Nielsen transformation into what we will call a Nielsen reduced system.

Choose a total order of the transversals R_1, R_{-1}, and order products $\ell_1 t^{\epsilon_1} \ell_2 t^{\epsilon_2} \cdots \ell_m t^{\epsilon_m}$ by using the lexicographical order on the sequences $(\ell_1, \ell_2, \ldots, \ell_m)$. Next we extend this order to the set of pairs $\{g, g^{-1}\}$, $g \in G$, where the notation is chosen such that the leading half of g precedes that of g^{-1} with respect to the above ordering.

Let $\{g, g^{-1}\} \prec \{h, h^{-1}\}$ if either $|g| < |h|$ or $|g| = |h|$ and the leading half of g strictly precedes that of h or $|g| = |h|$ and the leading halves of g and h coincide while the leading half of g^{-1} precedes that of h^{-1}. Hence if $\{g, g^{-1}\} \prec \{h, h^{-1}\}$ and $\{h, h^{-1}\} \prec \{g, g^{-1}\}$ then at most the kernels of g and h may be different. A finite system $\{g_1, g_2, \ldots, g_n\}$ in G is called *shorter* than a system $\{h_1, h_2, \ldots, h_m\}$ if $\{g_i^{\epsilon(g_i)}, g_i^{-\epsilon(g_i)}\} \prec \{h_i^{\epsilon(h_i)}, h_i^{-\epsilon(h_i)}\}$ holds for $i = 1, 2, \ldots, m$ and for at least one $j \in \{1, 2, \ldots, m\}$ we do not have $\{h_i^{\epsilon(h_i)}, h_i^{-\epsilon(h_i)}\} \prec \{g_i^{\epsilon(g_i)}, g_i^{-\epsilon(g_i)}\}$, $\epsilon(g_i), \epsilon(h_i) = \pm 1$ (here the leading half of $g^{\epsilon(g_i)}$ and $h^{\epsilon(g_i)}$ precedes that of $g^{-\epsilon(g_i)}$ and $h^{-\epsilon(g_i)}$, respectively). A finite system $\{g_1, g_2, \ldots, g_m\} \subset G$ is said to be *Nielsen reduced* or *minimal* with respect to $\prec$ if $\{g_1, g_2, \ldots, g_m\}$ cannot be carried by a Nielsen transformation into a system $\{h_1, h_2, \ldots, h_m\}$ with $h_i = 1$ for some $i \in \{1, 2, \ldots, m\}$ or there is no system Nielsen equivalent to $\{g_1, g_2, \ldots, g_m\}$ which is shorter.

If the group G is countable, then each finite system, as in the case of free products with amalgamation, can be carried by a Nielsen transformation to a minimal system. In general for a given finite system a suitable order can always be chosen so that a finite system can be carried by a Nielsen transformation to a minimal system.

The following theorem is a slightly refined summary of the results of Peczynski and Reiwer [210].

Theorem 1.6.32. *Let* $G = \langle H, t \mid t^{-1} A_1 t = A_{-1} \rangle$. *If* $\{x_1, x_2, \ldots, x_m\}$ *is a finite system in* G *then there is a Nielsen transformation from* $\{x_1, x_2, \ldots, x_m\}$ *to a system* $\{y_1, y_2, \ldots, y_m\}$ *for which one of the following holds:*

(a) $y_i = 1$ *for some* $i \in \{1, 2, \ldots, m\}$.

(b) *Each* $w \in \langle y_1, y_2, \ldots, y_m \rangle$ *can be written as* $w = \prod_{i=1}^{q} y_{\nu_i}^{\epsilon_i}$, $\epsilon_i = \pm 1$ *and* $\epsilon_i = \epsilon_{i+1}$ *if* $\nu_i = \nu_{i+1}$ *and* $|y_{\nu_i}| \leq |w|$ *for* $i = 1, 2, \ldots, q$.

(c) *Some conjugate of H contains p, $p \geq 1$, of the y_i and some product of these y_i is conjugate to a non-trivial element of A_1 or A_{-1}.*

The Nielsen transformation can be chosen in finitely many steps such that $\{y_1, y_2, \ldots, y_m\}$ is shorter than $\{x_1, x_2, \ldots, x_m\}$ or the length of the elements of $\{x_1, x_2, \ldots, x_m\}$ are preserved.

The main tool in the proof of this theorem, which is not included in the work of Peczynski and Reiwer, is the following lemma. This is proved in an analogous fashion to the corresponding result in the free product with amalgamation case. We leave the proof as an exercise.

Lemma 1.6.33. *Let $x, y, z \in G$, $x \neq 1 \neq z$ with $|xy| \geq |x|, |y|$ and $|yz| \geq |z|, |y|$. Moreover, let neither xy precede x nor yz precede z.*

Then the following hold.

1. *If $|xyz| \leq |x| - |y| + |z|$ then y is conjugate to an element of H.*
2. *If $|x|, |z| > |y|$ and $|xyz| < |x| - |y| + |z|$ then y is conjugate to an element of A_ϵ, $\epsilon = \pm 1$.*
3. *If $V \subset G$ is a finite Nielsen reduced system, $x, y, z \in V \cup V^{-1}$ and $|xyz| < |x| - |y| + |z|$ then y is conjugate to an element of A_ϵ, $\epsilon = \pm 1$, or $x = y = z$.*

We note that if $|g^n| < |g|$ for $g \in G$ and $n \in \mathbb{N}$ then g is conjugate to an element of the base H and g^n is conjugate in G to an element of A_1 or A_{-1}. If A_1 and A_{-1} are malnormal in H and $x, y, z \in G$ with $\{x, y, z\}$ Nielsen reduced, $y \in A_1$ and $|xyz| < |x| - |y| + |z|$ then the symmetric form of x must end with t^{-1} and the symmetric form of z must begin with t. If we continue that way we directly obtain the following.

Lemma 1.6.34. *Let $G = \langle H, t \mid t^{-1} A_1 t = A_{-1} \rangle$ with A_1 and A_{-1} malnormal in H. Suppose $\langle a, b \rangle$ is a non-cyclic two-generator subgroup of G.*

Then $\langle a, b \rangle$ is conjugate to $\langle a', b' \rangle$ satisfying one of the following conditions.

(a) *The subgroup $\langle a', b' \rangle$ is a free product of cyclics.*
(b) *The subgroup $\langle a', b' \rangle$ is contained in H.*
(c) *The element a' has the form $th_1 th_2 \cdots th_n$, $n \geq 1$, $h_j \in H$ for $j = 1, 2, \ldots, n$, $b' \in A_1$ and $a'^{-1} b' a' \in H$.*

Corollary 1.6.35. *Suppose G is a two-generator subgroup with both A_1 and A_{-1} are malnormal in H. Then any system $\{x, y\}$ of generators of G is Nielsen equivalent to a system $\{th, b\}$ where $h \in H$ and $b \in K_1$.*

We now present some application of the Nielsen method in HNN groups. More applications will follow in the next section on one-relator groups. First of all, also in the next section we need the following.

Lemma 1.6.36. *Let $G = \langle a_1, a_2, \ldots, a_n \mid R = 1 \rangle$, $n \geq 2$, be a one-relator group. Then G has a presentation $G = \langle t, b_1, b_2, \ldots, b_{n-1} \mid R' = 1 \rangle$ where $\sigma_t(R') = 0$, $\sigma_t(R')$ the exponent sum on t in R'.*

Proof. If either $\sigma_{a_1}(R) = 0$ or $\sigma_{a_2}(R) = 0$ then there is nothing to show. If neither $\sigma_{a_1}(R)$ nor $\sigma_{a_2}(R)$ is 0 then we proceed by induction on $|\sigma_{a_1}(R)| + |\sigma_{a_2}(R)|$. Suppose that $0 < |\sigma_{a_1}(R)| \leq |\sigma_{a_2}(R)|$. Then G has a presentation

$$G = \langle t, a_2, a_3, \ldots, a_n \mid R' = 1\rangle$$

with $t = a_1 a_2^{\epsilon}$ and $R' = R(ta_2^{-\epsilon}, a_2, a_3, \ldots, a_n)$, $\epsilon = \pm 1$. For $\epsilon = 1$ or $\epsilon = -1$ we have $|\sigma_{a_2}(R')| < |\sigma_{a_2}(R)|$. □

We remark that the transition from $\{a_1, a_2, \ldots, a_n\}$ to $\{t, b_1, b_2, \ldots, b_{n-1}\}$ is done by a Nielsen transformation with respect to $R(a_1, a_2, \ldots, a_n)$.

Theorem 1.6.37 ([214]). *Let $G = \langle t, a \mid P^n = 1\rangle$, $n \geq 2$, with $P \neq 1$ a non-primitive, non-proper power element in the free group on t and a. If $\{x, y\}$ is a generating pair of G then $\{x, y\}$ is Nielsen equivalent to $\{t, a\}$. Hence, the isomorphism problem for G is solvable, that is, we can decide algorithmically in finitely many steps whether or not an arbitrary one-relator group is isomorphic to G.*

Sketch of the proof. We note first that the solution to the isomorphism problem uses the Whitehead algorithm (see Section 1.1) to decide if in a free group two elements are congruent via a Nielsen transformation α of the free group.

We may assume that P is cyclically reduced and $\sigma_t(P) = 0$ for the exponent sum of t in P (see Lemma 1.6.36).

Let $a_i = t^{-i}at^i$, $i \in \mathbb{Z}$. Let Q be the word expressing P in terms of the a_i. Then G is an HNN group $G = \langle a_m, a_{m+1}, \ldots, a_M, t \mid Q^n = 1, t^{-1}a_i t = a_{i+1}$ for $i = m, m+1, \ldots, M-1\rangle$, where m, M represent, respectively, the least and greatest value of i for which a_i occurs in Q.

By the spelling Theorem 1.7.9 of Newman [202] we see that the associated subgroups $A_1 = \langle a_m, \ldots, a_{M-1}\rangle$ and $K_{-1} = \langle a_{m+1}, \ldots, a_m\rangle$ are malnormal in the base $H = \langle a_m, \ldots, a_M \mid Q^n = 1\rangle$. We remark that we prove Newman's spelling theorem in the next section. Now applying Corollary 1.6.35 we see that a generating pair $\{x, y\}$ for G is Nielsen equivalent to a system $\{tg, h\}$, $g \in H$, $h \in K_1$. The pair $\{tg, h\}$ is then Nielsen equivalent to $\{t, a\}$ (for the difficult technical details see Pride's paper [214]). The rest of the theorem follows from Whitehead's algorithm. □

In this connection the following is worth to mention. Let G be a two-generator group, and let $\{a, b\}$ be a fixed but arbitrary generating pair. We say that G has the *commutator-generator property* if two elements x and y of G generate G if and only if $[x, y]$ is conjugate to $[a, b]$ or $[a, b]^{-1}$.

It is well known that the groups

(1) $F_2 = \langle a, b \mid \rangle$,
(2) $\langle a, b \mid [a, b]^n = 1\rangle$, $n \geq 2$, and
(3) $\langle s_1, s_2, s_3, s_4 \mid s_1^2 = s_2^2 = s_3^2 = s_4^q = s_1 s_2 s_3 s_4 = 1\rangle$ with $q \geq 2$, q odd

have the commutator-generator property. Hill and Pride [130] showed the following fact:

Let $G = \langle a, b \mid R^n = 1\rangle$ where $n \geq 5$ and R is a non-empty cyclically reduced word which is not a power of a primitive element in $\langle a, b \mid \rangle$. Then G has the commutator-generator property.

It should be mentioned that this does not hold for $2 \leq n \leq 4$ (see [130]).

As a consequence of Theorem 1.6.37 we then see that if G is as in Theorem 1.6.37 then G is Hopfian and the automorphism group of G is finitely presented (see McCool's work described in Section 1.2). Another direct application from Theorem 1.6.37 and Lemma 1.6.34 is as follows.

Theorem 1.6.38. *Let $G = \langle a_1, a_2, \dots, a_m \mid P^n = 1\rangle$, $m \geq 2$, $n \geq 2$, and P cyclically reduced. Then every two-generator subgroup of G is either a free product of two cyclic groups or is a one-relator group with torsion.*

A proof with all the details is given in [215].

Let $G = \langle a_1, a_2, \dots, a_n \mid R = 1\rangle$, $n \geq 2$, $R \neq 1$ and R not primitive in $\langle a_1, a_2, \dots, a_n \mid \rangle$. If $\alpha \in \operatorname{Aut}(F_n)$ is a Nielsen transformation then we get

$$\begin{aligned} G &= \langle \alpha(a_1), \alpha(a_2), \dots, \alpha(a_n) \mid \alpha^{-1}(R(\alpha(a_1), \alpha(a_2), \dots, \alpha(a_n))) = 1\rangle \\ &= \langle b_1, b_2, \dots, b_n \mid \alpha^{-1}(R(b_1, b_2, \dots, b_n)) = 1\rangle. \end{aligned}$$

We say that G has exactly one Nielsen class, if each generating system $\{x_1, x_2, \dots, x_n\}$ can be transferred by a Nielsen transformation to $\{a_1, a_2, \dots, a_n\}$. If G has exactly one Nielsen class then we may effectively decide if a one-relator group G is isomorphic to or not.

Examples 1.6.39. The following groups are one-relator groups with exactly one Nielsen class:

1. $G = \langle a_1, b_1, a_2, b_2, \dots, a_g, b_g \mid \prod_{i=1}^{g} [a_i, b_i] = 1\rangle$, $g \geq 1$;
2. $G = \langle a_1, a_2, \dots, a_g \mid a_1^2 a_2^2 \cdots a_g^2 = 1\rangle$, $g \geq 1$;
3. $G = \langle a_1, a_2, \dots, a_n, b_1, b_2, \dots, b_m \mid (UV)^\gamma = 1\rangle$, $n \geq 1$, $m \geq 1$, $\gamma \geq 2$, $U = U(a_1, a_2, \dots, a_n) \neq 1$ and not primitive in $\langle a_1, a_2, \dots, a_n \mid \rangle$, $V = V(b_1, b_2, \dots, b_m) \neq 1$ and not primitive in $\langle b_1, b_2, \dots, b_m \mid \rangle$;
4. $G = \langle a, b \mid R^m = 1\rangle$, $m \geq 2$, R not a proper power and not primitive in $\langle a, b \mid \rangle$; and
5. $G = \langle a, b, t \mid a[b, a][b, t] = 1\rangle$ (see [99]).

This last example is of special interest. The class of groups

$$G_{i,j} = \langle a, b, t \mid a[b^i, a][b^j, t] = 1\rangle$$

for positive integers i, j was introduced by G. Baumslag. The $G_{i,j}$ are *parafree*, that is, they share many properties with the free group F_2 of rank 2.

In particular:

(1) If $\gamma_n G_{i,j}$ are the terms of the lower central series of $G_{i,j}$ then $G_{i,j}/\gamma_n G_{i,j} \cong F_2/\gamma_n F_2$ for all n;

(2) $G_{i,j}$ has a normal subgroup with an infinite cyclic quotient group;
(3) $G_{i,j}$ is 2-free but not free; and
(4) $G_{i,j}$ is hyperbolic (we discuss hyperbolic groups in Chapter 3).

The group $G_{i,j}$ can be expressed as a conjugacy pinched one-relator group

$$G_{i,j} = \langle a, b, t \mid t^{-1}a[b^i, a]b^j t = b^j \rangle.$$

If in addition $j = 1$ then $\langle b \rangle$ and $\langle a[b^i, a]b \rangle$ are malnormal in $\langle a, b \mid \rangle$, and hence we may apply the techniques developed in [101]. The main result is the following theorem.

Theorem 1.6.40 ([99]). 1. *If i and k are prime numbers then $G_{i,1} \cong G_{k,1}$ if and only if $i = k$.*
2. *$G_{i,1}$ is Hopfian, and the isomorphism problem for $G_{i,1}$ is solvable.*
3. *$G_{1,1}$ is not isomorphic to $G_{i,j}$ with $i \geq 2$.*

In fact, the main purposes in [101] was to consider the 3-generator subgroups in cyclically pinched one-relator groups.

Theorem 1.6.41. *Let $G = \langle a_1, a_2, \ldots, a_n, t \mid t^{-1}Ut = V \rangle$ with $n \geq 1$ and $U = U(a_1, a_2, \ldots, a_n)$, $V = V(a_1, a_2, \ldots, a_n)$ non-trivial elements of the free groups F on $a_1, a_2, \ldots, a_n$ with neither U nor V a proper power in F. Suppose further that U is not conjugate in F to either V or V^{-1}. Let $H = \langle x_1, x_2, x_3 \rangle \subset K$. Then H is a free group of rank at most 3 or H has a presentation with one defining relation for $\{x_1, x_2, x_3\}$.*

For a proof we refer to [101]. Instead we want to describe the two-generator subgroups in conjugacy pinched one-relator groups in detail.

Theorem 1.6.42. *Let $G = \langle a_1, a_2, \ldots, a_n, t \mid t^{-1}Ut = V \rangle$ with $n \geq 1$ and $U = U(a_1, a_2, \ldots, a_n)$, $V = V(a_1, a_2, \ldots, a_n)$ non-trivial elements in the free group $F = \langle a_1, a_2, \ldots, a_n \mid \rangle$ with neither U nor V a proper power in F. Suppose $\langle x, y \rangle$ is a two-generator subgroups of G. Then*
(i) *$\langle x, y \rangle$ is free of rank 2, or*
(ii) *$\langle x, y \rangle$ is Abelian, or*
(iii) *$\langle x, y \rangle$ has a presentation $\langle a, b \mid a^{-1}ba = b^{-1} \rangle$, the Klein bottle group.*

We remark that $\langle U \rangle$ and $\langle V \rangle$ are malnormal in F because they are both not a proper power. Hence, we may apply Lemma 1.6.34. Furthermore, we need the following.

Lemma 1.6.43. *Let $G = \langle a, t \mid t^{-1}at = a^{-1} \rangle$. Let H be any subgroup of G. Then H is either Abelian or isomorphic to G. Furthermore, if H is Abelian then $H \cong \mathbb{Z}$ or $H \cong \mathbb{Z} \oplus \mathbb{Z}$. In particular every subgroup has two generators or less.*

Proof. From $t^{-1}at = a^{-1}$ we obtain $t^{-1}at = tat^{-1}$ or $t^{-2}at^2 = a$. Thus t^2 is central and so t^m is central if m is even. Furthermore, from $t^{-1}at = a^{-1}$ it is clear that every element of G can be written in the form $t^m a^k$ for integers m, k.

Now let H be any subgroup of G. If H is a subgroup of $\langle a\rangle$ then H is cyclic. Thus we may assume that H contains an element of the form $t^m a^k$ with $m \neq 0$. Let $d > 0$ be the minimal positive power of t which appears in the representation of elements of H of the form $t^m a^k$. Thus there is an integer j such that $t^d a^j \in H$.

Let $t^n a^k$ be any element of H. We claim that $n = qd$ for some $q \in \mathbb{Z}$. From the division algorithm $n = qd + r$ with q, r integers and $0 \le r < d$. Now $(t^d a^j)^q = t^{qd}a^k$ for some k (using $t^{-1}at = a^{-1}$). Thus $(a^{-k}t^{-qd})(t^{qd+r}a^k) = t^r a^v \in H$. Since d is minimal we get $r = 0$ and $n = qd$.

Hence every element of H has the form $t^{qd}a^v$. If every element of H is a power of $t^d a^j$ then H is cyclic. Suppose that H is non-cyclic. Then H contains two element $t^{qd}a^h$ and $t^{qd}a^v$ with $v \neq h$. Then H contains non-trivial powers of a since $(a^{-h}t^{-qd})(t^{qt}a^v) = a^{v-h}$. Now $H \cap \langle a\rangle$ is cyclic, say $\langle a^s\rangle$. Then a^{v-h} is a power of a^s, and we see that H is generated by $t^d a^j$ and a^s. Let $x = t^d a^j, y = a^s$, then $H = \langle x, y\rangle$. If d is even then t^d is central and x and y commute. In this case $H = \mathbb{Z} \oplus \mathbb{Z}$.

If d is odd then $t^d a^j = a^{-j}t^d$, and we obtain $x^{-1}yx = y^{-1}$. Thus H is a homomorphic image of G under the map $t \mapsto x$, $a \mapsto y$. Let $g = t^m a^k$ be in the kernel of this map. Then $t^m a^k \mapsto x^m y^k = (t^d a^j)^m (a^s)^k = t^{md}a^v = 1$. This is possible only if $m = 0$. Then $g = a^k$ and $g \mapsto y^k = a^{sk} = 1$, which is possible only if $k = 0$. Thus $g = 1$, the kernel is trivial and H is isomorphic to G. □

There is a nice topological proof of Lemma 1.6.43. Note that G is the fundamental group of a Klein bottle. If H is a subgroup of G then H corresponds to the fundamental group of a covering surface T of the Klein bottle. Thus T is homotopically a circle, a torus or a Klein bottle. In the first case $H = \mathbb{Z}$, in the second $H = \mathbb{Z} \oplus \mathbb{Z}$, while in the third $H = G$. We consider covering space theory later in this book.

Now we give the proof of Theorem 1.6.42.

Proof. First if $U = g^{-1}V^e g$ for some $g \in \langle a_1, a_2, \ldots, a_n\rangle$ and $e = \pm 1$, then the substitution $t' = g^t$ gives us $t^{-1}Ut = t'^{-1}V^e t'$. Thus we can assume that $U = V^e, e = \pm 1$ if U is conjugate in $\langle a_1, a_2, \ldots, a_n\rangle$ for V or V^{-1}.

Now $G = \langle a_1, a_2, \ldots, a_n, t \mid t^{-1}Ut = V^\epsilon, \epsilon = \pm 1\rangle$ is an HNN extension with base $F = \langle a_1, a_2, \ldots, a_n \mid \rangle$, free part $\langle t\rangle$ and associated subgroups $A_1 = \langle U\rangle$ and $A_{-1} = \langle V\rangle$. Since $\langle U\rangle$ and $\langle V\rangle$ are malnormal in F we may apply Lemma 1.6.34.

Let $H = \langle x, y\rangle$ be a two-generator subgroup of G. From Lemma 1.6.34 we see that H is

(1) A free product of cyclics, or
(2) Abelian, or
(3) Conjugate to $\langle x', y'\rangle$ with $x' = th_1th_2 \cdots th_r$, $r \ge 1$, $h_j \in F$ for $j = 1, 2, \ldots, r$, $y' \in A_1$ and $x'^{-1}y'x' \in F$.

Case (1) and (2) are handled in the statement of the theorem, so we can assume case (3): $x = th_1th_2 \cdots th_r, r \geq 1, h_j \in F$ for $j = 1, 2, \ldots, r, y \in A_1 = \langle U \rangle$ and $x^{-1}yx \in F$. Let $y \in U^k$ for some $k \in \mathbb{Z} \setminus \{0\}$. Now $\langle x, y \rangle = \langle x, U^k \rangle \subset \langle x, U \rangle$. If $\langle x, U \rangle$ where free Abelian then $\langle x, y \rangle$ as a subgroup would be also. Then from Lemma 1.6.43 it suffices to show that $\langle x, U \rangle$ is free, is Abelian or has a presentation $\langle x, U \rangle = \langle a, b \mid a^{-1}ba = b^{-1} \rangle$. Therefore let $y = U$, $x = th_1th_2 \cdots th_r$ as above with $x^{-1}yx \in F$.

Now

$$x^{-1}yx = h_r^{-1}t^{-1} \cdots h_1^{-1}t^{-1}Uth_1 \cdots th_r = h_r^{-1}t^{-1} \cdots (h_1^{-1}Vh_1)t \cdots th_r \in F.$$

If $r \geq 2$ then from the normal form theorem we see that by $h_1^{-1}Vh_1 = U^e, e = \pm 1$. By our original reduction we then can assume that $V = U^e, e \pm 1$, and thus we can assume that $h_1 = 1$.

Similarly we may assume that $h_i = 1$ for $i < r$.

Hence $x = t^r h, r \geq 1$ for some $h \in F$. Because $y = U$ we may assume that $h = 1$ or $h \notin \langle U \rangle$. We claim now that $\langle x, y \rangle$ is free of rank 2 if $h \notin \langle U \rangle$ or if $V \neq U^e, e = \pm 1$.

First suppose $V \neq U^e, e = \pm 1$. Then $r = 1$ because otherwise $x^{-1}yx \notin F$. Now we may assume that $h = 1$ because $h^{-1}t^{-1}Uth = h^{-1}Vh$ and we may replace t by $t' = th$ and V by $V' = h^{-1}Vh$. Therefore it suffices to regard $x = t$ and $y = U$ if $U \neq V^e, e = \pm 1$. But $\langle t, U \rangle$ is free of rank 2 if $U \neq V^e, e = \pm 1$.

Now, suppose $U = V^e, e = \pm 1$, and let $h \in F \setminus U$. Then $\langle x, y \rangle = \langle t^r h, U \rangle$ is free of rank 2 because $h \notin \langle U \rangle$ and U is malnormal in F. In the remaining case where $U = V^e, e = \pm 1$, and $h \in \langle U \rangle$, we may assume that $h = 1$ and $\langle x, y \rangle = \langle t^r, U \rangle$. Now, $\langle x, y \rangle$ is a subgroup of $\langle t, U \rangle$, and $\langle t, U \rangle$ has a presentation $\langle t, U \rangle = \langle t, U \mid t^{-1}Ut = U^e \rangle, e = \pm 1$. If $e = 1$ then $\langle t, U \rangle$ is free Abelian of rank 2, and hence also $\langle x, y \rangle$. If $e = -1$ then $\langle t, U \rangle = \langle t, U \mid t^{-1}Ut = U^{-1} \rangle$, and the result follows from Lemma 1.6.43. □

In general, to work in one-relator groups with Nielsen transformations on tuples $\{x_1, x_2, \ldots, x_n\}$ is quite complicated, especially in torsion-free one-relator groups. For instance, Brunner [38] showed that $G = \langle a, b \mid a^{-1}b^{-1}aba^{-1}bab^{-2} = 1 \rangle$ possesses infinitely many Nielsen inequivalent one-relator presentations in two generators.

We conclude this section with the remark that we may consider equations in HNN extensions analogously as for free products with amalgamation. We only have to modify one definition. Let $G = \langle H, t \mid t^{-1}A_1t = A_{-1} \rangle$. We call an ordered set $U = \{u_1, u_2, \ldots, u_n\} \subset G$ *regular* if there is no Nielsen transformation from U to a system $U' = \{u_1', u_2', \ldots, u_n'\}$ in which one element is conjugate in G to an element of A_1.

With this modified definition we get the results analogously to Theorems 1.5.50, 1.5.51 and 1.5.52 (see [223] and [98]). To use the result in concrete HNN groups is more difficult than in free products with amalgamation. Anyway we have two little nice results. The first result is straightforward.

Theorem 1.6.44 ([88]). *Let $G = \langle H, t \mid t^{-1}A_1t = A_{-1}\rangle$. Suppose that A_1 and A_{-1} are both malnormal in H, that A does not contain an element of order 2 and that H satisfies the Lyndon property LZ.*

Then if $x^2y^2z^2 = 1$ in G and $U = \{x, y, z\}$ is regular then $\langle x, y, z\rangle$ is cyclic.

We present an example that shows that Theorem 1.6.44 is best possible in general.

Let $G = \langle x, y, z \mid x^2y^2z^2 = 1\rangle$. The equation $x^2y^2z^2 = 1$ holds trivially and G is noncyclic. Since the equation $x^2y^2z^2 = 1$ is equivalent to the equation $v^2[u, t^{-1}] = 1$ we may write G as an HNN group $G = \langle H, t \mid t^{-1}ut = v^2u\rangle$ with $H = \langle u, v \mid \rangle$. The element u is not conjugate in the base H to v^2u and both associated subgroups are malnormal in the base. However, the system $\{x, y, z\}$ is not regular showing that in Theorem 1.6.44 regularity is essential.

We could manage the regularity problem in one-relator groups with odd torsion for quadratic equations $w(x_1, x_2, x_3) = 1$.

Theorem 1.6.45 ([87]). *Let G be a one-relator group $G = \langle a, b, c, \ldots \mid P^m = 1\rangle$ with $m \geq 3$ and m odd and P a cyclically reduced word, not a proper power in the free group on $a, b, c, \ldots$. If $x^2y^2z^2 = 1$ in G then $\langle x, y, z\rangle$ is cyclic.*

We also should mention that there are first results concerning the rank problem and the Nielsen method in some different construction like graph of groups and groups acting on $\mathbb{R}$-trees. We just mention some of the respective references: [145, 146, 147, 100, 265, 148, 149, 268, 150, 71].

1.7 One-relator groups

One-relator groups have always played a fundamental role in infinite group theory, especially for the topics in this book. We already considered in part the use of the Nielsen method for one-relator groups. They also arise naturally in the study of low-dimensional topology, an aspect we consider later in the book. Here we present a kind of classical introduction to some main topics, most of them with proof. We especially prove those results we need later in the book. There is a massive amount of work done over the past decades and many of the important and beautiful results on one-relator groups we cannot touch upon in this book. To get a feeling for what has been accomplished we suggest the survey by Baumslag, Fine and Rosenberger [21].

1.7.1 Magnus's Freiheitssatz and finite subgroups of one-relator groups

We start with the following.

Theorem 1.7.1 (Magnus's Freiheitssatz). *Let $G = \langle a, b, c, \ldots \mid r = 1\rangle$ a one-relator group with defining relator r, where r is a non-trivial cyclically reduced word in the free group* *on $a, b, c, \ldots$. Let $T \subset \{a, b, c, \ldots\}$ be such that r cannot be written as a word in T. Then T is a basis for a free subgroup of G.*

Proof. The proof will be by induction on the length of r. The case that r contains only one generator is obvious.

We now assume that r contains at least two generators, say a and b. By Lemma 1.6.36 we may assume that the exponent sum $\sigma_a(r) = 0$. Also we may assume that r begins with b^ϵ, $\epsilon = \pm 1$, after a cyclic permutation if necessary. We define $t = a$, $b_i = t^i b t^{-i}$, $c_i = t^i c t^{-i}, \ldots$, with $i \in \mathbb{Z}$. The relator r belongs to the normal closure of $b, c, \ldots$ in $F = \langle t, b, c, \ldots \mid \rangle$.

Hence, r may be written as a cyclically reduced word s in the $b_i, c_i, \ldots$. Let m, M represent, respectively, the least and greatest value of i for which b_i occurs in s. Then

$$G = \langle t, b_m, \ldots, b_M, c_i, d_i, \ldots, i \in \mathbb{Z} \mid s = 1, t b_j t^{-1} = b_{j+1}, t c_i t^{-1} = c_{i+1}, \ldots, \\ j = m, \ldots, M-1, i \in \mathbb{Z} \rangle.$$

Let $H = \langle b_m, \ldots, b_M, c_i, \ldots, i \in \mathbb{Z} \mid s = 1 \rangle$. By the induction hypothesis the groups $A_{-1} = \langle b_m, \ldots, b_{M-1}, c_i, \ldots, i \in \mathbb{Z} \rangle$ and $A_1 = \langle b_{m+1}, \ldots, b_M, c_i, \ldots, i \in \mathbb{Z} \rangle$ are free on the given generators. Especially the map

$$b_j \mapsto b_{j+1}, \quad c_i \mapsto c_{i+1}, \quad m \le j < M, i \in \mathbb{Z}$$

defines an isomorphism $i : A_{-1} \to A_1$. Hence $G = \langle H, t \mid t^{-1} A_1 t \overset{i}{=} A_{-1} \rangle$. Let first $T \subset L = \{b, c, d, \ldots\}$.

Since t occurs in r with $\sigma_t(r) = 0$, then at least one generator of H occurs with index $\neq 0$ in s. By the induction hypothesis L generates a free subgroup of H with basis L, and hence $\langle L \rangle$, and also $\langle T \rangle$, is a free subgroup of G. Now, let $T \subset L = \{t, c, d, \ldots\}$, that is, the omitted generator is b. Let w be a non-trivial, freely reduced word in $\{t, c, d, \ldots\}$. If $\sigma_t(w) \neq 0$, then $w \neq 1$ in G because each word which is 1 in G is freely equivalent to a product of conjugates of r^ϵ, $\epsilon = \pm 1$. Now, let $\sigma_t(w) = 0$. Then we may write w as a word in the set $J = \{c_i, d_i, \ldots, i \in \mathbb{Z}\}$ analogously to above when we rewrite r. This gives a freely reduced non-trivial wort w^*. By the induction hypothesis J generates freely a free subgroup of H. Hence, $w^* \neq 1$ in H. Since $w^* = w$ in G we have $w \neq 1$ in G. Hence, $\langle T \rangle$ is a free subgroup of G with basis T. □

Theorem 1.7.2 (Freiheitssatz for one-relator free groups with torsion). *Let*

$$G = \langle a_1, a_2, \ldots, a_n \mid a_1^{e_1} = a_2^{e_2} = \cdots = a_n^{e_n} = r^m(a_1, a_2, \ldots, a_m) = 1 \rangle$$

with $n \ge 2$, $m \ge 2$, $e_i = 0$ or $e_i \ge 2$ for $i = 1, 2, \ldots, n$, and $r(a_1, a_2, \ldots, a_n)$ a cyclically reduced word in the free product of $\langle a_1 \rangle, \langle a_2 \rangle, \ldots, \langle a_n \rangle$, which contains all $a_1, a_2, \ldots, a_n$. Then

$$\langle a_1, a_2, \ldots, a_{n-1} \rangle = \langle a_1 \mid a_1^{e_1} = 1 \rangle * \langle a_2 \mid a_2^{e_2} = 1 \rangle * \cdots * \langle a_{n-1} \mid a_{n-1}^{e_{n-1}} = 1 \rangle.$$

We know that we may embed free products of finitely many cyclic groups in $\mathrm{PSL}(2,\mathbb{C})$. Hence, Theorem 1.7.2 follows directly from

Theorem 1.7.3 ([81]). *Let A and B be groups which can be embedded in $\mathrm{PSL}(2,\mathbb{C})$. Let $m \geq 2$, R a cyclically reduced word in the free product $A * B$ of length ≥ 2 and $G = A * B/\langle\langle R^m\rangle\rangle$. Then there exists a homomorphism $\rho\colon G \to \mathrm{PSL}(2,\mathbb{C})$ such that*

$$A \to G \xrightarrow{\rho} \mathrm{PSL}(2,\mathbb{C}) \quad \text{and} \quad B \to G \xrightarrow{\rho} \mathrm{PSL}(2,\mathbb{C})$$

are injective and $\rho(R)$ has order m. Especially $A \to G$, $B \to G$ are injective.

Proof. Let without loss of generality $R = a_1b_1a_2b_2\cdots a_kb_k$ with $a_i \in A$, $b_i \in B$. Since R is cyclically reduced of length ≥ 2 we may assume that all $a_i \neq 1$ and all $b_i \neq 1$.

Choose injective homomorphisms

$$\rho_A\colon A \to \mathrm{PSL}(2,\mathbb{C}) \quad \text{and} \quad \rho_B\colon B \to \mathrm{PSL}(2,\mathbb{C})$$

with

$$\rho_A(a_i) = \pm 1\begin{pmatrix} * & x_i \\ * & * \end{pmatrix}, \quad \rho_B(b_i) = \pm\begin{pmatrix} * & y_i \\ * & * \end{pmatrix} \quad \text{for } i = 1,2,\ldots,k.$$

We may assume that all $x_i \neq 0$ and all $y_i \neq 0$. This we see as follows. Let $\rho_A(a_i) = A_i = \pm\left(\begin{smallmatrix} \alpha_i & \beta_i \\ \gamma_i & \delta_i \end{smallmatrix}\right) \neq \pm\left(\begin{smallmatrix} 1 & 0 \\ 0 & 1 \end{smallmatrix}\right)$. Then $\left(\begin{smallmatrix} 1 & y \\ 0 & 1 \end{smallmatrix}\right)A_i\left(\begin{smallmatrix} 1 & -y \\ 0 & 1 \end{smallmatrix}\right) = \pm\left(\begin{smallmatrix} * & g_i \\ * & * \end{smallmatrix}\right)$ with $g_i = \beta_i + (\delta_i - \alpha_i)y - \gamma_i y^2$, $y \in \mathbb{C}$.

We can have $g_i = 0$ only for finitely many y. Hence we may choose y so that $g_i \neq 0$ for $i = 1,2,\ldots,k$. Here we replace ρ_A by the homomorphism which we get by a following conjugation with $\pm\left(\begin{smallmatrix} 1 & y \\ 0 & 1 \end{smallmatrix}\right)$, if necessary. So we may assume $x_i \neq 0$ for all $i = 1,2,\ldots,k$.

Analogously we may assume $y_i \neq 0$ for all $i = 1,2,\ldots,k$.

Let $w \in \mathbb{C}$ and ρ_A^w the homomorphism we get from ρ_A and the following conjugation with $\pm\left(\begin{smallmatrix} 1 & 0 \\ w & 1 \end{smallmatrix}\right)$, that is, $\rho_A^w(a) = \left(\begin{smallmatrix} 1 & 0 \\ w & 1 \end{smallmatrix}\right)\rho_A(a)\left(\begin{smallmatrix} 1 & 0 \\ -w & 1 \end{smallmatrix}\right)$. We consider w as a variable. Let

$$T(w) = \pm\operatorname{tr}(\rho_A^w(a_1)\rho_B(b_1)\cdots\rho_A^w(a_k)\rho_B(b_k)).$$

We note that $T(w)$ is a polynomial in w of degree $2k$, and the highest coefficient is

$$\pm x_1y_1x_2y_2\cdots x_ky_k \neq 0.$$

By the fundamental theorem of algebra there exists a $w_0 \in \mathbb{C}$ with $T(w_0) = 2\cos(\frac{\pi}{m})$. Hence $\rho_A^{w_0}(a_1)\rho_B(b_1)\cdots\rho_A^{w_0}(a_k)\rho_B(b_k)$ has order m in $\mathrm{PSL}(2,\mathbb{C})$. We define now a homomorphism $\rho\colon G \to \mathrm{PSL}(2,\mathbb{C})$ via $\rho|_A = \rho_A^{w_0}$ and $\rho|_B = \rho_B$.

Note that $\rho(R)$ has trace $\pm T(w_0) = \pm 2\cos(\frac{\pi}{m})$, and hence the order m. Also note that $\rho_A^{w_0}$ and ρ_B are injective by construction. □

We give some examples for possible groups A as in Theorem 1.7.3:

(1) Free products of finitely many cyclic groups.

(2) Fuchsian groups, orientable surface groups, non-orientable surface groups of genus ≥ 3, Kleinian groups.
(3) $\mathbb{Z}^n$, $n \in \mathbb{N}$.
(4) F_2/F_2'' with F_2 free of rank 2 and $F_2'' = [F_2', F_2']$ (free metabelian group of rank 2), and
(5) $G = \langle a_1, a_2, \ldots, a_m, b_1, b_2, \ldots, b_n \mid (UV)^\gamma = 1\rangle$, $\gamma \geq 1$, $m \geq 1$, $n \geq 1$, $1 \neq U = U(a_1, a_2, \ldots, a_m)$ and $1 \neq V = V(b_1, b_2, \ldots, b_n)$ such that either
   (i) $\gamma \geq 2$, or
   (ii) $\gamma = 1$ and at most one of U and V is a proper power in the respective free group.
   The groups in (ii) are the hyperbolic cyclically pinched one-relator groups. This we show later in the book.
(6) Let $G = \langle a_1, a_2, \ldots, a_m, b_1, b_2, \ldots, b_n \mid a_1^{e_1} = a_2^{e_2} = \cdots = a_m^{e_m} = b_1^{f_1} = b_2^{f_2} = \cdots = b_n^{f_n} = uv = 1\rangle$ where $1 \leq m, n$, $e_i = 0$ or $e_i \geq 2$ for $i = 1, 2, \ldots, m$, $f_j = 0$ or $f_j \geq 2$ for $j = 1, 2, \ldots, n$, $u = u(a_1, a_2, \ldots, a_m)$ cyclically reduced and of infinite order in the free product on $\langle a_1\rangle, \langle a_2\rangle, \ldots, \langle a_m\rangle$ and $v = v(b_1, b_2, \ldots, b_n)$ is cyclically reduced and of infinite order in the free product on $\langle b_1\rangle, \langle b_2\rangle, \ldots, \langle b_n\rangle$ unless u is a proper power or a product of two elements of order 2 and v is also a proper power or a product of two elements of order 2.
   These are the hyperbolic groups of F-type. This we show later in the book. Concerning the faithful representation in $\mathrm{PSL}(2, \mathbb{C})$ (in fact in $\mathrm{PSL}(2, \mathbb{R})$ see [84]).
(7) Finitely generated fully residually free groups.
(8) $H_1 *_A H_2$ where H_1, H_2 admit faithful representations in $\mathrm{PSL}(2, \mathbb{C})$; H_1, H_2 have cyclic centralizers and $A = \langle u_1\rangle = \langle u_2\rangle$ with $\langle u_i\rangle$ maximal cyclic subgroups in H_i, $i = 1, 2$.
(9) $\langle H, t \mid t^{-1}wt = w\rangle$ where H is torsion free and finitely generated, has cyclic centralizers, has a faithful representation in $\mathrm{PSL}(2, \mathbb{C})$, and $\langle w\rangle$ is a maximal cyclic subgroup in H.

For (7)–(9) see [78]. We now give an extension of Theorem 1.7.3 We call two elements $A, B \in \mathrm{PSL}(2, \mathbb{C})$ *irreducible* if $\mathrm{tr}[A, B] \neq 2$. Equivalently, A and B, considered as linear fractional transformations, have no common fixed point.

Theorem 1.7.4. *Let $G = H_1 *_A H_2$ with $A = \langle a\rangle$ cyclic. Let $R \in G \setminus A$ be given in reduced form $R = a_1 b_1 a_2 b_2 \cdots a_k b_k$ with $k \geq 1$ with $a_i \in H_1 \setminus A$ and $b_i \in H_2 \setminus A$ for $i = 1, 2, \ldots, k$.*

Suppose there exists a representation $\varphi: G \to \mathrm{PSL}(2, \mathbb{C})$ such that $\varphi|_{H_1}$ and $\varphi|_{H_2}$ are injective and all pairs $\{\varphi(a_i), \varphi(a)\}$ and $\{\varphi(b_i), \varphi(a)\}$ are irreducible. Then the group $H := G/\langle\langle R^m\rangle\rangle$ with $m \geq 2$ has a representation $\rho: H \to \mathrm{PSL}(2, \mathbb{C})$ such that

$$H_1 \to H \xrightarrow{\rho} \mathrm{PSL}(2, \mathbb{C}) \quad \text{and} \quad H_2 \to H \xrightarrow{\rho} \mathrm{PSL}(2, \mathbb{C})$$

are injective and $\rho(R)$ has order m. Especially $H_1 \to H$, $H_2 \to H$ are injective.

Proof. Let $\varphi: G \to \mathrm{PSL}(2, \mathbb{C})$ given with $\varphi|_{H_1}$, $\varphi|_{H_2}$ injective and all pairs $\{\varphi(a_i), \varphi(a)\}$, $\{\varphi(b_i), \varphi(a)\}$ irreducible. We may assume that $\varphi(a) = \pm\left(\begin{smallmatrix} s & 0 \\ 0 & s^{-1} \end{smallmatrix}\right)$ or $\varphi(a) = \pm\left(\begin{smallmatrix} 1 & 1 \\ 0 & 1 \end{smallmatrix}\right)$. Let first

$\varphi(a) = \pm\left(\begin{smallmatrix} s & 0 \\ 0 & s^{-1} \end{smallmatrix}\right)$. Choose $T = \pm\left(\begin{smallmatrix} t & 0 \\ 0 & t^{-1} \end{smallmatrix}\right)$, $t \in \mathbb{C}$. We consider t as a variable whose value in $\mathbb{C}$ has to be determined.

Define

(1) $\rho(h_1) = \varphi(h_1)$, $h_1 \in H_1$; and

(2) $\rho(h_2) = T\varphi(h_2)T^{-1}$, $h_2 \in H_2$.

Since T commutes with $\varphi(a)$ for each t, this map extends to a representation $\rho: H \to \mathrm{PSL}(2,\mathbb{C})$ with the desired properties, if there is a $t \in \mathbb{C}$ such that $\rho(R)$ has order m. We define

$$f(t) = \pm\operatorname{tr}(\varphi(a_1)T\varphi(b_1)T^{-1}\cdots\varphi(a_k)T\varphi(b_k)T^{-1}).$$

Note that $f(t)$ is a Laurent polynomial of degree $2k$ in t and t^{-1} since the coefficients of t^{2k} and t^{-2k} are unequal 0 because the pairs $\{\varphi(a_1),\varphi(a)\}$ and $\{\varphi(b_i),\varphi(a)\}$ are irreducible. By the fundamental theorem of algebra there exists a $t_0 \in \mathbb{C}$ with $f(t_0) = 2\cos(\frac{\pi}{m})$. With this t_0 we see that $\rho(R)$ has order m. Hence ρ has all desired properties.

Now, let $\varphi(a) = \pm\left(\begin{smallmatrix} 1 & 1 \\ 0 & 1 \end{smallmatrix}\right)$. In this case we take $T = \pm\left(\begin{smallmatrix} 1 & t \\ 0 & 1 \end{smallmatrix}\right)$ with $t \in \mathbb{C}$ considered as a variable. Since T commutes with $\varphi(a)$ we get the desired representation in an analogously manner. □

We now continue with the more general aspects of the theory. We remark that it was not necessary to work in the proof of Theorem 1.7.1 with all the infinitely many $c_i, d_i, \ldots$. We may choose also for the $c_i, d_i, \ldots$ (if they occur in s) the least and the greatest value of i.

Example 1.7.5. Let $G = \langle t,b,c \mid b^2t^{-1}c^2b^2tc^2 = 1\rangle$. Then we get

$$\begin{aligned} G &= \langle t, b_{-1}, b_0, c_i, i \in \mathbb{Z} \mid b_0^2c_{-1}^2b_{-1}^2c_0^2 = 1, tb_{-1}t^{-1} = b_0, tc_it^{-1}, i \in \mathbb{Z}\rangle \\ &= \langle t, b_{-1}, b_0, c_{-1}, c_0 \mid b_0^2c_{-1}^2b_{-1}^2c_0^2 = 1, tb_{-1}t^{-1} = b_0, tc_{-1}t^{-1} = c_0\rangle. \end{aligned}$$

The associated subgroups are then $\langle b_0, c_{-1}, c_0\rangle$ and $\langle b_1, c_{-1}, c_0\rangle$. The advantage is, if the one-relator group is finitely generated (this is not an essential restriction because the relator contains only finitely many of the generators), then also the base group in the description is finitely generated and has an HNN extension.

This gives inductively the following theorem.

Theorem 1.7.6. *Let $G = \langle a_1, a_2, \ldots, a_n \mid r = 1\rangle$, $n \geq 1$ and r cyclically reduced. Suppose that L arises from $\{a_1, a_2, \ldots, a_n\}$ by omission of one generator which occurs in r. Then the generalized word problem is solvable for $K = \langle L\rangle$ in G.*

Proof. This is clear for $n = 1$. Now, let $n \geq 2$. If r contains only one generator then G is the free product of a free group of rank $n-1$ and a cyclic group which is finite or trivial.

For this the generalized word problem for K in G is solvable using the normal form theorem for free products.

Now suppose that r contains at least two generators. Now the statement follows inductively as in the proof of Theorem 1.7.1 and the normal form theorem, Theorem 1.6.4, for HNN extensions (or Britton's lemma, Corollary 1.6.5). □

Corollary 1.7.7. *Let $G = \langle a_1, a_2, \ldots, a_n \mid r = 1\rangle$, $n \geq 1$, a one-relator group. Then G has a solvable word problem.*

Proof. We may assume that r is cyclically reduced. The proof follows from Theorem 1.7.6 and Corollary 1.6.7 on the solvability of the word problem in HNN groups. □

Theorem 1.7.8. *Let $G = \langle a, b, c, \ldots \mid r = 1\rangle$, r cyclically reduced. Then*
1. *G is torsion-free if r is not a proper power in the free group $\langle a, b, c, \ldots \mid \rangle$; and*
2. *If $r = u^n$, $n \geq 2$, where u is not a proper power in the free groups $\langle a, b, c, \ldots \mid \rangle$, then u has order n in G, and all elements of G of finite order are conjugate to a power of u. Moreover, every finite subgroup of G is conjugate to a subgroup of $\langle u \rangle$.*

Proof. If r contains only one of the generators then the theorem holds by the torsion theorem for free products because an element of finite order is contained in a conjugate of one factor. Now assume that r contains at least two generators, say a and b. By Lemma 1.6.36 we may assume that $\sigma_a(r) = 0$ for the exponent sum of a in r. Then we may write G as an HNN extension with stable letter $t = a$ and base $H = \langle b_m, \ldots, b_M, c_i, \ldots, i \in \mathbb{Z} \mid s = 1\rangle$ as in the proof of Theorem 1.7.1. Then s is a nth power if and only if r is a nth power. By the induction hypothesis H, and so G, is torsion-free if r is not a proper power. If $r = u^n$, $n \geq 2$, then $s = v^n$, v the rewriting form of u. Now the result follows inductively from the torsion theorem, Theorem 1.6.10, for HNN extensions (recall that $u = v$ in G). □

1.7.2 Newman's spelling theorem

Next we prove the following theorem:

Theorem 1.7.9 (Newman's spelling theorem). *Let $G = \langle t, b, c, \ldots \mid r^n = 1\rangle$ where r is cyclically reduced and $n \geq 2$. Suppose $w = v$ in G where w is a freely reduced word on $\{t, b, c, \ldots\}$ and v omits a generator occurring in w. Then w contains a subword S such that S is also a subword of r^ϵ, $\epsilon = \pm 1$, and the length of S is greater than $\frac{n-1}{n}$ times the length of r^n.*

Proof. Suppose that an equation $w = v$ holds in G where v omits some generator, say t, which occurs in w but not in r. Then $G = K * \langle t\rangle$, where K is the subgroup of G generated by all the given generators except t, and $v \in K$.

Let $w = w_1 w_2 \cdots w_n$ where the w_i are alternately from K and $\langle t\rangle$. Since w contains t some w_j is a non-trivial power of t. By the normal form theorem for free products, the only way the equation $w = v$ can hold is if some $w_i = 1$ in K. But if the conclusion holds for the equation $w_i = 1$ it holds for the equation $w = v$.

Thus we need consider only equations in which the omitted generator occurs in r^n. The proof is by induction in the length of r. If r involves only one generator the result holds. Therefore we may assume that r involves at least two generators, say t and b. Again we may assume that $\sigma_t(r) = 0$ for the exponent sum of t in r. As before $G = \langle H, t \mid t^{-1}A_1 t = A_{-1}\rangle$ with $H = \langle b_m, \ldots, b_M, c_i, \ldots, i \in \mathbb{Z} \mid s^n = 1\rangle$ with $A_{-1} = \langle b_m, \ldots, b_{M-1}, c_i, \ldots, i \in \mathbb{Z}\rangle$ and $A_1 = \langle b_{m+1}, \ldots, b_M, c_i, \ldots, i \in \mathbb{Z}\rangle$. Note that A_{-1} and A_1 are free on the given generators.

Suppose w is freely reduced on $t, b, c, \ldots$ and that $w(t, b, \ldots) = v(b, \ldots)$ where t occurs in w but not in v. The equation $w(t, b, \ldots) = v(b, \ldots)$ implies that w can be reduced to a t-free word by t-reductions (see Section 1.6). Starting with the original word w perform, whenever possible, t-reductions. Suppose we reach a word w' to which no such t-reductions can be applied but w' is not t-reduced. Then w' contains a subword $t^{\epsilon}ut^{-\epsilon}$, $\epsilon = \pm 1$, where u is t-free and u itself is not in A_1 or A_{-1} but u is equal to a word z in A_1 or A_{-1} (depending on ϵ). Thus z omits a generator of H occurring in u, and $u = z$ in H. By the induction hypothesis, u contains a subword Q of $s^{\epsilon n}$, $\epsilon = \pm 1$, of the required length. Since we have only used t-reductions to obtain w', we can recover w from w' by replacing each $b_i, \ldots$ by $t^i b t^{-i}, \ldots$, and freely reducing, canceling only t-symbols. Note that the part S of u which is recovered from Q will contain a subword of $r^{\epsilon n}$, $\epsilon = \pm 1$, of the desired length even if we disregard any occurrences of t^{ϵ}, $\epsilon = \pm 1$, at the beginning or end of S.

If w can be reduced to a t-free word w^* by t-reductions, then w^* must contain a generator with a non-zero subscript and $w^* = v$ in H. By the induction hypothesis w^* must contain a subword Q of $r^{\epsilon n}$, $\epsilon = \pm 1$, of the desired length. As before this implies that w contains a suitable subword S of $r^{\epsilon n}$, $\epsilon = \pm 1$, and, moreover, S can be chosen so that it does not begin or end with t^{ϵ}, $\epsilon = \pm 1$.

Next suppose that an equation $w(t, b, c, \ldots) = v(t, c, \ldots)$ holds in G where b occurs in w but not in v. By Britton's lemma we have $\sigma_t(w) = \sigma_t(v) = \alpha$. Hence $wt^{-\alpha} = vt^{-\alpha}$. Since $vt^{-\alpha}$ has exponent sum zero on t and does not contain any b_j-symbol, $vt^{-\alpha}$ can be rewritten as a t-free word v^* by t-reductions. The equation $wt^{-\alpha} = v^*$ holds in G. As in the previous argument, if $wt^{-\alpha}$ cannot be reduced to a t-free word by t-reductions we conclude that $wt^{-\alpha}$ must contain a subword S of $r^{\epsilon n}$, $\epsilon = \pm 1$, of the desired length where S does not begin or end with t^{ϵ}, $\epsilon = \pm 1$. The latter condition ensures that S is a subword of w. If $wt^{-\alpha}$ can be reduced to a t-free word w^* by t-reductions, then some b_i occurs in w^*. Since $w^* = v^*$ in H, the induction hypothesis says that w^* contains a subword Q of $s^{\epsilon n}$, $\epsilon = \pm 1$ of the required length. As before, w must contain a subword S of $r^{\epsilon n}$, $\epsilon = \pm 1$, of the required length where S does not begin or end with t^{ϵ}, $\epsilon = \pm 1$. □

The condition in Theorem 1.7.9 is exactly a condition we have in hyperbolic groups which we consider later in the book. Hence we have

Corollary 1.7.10. 1. *All one-relator groups with torsion are hyperbolic.*
2. *All one-relator groups with torsion have solvable word problem.*
3. *All one-relator groups with torsion have solvable conjugacy problem.*

These properties in Corollary 1.7.10 will be considered later in the book in general for hyperbolic groups.

The conjugacy problem for one-relator groups in general is still unsolved. There is another direct consequence of Newman's spelling theorem. Let $G = \langle t, b, c, \ldots \mid r^n = 1\rangle$, $n \geq 2, 1 \neq r$, cyclically reduced and not a proper power. Suppose without loss of generality that $\sigma_t(r) = 0$ and that b occurs in r. Again we define $b_i = t^i b t^{-i}$, $c_i = t^i c t^{-i}, \ldots, i \in \mathbb{Z}$. Let m, M represent the least and greatest value of i for which b_i occurs in s, the word we get if we rewrite r in the $b_i, c_i, \ldots$. Then $G = \langle H, t \mid t^{-1}A_1 t = A_{-1}\rangle$ where again $H = \langle b_m, \ldots, b_M, c_i, \ldots, i \in \mathbb{Z} \mid s^n = 1\rangle$, $A_1 = \langle b_{m+1}, \ldots, b_M, c_i, \ldots, i \in \mathbb{Z}\rangle$ and $A_{-1} = \langle b_m, \ldots, b_{M-1}, c_i, \ldots, i \in \mathbb{Z}\rangle$. Then A_1 and A_{-1} are malnormal in H.

1.7.3 Normal closures of single elements in free groups

Here we discuss the normal closures of single elements in free groups.

Theorem 1.7.11. *Let F be a free group and $r_1, r_2 \in F$.*
1. *If $\langle\langle r_1\rangle\rangle = \langle\langle r_2\rangle\rangle$ then r_1 is conjugate in F to r_2^ϵ, $\epsilon = \pm 1$.*
2. *If, for some positive n, r_1^n is in the normal closure of r_2^n, then r_1 is in the normal closure of r_2.*

Proof. It is enough to prove 1. We remark that 2. can also obtained as a consequence of Theorem 1.7.8.

We now prove 1. Let $r, s \in F = F(X)$ with $N = \langle\langle r\rangle\rangle = \langle\langle s\rangle\rangle$. We may assume that r and s are non-trivial and cyclically reduced. Each is a consequence of the other, and therefore r and s contain the same generators by Theorem 1.7.1. The case of just one generator is clear. Suppose that r and s both contain at least two generators. We argue, as before, inductively on the length of r. Let $X = \{t, b, c, \ldots\}$ and r and s may contain both t and b. Without loss of generality we may assume that $\sigma_t(r) = 0$. Then r is contained in the normal closure F_1 of $b, c, \ldots$ in F.

The normal closure F_1 is generated by the set X_1 of the elements $b_i = t^i b t^{-i}$, $c_i = t^i c t^{-i}, \ldots, i \in \mathbb{Z}$. The Nielsen method shows that F_1 is freely generated by X_1. Further, N is the normal closure of $S_1 = \{s_i, i \in \mathbb{Z}\}$, $s_i = t^i s t^{-i}$ in F_1. Let $r_i = t^i r t^{-i}$. Some r_p contains $b = b_0$ with least and a b_m with greatest value of i. Analogously some s_q contains $b = b_0$ with the least and a b_n with greatest value of i.

The Freiheitssatz 1.7.1 now shows that $m = n$ and that r_p is a consequence of s_q and s_q is a consequence of r_p. Hence, r_p and s_q have the same normal closure in F_1. By the induction hypothesis r_p is in F_1 conjugate to s_q^ϵ, $\epsilon = \pm 1$. Now r_p is in F conjugate to r and s_q is in F conjugate to s. Hence r is in F conjugate to s^ϵ, $\epsilon = \pm 1$. □

We get now again

Corollary 1.7.12. *Let $G = \langle a_1, a_2, \dots, a_n \mid r = 1\rangle$, $n \geq 2$. If G is a free group, then $r = 1$ or r is a primitive element in $\langle a_1, a_2, \dots, a_n \mid \rangle$ and G is free of rank $n-1$.*

Proof. Let $r \neq 1$ in $\langle a_1, a_2, \dots, a_n \mid \rangle = F$. Let G be a free group. By Theorem 1.7.1 we have $\operatorname{rk}(G) \leq n-1$. Abelianization shows $\operatorname{rk}(G) \geq n-1$. By the theorem of Grushko F has a basis $\{y_1, y_2, \dots, y_n\}$ such that the images $y_1, y_2, \dots, y_{n-1}$ form a basis of G.

From $\langle\langle r\rangle\rangle = \langle\langle y_n\rangle\rangle$ we see that r is conjugate to y_n^{ϵ}, $\epsilon = \pm 1$, that is, r primitive in F. □

We remark that the Magnus theorem, Theorem 1.7.11, also holds, respectively, for elementary free groups (see [78] for the definition).

Theorem 1.7.13. *Let G be an elementary free group and r_1, r_2 elements of G. Then if $\langle\langle r_1\rangle\rangle = \langle\langle r_2\rangle\rangle$ then r_1 is conjugate to either r_2 or r_2^{-1}.*

For a proof see [78]. This fact holds especially for orientable surface groups of genus $g \geq 2$ and non-orientable surface groups of genus ≥ 4. These special cases go back to Bogopolski [29] and Howie [133].

Shenitzer [240] determined the possibilities for the free decomposition of a one-relator group. He observes first that if w is an element of a free group F, of minimal length under the $\operatorname{Aut}(F)$ and if α is a Whitehead transformation other than a permutation of the letters such that $|\alpha(w)| = |w|$, then α cannot introduce into $\alpha(w)$ any generator that does not already occur in w.

Using Whitehead's result (see Section 1.1) we get the following theorem.

Theorem 1.7.14. *If r_1 and r_2 are elements of a free group F, of minimal length under* $\operatorname{Aut}(F)$*, and equivalent under the action of* $\operatorname{Aut}(F)$*, then the number of generators that actually occur in r_1 is the same as for r_2.*

Corollary 1.7.15. *Let $G = \langle x_1, x_2, \dots, x_n \mid r = 1\rangle$ where r is of minimal length under* $\operatorname{Aut}(F)$*, $F = \langle x_1, x_2, \dots, x_n \mid \rangle$ and contains exactly the generators $x_1, x_2, \dots, x_k$ for some k, $0 \leq k < n$. Then $G \cong G_1 * G_2$ where $G_1 = \langle x_1, x_2, \dots, x_k \mid r = 1\rangle$ is freely indecomposable and G_2 is free with basis $\{x_{k+1}, x_{k+2}, \dots, x_n\}$.*

In 1967 Baumslag conjectured that all one-relator groups with torsion are residually finite. Over the years there were many partial results on the conjecture using particular forms of the relator. However the complete conjecture was settled affirmatively by Wise [275]. An overview of Wise's methods can be found in [21].

Theorem 1.7.16. *If G is a one-relator group with torsion, then G is residually finite.*

We remark that recently Louder and Wilton have shown that one-relator groups with torsion are coherent [173].

The general case, that is, especially the case of torsion-free one-relator groups was been recently handled by Jaikin-Zapirain and Linton [137]. Hence we have that one-relator groups are coherent. This confirmed G. Baumslag's respective conjecture.

1.7.4 The surface group conjecture and property IF

Another famous conjecture concerning one-relator groups is the surface group conjecture. This conjecture was known a long time as a question by several people including Zieschang and Rosenberger. Melnikov proposed 1980 in the Kourovka notebook: Suppose that G is a residually finite, non-free, non-cyclic one-relator group such that every subgroup of finite index is again a one-relator group. Must G be a surface group?

As asked by Melnikow the answer must be no. The Baumslag–Solitar group $\mathrm{BS}(1,n)$, $|n|>1$, also satisfies the hypothesis of Melnikov's question.

Recall, a Baumslag–Solitar group $\mathrm{BS}(m,n)$ is a one-relator group with a presentation

$$\mathrm{BS}(m,n) = \langle a,b \mid a^m = ba^nb^{-1}\rangle$$

with $1 \le |m|,|n|$. If $|m| = |n| = 1$ then $\mathrm{BS}(m,n)$ is a surface group.

If $1 < |m|,|n|$ then G contains a free Abelian group of rank 2 of infinite index.

If $|n| = 1$ and $|m| > 1$ then $\mathrm{BS}(1,m)$ contains a non-free, non-finitely generated subgroup of infinite index, namely $\mathbb{Z}[\frac{1}{m}]$, the kernel of the map onto $\mathbb{Z}$. It turned out that a property on subgroups of infinite index seems to be more intrinsic than the residual finiteness.

Definition 1.7.17. A group G satisfies *property IF* if every subgroup of infinite index is free.

Hence our favorite version of the surface group conjecture is the following.

Surface Group Conjecture. *Suppose that G is a non-free, non-cyclic one-relator group such that every subgroup of finite index is again a one-relator group and every subgroup of infinite index is a free group. Then G is a surface group.*

Over the years there exist several partial results concerning the surface group conjecture. Fine, Kharlampovich, Myasnikov, Remeslennikov and Rosenberger [82] started to consider fully residually free groups with respect to the surface group conjecture. Using the structure theorem for fully residually free groups in terms of the JSJ decomposition (see [78]) they get the following result.

Theorem 1.7.18. *Suppose that G is a finitely generated fully residually free group with property IF. Then G is either free or a cyclically pinched one-relator group or a conjugacy pinched one-relator group.*

Moreover, G is either hyperbolic or free Abelian on rank 2.

Corollary 1.7.19. *Suppose that G is a finitely generated fully residually free group with property IF. Then G is either free or every subgroup of finite index is freely indecomposable and hence a one-relator group.*

Recall that a group is said to be freely indecomposable if it cannot be written as a free product of two non-trivial subgroups. Using a result of Wilton [274] combined with the Karrass–Solitar subgroup theorem for free products of two groups with an amalgamated subgroup [151], Ciobanu, Fine and Rosenberger [54] solved the surface group conjecture affirmatively for fully residually free groups, cyclically pinched one-relator groups and conjugacy pinched one-relator groups.

Theorem 1.7.20. *Suppose that G is a finitely generated non-free freely indecomposable fully residually free group with property IF. Then G is a surface group.*

Thus fully residually free and property IF essentially characterize surface groups.

Theorem 1.7.21. *A non-free group G with rank $\mathrm{rk}(G) \geq 4$ is a surface group if and only if G is finitely generated, freely indecomposable, fully residually free and it satisfies property IF.*

Recall that the non-orientable surface group $\langle x, y, z \mid x^2y^2z^2 = 1\rangle$ is not fully residually free.

Theorem 1.7.22. *Let G be a cyclically pinched or a conjugacy pinched one-relator group satisfying property IF. Then G is free or it is a surface group.*

Remark 1.7.23. In [54] this was stated a bit differently. But this a bit more general statement follows from the above observation about $\mathrm{BS}(1, n)$.

Recently, Gardam, Kielak and Logan [106] proved the following for two-generator one-relator groups. Let $G = \langle a, b \mid r = 1\rangle$ be a non-free two-generator one-relator group with property IF. Then G is a surface group, that is here, G is free Abelian of rank 2 or the Klein bottle group $\mathrm{BS}(1, -1)$.

The surface group conjecture is related to a question of Gromov, which asks if a one-ended hyperbolic group must contain a surface subgroup. There are many efforts being made considering Gromov's questions.

So far, many authors considered the Baumslag doubles (see for instance [161] and the references mentioned there). Recall that a Baumslag double is an amalgamated product of the form $G = \langle F, \bar{F} \mid w = \bar{w}\rangle$ where F is a finitely generated free group of rank ≥ 2, $\overline{F}$ is an isomorphic copy of F, w is a non-trivial word in F and $\overline{w}$ is its copy in $\overline{F}$.

The resulting Baumslag double is hyperbolic if and only if w is not a proper power (for a proof of this fact see Subsection 3.5.4). Often we allow also rank $\mathrm{rk}(F) = 1$ to occur for a Baumslag double. But if $\mathrm{rk}(F) = 1$ then G is hyperbolic if and only if G is infinite cyclic. Hence, we assume here rank $\mathrm{rk}(F) \geq 2$.

An orientable surface group of genus 2 contains an orientable surface group of any finite genus as a subgroup. Hence, it is worth to mention the following.

Theorem 1.7.24 (See also Theorem 1.5.62). *Let $G = \langle F, \bar{F} \mid w = \bar{w}\rangle$ be a hyperbolic Baumslag double. Then G contains a hyperbolic orientable surface group of genus 2 if and only if w is a commutator, that is, $w = [U, V]$ for some elements $U, V \in F$.*

Further, G contains a non-orientable surface group of genus 4 if and only if $w = X^2Y^2$ for some $X, Y \in F$.

Corollary 1.7.25. *Let $G = \langle F, \bar{F} \mid w = \bar{w} \rangle$ be a hyperbolic Baumslag double. Then G contains orientable surface groups of all finite genuses if and only if w is a commutator.*

1.7.5 Torsion-free normal subgroups of finite index, the Tits alternative and decompositions of one-relator groups

We discuss a series of important results for one-relator groups.

Theorem 1.7.26. *Let $G = \langle a_1, a_2, \ldots, a_n \mid r^m = 1 \rangle$, $m \geq 2$, $n \geq 2$, r cyclically reduced and not a proper power in $\langle a_1, a_2, \ldots, a_n \mid \rangle$.*

Then G contains a torsion-free normal subgroup of finite index, that is, G is virtually torsion-free.

Proof. We use a theorem of Selberg [238] in the special case of $\mathrm{PSL}(2, \mathbb{C})$:

Each finitely generated subgroup of $\mathrm{PSL}(2, \mathbb{C})$ contains a torsion-free normal subgroup of finite index. Now, we may assume that r contains all generators $a_1, a_2, \ldots, a_n$. Let $\rho: G \to \mathrm{PSL}(2, \mathbb{C})$ a homomorphism as in the proof of Theorem 1.7.2 with $\rho|_{\langle a_1, a_2, \ldots, a_{n-1} \mid \rangle}$ and $\rho|_{\langle a_n \rangle}$ injective and $\rho(r)$ of order m. Then $\rho(G)$ is a finitely generated subgroup of $\mathrm{PSL}(2, \mathbb{C})$ and, hence, contains a torsion-free normal subgroup H of finite index. Note that $\rho^{-1}(H)$ is a normal subgroup of finite index in G. We claim that $\rho^{-1}(H)$ is torsion-free.

Let $g \in \rho^{-1}(H)$ with $g = 1$ or of finite order ≥ 2. We have $g \in \ker(\rho)$ because H is torsion-free. Furthermore, $g = 1$ or g is conjugate to a non-trivial power of r by Theorem 1.7.8. Since $\rho(r)$ has order m we must have $g = 1$. Hence, $\rho^{-1}(H)$ is torsion-free. □

Although the general development of group homology and Euler characteristic and their use are beyond the scope of this book we want to make some remarks. We refer the reader to the book [37] for the necessary terminology and a complete development. We note, however, that a group G is of type FL if $\mathbb{Z}$ admits a finite free resolution over the group ring $\mathbb{Z}G$ and is of type WFL if G is virtually torsion-free and every torsion-free subgroup of finite index is of type FL. Now, let G be as in Theorem 1.7.26. G is virtually torsion-free and $\mathbb{Z}$ admits a finite free resolution over the group ring $\mathbb{Z}G$ [176]. Hence, G is of type WFL and has the virtual cohomological dimension $\mathrm{vcd}(G) \leq 2$ (the bound 2 has the reason that there is a two-dimensional CW-complex, with G as fundamental group, which has a contractible universal covering space, a subject we discuss later in the book). Now Chiswell [53] showed that G has rational Euler characteristic $\chi(G) = 1 - n + \frac{1}{m}$, and if H is a subgroup of G of finite index $[G : H]$, then $\chi(H)$ is defined with $\chi(H) = [G : H]\chi(G)$.

Results of Howie and Duncan [131, 132] and [135] allow us to extend Theorem 1.7.26 to further one-relator products of cyclic.

Theorem 1.7.27. *Let*

$$G = \langle a_1, a_2, \ldots, a_n \mid a_1^{e_1} = a_2^{e_2} = \cdots = a_n^{e_n} = 1 = R^m(a_1, a_2, \ldots, a_n) = 1 \rangle$$

with $m \geq 3$, $n \geq 2$ *and* $e_i = 0$ *or* $e_i \geq 2$ *for* $i = 1, 2, \ldots, n$ *and* $R(a_1, a_2, \ldots, a_n)$ *a cyclically reduced word in the free product on* $a_1, a_2, \ldots, a_n$ *which involves all the* a_i.

If $m \geq 4$ *then suppose that* $R(a_1, a_2, \ldots, a_n)$, *is not conjugate in the free product on* $a_1, a_2, \ldots, a_n$ *to a word of the form* xy *of orders* $p \geq 2$, $q \geq 2$, *respectively, where* $\frac{1}{m} + \frac{1}{p} + \frac{1}{q} > 1$.

If $m = 3$ *then suppose that no letter (with respect to the free product on* $a_1, a_2, \ldots, a_n$*) appearing in* $R(a_1, a_2, \ldots, a_n)$ *has order 2. Then:*

(1) G *is virtually torsion-free.*

(2) G *is of homological type WFL with* $\operatorname{vcd}(G) \leq 2$, *and* G *has a rational Euler characteristic given by*

$$\chi(G) = 1 - n + \sum_{i=1}^{n} \beta_i + \frac{1}{m},$$

where $\beta_i = 0$ *if* $e_i = 0$ *and* $\beta_i = \frac{1}{e_i}$ *if* $e_i \geq 2$. *Furthermore, if the index* $[G : H]$ *is finite for* $H < G$ *then* $\chi(H)$ *is defined and* $\chi(H) = [G : H]\chi(G)$ *(Riemann–Hurwitz formula).*

For a proof of this theorem, based on Howie's work, see [93]. In this book there are also, respectively, results for generalized tetrahedron groups, cyclically pinched one-relator groups, and groups of F-type.

This type of reduction to linear groups gives many other results. Besides the results by Malcev and Selberg there have to be mentioned two more results. We first consider Tits's theorem, see [258]: Let G be a finitely generated subgroup of $\mathrm{GL}(n, K)$, $n \geq 2$, K a commutative field. Then G contains a free subgroup of rank 2 or a solvable subgroup of finite index.

In this spirit we give the following definition for finitely generated groups.

Definition 1.7.28. A finitely generated group G satisfies the *Tits alternative* if it either contains a non-Abelian free subgroup or a solvable subgroup of finite index.

We remark some terminology concerning the solvability of a group.

Remark 1.7.29. We remark that a group which contains a solvable subgroup of finite index is often just called *virtually solvable* or *solvable-by-finite*. The term solvable-by-finite is also used in the literature to coin groups that contain a normal solvable subgroup of finite index. By virtue of Theorem 1.2.22 these two notions are equivalent for finitely generated groups. In this book, we choose to use the term virtually solvable throughout.

There are many research activities concerning the Tits alternative for finitely generated groups by many people including Fine, Howie and Rosenberger for one-relator products of cyclics, generalized tetrahedron groups, groups of F-type, one-relator groups, automorphism groups and other types of groups. Using the essential representations we easily get the following. We leave the details as an exercise.

Theorem 1.7.30. *Let*

$$G = \langle a_1, a_2, \ldots, a_n \mid a_1^{e_1} = a_2^{e_2} = \cdots = a_n^{e_n} = R^m(a_1, a_2, \ldots, a_n) = 1\rangle$$

with $m \geq 2$, $e_i = 0$ or $e_i \geq 2$ for $i = 1, 2, \ldots, n$ and $R(a_1, a_2, \ldots, a_n)$ a cyclically reduced word in the free product on $a_1, a_2, \ldots, a_n$ involving all $a_1, a_2, \ldots, a_n$.

If $n \geq 3$ or $n = 2$, $e_1 = 0$ or $n = 2$, $e_2 = 0$ then the Tits alternative holds for G.

In a series of papers Howie, Rosenberger and others (see [93, 134], and the references cited therein) could show the following.

Theorem 1.7.31. *Let $G = \langle a, b \mid a^p = b^q = R^r(a, b) = 1\rangle$ where $R(a, b)$ is a non-trivial cyclically reduced word in the free product on a and b and $2 \leq p \leq q$. Then:*
1. *If $r \geq 3$ then the Tits alternative holds for G.*
2. *If $3 \leq p \leq q$, $r = 2$ then the Tits alternative holds for G.*

A group as in Theorem 1.7.31 is called a *generalized triangle group*.

The conjecture that the Tits alternative holds for a generalized triangle group is now reduced to the case $p = r = 2$, $3 \leq q \leq 5$ (the case $q \geq 6$ is handled already).

We remark that the complete list of the finite generalized triangle groups is given by Theorem 7.3.4.1 in [90].

We now consider Bass's theorem, see [9].

Let G be a finitely generated subgroup of $\mathrm{GL}(2, \mathbb{C})$. Then one of the following cases holds:
(i) There exists an epimorphism $f: G \to \mathbb{Z}$ with $f(u) = 0$ for all unipotent elements u of G.
(ii) We have $G = G_1 *_H G_2$ with $G_1 \neq H \neq G_2$, and each finitely generated unipotent subgroup of G is contained in a conjugate of G_1 or G_2.
(iii) The group G is conjugate to a group of triangular matrices $\left(\begin{smallmatrix} a & b \\ 0 & d \end{smallmatrix}\right)$ where a and d are roots of units.
(iv) The group G is conjugate to a subgroup of $\mathrm{GL}(2, A)$ where A is a ring of algebraic integers.

This easily gives the following theorem.

Theorem 1.7.32. *Suppose*

$$G = \langle a_1, a_2, \ldots, a_n \mid a_1^{e_1} = a_2^{e_2} = \cdots = a_n^{e_n} = R^m(a_1, a_2, \ldots, a_n) = 1\rangle.$$

If $m \geq 2$, $n \geq 3$ and $e_i = 0$ or $e_i \geq 2$ for all $i = 1, 2, \ldots, n$, then G is a non-trivial free product with amalgamation.

Proof. We may assume that each $e_i \geq 2$. If not we introduce relations $a_i^{f_i} = 1$ with large enough $f_i \geq 2$ for each i with $e_i = 0$, and write $f_j = e_j$ if $e_j \geq 2$. The group $G^* = \langle a_1, a_2, \ldots, a_n \mid a_1^{f_1} = a_2^{f_2} = \cdots = a_n^{f_n} = R^m(a_1, a_2, \ldots, a_n) = 1\rangle$ is an epimorphic image of G. If G^* is a non-trivial free product with amalgamation then also G.

Hence, let $e_i \geq 2$ for $i = 1, 2, \ldots, n$. This condition indicates that G is generated by elements of finite order. We may assume that R is cyclically reduced in the free product on $a_1, a_2, \ldots, a_n$ and involves all the generators for if a generator is omitted in $R(a_1, a_2, \ldots, a_n)$ then G is a non-trivial free product because $n \geq 3$.

Since $m \geq 2$ and $n \geq 3$ there exists an irreducible essential representation of G in $\mathrm{PSL}(2, \mathbb{C})$ and therefore especially the space of representations of G in $(\mathrm{PSL}(2, \mathbb{C}))^n$ is non-empty. Here an irreducible representation means that in the image of G there are at least two elements which have no common fixed point considered as linear fractional transformations.

Hence, from the existence of irreducible essential representations for such one-relator products of cyclics we see that each of the $n-1$ matrices have two degrees of freedom with the trace and determinant being specified, while the final matrix has one degree of freedom, the trace, determinant and relator condition specified. Therefore the dimension of the character space is $2(n-1) + 1 - 3 = 2n - 4$. This is positive if $n \geq 3$. We may avoid case (iv) of Bass's theorem by choosing a suitable transcendental number for a matrix. We already avoided case (iii) by the irreducibility of the representation. Thus if $n \geq 3$, G splits as a non-trivial free product with amalgamation or an HNN group. However G is generated by elements of finite order so its abelianization G^{ab} is finite. Therefore G cannot be an HNN group and must therefore be a non-trivial free product with amalgamation. □

Corollary 1.7.33. *Let $G = \langle a_1, a_2, \ldots, a_n \mid R^m = 1 \rangle$, $n \geq 3$, $m \geq 2$ be a one-relator group with torsion. Then G is decomposable as a non-trivial free product with amalgamation.*

Benyash–Krivets [26] refined the representation arguments deeply and showed the following.

Theorem 1.7.34. *Let $G = \langle a, b \mid R^m = 1 \rangle$, $m \geq 2$. Then G is decomposable as a non-trivial free product with amalgamation.*

We now want to show how to find special free subgroups in one-relator groups with torsion by taking powers of one generator. The idea goes back to Mendelsohn and Ree [190].

We consider one-relator groups of form $G = \langle a, b \mid r^m = 1 \rangle$ with $m \geq 2$ and r not a conjugate of a power of a or b. We construct essential representations of G in $\mathrm{PSL}(2, \mathbb{C})$ such that the image $\langle A, B \rangle$ has free subgroups of the form $\langle A, B^n \rangle$.

Definition 1.7.35. Define

$$\mathcal{D} := \{z \in \mathbb{C} \mid |\mathrm{Re}(z)| \leq 1\} \text{ and } \mathcal{D}' := \{z \in \mathbb{C} \mid |\mathrm{Re}(z)| \geq 1\}.$$

Lemma 1.7.36. 1. *We have that $z \in \mathcal{D}'$ implies $|z^{-1} - \frac{1}{2}| \leq \frac{1}{2}$ or $|z^{-1} + \frac{1}{2}| \leq \frac{1}{2}$, and,*
2. *if $|z - \frac{1}{2}| \geq \frac{1}{2}$ and $|z + \frac{1}{2}| \geq \frac{1}{2}$ then $z^{-1} \in \mathcal{D}$.*

We leave the calculation as an exercise.

Now, let $B_\lambda := \pm\left(\begin{smallmatrix} 1 & 0 \\ \lambda & 1 \end{smallmatrix}\right) \in \mathrm{PSL}(2,\mathbb{C})$ and $B_\lambda(z)$ the linear fractional transformation $z \to \frac{z}{\lambda z+1}$ related to B_λ. Then $(B_\lambda)^n = \pm\left(\begin{smallmatrix} 1 & 0 \\ n\lambda & 1 \end{smallmatrix}\right)$, and we get

Lemma 1.7.37. *Let $\lambda \in \mathbb{C}$ with $|\lambda| \geq 1$, $|\lambda - 1| \geq 1$ and $|\lambda + 1| \geq 1$. Let B_λ be as above and $n \in \mathbb{Z} \setminus \{0\}$. Then $z \in \mathcal{D}'$ implies $B_\lambda^n(z) \in \mathcal{D}$.*

Proof. We have

$$B_\lambda^n = \pm\begin{pmatrix} 0 & 1 \\ 1 & 0 \end{pmatrix}\begin{pmatrix} 1 & n\lambda \\ 0 & 1 \end{pmatrix}\begin{pmatrix} 0 & 1 \\ 1 & 0 \end{pmatrix}.$$

Using this factorization we get $B_\lambda^n(z) = z_2^{-1}$ where $z_2 = z_1 + n\lambda$ and $z_1 = z^{-1}$. Since $|n| \geq 1$, as λ also $n\lambda$ satisfies the above inequalities. Let $z \in \mathcal{D}'$. Then $|z_1 - \frac{1}{2}| \leq \frac{1}{2}$ or $|z_1 + \frac{1}{2}| \leq \frac{1}{2}$.

If $|z_1 - \frac{1}{2}| \leq \frac{1}{2}$ then

$$\left|z_2 - \frac{1}{2}\right| = \left|z_1 + n\lambda - \frac{1}{2}\right| \geq |n\lambda| - \left|z_1 - \frac{1}{2}\right| \geq 1 - \frac{1}{2} = \frac{1}{2},$$

and in the same manner

$$\left|z_2 + \frac{1}{2}\right| = \left|z_1 + n\lambda + \frac{1}{2}\right| \geq |n\lambda + 1| - \left|z_1 - \frac{1}{2}\right| \geq 1 - \frac{1}{2} = \frac{1}{2}.$$

If $|z_1 + \frac{1}{2}| \leq \frac{1}{2}$ we get analogously $|z_2 - \frac{1}{2}| \geq \frac{1}{2}$ and $|z_2 + \frac{1}{2}| \geq \frac{1}{2}$. Hence, $B_\lambda^n(z_2)$ satisfies the conditions in part (b) of Lemma 1.7.36.

Therefore $B_\lambda^n(z) \in \mathcal{D}$. □

Theorem 1.7.38. *Let $\lambda \in \mathbb{C}$ with $|\lambda| \geq 1$, $|\lambda - 1| \geq 1$ and $|\lambda + 1| \geq 1$. Then B_λ and $A = \pm\left(\begin{smallmatrix} 1 & 2 \\ 0 & 1 \end{smallmatrix}\right)$ generate a free subgroup of* PSL(2, ℂ).

Proof. Assume there is a relation

$$W = B_\lambda^{n_r} A^{m_r} \cdots B_\lambda^{n_1} A^{m_1} = \pm E_2 = \pm\begin{pmatrix} 1 & 0 \\ 0 & 1 \end{pmatrix},$$

where, without loss of generality, all $n_i, m_i \neq 0$. Then $W(z) = z$ for all $z \in \mathbb{C}$.

We define $z_1 = A^{m_1}(0)$, $z_1' = B^{n_1}(z_1)$ and for $k \leq r-1$ inductively $z_k = A^{m_k}(z_{k-1}')$, $z_k' = B^{n_k}(z_k)$. Since W is a relation, we must have

$$z_r := A^{-m_r} B_\lambda^{-n_r}(0) = z_{r-1}'.$$

Now $z_1 = 2m_1$ which gives $|\mathrm{Re}(z_1)| \geq 1$, that is, $z_1 \in \mathcal{D}'$. Then $z_1' = B^{n_1}(z_1) \in \mathcal{D}$ by Lemma 1.7.37. By definition we have $z_2 = z_1' + 2m_2$, that is, $|\mathrm{Re}(z_2)| \geq |2m_2| - |\mathrm{Re}(z_1')| \geq 2 - 1 = 1$, that is $z_2 \in \mathcal{D}'$ and as before $z_2' \in \mathcal{D}$. Inductively we get $z_{r-1}' \in \mathcal{D}$.

On the other side $z_r = -2m_r$, that is, $|\mathrm{Re}(z_r)| \geq 2$. Hence, $z_r \notin \mathcal{D}$ and therefore $z_r \neq z_{r-1}'$ which gives a contradiction. This means that W is not a relator, which shows that $\langle A, B_\lambda \rangle$ is free of rank 2. □

Theorem 1.7.39. *Let $G = \langle a, b \mid r^m = 1\rangle$, $m \geq 2$, $1 \neq r$ not a conjugate to a power of a or b. Then there exists a $n_0 \in \mathbb{N}$ such that $\langle a, b^n\rangle$ is free of rank 2 for all $n \geq n_0$.*

Proof. We look for a suitable $\lambda \in \mathbb{C}$ such that $a \mapsto A = \pm\begin{pmatrix} 1 & 2 \\ 0 & 1 \end{pmatrix}$, $b \mapsto B_\lambda = \pm\begin{pmatrix} 1 & 0 \\ \lambda & 1 \end{pmatrix}$ is an essential representation $G \to \mathrm{PSL}(2, \mathbb{C})$. We consider λ as a variable and r as a word in a and b. Then r is represented by a matrix $R_\lambda = \pm\begin{pmatrix} p(\lambda) & q(\lambda) \\ s(\lambda) & t(\lambda) \end{pmatrix}$ with polynomials $p, q, s, r \in \mathbb{Z}[\lambda]$ and $\det(S_\lambda) = 1$.

If $\mathrm{tr}(S_\lambda) = 2\cos(\frac{\pi}{m})$, then S_λ has the eigenvalues $e^{\frac{\pi i}{m}}$ and $e^{-\frac{\pi i}{m}}$. Hence we look for a solution of the polynomial equation

$$p(\lambda) + t(\lambda) = 2\cos\left(\frac{\pi}{m}\right).$$

If there is a solution $\lambda_0 \neq 0$ then the matrix S_{λ_0} is diagonalizable. Hence, there exists a regular P with

$$S_{\lambda_0} = \pm P^{-1}\begin{pmatrix} e^{\frac{\pi i}{m}} & 0 \\ 0 & e^{-\frac{\pi i}{m}} \end{pmatrix} P,$$

that is, $S_{\lambda_0}^m = \pm E_2$, and $a \mapsto A$, $b \mapsto B_{\lambda_0}$ defines an essential representation of G in $\mathrm{PSL}(2, \mathbb{C})$.

We now choose $n_0 \in \mathbb{N}$ such that $|n_0\lambda_0| \geq 1$, $|n_0\lambda_0 - 1| \geq 1$ and $|n_0\lambda_0 + 1| \geq 1$. Then $\langle A, B_{\lambda_0}^{n_0}\rangle = \langle A, B_{n_0\lambda_0}\rangle$ is free of rank 2 and so $\langle A, B_\lambda\rangle$ for $|\lambda| \geq |n_0\lambda_0|$. We have to show that the equation $p(\lambda) + t(\lambda) = 2\cos(\frac{\pi}{m})$ has a solution $\lambda \in \mathbb{C} \setminus \{0\}$. Since r is not a conjugate of a power of a or b we see that r contains a and b, and is cyclically reduced. Let, without loss of generality, $r = a^{r_1}b^{s_1}a^{r_2}b^{s_2}\cdots a^{r_k}b^{s_k}$, all $r_i, s_i \neq 0$. By induction we get

$$r \mapsto \pm\begin{pmatrix} 1 + 2\lambda P(\lambda) & 2Q(\lambda) \\ \lambda S(\lambda) & 1 + 2\lambda T(\lambda) \end{pmatrix}$$

with $P, Q, S, T \in \mathbb{Z}[\lambda]$ with P, Q, S of degree $k - 1$, and T of degree $k - 2$ for $k \geq 2$ and $T(\lambda) = 0$ for $k = 1$.

We need a solution $\lambda_0 \neq 0$ of the equation

$$\lambda \cdot (P(\lambda) + T(\lambda)) + \left(1 - \cos\left(\frac{\pi}{m}\right)\right) = 0.$$

Now $(1 - \cos(\frac{\pi}{m})) \neq 0$ and we have a solution $\lambda_0 \neq 0$. □

Remark 1.7.40. Using an embedding of the free group of rank $n \geq 2$ in a free group of rank 2. We get directly the following theorem.

Theorem 1.7.41. *Let $G = \langle a_1, a_2, \ldots, a_k \mid r^m = 1\rangle$, $k \geq 2$, $m \geq 2$, r cyclically reduced in $\langle a_1, a_2, \ldots, a_k \mid \rangle$ and r contains all $a_1, a_2, \ldots, a_k$. Then there is a $n_0 \in \mathbb{N}$ such that $\langle a_1^n, a_2, \ldots, a_k\rangle$ is free of rank k for all $n \geq n_0$.*

We now show that this type of phenomenon and property holds in a more general context. Much depends on the traces of generators of subgroups of $SL(2, \mathbb{C})$. If $A \in SL(2, \mathbb{C})$ with $A \neq \pm E$, $E = \left(\begin{smallmatrix} 1 & 0 \\ 0 & 1 \end{smallmatrix}\right)$, then
1. A is *hyperbolic* if $tr(A) \in \mathbb{R}$ and $|tr(A)| > 2$.
2. A is *parabolic* if $tr(A) \in \mathbb{R}$ and $|tr(A)| = 2$.
3. A is *elliptic* if $tr(A) \in \mathbb{R}$ and $|tr(A)| < 2$.
4. A is *loxodromic* if $tr(A) \notin \mathbb{R}$.

Furthermore, A has finite order if and only if $tr(A) = \pm 2\cos(\frac{q\pi}{p})$ for some $p, q \in \mathbb{N}$, $1 \leq q < p$, $\gcd(p, q) = 1$; and if A is loxodromic then there exists a natural number n such that $|tr(A^n)| > 2$.

Now, we repeat the definition of the Chebyshev polynomials $S_n(t)$ recursively by $S_0(t) = 0$, $S_1(t) = 1$ and $S_n(t) = tS_{n-1}(t) - S_{n-2}(t)$ for $n \geq 2$ and $S_n(t) = -S_{-n}(t)$ if $n < 0$. These polynomials satisfy the following identities:
(1) $S_{n+m}(t) = S_n(t)S_{m+1}(t) - S_m(t)S_{m-1}(t)$,
(2) $S_n^2(t) - S_{n+1}(t)S_{n-1}(t) = 1$,
(3) $S_{nm}(t) = S_m(S_{n+1}(t) - S_{n-1}(t))S_n(t)$,
(4) $S_n(t + \frac{1}{t}) = \frac{1-t^{2n}}{(1-t^2)t^{n-1}}$,

for $n, m \in \mathbb{N}$.

If $0 \leq x < 2$ then there is $\theta \in \mathbb{R}$, $0 < \theta \leq \frac{\pi}{2}$ with $x = 2\cos(\theta)$ and we have $S_n(x) = \frac{\sin(n\theta)}{\sin(\theta)}$, $n \geq 0$. If especially $x = 2\cos(\frac{\pi}{p})$, $p \in \mathbb{N}$, $p \geq 4$, then $S_n(x) > 1$ for $1 < n < p-1$. If $x = 2$, then $S_n(x) = n$ for $n \in \mathbb{N}$. If $x > 2$, then there is a $\theta \in \mathbb{R}$ with $x = 2\cosh(\theta)$ and we have

$$S_n(x) = \frac{\sinh(n\theta)}{\sinh(\theta)} = \frac{\alpha^n - \alpha^{-n}}{\sqrt{x^2 - 4}}$$

for $n \geq 0$ where $\alpha = \frac{1}{2}(x + \sqrt{x^2 - 4})$. Hence, if $x > 2$ then

$$\lim_{n\to\infty} \frac{S_{n+1}(x)}{S_n(x)} = \alpha.$$

Now let $A, B \in SL(2, \mathbb{C})$, and let $x = tr(A)$, $y = tr(B)$ and $z = tr(AB)$. We then have the identities
(1) $tr(AB^{-1}) = tr(A) \cdot tr(B) - tr(AB) = xy - z$;
(2) $tr([A, B]) = x^2 + y^2 + z^2 - xyz - 2$;
(3) $tr([A^n, B^m]) - 2 = S_n^2(x)S_m^2(y)(tr([A, B]) - 2)$ for $n, m \geq 0$; and
(4) $A^n = S_n(x)A - S_{n-1}(x)E$ for $n \geq 0$.

These identities have the following consequences. Let $G = \langle A, B\rangle < SL(2, \mathbb{C})$ and $w \in G$. Then $tr(w)$ is a polynomial in x, y, z over $\mathbb{Z}$. Especially, if $x, y, z \in \mathbb{R}$, then $tr(w) \in \mathbb{R}$.

We also use the following result which is quite useful and the proof of which is computational (we leave it as an exercise).

Lemma 1.7.42. *Let A, B, C, X be four elements of* $\mathrm{SL}(2,\mathbb{C})$ *with* $AB = C$. *For A we write* $A = \left(\begin{smallmatrix} a_1 & a_2 \\ a_3 & a_4 \end{smallmatrix}\right)$ *and use similar notation for B, C and X. Let*

$$\vec{x} = (x_1, x_2, x_3, x_4)^t, \quad \vec{r} = (\mathrm{tr}(X), \mathrm{tr}(AX), \mathrm{tr}(BX), \mathrm{tr}(X))^t$$

and

$$M = \begin{pmatrix} 1 & 0 & 0 & 1 \\ a_1 & a_3 & a_2 & a_4 \\ b_1 & b_3 & b_2 & b_4 \\ c_1 & c_3 & c_2 & c_4 \end{pmatrix} \in M(4,\mathbb{C}).$$

Then $M\vec{x} = \vec{r}$ *and* $\det(M) = \mathrm{tr}[A,B] - 2$.

Hence, we have $\det(M) = 0$ if and only if $\mathrm{tr}([A,B]) = 2$ which is equivalent to the property that A and B, considered as linear fractional transformation, have at least one common fixed point. Based on representation theory, we call the subgroup $\langle A,B\rangle$ *reducible* if $\mathrm{tr}([A,B]) = 2$, and *irreducible* if $\mathrm{tr}([A,B]) \neq 2$. If $\langle A,B\rangle$ is irreducible, that is, if M is invertible, then $\vec{x} = M^{-1}\vec{r}$, and we get X completely from M and $\vec{r}$.

A key lemma for the following considerations follows.

Lemma 1.7.43. *Let* $A,B \in \mathrm{SL}(2,\mathbb{R})$ *with* $|\mathrm{tr}(A)| \leq 2$. *Then* $\mathrm{tr}([A,B]) \geq 2$.

Proof. We may assume, after a possible conjugation, that $A = \left(\begin{smallmatrix} 1 & \lambda \\ 0 & 1 \end{smallmatrix}\right)$ or $A = \left(\begin{smallmatrix} t & s \\ -s & t \end{smallmatrix}\right)$ where $t = \cos(\theta)$, $s = \sin(\theta)$. Let $B = \left(\begin{smallmatrix} a & b \\ c & d \end{smallmatrix}\right)$. Note that $a^2 + b^2 + c^2 + d^2 \geq 2$. Then depending on the case

$$\mathrm{tr}([A,B]) = 2(ad - bc) + \lambda^2c^2 = 2 + \lambda^2c^2 \geq 2$$

or

$$\mathrm{tr}([A,B]) = 2t^2(ad - bc) + s^2(a^2 + b^2 + c^2 + d^2) \geq 2(t^2 + s^2) = 2. \qquad \square$$

Note that from Lemma 1.7.42 if $A,B \in \mathrm{SL}(2,\mathbb{R})$ with $0 \leq \mathrm{tr}(A), \mathrm{tr}(B)$ and $\mathrm{tr}([A,B]) < 2$ then automatically both $|\mathrm{tr}(A)| > 2$ and $|\mathrm{tr}(B)| > 2$.

Moreover, $\mathrm{tr}([A,B]) < 2$ is equivalent to the fact that the fixed points of A and B, considered as linear fractional transformations, are all distinct and separate from each other.

Lemma 1.7.44. *Let* $A,B \in \mathrm{SL}(2,\mathbb{C})$ *with* $x = \mathrm{tr}(A)$, $y = \mathrm{tr}(B)$, $z = \mathrm{tr}(AB)$ *such that* $x,y,z \in \mathbb{R}$ *and* $\mathrm{tr}([A,B]) < 2$. *If* $|\mathrm{tr}(A)| < 2$ *then also* $|\mathrm{tr}(B)|$, $|\mathrm{tr}(AB)| < 2$, *that is, if* $|x| < 2$, *then also* $|y| < 2$ *and* $|z| < 2$.

Proof. Without any loss of generality, let $0 \leq x \leq y \leq z$. After a suitable conjugation in $\mathrm{SL}(2,\mathbb{C})$ we may assume that

$$A = \begin{pmatrix} 0 & 1 \\ -1 & x \end{pmatrix}, \quad B = \begin{pmatrix} y & -\rho^{-1} \\ \rho & 0 \end{pmatrix}.$$

Then $z = \mathrm{tr}(AB) = \rho + \rho^{-1}$. Suppose that $z \geq 2$. Then $\rho \in \mathbb{R}$ and therefore $A, B \in \mathrm{SL}(2, \mathbb{R})$. But this contradicts Lemma 1.7.43. Hence $z < 2$ and $y < 2$ because $y \leq z$. □

Corollary 1.7.45. *Let* $G = \langle A, B\rangle < \mathrm{SL}(2, \mathbb{C})$ *with* $x = \mathrm{tr}(A) \in \mathbb{R}$, $y = \mathrm{tr}(B) \in \mathbb{R}$, $z = \mathrm{tr}(AB) \in \mathbb{R}$. *Let* $\mathrm{tr}([A, B]) \neq 2$ *and further, if* $\mathrm{tr}([A, B]) < 2$ *then let* $|x|, |y|, |z| \geq 2$. *Then there exist an* $N \in \mathrm{SL}(2, \mathbb{C})$ *with* $NGN^{-1} \subset \mathrm{SL}(2, \mathbb{R})$.

Proof. We may assume again that $0 \leq x \leq y \leq z$.
Case 1: Let $z \geq 2$.

Then G is a conjugate in $\mathrm{SL}(2, \mathbb{C})$ to $\tilde{G} = \langle (\begin{smallmatrix} 0 & 1 \\ -1 & \lambda \end{smallmatrix}), (\begin{smallmatrix} y & -\rho^{-1} \\ \rho & 0 \end{smallmatrix})\rangle$ with $z = \rho + \rho^{-1}$. Since $z \geq 2$ we have $\tilde{G} < \mathrm{SL}(2, \mathbb{R})$.
Case 2: Let $z < 2$.

Then $\mathrm{tr}([A, B]) > 2$ by Lemma 1.7.43. Also, at most one of x, y, z is 0 because if we would have $x = y = 0$ then $2 < \mathrm{tr}([A, B]) = z^2 - 2$, that is, $z > 2$ giving a contradiction. Hence, let $0 < y \leq z < 2$.

We construct a generating pair $\{U, V\}$ of G, which is Nielsen equivalent to $\{A, B\}$, such that $0 \leq \mathrm{tr}(U) \leq \mathrm{tr}(V) \leq \mathrm{tr}(UV)$ and $2 \leq \mathrm{tr}(UV)$. We use, as in Section 1.3, the action of the Nielsen transformations on trace triples, but this time not in a trace minimizing manner. We write $(U, V) \overset{N}{\sim} (A, B)$ if there is a (regular) Nielsen transformation from (A, B) to (U, V). Let

$$E_G = \{(UV) \mid (U, V) \overset{N}{\sim} (A, B)\} \quad \text{and} \quad L_G = \{\mathrm{tr}(U), \mathrm{tr}(V), \mathrm{tr}(UV) \mid (U, V) \in E_G\}.$$

Starting from the triple $(\mathrm{tr}(A), \mathrm{tr}(B), \mathrm{tr}(AB)) = (x, y, z) \in L_G$ we get all triples $(u, v, w) \in L_G$ by repeated application of the following transformation

$$\begin{aligned} (u, v, w) &\mapsto (v, u, w), \\ (u, v, w) &\mapsto (w, u, v), \quad \text{and} \\ (u, v, w) &\mapsto (u, v, uv - w). \end{aligned}$$

This follows from the fact that $\mathrm{tr}(AB^{-1}) = \mathrm{tr}(A) \cdot \mathrm{tr}(B) - \mathrm{tr}(AB)$. We start with (x, y, z), $0 \leq x$, $0 < y \leq z < 2$. Because $y < 2$ we have $z > \frac{1}{2}xy$, that is,

$$z = \frac{1}{2}xy + \frac{1}{2}\sqrt{x^2y^2 + 4b - 4x^2 - 4y^2},$$

where $b := x^2 + y^2 + z^2 - xyz > 4$.

Fact 1. $zy - x > x$.

Proof. Assume that $zy - x \leq x$. Then

$$2x \geq zy = \frac{1}{2}xy^2 + \frac{1}{2}y\sqrt{x^2y^2 + 4b - 4x^2 - 4y^2},$$

that is, $x^2(4 - y^2) \geq y^2(b - y^2) > 0$ from which it follows that $x^2 > y^2$ because $4 < b$ and $0 < y < 2$. This contradicts $x \leq y$. Therefore $zy - x > x$. □

Fact 2. $zy - x > y$.

Proof. Assume that $zy - x \leq y$. By Fact 1 we have $0 \leq x < zy - x$. We consider the triple (x', y, z) with $x' = zy - x$. We have $0 < x' \leq y \leq z$. By Fact 1 again $x' < zy - x'$, that is, $zy - x < zy - (zy - x) = x$. This contradicts $x < zy - x$. Therefore $zy - x > y$. □

For $zy - x \leq z$ we consider the triples (x, y, z) with $x \leq y \leq z$ and $(y, yz - x, z)$ with $y \leq yz - x \leq z$. For $zy - x > z$ we consider the triples (x, y, z) with $x \leq y \leq z$ and $(y, z, zy - x)$ with $y \leq z < yz - x$.

In either case

$$x + y + z < y + yz - x + z = y + z + yz - x.$$

We now obtain a sequence (x_n, y_n, z_n) of elements of L_G where $(x_1, y_1, z_1) = (x, y, z)$ and

$$(x_{n+1}, y_{n+1}, z_{n+1}) = \begin{cases} (y_n, y_n z_n - x_n, z_n) & \text{if } y_n z_n - x_n \leq z_n, \\ (y_n, z_n, y_n z_n - x_n) & \text{if } y_n z_n - x_n > z_n, \end{cases}$$

if $n \geq 1$. In both cases

$$x_n + y_n + z_n < x_{n+1} + y_{n+1} + z_{n+1}.$$

Fact 3. *The sequence $(a_n)_{n \in \mathbb{N}}$ with $a_n = x_n + y_n + z_n$ is unbounded.*

Proof. Assume that (a_n) is bounded. Then (a_n) converges, and so, consequently, does each of the sequences (x_n), (y_n), (z_n). Let

$$x_0 = \lim_{n \to \infty} x_n, \quad y_0 = \lim_{n \to \infty} y_n,$$
$$z_0 = \lim_{n \to \infty} z_n, \quad a_0 = \lim_{n \to \infty} a_n.$$

We have $x_n \leq x_0, y_n \leq y_0, z_n \leq z_0$ and $a_n \leq a_0$ for all $n \in \mathbb{N}$. □

Moreover, we have

(i) $0 < x_0 \leq y_0 \leq z_0$;

(ii) $x_0^2 + y_0^2 + z_0^2 - x_0 y_0 z_0 = b > 4$; and

(iii) $y_0 z_0 - x_0 \geq y_0$.

The case $y_0z_0 - x_0 > y_0$ cannot occur, because otherwise there exists some y_n with $y_n > y_0$. Therefore $y_0z_0 - x_0 = y_0$.

Similarly the case $y_0z_0 - x_0 > x_0$ cannot occur. Therefore $y_0z_0 - x_0 = x_0$, from which it follows that $x_0 = y_0$ and $x_0z_0 = 2x_0$, that is, $z_0 = 2$, because $x_0 = y_0 \neq 0$. Hence

$$4 < b = x_0^2 + y_0^2 + z_0^2 - x_0y_0z_0 = 2x_0^2 + 4 - 2x_0^2 = 4,$$

which is a contradiction. Therefore (a_n) is unbounded. Consequently each of the sequences (x_n), (y_n), (z_n) is unbounded. That means there is a generating pair $\{U, V\}$ for G with $\operatorname{tr}(U) > 2$, $\operatorname{tr}(V) > 2$, and $(U, V) \overset{N}{\sim} (A, B)$. Now we are in Case 1 with U and V, which proves Corollary 1.7.45. □

Definition 1.7.46. Let G be a subgroup of $\mathrm{SL}(2, \mathbb{C})$.
1. We say that G is *discrete* if it does not contain any convergent sequence of pairwise distinct elements.
2. We say that G is *elementary* if the commutator of any two elements of infinite order has trace 2; or equivalently, G is elementary if any two elements of infinite order, regarded as linear fractional transformations, have at least one common fixed point.
3. We say that G is *elliptic* if each of its elements $A \neq \pm E$ is elliptic.

The elementary subgroups of $\mathrm{SL}(2, \mathbb{C})$ are well known and easily dealt with (see [104]). Prominent examples of elementary subgroups of $\mathrm{SL}(2, \mathbb{C})$ modulo $\pm E$ are the cyclic groups, the dihedral groups D_n, $2 \leq n \leq \infty$, the alternating groups A_4 and A_5 and the symmetric group S_4.

We now describe the main theorem for this section.

Theorem 1.7.47. *Let G be a non-elementary and non-elliptic subgroup of* $\mathrm{SL}(2, \mathbb{C})$. *Suppose that* $\operatorname{tr}(U) \in \mathbb{R}$ *for all $U \in G$. Then there exists an $N \in$* $\mathrm{SL}(2, \mathbb{C})$ *with $NGN^{-1} <$* $\mathrm{SL}(2, \mathbb{R})$.

Proof. Since G is non-elliptic there exists an $1 \neq A \in G$ with $|\operatorname{tr}(A)| \geq 2$. Especially A is of infinite order. Further, since G is also non-elementary there exists a $B \in G$ with $\operatorname{tr}([A, B]) \neq 2$. Let $x = \operatorname{tr}(A)$, $y = \operatorname{tr}(B)$, $z = \operatorname{tr}(AB)$, and without loss of generality, $x \geq 2$, $0 \leq y \leq z$.

If $\operatorname{tr}([A, B]) < 2$ then $y, z \geq 2$ by Lemma 1.7.44 because $x \geq 2$. If $\operatorname{tr}([A, B]) > 2$ then we may assume that also $y \geq 2$, eventually after replacing $\{A, B\}$ by a Nielsen equivalent pair (see proof of Corollary 1.7.45). By Corollary 1.7.45 there exists a $N \in \mathrm{SL}(2, \mathbb{C})$ with $NHN^{-1} \subset \mathrm{SL}(2, \mathbb{R})$, where $H = \langle A, B \rangle$. We may choose N in such a way, eventually after a suitable conjugation, that

$$A_1 = NAN^{-1} = \begin{pmatrix} 0 & 1 \\ -1 & x \end{pmatrix} \quad \text{and} \quad B_1 = NBN^{-1} = \begin{pmatrix} a & b \\ c & d \end{pmatrix},$$

$a, b, c, d \in \mathbb{R}$, $ad - bc = 1$, $a + d = y$.

We now consider the group $\tilde{G} = NGN^{-1}$. Also, all elements of $\tilde{G}$ have real traces. Let $D = \begin{pmatrix} e & f \\ g & h \end{pmatrix} \in \tilde{G}$. The conditions $\mathrm{tr}(D) = r_1 \in \mathbb{R}$, $\mathrm{tr}(A_1 D) = r_2 \in \mathbb{R}$, $\mathrm{tr}(B_1 D) = r_3 \in \mathbb{R}$, $\mathrm{tr}(A_1 B_1 D) = r_4 \in \mathbb{R}$ lead to the system of linear equations $M\vec{x} = \vec{r}$ where

$$\vec{x} = (e, f, g, h)^t, \quad \vec{r} = (r_1, r_2, r_3, r_4)^t$$

and

$$M = \begin{pmatrix} 1 & 0 & 0 & 1 \\ 0 & -1 & 1 & x \\ a & c & b & d \\ c & cx - a & d & dx - b \end{pmatrix} \in M(4, \mathbb{R})$$

(see Lemma 1.7.42).

Since $\det(M) = \mathrm{tr}([A, B]) - 2 \neq 0$ we have $\vec{x} \in \mathbb{R}^4$ and hence $D \in \mathrm{SL}(2, \mathbb{R})$.

Therefore $\tilde{G} < \mathrm{SL}(2, \mathbb{R})$. □

We remark that the conjugacy factor N in Theorem 1.7.47 is completely determined by two matrices A, B with $|\mathrm{tr}(A)| \geq 2$ and $\mathrm{tr}([A, B]) \neq 2$.

Corollary 1.7.48. *Let G be a non-elementary subgroup of $\mathrm{SL}(2, \mathbb{C})$ with $\mathrm{tr}(U) \in \mathbb{R}$ for all $U \in G$, and suppose that there exist $A, B \in G$ with $|\mathrm{tr}(A)| < 2$ and $\mathrm{tr}([A, B]) < 2$. Then G is elliptic.*

Proof. We assume that G is not elliptic. Then, by Theorem 1.7.47, there exists a $N \in \mathrm{SL}(2, \mathbb{C})$ with $\tilde{G} = NGN^{-1} \subset \mathrm{SL}(2, \mathbb{R})$. Let $R = NAN^{-1}$, $S = NBN^{-1}$. We have $|\mathrm{tr}(R)| < 2$ and $\mathrm{tr}([R, S]) < 2$. But this contradicts Lemma 1.7.43. □

We now describe Majeed's observation (see [186]).

Theorem 1.7.49. *Let $\langle A, B\rangle$ be an irreducible subgroup of $\mathrm{SL}(2, \mathbb{C})$ with $\mathrm{tr}(A) = x$, $\mathrm{tr}(B) = y$. Suppose that $|x| > 2$, $|y| > 2$. Then $\langle A^n, B^n\rangle$ is free and discrete for some sufficiently large n.*

Proof. We may assume, after a suitable conjugation and inversion that

$$A = \begin{pmatrix} \lambda & 0 \\ \alpha & \lambda^{-1} \end{pmatrix}, \quad B = \begin{pmatrix} \mu & \beta \\ 0 & \mu^{-1} \end{pmatrix},$$

with $|\lambda| > |\lambda^{-1}|$, $|\mu| > |\mu^{-1}|$. We consider A and B as linear fractional transformations A and B acting on $\mathbb{C} \cup \{\infty\}$, respectively. Then

$$A(z) = \frac{\lambda z}{\alpha z + \lambda^{-1}}, \quad B(z) = \frac{\mu z + \beta}{\mu^{-1}},$$

and

$$A^n(z) = \frac{\lambda^n z}{\alpha'(\lambda^n - \lambda^{-n})z + \lambda^{-n}}, \qquad B^n(z) = \frac{\mu^n z + \beta'(\mu^n - \mu^{-n})}{\mu^{-n}},$$

where $\alpha' = \frac{\alpha}{\lambda-\lambda^{-1}}$, $\beta' = \frac{\beta}{\mu-\mu^{-1}}$. If $z \neq 0$, then $A^n(z) \to \frac{1}{\alpha'}$ as $n \to \infty$, and if $z \neq \frac{1}{\alpha'}$ and $z \neq \infty$, then $A^{-n}(z) \to 0$ as $n \to \infty$. Similarly, if $z \neq 0$, then $B^n(z) \to \infty$ as $n \to \infty$, and if $z \neq -\beta'$ and $z \neq \infty$, then $B^{-n}(z) \to -\beta'$ as $n \to \infty$.

Now choose disjoint open sets O_1, O_2, U_1, U_2 of $\mathbb{C} \cup \{\infty\}$ such that $0 \in O_1$, $\frac{1}{\alpha'} \in D_2$, $\infty \in U_1$, $-\beta' \in U_2$. Such sets always exist, for example, for suitable small $\epsilon > 0$, take

$$\begin{aligned} O_1 &= \{z \in \mathbb{C} \cup \{\infty\} \mid |z| < \epsilon\}, \\ O_2 &= \left\{z \in \mathbb{C} \cup \{\infty\} \,\middle|\, \left|z - \frac{1}{\alpha'}\right| < \epsilon\right\}, \\ U_1 &= \left\{z \in \mathbb{C} \cup \{\infty\} \,\middle|\, \left|\frac{1}{z}\right| < \epsilon\right\}, \\ U_2 &= \{z \in \mathbb{C} \cup \{\infty\} \mid |z + \beta'| < \epsilon\}, \end{aligned}$$

where $\frac{1}{\alpha'} \neq -\beta'$, for otherwise $\alpha\beta = -(\lambda - \lambda^{-1})(\mu - \mu^{-1})$, that is, $\operatorname{tr}([A,B]) = 2$, which is a condition for reducibility of $\langle A, B\rangle$. Put $O = O_1 \cup O_2$, $U = U_1 \cup U_2$. Let $z \in U$; if $z \neq \infty$, then

$$\begin{aligned} A^n(z) &\to \frac{1}{\alpha'} \in O \quad \text{and} \\ A^{-n}(z) &\to 0 \in O \quad \text{as } n \to \infty \end{aligned}$$

and, if $z = \infty$, then $A^n(\infty) = A^{-n}(\infty) = \frac{1}{\alpha'}$. Hence, $A^{\epsilon n}(U) \subset O$ for sufficiently large n, $\epsilon = \pm 1$.

Analogously, let $z \in O$; if $z \neq 0$, then $B^n(z) \to \infty \in U$ and $B^{-n}(z) \to -\beta'$ as $n \to \infty$ and, if $z = 0$, then $B^n(0) = \infty$, $B^{-n}(0) = -\beta'$.

Hence, $B^{\epsilon n}(O) \subset U$ for sufficiently large n, $\epsilon = \pm 1$.

Therefore $\langle A^n, B^n\rangle$ is a free and discrete group on the two generators A^n and B^n for sufficiently large n by the ping-pong argument. First of all, A and B have infinite order. Hence, let, after a suitable conjugation,

$$w = (A^n)^{r_1}(B^n)^{s_1}(A^n)^{r_2}(B^n)^{s_2}\cdots(A^n)^{r_k}(B^n)^{s_k}(A^n)^{r_{k+1}},$$

where $k \geq 1$ and $r_1, r_2, \ldots, r_{k+1}, s_1, s_2, \ldots, s_k$ nonzero integers. If $z \in U$ then $w(z) \in O$ which gives $w \neq 1$ and hence the statement because $O \cap U = \emptyset$. □

Theorem 1.7.50. *Let $A, B \in \mathrm{SL}(2,\mathbb{C})$ and G be the subgroup of $\mathrm{SL}(2,\mathbb{C})$ generated by A and B. Let G be non-elementary and non-elliptic. Then there is a generating pair $\{U, V\}$ for G which is Nielsen equivalent to $\{A, B\}$ such that $\langle U^n, V^n\rangle$ is a discrete free group of rank 2 for some sufficiently large n.*

Proof. Let, as usual, $x = \operatorname{tr}(A), y = \operatorname{tr}(B)$ and $z = \operatorname{tr}(AB)$. We note that $\operatorname{tr}([A,B]) = x^2 + y^2 + z^2 - xyz - 2 \neq 2$ and that at most one of x, y, z is zero because G is non-elementary.

Case 1: Two of x, y and z are non-real.

Let, without loss of generality $x, y \notin \mathbb{R}$ (if, for instance, $x \in \mathbb{R}$ then $y, z \notin \mathbb{R}$ and we consider the generating pair $\{B, AB\}$ for G). Then A and B are loxodromic, and there exists a positive integer q such that $|\operatorname{tr}(A^q)| > 2$ and $|\operatorname{tr}(B^q)| > 2$. By Theorem 1.7.49 we therefore have an $n \in \mathbb{N}$ such that $\langle A^n, B^n \rangle$ is a discrete free group of rank 2.

Case 2: Two of x, y and z are real and one of them is non-real.

Let, without loss of generality, $x \in \mathbb{R}$. We consider subcases.

(a) Let $y \in \mathbb{R}, y \neq 0$ and $z \notin \mathbb{R}$.
Then AB is loxodromic. Also BAB is loxodromic because $y \neq 0$, $\operatorname{tr}(BAB) = yz - x$ and at most one of x, y and z is zero.
We regard AB and BAB, and the proof goes through as in Case 1.

(b) Let $y = 0$ and $z \notin \mathbb{R}$.
Then $x \neq 0$ and AB is loxodromic. Also ABA is loxodromic because $\operatorname{tr}(ABA) = xz$.
We now regard AB and ABA, and the proof goes through as in Case 1.

(c) Let $y \notin \mathbb{R}$, $z \in \mathbb{R}$ and $z \neq 0$. Then B is loxodromic. Also AB^2 is loxodromic because $z \neq 0$, $\operatorname{tr}(AB^2) = yz - x$. We now regard AB^2 and B, and the proof goes through as in Case 1.

(d) Let $y \notin \mathbb{R}$ and $z = 0$.
Then $x \neq 0$, and B is loxodromic. Also AB^{-1} is loxodromic because we have $\operatorname{tr}(AB^{-1}) = xy$. Now we regard AB^{-1} and B, and the proof goes through as in Case 1.

Case 3: $x, y, z \in \mathbb{R}$.

By Corollary 1.7.45 we may assume, after a suitable conjugation, that $G \subset \operatorname{SL}(2, \mathbb{R})$. Let, without loss of generality $0 \leq x, y, z$. If $\operatorname{tr}([A,B]) < 2$ then $x, y > 2$ from Lemma 1.7.43, and the result follows from Theorem 1.7.49. Now, let $\operatorname{tr}([A,B]) > 2$.

The construction in the proof of Corollary 1.7.45 leads to a generating pair $\{U, V\}$ of G which is Nielsen equivalent to $\{A, B\}$ and with $\operatorname{tr}(U) > 2$ and $\operatorname{tr}(V) > 2$, and the result follows from Theorem 1.7.49. □

Remarks 1.7.51. 1. This result is best possible because a two-generator subgroup of $\operatorname{SL}(2, \mathbb{C})$ which is elementary or elliptic does not contain a non-Abelian discrete free subgroup.

2. Let P denote the natural map: $\operatorname{SL}(2, \mathbb{C}) \to \operatorname{PSL}(2, \mathbb{C})$. We use, with slight ambiguity, the term $\operatorname{tr}(A)$ for the trace of $A \in \operatorname{PSL}(2, \mathbb{C})$, and also the term E for the identity element in $\operatorname{PSL}(2, \mathbb{C})$.
We adapt the above definitions accordingly for a subgroup G of $\operatorname{PSL}(2, \mathbb{C})$. We first get the next corollary.

Corollary 1.7.52. *Let H be any two-generator group and $f: H \twoheadrightarrow G$, with $G \subset \operatorname{SL}(2, \mathbb{C})$ or $G \subset \operatorname{PSL}(2, \mathbb{C})$, be an epimorphism of H onto G such that G is non-elementary and* *non-elliptic. Then there is a generating pair* $\{U, V\}$ *for* H *such that* $\langle U^n, V^n\rangle$ *is a free group of rank 2 for some sufficiently large* n.

3. Let H be any group. We recall that we call a cardinal number r the rank $\operatorname{rk}(H)$ of H if H can be generated by a generating system X with cardinal number r but not by a generating system Y with a cardinal number s less than r. We call a generating system X of H a *minimal generating system* of H if X has the cardinal number $\operatorname{rk}(H)$. The constructions in this section easily give the following.

   Theorem 1.7.53. *Let* G *be a non-elementary and non-elliptic subgroup of* $\mathrm{SL}(2,\mathbb{C})$. *Then* G *can be generated by a minimal generating system which contains either only hyperbolic matrices or only loxodromic matrices.*

   We leave the proof as an exercise.
4. A *Fuchsian group* is here a non-elementary discrete subgroup F of $\mathrm{PSL}(2,\mathbb{R})$ or a conjugate of such a group in $\mathrm{PSL}(2,\mathbb{C})$. In [92] Rosenberger and Fine give a complete classification of all possibilities for generating pairs for two-generator Fuchsian groups. As a corollary we have the following.

   Corollary 1.7.54. 1. *Suppose* $G = \langle U, V\rangle \subset \mathrm{PSL}(2,\mathbb{R})$ *is a Fuchsian group. Then*

$$|\operatorname{tr}([U,V]) - 2| \geq 2 - 2\cos\left(\frac{\pi}{7}\right).$$

   2. *Let* G *be a non-elementary subgroup of* $\mathrm{PSL}(2,\mathbb{R})$. *Then* G *is discrete if and only if each cyclic subgroup of* G *is discrete.*

   In this connection we want to mention Jørgensen's result [139].

Theorem 1.7.55. 1. *Let* $\langle A, B\rangle \subset \mathrm{PSL}(2,\mathbb{C})$ *be a discrete non-elementary group. Then*

$$|(\operatorname{tr}(A))^2 - 4| + |\operatorname{tr}([A,B]) - 2| \geq 1.$$

2. *Let* G *be a non-elementary subgroup of* $\mathrm{PSL}(2,\mathbb{C})$. *Then* G *is discrete if and only if each two-generator subgroup of* G *is discrete.*

1.7.6 RG type, CT, and CSA properties in one-relator groups

In the previous sections commutative transitive (CT), conjugately separated Abelian (CSA) and restricted Gromov (RG) groups appeared in various contexts. In the final subsection of this chapter, we consider their relations in one-relator groups.

Theorem 1.7.56. *Let* $G = \langle a_1, a_2, \dots, a_n \mid R^m = 1\rangle$, $m \geq 2$, $n \geq 2$, *be a one-relator group with torsion. Then* G *is a CT group.*

Proof. From Theorem 1.6.38 and Corollary 1.7.52 we see that G has a trivial center and every Abelian subgroup is locally cyclic. Hence, G is a CT group. □

Theorem 1.7.57. *Let $G = \langle a_1, a_2, \dots, a_n \mid R^m = 1\rangle$, $m \geq 2$, $n \geq 2$, be a one-relator group with torsion. Then the following are equivalent:*
(1) *The group G is CSA.*
(2) *The group G is RG.*
(3) *The group G does not contain a copy of the infinite dihedral group $\mathbb{Z}_2 * \mathbb{Z}_2$.*
(4) *The group G satisfies the Lyndon property LZ, that is, if $x^2y^2z^2 = 1$ in G, then $\langle x, y, z\rangle$ is cyclic.*

Proof. From Theorem 1.5.57, 1.6.38 and Corollary 1.7.52 we know that G is RG if and only if it does not contain a copy of $\mathbb{Z}_2 * \mathbb{Z}_2$. From [115] we know that G is CSA if and only if it does not contain a copy of $\mathbb{Z}_2 * \mathbb{Z}_2$. From Theorem 1.6.44 we know that G satisfies LZ if and only if it does not contain a copy of $\mathbb{Z}_2 * \mathbb{Z}_2$. □

For a torsion-free one-relator group the situation is different. A torsion-free one-relator group fails to be a CSA group if and only if it contains a copy of some non-Abelian Baumslag–Solitar group $BS(1,n) = \langle x, y \mid yxy^{-1} = x^n\rangle$, $n \neq 1, 0$, or a copy of the group $F_2 \oplus \mathbb{Z}$, the direct product of the free group F_2 of rank 2 and $\mathbb{Z}$ (see [115]). But $F_2 \oplus \mathbb{Z}$ contains a copy of $BS(1,1)$. Hence, a non-Abelian torsion-free one-relator group is a CSA group if and only if it does not contain a copy of $BS(1,n)$ with $n \in \mathbb{Z} \setminus \{-1, 0, 1\}$.

We remark that $BS(1,n)$ is a CT group if $n \neq -1, 0$ (see [168]). If $n > 1$ then this also follows from the fact that

$$y \mapsto \pm\begin{pmatrix} \sqrt{n} & 0 \\ 0 & \frac{1}{\sqrt{n}} \end{pmatrix}, \quad a \mapsto \pm\begin{pmatrix} 1 & 1 \\ 0 & 1 \end{pmatrix}$$

defines a faithful representation of $BS(1,n)$, $n > 1$, into $PSL(2,\mathbb{R})$, and subgroups of $PSL(2,\mathbb{R})$ are CT groups.

Theorem 1.7.58. *Let G be a torsion-free one-relator group. Then G is CT if and only if it does not contain a copy of $F_2 \oplus \mathbb{Z}$ or a copy of $BS(1,-1)$.*

Proof. Suppose G is CT. Then every subgroup of G is CT, so $F_2 \oplus \mathbb{Z}$ is not isomorphic to a subgroup of G because in

$$F_2 \oplus \mathbb{Z} = \langle a, b \mid \,\rangle \oplus \langle c \mid \,\rangle$$

we have $[a,c] = [b,c]$ but $[a,b] \neq 1$. Also the Klein bottle group $BS(1,-1) = \langle x, y \mid yxy^{-1} = x^{-1}\rangle$ is not CT because $[x,y^2] = [y^2,y] = 1$ but $[x,y] \neq 1$. Hence, G does not contain a copy of $BS(1,-1)$. Conversely, assume now that G does not contain subgroups isomorphic to $F_2 \oplus \mathbb{Z}$ or $BS(1,-1)$. We claim that G is CT.

Let $a_1, a_2, b \in G \setminus \{1\}$ such that $[a_1, b] = 1 = [a_2, b]$. Put $H = \langle a_1, a_2 \rangle$. Then H is either cyclic or contains a free group F_2 of rank 2 or is isomorphic to some $\mathrm{BS}(1, n)$, $n \neq 1$ (see [177, Chapter 5] or [115]).

If H is cyclic or is isomorphic to $\mathrm{BS}(1, 1)$, then H is Abelian and hence $[a_1, a_2] = 1$. We have to consider the following two cases:

(1) H contains a free group of rank 2.
(2) $H \cong \mathrm{BS}(1, n)$, $n \in \mathbb{Z} \setminus \{-1, 0, 1\}$.

(1) $F_2 < H$. Then $\langle b \rangle \cap F = \{1\}$ because $[b, a_1] = [b, a_2] = 1$. Therefore $\langle b, F_2 \rangle \cong \mathbb{Z} \oplus F_2$, a contradiction. So this case cannot occur.
(2) $H \cong \mathrm{BS}(1, n)$, $n \in \mathbb{Z} \setminus \{-1, 0, 1\}$. The center $Z(\mathrm{BS}(1, n))$ is trivial and $\mathrm{BS}(1, n)$ is CT (see [167]). Hence, especially $[b, a] = 1$ for each $a \in \mathrm{BS}(1, n)$ and $\langle b \rangle \cap \mathrm{BS}(1, n) = \{1\}$, that is, $\langle b, \mathrm{BS}(1, n) \rangle \cong \mathbb{Z} \oplus \mathrm{BS}(1, n)$. Note that $Z \oplus \mathrm{BS}(1, n)$ cannot contain a free group of rank 2, and also cannot be Abelian. Therefore $Z \oplus \mathrm{BS}(1, n)$ is isomorphic to some $\mathrm{BS}(1, m)$, $m \neq 0$. Since the center $Z(\mathrm{BS}(1, n))$ is trivial and the center $Z(\mathbb{Z} \oplus \mathrm{BS}(1, n))$ is non-trivial, we must have $m = 1$ or $m = -1$. We cannot have $m = 1$ because $B(1, 1)$ is Abelian and $\mathbb{Z} \oplus \mathrm{BS}(1, n)$ is not. Therefore we have $m = -1$, that is, $\mathbb{Z} \oplus \mathrm{BS}(1, n) \cong \mathrm{BS}(1, -1)$. This cannot happen by Lemma 1.6.43 because a subgroup of $\mathrm{BS}(1, -1)$ is either Abelian or isomorphic to $\mathrm{BS}(1, -1)$. So, also case (2) cannot occur. Hence, altogether G is a CT group. □

Remarks 1.7.59. 1. The groups $F_2 \oplus \mathbb{Z}$ and $\mathrm{BS}(1, -1)$ both contain a copy of $\mathbb{Z}^2 = \mathbb{Z} \oplus \mathbb{Z}$. Hence we have the next corollary.

Corollary 1.7.60. *If a torsion-free one-relator group does not contain a copy of $\mathbb{Z}^2$ then it is CT.*

2. In [80] Fine, Gaglione, Rosenberger and Spellman examine in detail the relationship between CT and CSA.

Theorem 1.7.61. *Let G be a torsion-free one-relator group. Then G is an RG group if and only if G does not contain a copy of one of the Baumslag groups*

$$\mathrm{BS}(1, m) = \langle x, y \mid yxy^{-1} = x^m \rangle, \quad m \neq 0.$$

We give some hints for the proof. Certainly if G is an RG group then G does not contain a copy of one of the $\mathrm{BS}(1, m)$, $m \neq 0$. Suppose now that G does not contain a copy of one of the $\mathrm{BS}(1, m)$, $m \neq 0$.

If G is cyclic or free then G is an RG group. Hence, let G be not cyclic and not free, that is, $G = \langle a, b, c, \ldots \mid R = 1 \rangle$ with R cyclically reduced and not a proper power. We may assume that $\sigma_a(R) = 0$. Let $a = t$.

As usual we may write G as an HNN extension $G = \langle H, t \mid t^{-1} K_1 t = K_{-1} \rangle$ with $H = \langle b_m, b_{m+1}, \ldots, b_M, c_1, d_1, \ldots \mid S = 1 \rangle$, m the least and M the greatest value of i

for which b_i occurs in S, $K_1 = \langle b_m, b_{m+1}, \ldots, b_{M-1}, c_i, d_i, \ldots \mid \rangle$ and as before $K_{-1} = \langle b_{m+1}, \ldots, b_M, c_i, d_i \ldots \mid \rangle$.

Let $x, y \in G$ which do not generate a cyclic group. We do a Nielsen cancellation on $\{x, y\}$. We may assume that one of the three following cases holds:

(1) $x, y \in H$.

(2) x and y generate a free group.

(3) $x \in K_\delta$, $\delta = \pm 1$, and y has a symmetric form

$$y = \ell_1 t^{\epsilon_1} \ell_2 t^{\epsilon_2} \cdots \ell_p t^{\epsilon_p} k_y t^{\mu_p} r_p \cdots t^{\mu_1} r_1$$

as described in Section 1.6 with $|y| \geq 1$. If (1) holds then by induction on the length of the relator x^k and y^k generate a free group of rank 2 for a suitable k. If (2) holds then $\langle x, y \rangle$ is already free.

Hence, we have to consider case (3). The proof in this case is a long combinatorial one. But essentially the proof is reduced to situations like this:

$z \in K_{-1} \setminus K_1$ not a proper power in K_{-1}, $w \in K_1 \setminus K_{-1}$ not a proper power in K_1 with $z^\alpha = w^\beta$, $\alpha, \beta \geq 2$ minimal with this property.

Then $\langle z^\alpha, zw \rangle$ is non-cyclic and Abelian, and hence, a copy of BS(1, 1). But such a situation leads to a contradiction because G does not contain a copy of BS(1, 1). For the technical details see [85] and [86].

As a summary we get the following.

Theorem 1.7.62. *Let G be a torsion-free one-relator group which does not contain a copy of* BS(1, 1). *Then the following are equivalent:*

(1) *The group G is RG.*

(2) *The group G is CSA.*

(3) *The group G does not contain a copy of the* BS(1, m) $= \langle x, y \mid yxy^{-1} = x^m \rangle$ *with* $m \in \mathbb{Z} \setminus \{-1, 0, 1\}$.

2 Covering spaces

2.1 Fundamental groups and CW complexes

We expect the reader to be familiar with topology and discrete mathematics but we will explicitly mention many of the relevant concepts. All definitions and details can be found in [48]. We also refer to [105] and [196] as introductory text books for algebraic topology. In the following X, Y and Z are always topological spaces and all maps $f: X \to Y$ are assumed to be continuous, also we let I denote the unit interval $[0, 1]$.

2.1.1 Paths, homotopy equivalences and fundamental groups

Definition 2.1.1. 1. A *path* in X is a continuous map $\sigma: I \to X$. The image $\sigma(I)$ is the *support* of σ.
2. A space X is *path connected* if any two points of X are connected by a path, that is, if $a, b \in X$ then there is a path $\sigma: I \to X$ with $\sigma(0) = a$, the *start point* of σ, and $\sigma(1) = b$, the *end point* of σ.
3. If σ is a path, then σ^{-1} is defined by $\sigma^{-1}(t) = \sigma(1 - t)$.

If we talk about a path then we often have its support in mind. There will be no misunderstanding if we just say path for the support of a path. In this sense, each path is connected. Recall that any path connected space X is also connected. If X is locally path connected then an open subset E is path connected if and only if E is connected. If X is path connected and $f: X \to Y$ with $Y = f(X)$, then $f(X)$ is path connected.

Definition 2.1.2. 1. For maps $f, g: X \to Y$ a *homotopy* from f to g is a family of continuous maps $f_t: X \to Y$ for all $t \in I$ such that $f_0 = f$ and $f_1 = g$, and the maps $(x, t) \mapsto f_t(x)$ are continuous from $X \times I$ to Y. Two maps $f_0, f_1: X \to Y$ are called *homotopic*, written $f_0 \simeq f_1$, if there is a homotopy from f_0 to f_1—or equivalently—a homotopy from f_1 to f_0 (see below).
In other words: Two maps $f, g: X \to Y$ are homotopic if there exists a continuous map $F: X \times I \to Y$ with $F(x, t) = f_t(x)$ such that $F(x, 0) = f_0(x) = f(x)$ and $F(x, 1) = f_1(x) = g(x)$ for all $x \in X$. We also call F a homotopy from f to g.
2. If all $f_t: X \to Y$ are embeddings then the homotopy is called an *isotopy* and f_0 is *isotopic* to f_1, written $f_0 \cong f_1$.
3. A map $f: X \to Y$ is called *nullhomotopic* if it is homotopic to a constant map. A space X is called *contractible* if the identity map of X is nullhomotopic.

Being homotopic is an equivalence relation in the set of maps $X \to Y$ which can easily be deduced from the following fact: If $f \simeq g: X \to Y$ and $f' \simeq g': Y \to Z$ then $f \circ f' \simeq g \circ g': X \to Z$. Using the notion of a homotopy we can give another characterization of path connectedness: A space X is path connected if and only if each map

https://doi.org/10.1515/9783111340043-002

$\{-1, 1\} \to X$ is nullhomotopic. Here, of course, $\{-1, 1\}$ can be replaced by any discrete space with two points.

Definition 2.1.3. A map $f: X \to Y$ is a *homotopy equivalence* if there exists a map $g: Y \to X$ such that $g \circ f \simeq \mathrm{id}_X$ and $f \circ g \simeq \mathrm{id}_Y$; g is then called the *homotopy inverse* of f. Two spaces X and Y are *homotopy equivalent*, or of the same *homotopy type*, written $X \simeq Y$, if there exists a homotopy equivalence $f: X \to Y$.

Of course, being homotopy equivalent is an equivalence relation.

Definition 2.1.4. 1. Let $\omega_i: I \to X$, $i = 1, 2$, be two paths with $\omega_1(1) = \omega_2(0)$. Then

$$\omega : I \to X, \quad \omega(t) = \begin{cases} \omega_1(2t) & \text{if } 0 \le t \le \frac{1}{2}, \\ \omega_2(2t-1) & \text{if } \frac{1}{2} \le t \le 1 \end{cases}$$

is called the *product* of ω_1 and ω_2. We write $\omega = \omega_1\omega_2$.
2. Let $\omega, \omega': I \to X$ be paths with the same start points and the same end points, that is, $\omega(0) = \omega'(0)$ and $\omega(1) = \omega'(1)$. Then ω and ω' are *homotopic* if there exists a homotopy $\omega_t: I \to X$ with $\omega_0 = \omega$ and $\omega_1 = \omega'$ such that $\omega_t(0) = \omega(0)$ and $\omega_t(1) = \omega(1)$ for all $t \in I$. We write $\omega \simeq \omega'$.

Please note that the composition of paths $\omega_1\omega_2$ is read from left to right; we go first through ω_1 and continue to go through ω_2 (both at double speed), in contrast to the composition of maps $g \circ f$ where we first apply f and then g. We also remark that the composition of paths is a different concept than the usual composition of maps and that a homotopy of paths is a stronger notion than just a homotopy of maps.

Lemma 2.1.5. 1. *If $\omega_i \simeq \omega_i'$, $i = 1, 2$, then $\omega_1\omega_2 \simeq \omega_1'\omega_2'$.*
2. *For ω_1, ω_2, ω_3 we have $(\omega_1\omega_2)\omega_3 \simeq \omega_1(\omega_2\omega_3)$.*
3. *For some $x \in X$ we define $c_x: I \to X$ to be the constant path $c_x(t) = x$ for all $t \in I$. Then $c_x\omega \simeq \omega$ and $\omega c_x \simeq \omega$ if ω is a path with start point x and end point x, respectively.*

Proof. 1. Let u_t be the homotopy from ω_1 to ω_1' and v_t be the homotopy from ω_2 to ω_2'. Then $u_t\omega_2$ is a homotopy from $\omega_1\omega_2$ to $\omega_1'\omega_2$, and $\omega_1'v_t$ is a homotopy from $\omega_1'\omega_2$ to $\omega_1'\omega_2'$. Hence $\omega_1\omega_2 \simeq \omega_1'\omega_2 \simeq \omega_1'\omega_2'$.
2. We define the homotopy

$$h_\tau(t) = \begin{cases} t(1+\tau) & \text{if } 0 \le t \le \frac{1}{4}, \\ t + \frac{\tau}{4} & \text{if } \frac{1}{4} \le t \le \frac{1}{2}, \\ t + \frac{\tau}{2}(1-t) & \text{if } \frac{1}{2} \le t \le 1 \end{cases}$$

for $\tau \in I$ which deforms the partition depicted in Figure 2.1(a) into the one depicted in 2.1(b).

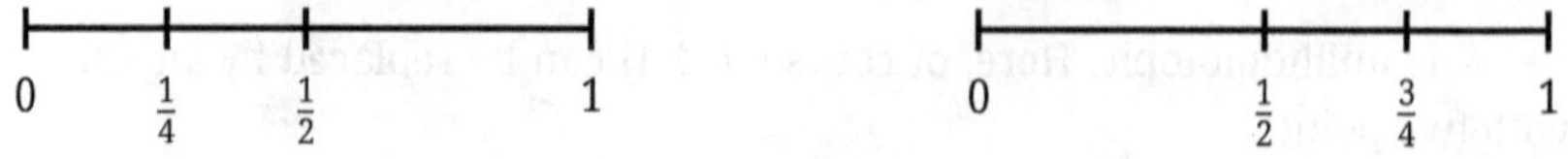

(a) First partition.

(b) Second partition.

Figure 2.1: Partitions in the proof of Lemma 2.1.5.

From Definition 2.1.4 we get

$$\omega_0(t) := (\omega_1(\omega_2\omega_3))(t) = \begin{cases} \omega_1(2t) & \text{if } 0 \le t \le \frac{1}{2}, \\ \omega_2(4t-2) & \text{if } \frac{1}{2} \le t \le \frac{3}{4}, \\ \omega_3(4t-3) & \text{if } \frac{3}{4} \le t \le 1, \end{cases}$$

and

$$\omega_0'(t) := ((\omega_1\omega_2)\omega_3)(t) = \begin{cases} \omega_1(4t) & \text{if } 0 \le t \le \frac{1}{4}, \\ \omega_2(4t-1) & \text{if } \frac{1}{4} \le t \le \frac{1}{2}, \\ \omega_3(2t-1) & \text{if } \frac{1}{2} \le t \le 1. \end{cases}$$

Then $\omega_0 \circ h_\tau$, $0 \le \tau \le 1$, is a homotopy from ω_0 to ω_0', as the respective start and end points stay fixed.

3. We define the homotopy

$$k_\tau(t) = \begin{cases} (1-\tau)t & \text{if } 0 \le t \le \frac{1}{2}, \\ (1+t)t - \tau & \text{if } \frac{1}{2} \le t \le 1 \end{cases}$$

for all $\tau \in I$.
From Definition 2.1.4 we get

$$\omega'(t) := c_x\omega(t) = \begin{cases} x & \text{if } 0 \le t \le \frac{1}{2}, \\ \omega(2t-1) & \text{if } \frac{1}{2} \le t \le 1. \end{cases}$$

Then $\omega \circ k_\tau$, $0 \le \tau \le 1$, is a homotopy from ω to $c_x\omega$ during which the start and end point of ω stay fixed. Analogously we proceed for ωc_x. □

Definition 2.1.6. Let $x_0 \in X$. A *loop* or *closed path* based at x_0 is a map $\sigma: I \to X$ such that the start and end point of σ are equal to x_0, that is, $\sigma(0) = \sigma(1) = x_0$.

According to Definition 2.1.4 we will call two loops $\sigma, \sigma': I \to X$ (based at the same basis point x_0) *homotopic* if there exists a homotopy $\sigma_t: I \to X$ with $\sigma_0 = \sigma$, $\sigma_1 = \sigma'$ and $\sigma_t(0) = \sigma_t(1) = x_0$ for all $t \in I$.

Now let $x_0 \in X$ and let $\pi_1(X, x_0)$ denote the set of all homotopy classes of loops based at x_0. If ω is a loop based at x_0 then let $[\omega]$ denote its homotopy class. We define the multiplication $[\omega][\omega'] = [\omega\omega']$ on $\pi_1(X, x_0)$.

Theorem 2.1.7. *The set $\pi_1(X, x_0)$ together with the above multiplication is a group.*

Proof. The set $\pi_1(X, x_0)$ is non-empty since it always contains the constant path c_{x_0}. Now let ω, ω' be loops based at x_0. By Lemma 2.1.5 the multiplication is well defined and associative, and there exists a neutral (or unit) element $1 = [c_{x_0}]$. It only remains to find an inverse for $[\omega]$. We claim that $[\omega^{-1}][\omega] = [\omega^{-1}\omega] = 1$. Define $\omega' = \omega^{-1}\omega$. Then

$$\omega'(t) = \begin{cases} \omega^{-1}(2t) & \text{if } 0 \le t \le \frac{1}{2}, \\ \omega(2t-1) & \text{if } \frac{1}{2} \le t \le 1, \end{cases} = \begin{cases} \omega(1-2t) & \text{if } 0 \le t \le \frac{1}{2}, \\ \omega(2t-1) & \text{if } \frac{1}{2} \le t \le 1. \end{cases}$$

We define the homotopy

$$h_\tau(t) = \begin{cases} \omega(1-2t(1-\tau)) & \text{if } 0 \le t \le \frac{1}{2}, \\ \omega(\tau + (2t-1)(1-\tau)) & \text{if } \frac{1}{2} \le t \le 1. \end{cases}$$

Then h_τ is a homotopy from $\omega^{-1}\omega$ to c_{x_0}. Analogously we proceed for $[\omega][\omega^{-1}] = 1$. □

The group $\pi_1(X, x_0)$ is called the *fundamental group, first homotopy group* or *path group* of X with respect to the basis point x_0. Direct computations of the fundamental group of topological spaces are rather complicated. In the following we will recall more sophisticated methods and structure theorems in the context of fundamental groups. We will postpone concrete examples to the next section. The next theorem shows in particular that the fundamental group of a path connected space is—up to isomorphism—independent of the choice of the basis point.

Theorem 2.1.8. *Let $x_0, x_1 \in X$ be points of the same path connected component of X. Let v be a path with start point x_0 and end point x_1. Then the map $\omega \mapsto v^{-1}\omega v$, where ω is a loop based at x_0, defines an isomorphism $\pi_1(X, x_0) \to \pi_1(X, x_1)$. Two distinct paths v, v' from x_0 to x_1 give isomorphisms which only differ by an inner automorphism of $\pi_1(X, x_1)$.*

Proof. From $\omega_1\omega_2 \mapsto v^{-1}\omega_1\omega_2 v \simeq v^{-1}\omega_1 v v^{-1}\omega_2 v$ follows that we get a homomorphism. By $\omega' \mapsto v\omega' v^{-1}$ we get an inverse homomorphism. The difference between two such isomorphisms is given by vv'^{-1} or $v^{-1}v'$ if $v^{-1}\omega v \mapsto v'^{-1}\omega v'$. □

We now explain how a map between topological spaces induces a homomorphism between fundamental groups.

Theorem 2.1.9. 1. *If $f: X \to Y$ is continuous, then $\omega \mapsto f \circ \omega$ defines a homomorphism $f_*: \pi_1(X, x_0) \to \pi_1(Y, f(x_0))$, $[\omega] \mapsto [f \circ \omega]$.*
2. *If $X \xrightarrow{f} Y \xrightarrow{g} Z$ then $(g \circ f)_* = g_* \circ f_*$.*
3. *If $f: X \to X$ is the identity then $f_*: \pi_1(X, x_0) \to \pi_1(X, x_0)$ is the identity.*
4. *If $f, g: X \to Y$ are homotopic via a homotopy h_t with $h_{t_1}(x_0) = h_{t_2}(x_0)$ for all $t_1, t_2 \in I$, then $f_* = g_*$.*

Proof. We have $f_*[\omega_1\omega_2] = [f \circ (\omega_1\omega_2)] = [f \circ \omega_1][f \circ \omega_2] = f_*[\omega_1]f_*[\omega_2]$; compare Figure 2.2. This shows the first assertion, 2. and 3. are obvious. For the last assertion observe that if $f \simeq g$ then certainly $f \circ \omega \simeq g \circ \omega$. □

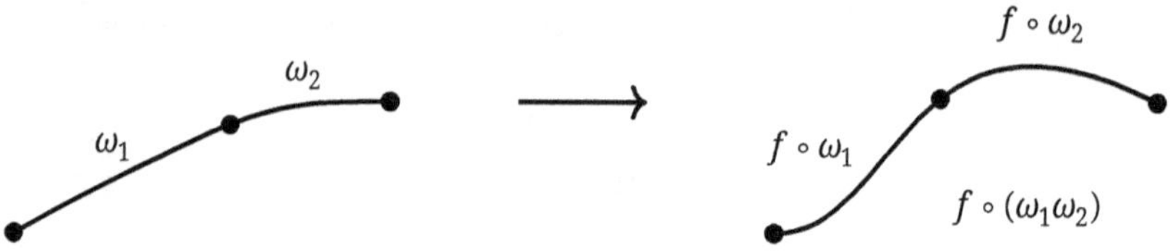

Figure 2.2: Visualization of the proof of Theorem 2.1.9.

We can interpret the map $f: X \to Y$ as a map between the pointed spaces (X, x_0) and $(Y, f(x_0))$. From this point of view Theorem 2.1.9 just says that we have a functor from the category of pointed spaces to the category of groups that maps pointed spaces to their fundamental groups and continuous maps to their induced homomorphisms. The next theorem shows how paths induced by homotopies yield isomorphisms of the respective fundamental groups.

Theorem 2.1.10. *Let $f_\tau: X \to Y$ be a homotopy. Let $x_0 \in X$ and $\gamma: I \to Y$ be the path $\tau \mapsto f_\tau(x_0)$. Via $\omega \mapsto \gamma^{-1}\omega\gamma$ we have an induced isomorphism $u: \pi_1(Y, f_0(x_0)) \to \pi_1(Y, f_1(x_0))$, compare Theorem 2.1.8. Then the diagram*

$$\begin{array}{ccc} \pi_1(X, x_0) & \xrightarrow{(f_0)_*} & \pi_1(Y, f_0(x_0)) \\ & \searrow^{(f_1)_*} & \downarrow u \\ & & \pi_1(Y, f_1(x_0)) \end{array}$$

is commutative.

Proof. Let $\omega: I \to X$ be a path with $\omega(0) = \omega(1) = x_0$. We have to show that $\gamma^{-1}(f_0 \circ \omega)\gamma \simeq f_1 \circ \omega$. Let $F: I \times X \to Y$ be defined by $F(t, \tau) = f_\tau(\omega(t))$. Then $F(t, 0) = f_0(\omega(t))$, $F(t, 1) = f_1(\omega(t))$ and $F(0, \tau) = F(1, \tau) = f_\tau(x_0) = \gamma(t)$. The part boundary of the square depicted in Figure 2.3 $I \times I$ is defined by $\gamma^{-1}(f_0 \circ \omega)$ and the roof (dashed in the figure) by $f_1 \circ \omega$. Hence, $\gamma^{-1}(f_0 \circ \omega)\gamma \simeq f_1 \circ \omega$. □

Figure 2.3: Boundary of the square $I \times I$.

We recall the notion of retracts and deformation retracts that help to gather geometric intuition of a topological space in order to simplify the calculations of the fundamental groups.

Definition 2.1.11. Let $A \subset X$ be a subspace and $i: A \hookrightarrow X$ the inclusion. Then A is a *retract* of X if there exists a map $r: X \to A$ with $r \circ i = \mathrm{id}_A$, that is, $r(x) = x$ for all $x \in A$; r is then called a *retraction*. Moreover, if $i \circ r \simeq \mathrm{id}_X$ then A is a *deformation retract* of X.

We discuss additional properties that maps induced by retractions have.

Theorem 2.1.12. 1. *Let $A \subset X$ be a retract, i the inclusion, r a retraction and $x_0 \in A$. Then $i_*: \pi_1(A, x_0) \to \pi_1(X, x_0)$ is injective.*
2. *If in addition $i_*(\pi_1(A, x_0))$ is a normal subgroup in $\pi_1(X, x_0)$ then $\pi_1(X, x_0)$ is the direct product $\pi_1(X, x_0) = i_*(\pi_1(A, x_0)) \oplus \ker(r_*)$.*
3. *There is an isomorphism $\pi_1(X \times Y, (x_0, y_0)) \cong \pi_1(X, x_0) \oplus \pi_1(Y, y_0)$.*
   Explicitly: Let $i: X \hookrightarrow X \times Y$, $x \mapsto (x, y_0)$ and $j: Y \hookrightarrow X \times Y$, $y \mapsto (x_0, y)$. Then $\pi_1(X \times Y, (x_0, y_0)) = i_*(\pi_1(X, x_0)) \oplus j_*(\pi_1(Y, y_0))$.

Proof. 1. Since $r \circ i = \mathrm{id}_A$ we see that $r_* \circ i_*$ is the identity of $\pi_1(A, x_0)$. Therefore i_* is injective.
2. Let $b \in \ker(r_*)$ and $a \in \mathrm{im}(i_*)$. Now $bab^{-1} \in \mathrm{im}(i_*)$, that is, there exist $a_1, a_2 \in \pi_1(A, x_0)$ with $i_*(a_1) = a$ and $i_*(a_2) = bab^{-1}$. Hence $a_2 = r_* \circ i_*(a_2) = r_*(bab^{-1}) = r_*(a) = r_* \circ i_*(a_1) = a_1$ because $b \in \ker(r_*)$. Therefore $a = bab^{-1}$ because i_* is injective. Furthermore, $\ker(r_*) \cap \mathrm{im}(i_*) = \{1\}$ which we can see as follows: $a = r_* \circ i_*(a) = r_*(i_*(a)) = 1$ for all $a \in \ker(r_*) \cap \mathrm{im}(i_*)$. Now $r_*(a(i_* \circ r_*(a))^{-1}) = r_*(a) r_*((i_* \circ r_*(a))^{-1}) = r_*(a)(r_*(a))^{-1} = 1$ which gives $a = i_* \circ r_*(a) \cdot b$ with $b \in \ker(r_*)$ and $a \in \pi_1(X, x_0)$. Hence $\ker(r_*) \cdot \mathrm{im}(i_*) = \pi_1(X, x_0)$.
3. Define the maps

$$X \underset{p}{\overset{i}{\rightleftarrows}} X \times Y \underset{q}{\overset{j}{\leftrightarrows}} Y$$

by $i(x) = (x, y_0)$, $p(x, y) = x$, $j(y) = (x_0, y)$ and $q(x, y) = y$.

We claim: $i_*(\pi_1(X, x_0)) = \ker(q_*: \pi_1(X \times Y, (x_0, y_0)) \to \pi_1(Y, y_0))$. This can be seen as follows: If $a \in i_*(\pi_1(X, x_0))$ then $a \in \ker(q_*)$ because $q \circ i: X \to Y$ is a constant map. Conversely, let $a \in \ker(q_*)$ and let $\omega: I \to X \times Y$ be a path which represents a. Since $q_*(a) = 1$ there exists a homotopy $\omega_t: I \to Y$ with $\omega_0 = q \circ \omega$ and $\omega_1(I) = y_0$. Let $\omega'_\tau: i \to X \times Y$ be defined by $\omega_tau'(t) = (p \circ \omega(t), \omega_\tau(t))$. Then $\omega'_0(t) = (p \circ \omega(t), q \circ \omega(t)) = \omega(t)$ and $\omega_1(t) = (p \circ \omega(t), y_0) = i \circ p \circ \omega(t)$, that is, $\omega \simeq i \circ p \circ \omega$, hence $a = i_* \circ p_*(a) \in \mathrm{im}(i_*)$.

From this claim and part 2 of the theorem we can now conclude: If we identify X with $X \times \{y_0\} \subset X \times Y$, then X is a retract of $X \times Y$, i is the inclusion and p the respective retraction, that is, $p \circ i(x, y_0) = (x, y_0)$. We know that $i_*(\pi_1(X, x_0))$ is normal in $\pi_1(X \times Y, (x_0, y_0))$ and that $\pi_1(X \times Y, (x_0, y_0)) = i_*(\pi_1(X, x_0)) \oplus \ker(p_*)$. For symmetric reasons we have $\ker(p_*) = j_*(\pi_1(Y, y_0))$. □

We are now able to proof the important fact that homotopy equivalent spaces have isomorphic fundamental groups. Put differently, the fundamental group is a homotopy invariant. In practice, this observation is often used to distinguish spaces by comparing their fundamental groups.

Theorem 2.1.13. 1. *If $f: X \to Y$ is a homotopy equivalence then $f_*: \pi_1(X, x_0) \to \pi_1(Y, f(x_0))$ is an isomorphism.*
2. *If A is a deformation retract and $i: A \hookrightarrow X$ the inclusion then $i_*: \pi_1(A, x_0) \to \pi_1(X, x_0)$, $x_0 \in A$, is an isomorphism.*

Proof. 1. Choose $g: Y \to X$ such that $g \circ f \simeq \mathrm{id}_X$ and $f \circ g \simeq \mathrm{id}_Y$. Let $u: \pi_1(X, x_0) \to \pi_1(X, g \circ f(x_0))$ and $v: \pi_1(Y, f(x_0)) \to \pi_1(Y, f \circ g \circ f(x_0))$ be the isomorphisms determined as in Theorem 2.1.10.
We get the commutative diagram

$$\begin{array}{ccc} \pi_1(X, x_0) & \xrightarrow{\ u\ } & \pi_1(X, g \circ f(x_0)) \\ {\scriptstyle f_*}\downarrow & \nearrow{\scriptstyle g_*} & \downarrow{\scriptstyle f'_*} \\ \pi_1(Y, f(x_0)) & \xrightarrow[v]{} & \pi_1(Y, f \circ g \circ f(x_0)) \end{array}$$

where we denote by f'_* the change of the basis point. Since u is an isomorphism, we see that f_* is injective and g_* is surjective. Since v is an isomorphism we see that g_* is injective and f'_* is surjective. Hence, g_* is an isomorphism and thus also f_*.
2. This follows from the first statement and the following fact: If $A \subset X$ is a deformation retract then $A \simeq X$ and $i: A \hookrightarrow X$ is a homotopy equivalence. □

Definition 2.1.14. A space X is called *simply connected* if it is path connected and if $\pi_1(X, x_0) = \{1\}$ for all $x_0 \in X$.

Note that we already know that it suffices if $\pi_1(X, x_0) = \{1\}$ for one $x_0 \in X$.

Theorem 2.1.15. *If X is contractible then X is simply connected.*

2.1.2 Simplicial and CW complexes

We now turn to simplices and simplicial complexes. We later generalize them to CW complexes.

Definition 2.1.16. We call k points $r_1, \dots, r_k \in \mathbb{R}^N$ with $k \le N+1$ *independent*, if they are not contained in a $(k-2)$-dimensional subspace. Let $\tau_0, \dots, \tau_q \in \mathbb{R}^N$ be $q+1$ independent points. The set

$$\sigma^q = \left\{ \tau \in \mathbb{R}^N \;\middle|\; \tau = \sum_{i=0}^{q} \lambda_i \tau_i \text{ with } \lambda_i \in \mathbb{R}, \lambda_i > 0 \text{ for all } i \text{ and } \sum_{i=0}^{q} \lambda_i = 1 \right\}$$

is called the *q-simplex* spanned by $\tau_0, \tau_1, \dots, \tau_q$ and we also write $\sigma^q = (\tau_0, \dots, \tau_q)$.

The set

$$\overline{\sigma^q} = \left\{ \tau \in \mathbb{R}^N \;\middle|\; \tau = \sum_{i=0}^{q} \lambda_i \tau_i \text{ with } \lambda_i \in \mathbb{R}, \lambda_i \geq 0 \text{ for all } i \text{ and } \sum_{i=0}^{q} \lambda_i = 1 \right\}$$

is called the *closed q-simplex* spanned by $\tau_0, \tau_1, \dots, \tau_q$ and we also write $\overline{\sigma^q} = [\tau_0, \dots, \tau_q]$.

We call q the *dimension* and $\tau_0, \tau_1, \dots, \tau_q$ the *vertices* of σ^q and $\overline{\sigma^q}$, respectively. The set $\partial\sigma^q = \overline{\sigma^q} \setminus \sigma^q$ is called the *boundary* of σ^q. The tuple $(\lambda_0, \dots, \lambda_q)$ are the *barycentric coordinates* of the point $\tau = \sum_{i=0}^{q} \lambda_i \tau_i$.

The barycentric coordinates are uniquely determined for fixed $\tau_0, \dots, \tau_q$. We observe that σ^0 is a point, σ^1 is a line segment without end points, σ^2 is a triangle without sides and σ^3 is a tetrahedron without side faces. We have $\partial\sigma^0 = \emptyset$.

Definition 2.1.17. Let σ and τ be simplices in $\mathbb{R}^N$. We call σ a *face* of τ if all vertices of σ are vertices of τ, we write $\sigma < \tau$. We call σ a *proper face* if $\sigma < \tau$ and $\sigma \neq \tau$.

We observe that for each simplex τ we have $\tau < \tau$ and a p-simplex has $\binom{p+1}{q+1}$ q-dimensional faces.

Definition 2.1.18. 1. A *simplicial complex* (in $\mathbb{R}^N$) is a finite set K of simplices (in $\mathbb{R}^N$) with the following properties:
   (a) If $\sigma \in K$ and $\tau \leq \sigma$ then $\tau \in K$.
   (b) If $\sigma, \sigma' \in K$ then $\sigma \cap \sigma' = \emptyset$ or $\sigma = \sigma'$.
   The set $|K| = \bigcup_{\sigma\in K} \sigma \subset \mathbb{R}^N$ is the *underlying topological space* or the *geometric realization* of K (with the induced topology of $\mathbb{R}^N$).
   Equivalently, we can define a simplicial complex K for closed simplices if we replace condition (b) by the following: The non-empty intersection of any two closed simplices $\overline{\sigma_1}, \overline{\sigma_2} \in K$ is a face of both $\overline{\sigma_1}$ and $\overline{\sigma_2}$.
2. A topological space X is a *polyhedron* or *triangulable* if there exists a simplicial complex K with $|K|$ homeomorphic to X, that is, $|K| \cong X$. If $f\colon |K| \to X$ is a homeomorphism then $f(\sigma) \subset X$, $\sigma \in K$, is a *curvilinear* simplex and $\{f(\sigma) \mid \sigma \in K\}$ a *curvilinear simplicial complex*. All notations from K are transferred to X via f. The pair (f, K) is a *triangulation* of X.

Examples 2.1.19. 1. The set T^n of face simplices of an n-simplex σ^n is a simplicial complex with $|T^n| = \overline{\sigma^n}$. The n-ball $D^n = \{x \in \mathbb{R}^n \mid \|x\| \leq 1\}$ is triangulable because $\overline{\sigma^n} \cong D^n$.
2. The set Σ^n of all proper face simplices of an $(n+1)$-simplex σ^{n+1} is a simplicial complex $|\Sigma^n| = \partial\sigma^{n+1}$. The n-sphere $S^n = \{x \in \mathbb{R}^{n+1} \mid \|x\| = 1\}$ is triangulable because $\partial\sigma^{n+1} \cong S^n$.
3. All compact surfaces are triangulable (theorem of Rado; see [281]). See Figure 2.4 for the torus.

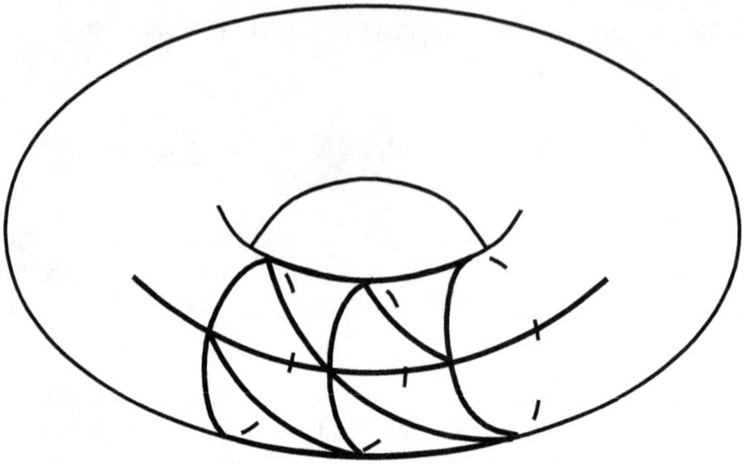

Figure 2.4: A triangulation of the torus.

4. The projective spaces $\mathbb{P}^n(T)$, $T = \mathbb{R}, \mathbb{C}$ or $\mathbb{H}$ are triangulable.
5. All compact 3-dimensional manifolds are triangulable (theorem of Moise; see [194]).
6. All compact differentiable manifolds are triangulable (theorem of Cairus; see [196]). This implies 4.

Remarks 2.1.20. 1. If K is a simplicial complex, then $K_0 \subset K$ is called a *simplicial subcomplex* if K_0 itself is a simplicial complex. Therefore $|K_0|$ is a closed subset of $|K|$.
2. If X is a polyhedron, then the subspace $X_0 \subset X$ is called *subpolyhedron* if there is a triangulation (f, K) of X and a subcomplex $K_0 \subset K$ with $f(|K_0|) = X_0$.

Triangulations can become quite large with respect to the minimal number of simplices needed to triangulate a space. Furthermore there are topological spaces that cannot be triangulated at all. We aim at CW complexes which provide a more flexible family of decompositions of topological spaces.

Definition 2.1.21. 1. A *q-dimensional* (topological) *cell* or *q-cell* is a topological space which is homeomorphic to the open q-ball $D^q \setminus S^{q-1}$.
2. A space *partitioned in cells* is a pair $K = (X, \zeta)$ where X is a topological space and ζ is a set of subspaces of X such that:
   (a) $X = \bigcup_{e \in \zeta} e$;
   (b) If $e, e' \in \zeta$ then $e \cap e' = \emptyset$ or $e = e'$; and
   (c) If $e \in \zeta$ then e is a cell.

   The space X is called the *underlying space*, written as $X = |K|$, ζ is the *cell decomposition* of X. The *dimension* of K is $\dim(K) = \sup\{\dim(e) \mid e \in \zeta\}$ (and can be finite or infinite). If $n \geq 0$ let $\zeta^n = \{e \in \zeta \mid \dim(e) \leq n\}$, $X^n = \bigcup_{e \in \zeta^n} e$. Then $K^n = (X^n, K^n)$ is a space partitioned in cells which is called the *n-dimensional skeleton* or *n-skeleton* of K. If $e \in \zeta$ then $\overline{e} \subset X$ is the *closed hull* of $e \subset X$ and $\partial e = \overline{e} \setminus e$ the *boundary* of the cell e.

Examples of cells are immediate: A 0-cell is a one point space, $D^q \setminus S^{q-1}$ is a q-cell, a q-simplex is a q-cell and $\mathbb{R}^q$ is a q-cell.

Examples 2.1.22. The following are examples of spaces partitioned in cells:
1. $(|K|, K)$ if K is a simplicial complex;
2. $(S^n, \{e^0, e^n\})$ where $e^0 \in S^n$ is any point and $e^n = S^n \setminus \{e^0\}$;
3. $(\mathbb{R}^n, \{\mathbb{R}^n\})$; and
4. (X, ζ) where X is a topological space and ζ the set of the one-point subspaces.

The last example shows that any space can be partitioned into cells and this partition does not provide new information. Hence, we will impose stronger (topological) conditions on cell complexes and we will do so by means of characteristic maps.

Definition 2.1.23. Let $e^q \subset X$ be a q-cell. A continuous map $F\colon D^q \to X$ is a *characteristic map* of e^q if F maps the open q-ball $D^n \setminus S^{n-1}$ homeomorphically onto e^q.

Examples 2.1.24. 1. Let K be a simplicial complex and $\sigma^q \in K$ a q-simplex. Each homeomorphism $F\colon D^q \to \overline{\sigma^q} \subset |K|$ is a characteristic map of σ^q.
2. There exist continuous maps $F\colon D^n \to S^n$ with $F(S^{n-1}) = e^0$ which maps $D^n \setminus S^{n-1}$ homeomorphically onto $e^n = S^n \setminus \{e^0\}$, each such map is a characteristic map of the cell e^n in $(S^n, \{e^0, e^n\})$.
3. The cell $\mathbb{R}^n$ in $(\mathbb{R}^n, \{\mathbb{R}^n\})$ has no characteristic map.
4. For each 0-cell there is exactly one characteristic map.
5. If a cell with positive dimension has a characteristic map then it has infinitely many.

Theorem 2.1.25. *Let X be Hausdorff and $F\colon D^q \to X$ be a characteristic map of $e^q \subset X$. Then $F(D^q) = \overline{e^q}$.*

Proof. We have $F(D^q) = F(\overline{D^q \setminus S^{q-1}}) \subset \overline{F(D^q \setminus S^{q-1})} = \overline{e^q}$ because F is continuous. Now $F(D^q)$ is compact as the image of the compact set D^q, and hence closed because X is Hausdorff. From $e^q \subset F(D^q)$ we therefore have $\overline{e^q} \subset F(D^q)$. □

Remark 2.1.26. The map $F\colon D^q \to \overline{e^q}$ is surjective. Hence $\overline{e^q}$ is a quotient space of D^q.

We can now give the definition of CW complexes.

Definition 2.1.27. A finite *CW complex* is a space partitioned in cells $K = (X, \zeta)$ with the following properties:
(a) X is Hausdorff and ζ is a finite set; and
(b) Each cell $e^q \in \zeta$ has a characteristic map $F\colon D^q \to X$ with $F(S^{q-1}) \subset X^{q-1} = \bigcup_{e \in \zeta^{q-1}} e$, where $\zeta^{q-1} = \{e \in \zeta \mid \dim(e) \leq q-1\}$ as above.

We consider the following examples of finite CW complexes.

Examples 2.1.28. 1. If K is a simplicial complex then $(|K|, K)$ is a finite CW complex.
2. The pair $(S^n, \{e^0, e^n\})$ is a finite CW complex.

3. Let X be arbitrary and ζ be the set of the one-point subspaces of X. Then (X,ζ) is a finite CW complex if and only if X is discrete with finitely many points.

We will consider many more examples later in the book.

Remark 2.1.29. Let (K,ζ) be a finite CW complex. Then we have the following.
Condition C: $\overline{e}$ is contained in the union of finitely many cells of ζ for each cell e.
(C stands for "closure-finite").
Condition W: A set $A \subset X$ is closed if and only if $A \cap \overline{e}$ is closed for all $e \in \zeta$.
(W stands for "weak topology").

If we replace the condition that ζ is finite by the conditions X and W then $K = (X,\zeta)$ is in general a CW complex. In the following, we will only consider finite CW complexes (unless stated otherwise).

Definition 2.1.30. Let Y be a topological space and $f\colon S^{q-1} \to Y$, $q \geq 1$, be a continuous map and $Y \sqcup D^q$ the topological sum of Y and D^q. We define an equivalence relation in $Y \sqcup D^q$ as follows:

$x \sim y$ if and only if one of the following conditions applies:

- $x = y$.
- $x \in S^{q-1}, y \in f(S^{q-1})$ and $y = f(x)$.
- $y \in S^{q-1}, x \in f(S^{q-1})$ and $x = f(y)$.
- $x, y \in S^{q-1}$ and $f(x) = f(y)$.

The set of the equivalence classes equipped with the quotient topology is a topological space, denoted by $Y \cup_f e^q$. Let Y be a subspace of X and $h\colon X \to Y \cup_f e^q$ be a homeomorphism with $h(y) = y$ for all $y \in Y$. We say that X arises from X by *attaching a q-cell* via the *attaching map f*.

Remarks 2.1.31. 1. The space $X \setminus Y$ is a q-cell.
If $i\colon D^q \hookrightarrow Y \sqcup D^q$ is the inclusion and $p\colon Y \sqcup D^q \to Y \cup_f e^q$ the identifying map, then $h \circ p \circ i\colon D^q \to X$ is a characteristic map.

2. We can describe $(Y \cup_f e^q, i, j)$ as a pushout

$$\begin{array}{ccc} S^{q-1} & \xrightarrow{\ f\ } & Y \\ {\scriptstyle i}\downarrow & & \downarrow{\scriptstyle j} \\ D^q & \xrightarrow[p\circ i_q]{} & Y \cup_f e^q \end{array}$$

of $(f, p \circ i_q)$. The universal property of $Y \cup_f e^q$ is as follows. The maps $j\colon Y \to Y \cup_f e^q$ and $p \circ i_q\colon D^q \to Y \cup_f e^q$ are continuous and make the above square commutative. Furthermore, for all triples (Z, g, h) consisting of a topological space Z and continuous maps $g\colon Y \to Z$ and $h\colon D^q \to Z$ such that the outer square in the diagram

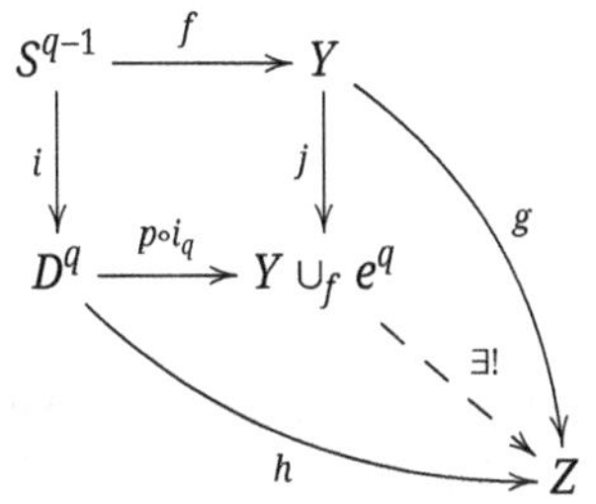

commutes, there exists a unique arrow $Y \cup_f e^q \to Z$ such that the two triangles commute.

3. Usually one attaches more than one cell to Y. Let $D_1^q, \dots, D_k^q$ be pairwise disjoint q-dimensional balls, $q \geq 1$, and $f_i: S_i^{q-1} \to Y$ be continuous maps, $i = 1, \dots, q$. Let $Y \cup_{f_1} e_1^1 \cup_{f_2} \cdots \cup_{f_k} e_k^q$ be the quotient space of the topological sum $Y \sqcup D_1^q \sqcup \cdots \sqcup D_k^q$ with respect to the equivalence relation: $x \sim y$ if and only if one of the following conditions applies:
   - $x = y$.
   - $x \in S^{q-1}, y \in f_i(S_i^{q-1})$ and $y = f_i(x)$ for one $i = 1, \dots, k$.
   - $y \in S_i^{q-1}, x \in f_i(S_i^{q-1})$ and $x = f_i(y)$ for one $i = 1, \dots, k$.
   - $x, y \in S_i^{q-1}$ and $f_i(x) = f_i(y)$ for one $i = 1, \dots, k$.

   Let Y be a subspace of X and $k: X \to Y \cup_{f_1} e_1^q \cup_{f_2} \cdots \cup_{f_k} e_k^q$ be a homeomorphism with $h(y) = y$ for all $y \in Y$. Then we say that X *arises from* Y *by attaching* k *q-cells.*

Theorem 2.1.32. 1. *Let $K = (Y, \zeta)$ be a finite CW complex. Then Y^q arises from Y^{q-1} by attaching finitely many q-cells.*
2. *If $\dim(K) = \max\{\dim(e) \in \zeta\} \leq q - 1$ and if X arises from Y by attaching finitely many q-cells, then $(X, \zeta \cup \zeta')$ is a finite CW complex where ζ' is the set of all cells from $X \setminus Y$.*

Proof. 1. Let $e_1^q, \dots, e_k^q$ be the q-cells of ζ such that $Y^q = Y^{q-1} \cup_{f_1} e_1^q \cup_{f_2} \cdots \cup_{f_k} e_k^q$. Let $F_i: D_i^q \to Y$ be a characteristic map of e_i^q, $i = 1, \dots, k$. Then

$$F_i(D_i^q) = F_i(D_i^q \setminus S_i^{q-1}) \cup F_i(S_i^{q-1}) \subset e_i^q \cup Y^{q-1} \subset Y^q,$$

and we may define a map $F: Y^{q-1} \sqcup D_1^q \sqcup \cdots \sqcup D_k^q \to Y^q$ via

$$F(x) = \begin{cases} x & \text{if } x \in Y^{q-1}, \\ F_i(x) & \text{if } x \in D_i^q. \end{cases}$$

Let $f_i: S_i^{q-1} \to Y^{q-1}$ be defined by $f_i(x) = F(x)$ for $x \in S_i^{q-1}$ and $i = 1, \dots, q$. In the diagram

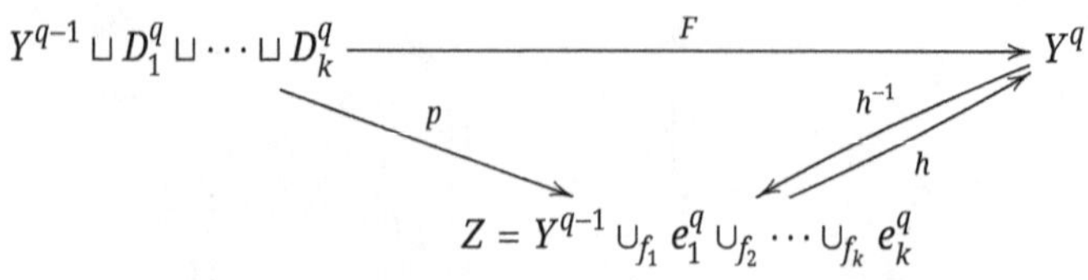

let p be the quotient map (identification map) and h be defined via $F = h \circ p$. Then h is bijective and $p = h^{-1} \circ F$. We show that h and h^{-1} are continuous which proves 1. For the continuity of h observe that if $A \subset Y^q$ is open then $F^{-1}(A) = p^{-1}(h^{-1}(A))$ is open because F is continuous. In the quotient topology $h^{-1}(A)$ is open if and only if the pre-image $p^{-1}(h^{-1}(A))$ is open. Hence $h^{-1}(A)$ is open, and h is continuous. For the continuity of h^{-1} it is enough to show that F is a closed map. But this is the case because the domain of F is compact and the range is Hausdorff.

2. $X = Y \cup_{f_1} e_1^q \cup_{f_2} \cdots \cup_{f_k} e_k^q$ for continuous maps $f_i \colon S_i^{q-1} \to Y$, $i = 1, \ldots, k$ (up to homeomorphism relative Y). Each e_i^q has a characteristic map $F_i \colon D_i^q \to X$ with $F_i(S_i^{q-1}) = f_i(S_i^{q-1}) \subset Y = Y^{q-1} = X^{q-1}$. By definition we only need to show that X is Hausdorff but this is already clear from the construction. □

We fix the following notation: If $K = (X, \zeta)$ is a CW complex with

$$\zeta = \{e_1^0, \ldots, e_{\alpha_0}^0, \ldots, e_1^q, \ldots, e_{\alpha_q}^q\}$$

then we write

$$X = e_1^0 \cup \cdots \cup e_{\alpha_0}^0 \cup \cdots \cup e_1^q \cup \cdots \cup e_{\alpha_q}^q.$$

Example 2.1.33. The *n-sphere* $S^n = e^0 \cup e^n$ is a finite CW complex. The space S^n arises from a one-point space $\{e^0\}$ by attaching an n-cell with the constant attaching map $S^{n-1} \to \{e^0\}$.

This gives immediately the following implication.

Corollary 2.1.34. *If $r < n$ then each map $f \colon S^r \to S^n$ is nullhomotopic.*

Proof. It is enough to show that each map $f \colon |\Sigma^r| \to |\Sigma^n|$ is nullhomotopic where Σ^s is the set of all proper faces of an $(s+1)$-simplex σ^{s+1} because $|\Sigma^s| \cong S^s$. Of course, Σ^r contains only simplices of a dimension less or equal to $r < n$, whereas Σ^n also contains n-simplices. Hence, f is not surjective. We now take S^n instead of $|\Sigma|$. Let $p \notin f(|\Sigma^r|)$, and without loss of generality, let $p = (0, \ldots, 0, 1)$. Take $h \colon S^n \setminus \{p\} \to \mathbb{R}^n$, $h(x_1, \ldots, x_{n+1}) = \frac{(x_1, \ldots, x_n)}{1 - x_{n+1}}$ which is a homeomorphism and set $h' = h^{-1}$. Then $g \colon |\Sigma^r| \to \mathbb{R}^n$ with $g = h \circ f$ is nullhomotopic and therefore also $f = h' \circ g$ is nullhomotopic. □

We now turn to orientable and non-orientable surfaces.

Examples 2.1.35. 1. Let $g \geq 1$, $g \in \mathbb{N}$. Divide $S^1 = \partial D^2$ in $4g$ equal positively directed arcs $a_1, b_1, c_1, d_1, \ldots, a_g, b_g, c_g, d_g$ and identify on S^1 the arcs a_1 with c_1^{-1}, b_1 with $d_1^{-1}, \ldots, a_g$ with c_g^{-1}, b_g with d_g^{-1}. Here the exponent -1 means that the respective edge points are opposite to the direction of the transversal.
This gives rise to a quotient space F_g of D^2: F_g is the orientable (compact) surface of genus g and is homeomorphic to a ball with g attached handles. The vertices of the arcs will in F_g be identified to a 0-cell e^0. In F_g we have 1-cells e_i^1 and $e_i'^1$ from $a_i = c_i^{-1}$ and $b_i = d_i^{-1}$, respectively. $D^2 \setminus S^1$ is a 2-cell, and $F_g = e^0 \cup e_1^1 \cup \cdots \cup e_g^1 \cup e_1'^1 \cup \cdots \cup e_g'^1 \cup e_2$ is a finite CW complex. The attaching map $D^2 \to F_g$ is the characteristic map of e^2. We finally define $F_0 = S^2 = e^0 \cup e^2$.
2. Let $g \geq 2$, $g \in \mathbb{N}$. Divide $S^1 = \partial D^2$ in $2g$ equal positively directed arcs $e_1, f_1, \ldots, e_g, f_g$ and identify e_1 with $f_1, \ldots, e_g$ with f_g. Then the quotient space N_g of D^2 is the non-orientable (compact) surface of genus g, and N_g is homeomorphic to a ball with g attached Moebius bands. N_g is a finite CW complex of the form $N_g = e^0 \cup e_1^1 \cup \cdots \cup e_g^1 \cup e_2$. We finally define $N_1 = \mathbb{P}^2(\mathbb{R})$.

In the following we will discuss projective spaces in detail. To this end, let $\mathbb{R}$, $\mathbb{C}$ and $\mathbb{H}$ be the field of real numbers, the complex numbers and the quaternions, respectively. Let T stand for $\mathbb{R}$, $\mathbb{C}$ or $\mathbb{H}$. The set of the ordered n-tuples $(t_1, \ldots, t_n)$ with $t_i \in T$ is a T-vector space, $n \geq 1$. Set

$$D^n(T) = \left\{ (t_1, \ldots, t_n) \in T^n \;\middle|\; \sum_{i=1}^{n} |t_i|^2 \leq 1 \right\}$$

and

$$S^n(T) = \left\{ (t_1, \ldots, t_n) \in T^n \;\middle|\; \sum_{i=1}^{n} |t_i|^2 = 1 \right\}.$$

Then $S^{n-1}(T) \subset D^n(T) \subset T^n$. If $T = \mathbb{R}$, then this reduces to $S^{n-1} \subset D^n \subset \mathbb{R}^n$, if $T = \mathbb{C}$, $t_\mu = x_\mu + iy_\mu$, then we identify $(t_1, \ldots, t_n) \in \mathbb{C}^n$ with $(x_1, y_1, \ldots, x_n, y_n) \in \mathbb{R}^{2n}$, and we get $S^{n-1}(\mathbb{C}) = S^{2n-1} \subset D^n(\mathbb{C}) = D^{2n} \subset \mathbb{C}^n = \mathbb{R}^{2n}$ and if $T = \mathbb{H}$, $t_r = x_r + iy_r + jz_r + ku_r$, and if we identify $(t_1, \ldots, t_n) \in \mathbb{H}^n$ with $(x_1, y_1, z_1, u_1, \ldots, x_n, y_n, z_n, u_n) \in \mathbb{R}^{4n}$, then we get $S^{n-1}(\mathbb{H}) = S^{4n-1} \subset D^n(\mathbb{H}) = D^{4n} \subset \mathbb{H}^n = \mathbb{R}^{4n}$.

Definition 2.1.36. For $(t_1, \ldots, t_n), (t_1', \ldots, t_n') \in T^n \setminus \{0, \ldots, 0\}$ let $(t_1, \ldots, t_n) \sim (t_1', \ldots, t_n')$ if and only if $t_r = \lambda t_r'$ for some $\lambda \in T$ and $r = 1, \ldots, n$. We denote the equivalence class of $(t_1, \ldots, t_n)$ by $[t_1, \ldots, t_n]$. The set $\mathbb{P}^{n-1}(T)$ of the equivalence classes is the $(n-1)$-dimensional projective space over T.

The function $q: S^{n-1}(T) \to \mathbb{P}^{n-1}(T)$, $q(t_1, \ldots, t_n) = [t_1, \ldots, t_n]$ is surjective. $\mathbb{P}^{n-1}(T)$ gets the quotient topology. We identify the subspace of all $[t_1, \ldots, t_n, 0] \in \mathbb{P}^n(T)$ with $\mathbb{P}^{n-1}(T)$. Then we get the following result.

Theorem 2.1.37. *Let $d = 1, 2, 4$ according to whether $T = \mathbb{R}, \mathbb{C}$ or $\mathbb{H}$. Then $\mathbb{P}^n(T)$ is a finite CW complex of the form $\mathbb{P}^n(T) = e^0 \cup e^d \cup e^{2d} \cup \cdots \cup e^{nd}$, and arises from $\mathbb{P}^{n-1}(T)$ by attaching an (nd)-cell.*

Proof. Define $F: D^n(T) \to \mathbb{P}^n(T)$ by $F(t_1, \ldots, t_n) = [t_1, \ldots, t_n, 1 - \sum_{r=1}^n |t_r|^2]$. Then $F(S^{n-1}(T)) \subset \mathbb{P}^{n-1}(T)$ and F maps $D^n(T) \setminus S^{n-1}(T)$ homeomorphically onto $\mathbb{P}^n(T) \setminus \mathbb{P}^{n-1}(T)$. Hence, $\mathbb{P}^n(T) \setminus \mathbb{P}^{n-1}(T) =: e^{nd}$ is a cell of dimension nd and with characteristic map F. Now the theorem follows by induction on n and the fact that $\mathbb{P}^0(T) = \{e^0\}$ is a one-point space. □

We now define the appropriate maps to compare CW and simplicial complexes.

Definition 2.1.38. Let $K_i = (X_i, \zeta_i)$, $i = 1, 2$, be finite CW complexes and $f: X_1 \to X_2$ be a continuous map. Then f is called *cellular* (with respect to the cell partitions ζ_1 and ζ_2) if $f(X_1^q) \subset X_2^q$ for all $q \geq 0$.

Remark 2.1.39. If ζ_i, $i = 1, 2$, is a simplicial complex and if f maps cells from ζ_1 onto cells of ζ_2, then f is called *simplicial*. Recall that we do not require from an arbitrary cellular map that it maps cells onto cells.

We would like to show the existence of cellular approximations. For this purpose, we need the following two lemmas.

Lemma 2.1.40. *Suppose that the space X arises from the space Y by attaching finitely many cells of dimension $> n$, and that $\varphi: (D^n, S^{n-1}) \to (X, Y)$ is continuous. Then there exists a homotopy $k_t: D^n \to X$ with $k_0 = \varphi$, $k_1(D^n) \subset Y$ and $k_t(x) = \varphi(x)$ for $x \in S^{n-1}$, $t \in I$.*

Recall that $\varphi: (D^n, S^{n-1}) \to (X, Y)$ means pairwise, that is, $\varphi(D^n) \subset X$ and $\varphi(S^{n-1}) \subset Y$.

Proof. We prove the lemma in four steps.

1. Let $f: (D^n, S^{n-1}) \to (\mathring{D}^k, \mathring{D}^k \setminus \{0\})$ be continuous and $n < k$.
   Recall that $\mathring{D}^k = D^k \setminus S^{k-1}$. Then there exists a homotopy $h_t: D^n \to \mathring{D}^k$ with $h_0 = f$, $h_1(D^n) \subset \mathring{D}^k \setminus \{0\}$ and $h_t(x) = f(x)$ for all $x \in S^{n-1}$ and $t \in I$.

   Proof. Let $f_1: S^{n-1} \to \mathring{D}^k \setminus \{0\}$ with $f_1(x) = f(x)$. Since $\mathring{D}^k \setminus \{0\} \cong S^{k-1}$ and $n - 1 < k - 1$ we see that f_1 is nullhomotopic by Corollary 2.1.34. Hence we may extend f_1 to $f_2: D^n \to \mathring{D}^k \setminus \{0\}$ (recall that if $f: X \to Y$ is nullhomotopic then for each of the maps $g: Y \to Z$ and $h: W \to X$ the maps $g \circ f$ and $f \circ h$ are nullhomotopic).
   Let $f_3: (D^n \times \{0\}) \cup (D^n \times \{1\}) \cup (S^{n-1} \times I) \to \mathring{D}^k$ be defined by $f_3(x, 0) = f(x)$, $f_3(x, 1) = f_2(x)$ and $f_3(y, t) = f(y)$ for $x \in D^n$, $y \in S^{n-1}$, $t \in I$. Let $g: (D^n \times I, (D^n \times \{0\}) \cup (D^n \times \{1\}) \cup (S^{n-1} \times I)) \to (D^{n+1}, S^n)$ be a homeomorphism of pairs. The restriction $f_3 \circ g^{-1}|_{S^n}: S^n \to \mathring{D}^k$ is nullhomotopic because $\mathring{D}^k$ is contractible. Hence, $f_3 \circ g^{-1}|_{S^n}$ can be extended to a map $f_4: D^{n+1} \to \mathring{D}^k$. Define $h_t: D^n \to \mathring{D}^k$, $h_t(x) = f_4 \circ g(x, t)$. Then $h_0 = f$, $f_4 \circ g(x, 1) = f_3(x, 1) = f(x) \in \mathring{D}^k \setminus \{0\}$, hence $h_1(D^n) \subset \mathring{D}^k \setminus \{0\}$ and $h_t(x) = f_4 \circ g(x, t) = f_3 \circ g^{-1} g(x, t) = f(x)$ for $x \in S^{n-1}$ and $t \in I$. □

2. Let K be a simplicial complex with subcomplex L, and let $\max\{\dim(\sigma) \mid \sigma \in K \setminus L\} < k$. Let $f: (|K|, |L|) \to (\mathring{D}^k, \mathring{D}^k \setminus \{0\})$ be continuous. Then there exists a homotopy $h_t: |K| \to \mathring{D}^k$ with $h_0 = f$, $h_1(|K|) \subset \mathring{D}^k \setminus \{0\}$ and $h_t(x) = f(x)$ for $x \in |L|$ and $t \in I$.

   Proof. We have $(\overline{\sigma}, \partial\sigma) \cong (D^n, S^{n-1})$, for each $\sigma \in K \setminus L$ with $\dim(\sigma) = n < k$. Hence step 2 follows from step 1 by induction on the number of simplices in $K \setminus L$. □

3. Without loss of generality let $X = Y \cup e_1^k \cup \cdots \cup e_m^k$ for some $k > n$. The general case follows then by induction on the dimension of the attached cells. Let $x_i \in e_i^k$ be a fixed element. Let $U = Y \cup (e_1^k \setminus \{x_1\}) \cup \cdots \cup (e_m^k \setminus \{x_m\})$. Then $\{U, e_1^k, \dots, e_m^k\}$ is an open cover of X, hence $\mathcal{U} = \{\varphi^{-1}(U), \varphi^{-1}(e_1^k), \dots, \varphi^{-1}(e_m^k)\}$ is an open cover of D^n, where φ is as in the statement of the theorem. Since $D^n \cong \overline{\sigma^n}$ there exists a simplicial complex K with $\dim(K) = n$ and $|K| \cong D^n$ (in the following we identify $|K|$ with D^n) such that each closed simplex of K is in a subset of $\mathcal{U}$. Let $K_0 = \{\sigma \in K \mid \varphi(\overline{\sigma}) \subset U\}$ and $K_i = \{\sigma \in K \mid \varphi(\overline{\sigma}) \subset e_i^k\}$ for $i = 1, \dots, m$.
   Then $K = K_0 \cup K_1 \cup \cdots \cup K_m$ and $K_i \cap K_j = \emptyset$ for $i \neq j$, $i \neq 0 \neq j$. Define $f_i: (|K_i|, |K_i \cap K_0|) \to (e_i^k, e_i^k \setminus \{x_k\})$ for $i = 1, \dots, m$ by $f_i(x) = \varphi(x)$ for $x \in |K_i|$. Now $\dim(K_i) \leq \dim(K) = n < k$ and $(e_i^k, e_i^k \setminus \{x_k\}) \cong (\mathring{D}^k, \mathring{D}^k \setminus \{0\})$. Hence, we get: There exists a homotopy $h_t^{(i)}: |K_i| \to e_i^k$ with $h_0^{(i)} = f_i$, $h_1^{(i)}(|K_i|) \subset e_i^k \setminus \{x_i\}$ and $h_t^{(i)}(x) = f_i(x) = \varphi(x)$ for $x \in |K_i \cap K_0|$. If $i \neq j$ we have $e_i^k \cap e_j^k = \emptyset$, hence, $K_i \cap K_j = \emptyset$ $(i \neq 0 \neq j)$. Therefore we get a homotopy $h_t': D^n \to X$,

$$h_t(x) = \begin{cases} \varphi(x) & \text{if } x \in |K_0|, \\ h_t^{(i)}(x) & \text{if } x \in |K_i|, i = 1, \dots, m. \end{cases}$$

   We have $h_0' = \varphi$, $h_t'(x) = \varphi(x)$ for $x \in S^{n-1}$, $t \in I$, because $S^{n-1} \subset |K_0|$, and $h_1'(D^n) \subset U$.

4. Since $D^k \setminus \{0\}$ can be deformed to S^{k-1} there exists a homotopy $h_t'': U \to U$ with $h_0'' = \mathrm{id}_U$, $h''(U) \subset Y$ (the characteristic map $F: D^k \to X$ causes $F(S^{k-1}) \subset Y$) and $h_t''(y) = y$ for $y \in Y$ and $t \in I$. Then

$$k_t(x) = \begin{cases} h_{2t}'(x) & \text{if } 0 \leq t \leq \frac{1}{2}, \\ h_{2t-1}''(x) & \text{if } \frac{1}{2} \leq t \leq 1, \end{cases}$$

   defines a homotopy $k_t: D^n \to X$ with the desired properties. □

2.1.3 Homotopy extension property and cellular approximation

Definition 2.1.41. We say that the pair of spaces (X, A) has the *homotopy extension property* if for each choice of $Y, f: X \to Y$ and $\psi: A \times I \to Y$ with $\psi(x, 0) = f|_A(x)$ there exists a homotopy $F: X \times I \to Y$ with $F(x, 0) = f_0(x) = f(x)$ and $F|_{A \times I} = \psi$.

Hence, if (X, A) has the homotopy extension property, then we may extend each homotopy of A which comes along with an $f: X \to Y$ to a homotopy of X.

Lemma 2.1.42. 1. *Let $A \subset X$ be closed. Then (X, A) has the homotopy extension property if and only if $(X \times \{0\}) \cup (A \times I)$ is a retract of $X \times I$.*
2. *Especially, (D^n, S^{n-1}) has the homotopy extension property.*

Proof. 1. Let $f: X \to Y$ and $\psi: A \times I \to Y$ be maps with $\psi(x, 0) = \psi_0(x) = f|_A(x)$. Let $g: (X \times \{0\}) \cup (A \times I) \to Y$ be defined by $g|_{X \times \{0\}}(x, 0) = f(x)$ and $g|_{A \times I} = \psi$. Then g is well defined. Since A is closed, we see that $A \times I$ is closed in $X \times I$. The same holds for $X \times \{0\}$. Since $g|_{X \times \{0\}}$ and $g|_{A \times I}$ are continuous it follows that g is continuous. If $r: X \times I \to (X \times \{0\}) \cup (A \times I)$ is a retraction, then $\psi = g \circ r$ is the desired extension. Now assume that (X, A) has the homotopy extension property. Take $Y = (X \times \{0\}) \cup (A \times I)$ and $g = \mathrm{id}_Y, f(x) = g(x, 0)$ and $\psi = g|_{A \times I}$. By the assumption we may extend g to a continuous map $r: X \times I \to Y$, and r is a retraction.
2. Take a point $p \in \mathbb{R}^{n-1}$ over $D^n \times I$ and project the inner points of $D^n \times I$ to the boundary with half rays starting in p; see Figure 2.5. □

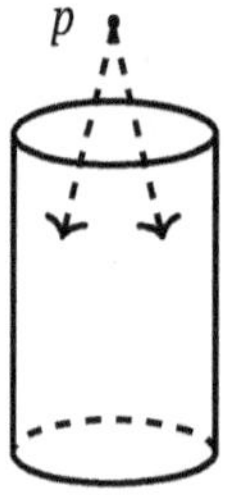

Figure 2.5: Visualization of the proof of Lemma 2.1.42.

Theorem 2.1.43 (Existence of a cellular approximation). *Let $K_i = (X_i, \zeta_i)$, $i = 1, 2$, be finite CW complexes and $f: X_1 \to X_2$ be a continuous map.*
1. *Then there exists a cellular map $g: X_1 \to X_2$ with $g \simeq f$. Each such map g is called a cellular approximation.*
2. *If f is cellular on X_1^k then we may choose g such that $g|_{X_1^k} = f|_{X_1^k}$.*

Proof. We prove the statement by induction on $\dim(X_1)$. If $\dim(X_1) = 0$ then X_1 is a discrete space with finitely many points, and it is enough to show that each point of X_2 can be connected by a path with a 0-cell of X_2. This follows from Theorem 2.1.32. Now X_2^q arises from X_2^{q-1} by attaching q-cells and the property that each cell e^q has a characteristic map $F: D^q \to X_2$ with $F(S^{q-1}) \subset X^{q-1}$. This ensures the existence of 0-cells in each connected component of X_2. Now, let $n \geq 1$, and the theorem hold for all $(n-1)$-dimensional finite CW complexes, and let $\dim(X_1) = n$. Then there exists a homotopy $h_t: X_1^{n-1} \to X_2$ with $h_0 = f|_{X_1^{n-1}}$ such that h_1 is cellular. Now the $(n-1)$-skeleton is a subcomplex of $K_1 = (X_1, \zeta_1)$. By the homotopy extension property we may extend h_t to a homotopy $k_t: X_1 \to X_2$ with $k_0 = f$. It is enough to show that k_1 has a cellular approximation. We write again f instead of k_1, so $f: (X_1, X_1^{n-1}) \to (X_2, X_2^{n-1})$ is cellular on X_1^{n-1}. We construct for each cell e_i^n, $i = 1, \ldots, r$, of $X_1 \setminus X_1^{n-1}$ a homotopy $h_t^{(i)}: X_1 \to X_2$ with $h_0^{(i)} = f|_{e_i^n}$, $h_1^{(i)}(e_i^n) \subset X_2^n$ and $h_t^{(i)}(x) = f(x)$ for $x \in X_1^{n-1}$ and $t \in I$. Then $h_t: X_1 \to X_2$, $h_t|_{e_i^n} = h_t^{(i)}$ for $i = 1, \ldots, r$ and $h_t(x) = f(x)$ for $(x,t) \in X_1^{n-1} \times I$, is a homotopy from f to the cellular map h_1.

Let e^n be one of the cells e_i^n, and let $F: (D^n, S^{n-1}) \to (X_1, X_1^{n-1})$ be a characteristic map of e^n. Define $\varphi: (D^n, S^{n-1}) \to (X_2, X_2^{n-1})$ by $\varphi(x) = f \circ F(x)$ for $x \in D^n$. From the above we get: There exists a homotopy $k_t: D^n \to X_2$ with $k_0 = \varphi$, $k_1(D^n) \subset X_2^n$ and $k_t(x) = f(x)$ for $x \in S^{n-1}$ and $t \in I$. Then $h_t: X_1 \to X_2$,

$$h_t(x) = \begin{cases} k_t \circ F^{-1}(x) & \text{if } x \in e_q^n, \\ f(x) & \text{if } x \in X_1^{n-1}, \end{cases}$$

is a homotopy with $h_0 = f$, $h_1(e_q^n) \subset X_2^n$ and $h_t(x) = f(x)$ for $x \in X^{n-1}$, $t \in I$. This gives the desired construction. □

Our last theorem in this paragraph shows that the fundamental group of a CW complex does only depend on the respective 2-skeleton.

Theorem 2.1.44. *Let $K = (X, \zeta)$ be a finite CW complex and $x_0 \in X^0$. Then $\pi_1(X, x_0) \cong \pi_1(X^2, x_0)$.*

Proof. The map $i: X^2 \to X$ induces a homomorphism $i_*: \pi_1(X^2, x_0) \to \pi_1(X, x_0)$. Now let $\omega: I \to X$ be a loop with $\omega(0) = \omega(1) = x_0$. By the theorem of the existence of a cellular approximation there exists a path $\omega': I \to X$, which is homotopic to ω, with $\omega'(0) = \omega'(1) = x_0$ and $\omega'(I) \subset X^1 \subset X^2$. Hence, $[\omega] = [\omega'] \in i_*(\pi_1(X^2, x_0))$, that is, i_* is surjective. Now, let $i_*[\omega_1] = i_*[\omega_2]$, $[\omega_1], [\omega_2] \in \pi_1(X^2, x_0)$. Then there exist loops $\omega_i: (I, \partial I) \to (X^1, \{x_0\})$ with $\omega_i \in [\omega_i]$, $i = 1, 2$. Furthermore, we have a map $f_0: D^2 \to X$ with $f_0(e^{\pi i t}) = \omega_1(t)$ and $f_0(e^{-\pi i t}) = \omega_2(t)$, $0 \le t \le 1$. It follows that $f_0(\partial D^2) \subset X^1$. By Theorem 2.1.43 we may deform f_0 with a homotopy to a map $f_1: D^2 \to X$ such that $f_1|_{\partial D^2} = f_0|_{\partial D^2}$ and $f_1(D^2) \subset X^2$. This means that ω_1 and ω_2 are already homotopic in X^2, that is, $[\omega_1] = [\omega_2]$ and i_* is injective. □

2.2 Covering spaces and the theorem of Seifert and van Kampen

Again, in this section $X, Y, Z, \tilde{X}, \tilde{Y}, \tilde{Z}, \ldots$ are always topological spaces and all maps $f: X \to Y$ are continuous, also let again I denote the unit interval $[0,1]$.

2.2.1 Covering maps and lifting properties

Definition 2.2.1. Given $\tilde{X}$, X and $p: \tilde{X} \to X$. Then p is called *covering map* or *covering projection* if each point of X has an open neighborhood U such that the following holds: There exists a family $(U_i)_{i \in I}$ of pairwise disjoint open sets in $\tilde{X}$ with $p^{-1}(U) = \dot{\bigcup}_{i \in I} U_i$ and $p|_{U_i}: U_i \to U$ is a homeomorphism for each $i \in I$. The space $\tilde{X}$ is called a *covering space* or just *covering* over X and X is called the *basis space* or just *basis* of the covering map p. We call a U with the above properties *elementary*.

Remarks 2.2.2. 1. The map p is surjective and a local homeomorphism.
2. If $\tilde{\tilde{X}} \to \tilde{X}$ and $q: \tilde{X} \to X$ are covering maps then also $q \circ p: \tilde{\tilde{X}} \to X$.

Examples 2.2.3. 1. The map $p: \mathbb{R} \to S^1$, $p(x) = (\cos(2\pi x), \sin(2\pi x))$ is a covering map. For $(1,0) \in S^1$ choose $U = \{(s,t) \in S^1 \mid s > 0\}$. Then $p^{-1}(U) = \dot{\bigcup}_{n\in\mathbb{Z}}(n - \frac{1}{4}, n + \frac{1}{4})$, and $p|_{(n-\frac{1}{4},n+\frac{1}{4})}: (n - \frac{1}{4}, n + \frac{1}{4}) \to U$ has the inverse given by $(s,t) \mapsto n + \frac{1}{2\pi}\arcsin(t)$. Now, let $z_0 = (\cos(2\pi x_0), \sin(2\pi x_0)) \in S^1$. We turn U around the angle $2\pi x_0$. The respective rotation Θ_{x_0} of S^1 around $2\pi x_0$ is given by the matrix $\left(\begin{smallmatrix} \cos(2\pi x_0) & -\sin(2\pi x_0) \\ \sin(2\pi x_0) & \cos(2\pi x_0) \end{smallmatrix}\right)$ and the associated translation $T_{x_0}: \mathbb{R} \to \mathbb{R}$ given by $T_{x_0}(x) = x + x_0$. We have the commutative diagram

$$\begin{array}{ccc} \mathbb{R} & \xrightarrow{T_{x_0}} & \mathbb{R} \\ {\scriptstyle p}\downarrow & & \downarrow{\scriptstyle p} \\ S^1 & \xrightarrow[\Theta_{x_0}]{} & S^1 \\ \cup\!\!\!\shortmid & & \cup\!\!\!\shortmid \\ (1,0) & & z_0 \end{array}$$

where T_{x_0} and Θ_{x_0} are homeomorphisms with $\Theta_{x_0}(1,0) = z_0$. As a suitable elementary neighborhood of z_0 choose $\Theta_{x_0}(U)$; $p^{-1}(\Theta_{x_0}(U)) = \dot{\bigcup}_{n\in\mathbb{Z}}(x_0 + n - \frac{1}{4}, x_0 + n + \frac{1}{4})$.
2. The map $z \mapsto z^n$, $n \in \mathbb{N}$, defines a covering map $p: S^1 \to S^1$.
3. The map $p: S^n \to \mathbb{P}^n(\mathbb{R}) = \mathbb{P}^n$, $p(x) = \{x, -x\}$, is a covering map.
4. The map $p: \mathbb{C} \to \mathbb{C} \setminus \{0\}$, $z \mapsto e^z$, is a covering map.

Remarks 2.2.4. 1. A covering map $p: \tilde{X} \to X$ is an open map and a local homeomorphism. But not each surjective, open map which is also a local homeomorphism is a covering map. Example: $x \mapsto (\cos(2\pi x), \sin(2\pi x))$ as a map $(-1,1) \to S^1$ is surjective, open and a local homeomorphism but not a covering map. There is no open neighborhood U of $(1,0) \in S^1$ such that $p^{-1}(U)$ is a suitable disjoint partition of open sets which are mapped by p homeomorphically onto U.
2. If $p: \tilde{X} \to X$ is a covering map then $p^{-1}(a)$, as a subspace, has the discrete topology for each $a \in X$.
3. If X is Hausdorff then also $\tilde{X}$ if $p: \tilde{X} \to X$ is a covering map.

Definition 2.2.5. Let $p: \tilde{X} \to X$ be a covering map. A *lifting* of $f: Y \to \tilde{X}$ is a function $\tilde{f}: Y \to \tilde{X}$ such that the following diagram commutes.

$$\begin{array}{ccc} & & \tilde{X} \\ & {\scriptstyle \tilde{f}}\nearrow & \downarrow{\scriptstyle p} \\ Y & \xrightarrow[f]{} & X \end{array}$$

Theorem 2.2.6 (Uniqueness of the lifting). *Let $p: \tilde{X} \to \tilde{X}$ be a covering map and $\varphi, \psi: Y \to \tilde{X}$ be liftings of the continuous map $f: Y \to X$. If Y is connected and if φ and ψ have the same value for one point of Y, then $\varphi = \psi$.*

Proof. We show that $Y_0 = \{y \in Y \mid \varphi(y) = \psi(y)\}$ and $Y \setminus Y_0$ are both open in Y. Then, if Y is connected and $Y_0 \neq \emptyset$, it follows that $Y_0 = Y$.

- The set Y_0 is open: Let $y_0 \in Y_0$ and U be an open neighborhood of $f(y_0)$ with a disjoint partition $p^{-1}(U) = \dot{\bigcup}_{i \in I} U_i$ in open U_i which are mapped by p homeomorphically onto U. We have $\varphi(y_0) = \psi(y_0) \in U_j$ for some $j \in I$. Let $p_j: U_j \to U$ with $p_j = p|_{U_j}$; p_j is a homeomorphism. $V = \varphi^{-1}(U_j) \cap \psi^{-1}(U_j)$ is an open neighborhood of y_0 with $\varphi(V) \subset U_j$, $\psi(V) \subset U_j$; see Figure 2.6.

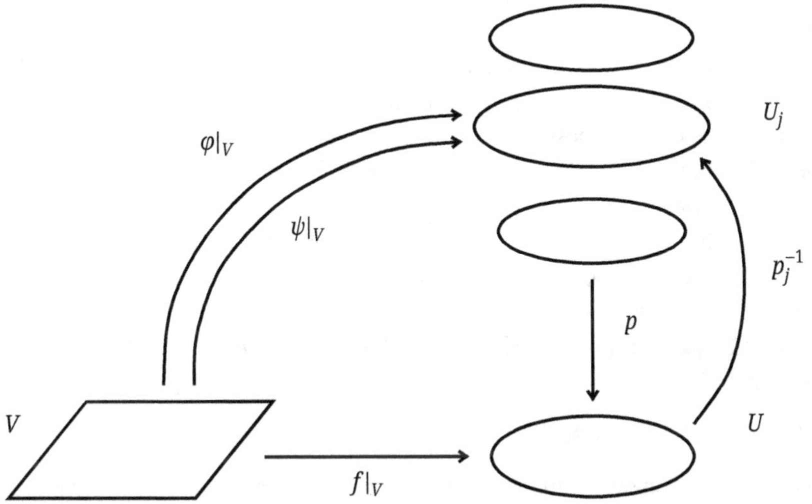

Figure 2.6: Visualization of the proof of Theorem 2.2.6.

  As $p \circ \varphi = f = p \circ \psi$ and p_j is a homeomorphism we get $f|_V = p \circ \varphi|_V = p_j \circ \varphi_V$ and $f|_V = p \circ \psi|_V = p_j \circ \psi_V$. Hence $\varphi_V = p_j^{-1} \circ f|_V = \psi_V$ and thus $V \subset Y_0$.
- The set $Y \setminus Y_0$ is open: Let $y_0 \in Y \setminus Y_0$. Choose U and $U_i, i \in I$ as above. From $\varphi(y_0) \neq \psi(y_0)$ we get $\varphi(y_0) \in U_j$ and $\psi(y_0) \in U_k$ with $j \neq k$. Let $V = \varphi^{-1}(U_j) \cap \psi^{-1}(U_k)$; V is a neighborhood of y_0. Then $\varphi(V) \subset U_j$, $\psi(V) \subset U_k$, and hence $V \subset Y \setminus Y_0$ because $U_j \cap U_k = \emptyset$. □

Theorem 2.2.7 (Path lifting property). *Let $p: \tilde{X} \to X$ be a covering map and $w: I \to X$ be a path, $\tilde{x}_0 \in \tilde{X}$ with $p(\tilde{x}_0) = w(0)$. Then there exists exactly one lifting of w to a path $\tilde{w}$ with $\tilde{w}(0) = \tilde{x}_0$.*

In order to prove Theorem 2.2.7 we need the following lemma.

Lemma 2.2.8 (Lemma of Lebesgue). *Let (E,d) be a compact metric space and $(O_i)_{i\in I}$ be an open cover of E. Then there exists a number $\delta > 0$ (the Lebesgue number) with the following property: If $A \subset E$ with $\delta(A) = \sup\{d(x,y) \mid x,y \in A\} < \delta$ then there exists a $j \in I$ with $A \subset O_j$.*

Proof. Assume there does not exist such a δ. Then, for each $n \in \mathbb{N}$, there is a non-empty $A_n \subset E$ with $\delta(A_n) < \frac{1}{n}$, which is none of the O_i. Let $y_n \in A_n$. Without loss of generality, let $(y_n)_{n\in\mathbb{N}}$ be convergent. Let $y = \lim_{n\to\infty} y_n$, then $y \in O_j$ for some $j \in I$ and $B(y,\varepsilon) \subset O_j$ for some $\varepsilon > 0$. Choose $N \in \mathbb{N}$ such that $d(y,y_n) < \frac{\varepsilon}{2}$ and $\frac{1}{n} < \frac{\varepsilon}{2}$ for all $n \geq N$. For these n we have $A_n \subset O_j$, which gives a contradiction. □

We can now prove Theorem 2.2.7.

Proof of Theorem 2.2.7. The uniqueness follows from Theorem 2.2.6 because $[0,1]$ is connected. For the existence of $\tilde{w}$ let $\mathcal{U}$ be a covering of X with open sets U such that $p^{-1}(U)$ is disjointly decomposable into open sets which are homeomorphically mapped under p onto U. The sets $w^{-1}(U)$, $U \in \mathcal{U}$ form an open cover of $[0,1]$. Let $\delta > 0$ be a Lebesgue number of this cover. Choose $0 = t_0 < t_1 < \cdots < t_n = 1$ with $t_k - t_{k-1} < \delta$, $k = 1,\ldots,n$. Then each set $w([t_{k-1},t_k])$ is contained in some $U \in \mathcal{U}$. Define $\tilde{w} = \tilde{x}_0$. Suppose that $\tilde{w}(t)$ is already defined and continuous for all $t \in [0,t_{k-1}]$ such that $p \circ \tilde{w}(t) = w(t)$. We have $w(t_{k-1}) \in U$ for some $U \in \mathcal{U}$; $p^{-1}(U) = \dot{\bigcup}_{i\in I} U_i$ a disjoint partition with open U_i. Then $p_i = p|_{U_i}: U_i \to U$ is a homeomorphism and there exists a $j \in I$ with $\tilde{w}(t_{k-1}) \in U_j$.

Define $\tilde{w}(t) = p_j^{-1} \circ w(t)$ for $t \in [t_{k-1},t_k]$. Then $t \mapsto \tilde{w}(t)$, $t \in [0,t_k]$, is a lifting if $w|_{[0,t_k]}$, and $w|_{[0,t_k]}$ is certainly continuous. Now we get the desired lifting $\tilde{w}: I \to \tilde{X}$ by induction on k. □

See Figure 2.7 for a visualization of the proof.

Theorem 2.2.9 (Homotopy lifting property). *Let $p: \tilde{X} \to X$ be a covering map and $f: Y \to X$ continuous. Given a lifting $\tilde{f}: Y \to \tilde{X}$ of f and a homotopy $H: Y \times I \to X$ with $h_0(y) = H(y,0) = f(y)$. Then there exists exactly one lifting $\tilde{H}: Y \times I \to \tilde{X}$ of H which is a homotopy with $\tilde{H}(y,0) = \tilde{h}_0(y) = \tilde{f}(y)$.*

Figure 2.8 visualizes the situation of Theorem 2.2.9.

Proof. The uniqueness follows from Theorem 2.2.6: For each $y \in Y$ we have with $\tilde{f} \in \tilde{X}$ a start point over $f(y) = H(y,0)$ for the lifting $t \mapsto \tilde{H}(y,t)$ of the path $t \mapsto H(y,t)$.

For the existence we define $\tilde{H}: Y \times I \to \tilde{X}$ so that $t \mapsto \tilde{H}(y,t)$ is the lifting of the path $t \mapsto H(y,t)$ with start point $\tilde{f}(y)$ for each $y \in Y$. Then we have to show the continuous dependence of (y,t) for this path lifting. The proof is analogous to the proof of Theorem 2.2.7 (for more details see [48]). □

We fix the following notation: Let $A \subset Y$ and $f|_A = g|_A$ for maps $f,g: X \to Y$. Then we call f and g *homotopic relative* A, written $f \simeq g$ rel A, if there is a homotopy f_t from f to g such that $f_t(a) = f(a) = g(a)$ for all $a \in A$ and all $t \in I$. In this case, f_t (and the related F) is called a *homotopy* from f to g *relative* A.

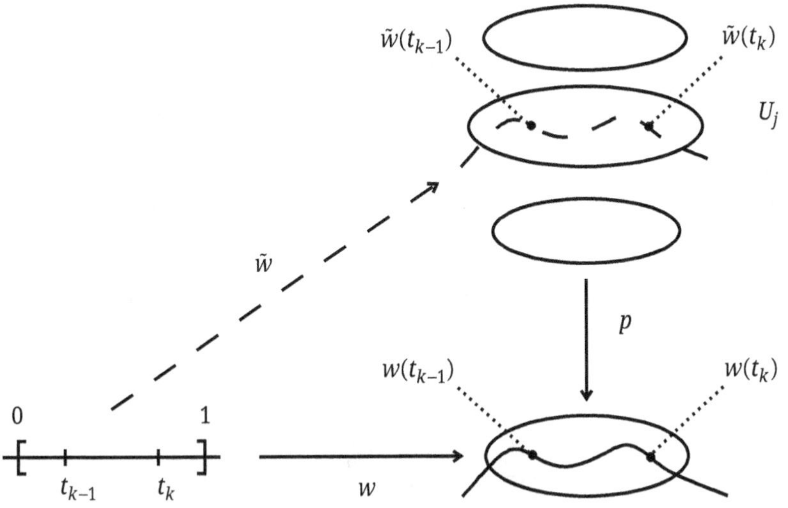

Figure 2.7: Visualization of the proof of Theorem 2.2.7.

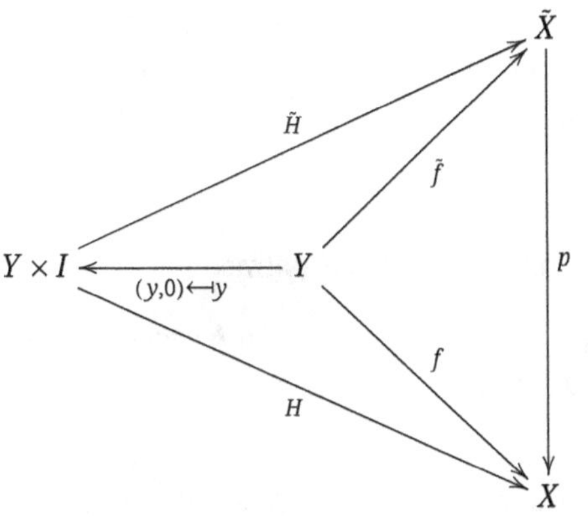

Figure 2.8: Homotopy lifting property.

Theorem 2.2.10. *Let $p: \tilde{X} \to X$ be a covering map and $w_0, w_1: I \to \tilde{X}$ be paths in $\tilde{X}$ with the same start point $\tilde{x}_0$. Suppose that $p \circ w_0$ and $p \circ w_1$ have the same end point. If $p \circ w_0 \simeq p \circ w_1 \text{ rel } \{0,1\}$ then $w_0 \simeq w_1 \text{ rel } \{0,1\}$. Especially, w_1 and w_2 have the same endpoints.*

Proof. There exists a homotopy $F: I \times I \to X$ with $F(t,0) = p \circ w_0(t)$, $F(t,1) = p \circ w_1(t)$, $F(0,\tau) = p \circ w_0(0) = p(\tilde{x}_0)$ and $F(1,\tau) = p \circ w_0(1)$. As in Theorem 2.2.7 and Lemma 2.2.8 there exist numbers $0 = t_0 < t_1 < \cdots < t_m = 1$ and $0 = \tau_0 < \tau_1 < \cdots < \tau_n = 1$ such that $F([t_i, t_{i+1}] \times [\tau_j, \tau_{j+1}])$ is an elementary neighborhood. We want to show that there is a unique map $G: I \times I \to X$ with $p \circ G = F$ and $G(0,0) = \tilde{x}_0$. First G will be defined on $[0, t_1] \times [0, \tau_1]$ which is possible in a unique manner. Then we extend G on $[t_1, t_2] \times [0, \tau_1]$ and so on, as long as G is explained on $I \times [0, \tau_1]$. Then we continue to explain G on $I \times [\tau_1, \tau_2]$

and so on. Note that G is uniquely determined by Theorem 2.2.6. From Theorem 2.2.9 we get $G(t,0) = w_0(t)$, $G(0,\tau) = \tilde{x}_0$, $G(t,1) = w_1(t)$ and $G(\{1\} \times I) = \{w_0(1)\}$. □

Corollary 2.2.11. *If $p: \tilde{X} \to X$ is a covering map then the sets $p^{-1}(x)$, $x \in X$ have the same cardinality.*

Corollary 2.2.12. *Let $p: \tilde{X} \to X$ be a covering map, $\tilde{x}_0 \in X$ and $x_0 = p(\tilde{x}_0)$. Then $p_*: \pi_1(\tilde{X}, \tilde{x}_0) \to \pi_1(X, x_0)$ is injective.*

With the theory of covering spaces we can now calculate the important example of the fundamental group of the 1-sphere.

Theorem 2.2.13. *The fundamental group $\pi_1(S^1, 1)$ is isomorphic to the group of integers, that is, $\pi_1(S^1, 1) \cong \mathbb{Z}$.*

Proof. The map $x \mapsto (\cos(2\pi x), \sin(2\pi x))$ defines a covering map $p: \mathbb{R} \to S^1$. Let $w: I \to S^1$ be a loop with $w(0) = w(1) = 1$. Let $\tilde{w}: U \to \mathbb{R}$ be the lifting of w, that is, $p \circ \tilde{w} = w$, with $\tilde{w}(0) = 0$. Let $\tilde{w}(1) = n \in \mathbb{Z}$. The map $t \mapsto (1-\tau)\tilde{w}(t) + \tau tn$ defines a homotopy rel $\{0,1\}$ from $\tilde{w}$ to a path $\tilde{\sigma}: I \to \mathbb{R}$, $t \mapsto tn$. We get $w \simeq p \circ \tilde{\sigma}_n =: \sigma_n$ rel $\{0,1\}$ from $\tilde{w}$ to a path $\tilde{\sigma}_n: I \to \mathbb{R}$, $t \mapsto tn$. Furthermore, $\sigma_n = (\sigma_1)^n$ if $n > 0$ and $\sigma_n = (\sigma_{-1})^{-n}$ if $n < 0$ and $\sigma_0: T \to \{1\}$. For the homotopy class we get $[\sigma_n] = [\sigma_1]^n$. Furthermore, there is no σ_n, $n \neq 0$, homotopic to σ_0 because the lifting of σ_n, $n \neq 0$, in $\mathbb{R}$ is not a loop. Hence $\pi_1(S^1, 1) \cong \langle[\sigma_1]\rangle \cong \mathbb{Z}$. □

An immediate consequence of the theorem is that S^1 is not contractible.

Theorem 2.2.14. *Let $p: \tilde{X} \to X$ be a covering map, $x_0 \in X$. Let $\tilde{X}$ be path connected. The subgroups $p_*(\pi_1(\tilde{X}, \tilde{x}_0))$, where $\tilde{x}_0$ varies over $p^{-1}(x_0)$, form a conjugacy class of subgroups of $\pi_1(X, x_0)$.*

Proof. Let $\tilde{x}_0, \tilde{x}_1 \in p^{-1}(x_0)$ and γ be a path from $\tilde{x}_0$ to $\tilde{x}_1$ (recall that $\tilde{X}$ is path connected). Let $u: \pi_1(\tilde{X}, \tilde{x}_0) \to \pi_1(\tilde{X}, \tilde{x}_1)$ be defined by $\tilde{w} \mapsto \gamma^{-1}\tilde{w}\gamma$. Furthermore, let $v: \pi_1(X, x_0) \to \pi_1(X, x_0)$ be defined by $w \mapsto (p \circ \gamma^{-1}) w (p \circ \gamma)$. The diagram

$$\begin{array}{ccc} \pi_1(\tilde{X}, \tilde{x}_0) & \xrightarrow{u} & \pi_1(\tilde{X}, \tilde{x}_1) \\ \downarrow p_* & & \downarrow p_* \\ \pi_1(X, x_0) & \xrightarrow[v]{} & \pi_1(X, x_0) \end{array}$$

is commutative, that means the images in $\pi_1(X, x_0)$ are conjugate because $p \circ \gamma$ is a loop. Now let $c^{-1} p_*(\pi_1(\tilde{X}, \tilde{x}_0)) c < \pi_1(X, x_0)$. Let δ be a path from c. Then we may lift δ in $\tilde{X}$ to a $\tilde{\delta}: I \to \tilde{X}$ with $\tilde{\delta}(0) = \tilde{x}_0$. Let $\tilde{x}_1 = \tilde{\delta}(1)$, then $p \circ \tilde{\delta}(1) = x_0$ and $p_*(\pi_1(\tilde{X}, \tilde{x}_1)) = c^{-1} p_*(\pi_1(\tilde{X}, \tilde{x}_0)) c$. □

Theorem 2.2.15. *Let $p: (\tilde{X}, \tilde{x}_0) \to (X, x_0)$ be a covering map. Let $f: (Y, y_0) \to (X, x_0)$ with Y path connected and locally path connected. Then a unique lift of f exists if and only if $f_*(\pi_1(Y, y_0))$ is a subgroup of $p_*(\pi_1(\tilde{X}, \tilde{x}_0))$.*

Proof. First suppose a lift $\tilde{f}$ of f exists. Then $f = p \circ \tilde{f}$ so that $f_* = (p \circ \tilde{f})_*$. Note that we choose $\tilde{f}$ so that $\tilde{f}(y_0) = \tilde{x}_0$. Then for $\alpha = [a] \in \pi_1(Y, y_0)$ we have $f_*(\alpha) = (p \circ \tilde{f})_*(\alpha) = [p \circ \tilde{f} \circ a] = p_*([\tilde{f} \circ a]) \in p_*(\pi_1(\tilde{X}, \tilde{x}_0))$ since $\tilde{f} \circ a$ is a loop at $\tilde{x}_0$. We thus obtain the desired property. Conversely, suppose that $f_*(\pi_1(Y, y_0))$ is a subgroup of $p_*(\pi_1(\tilde{X}, \tilde{x}_0))$. For $y \in Y$, let γ be a path in Y from y_0 to y. Then $f \circ \gamma$ is a path from x_0 to $f(y)$, and by the unique path lifting property (Theorem 2.2.7) there exists a lift $\widetilde{f \circ \gamma}: X \to \tilde{X}$ of $f \circ \gamma$ starting at x_0. Then $p((\widetilde{f \circ \gamma})(1)) = f(y)$. Since we want a lift $\tilde{f}$ such that $p(\tilde{f}(y)) = f(y)$, we will define $\tilde{f}(y) = (\widetilde{f \circ \gamma})(1)$. Since the choice of γ was arbitrary we have to show that $\tilde{f}$ does not depend on this choice. To do this, let γ' be another path in Y from y_0 to y. We have to show that $(\widetilde{f \circ \gamma'})(1) = (\widetilde{f \circ \gamma})(1)$. Note that $\gamma\gamma'^{-1}$ is a loop at y_0. Now, $f_*(\gamma\gamma'^{-1})$ is in the image of $\pi_1(\tilde{X}, \tilde{x}_0)$ under p. Thus there is some $\alpha = [a] \in \pi_1(\tilde{X}, \tilde{x}_0)$ with $p_*([a]) = f_*([\gamma\gamma'^{-1}])$, which means that $p \circ a \simeq f \circ (\gamma\gamma'^{-1})$ through some loop homotopy h_t, that is a homotopy rel $\{0, 1\}$. We apply the homotopy lifting property to h_t (Theorem 2.2.9) to obtain a homotopy $\tilde{h}_t$ from one lift of these loops to another. Since a is a lift of $p \circ a$ we can choose this homotopy so that $\tilde{h}_0 = a$ is a loop at $\tilde{x}_0$. Now, by Theorem 2.2.10, it follows that h_t is a loop homotopy. This $\tilde{h}_1$, which is a lift of $f \circ (\gamma\gamma'^{-1})$, is a loop at $\tilde{x}_0$. But by the unique path lifting $\tilde{h}_1 = \widetilde{f \circ \gamma} \cdot (\widetilde{f \circ \gamma'})^{-1}$ since this is also a lift of h_1. So the first half of the loop $\tilde{h}_1$ is the path $\widetilde{f \circ \gamma}$ and the second half is $(\widetilde{f \circ \gamma'})^{-1}$. This means $(\widetilde{f \circ \gamma})(1) = (\widetilde{f \circ \gamma'})(1)$, so $\tilde{f}$ is well defined. It remains to show that $\tilde{f}$ is continuous. Let U be an elementary neighborhood of $f(y)$, and let $\tilde{U} \subset p^{-1}(U)$ with $\tilde{f}(y) \in \tilde{U}$. Then $p|_{\tilde{U}}$ is a homeomorphism onto U. Choose a path connected open neighborhood V of y such that $f(V) \subset U$. Fix a path γ from y_0 to y and for $y' \in C$ let μ be a path in V from y to y'. We then have a product $w = \gamma\mu$. This yields a path $f \circ w$ in X, which goes through $f \circ \gamma$ and $f \circ \mu$. By the unique path lifting property, this has a lift $\widetilde{f \circ w} = \widetilde{f \circ w} \widetilde{f \circ \mu}$. Since the image of $f \circ \mu$ is contained in U we have $\widetilde{f \circ \mu} = p|_{\tilde{U}}^{-1} \circ (f \circ \mu)$, so that the image of $\widetilde{f \circ \mu}$ is contained in $\tilde{U}$. By construction of $\tilde{f}$ it follows that $\tilde{f}(y') = (\widetilde{f \circ w})(1) = (\widetilde{f \circ \mu})(1) \in \tilde{U}$. Since this holds for all $y' \in V$ it follows that $\tilde{f}|_V$ is contained in $\tilde{U}$ which means that $\tilde{f}$ is continuous at y because $\tilde{f}|_V = p|_{\tilde{U}}^{-1} \circ f$. Since y was arbitrary, it follows that $\tilde{f}$ is continuous on Y and $\tilde{f}$ is unique by Theorem 2.2.6. □

2.2.2 The theorem of Seifert and van Kampen

We fix the following notation: We denote by $w(Y, y_0)$ the set of all loops $w: I \to Y$ with $w(0) = w(1) = y_0$. This is the *loop space* of Y at y_0, which is of independent interest in algebraic topology.

Theorem 2.2.16 (Seifert and van Kampen). *Let X_1, X_2 be open subspaces of X with $X = X_1 \cup X_2$ and $X_1 \cap X_2 =: X \neq \emptyset$. Let the spaces X_0, X_1, X_2 be path connected and $x_0 \in X_0$. We denote the inclusions as $i_1: X_0 \hookrightarrow X_1$, $i_2: X_0 \hookrightarrow X_2$, $j_1: X_1 \hookrightarrow X$ and $j_2: X_2 \hookrightarrow X$. Then the diagram*

$$\begin{array}{ccc} \pi_1(X_0, x_0) & \xrightarrow{(i_1)_*} & \pi_1(X_1, x_0) \\ {\scriptstyle (i_2)_*}\downarrow & & \downarrow{\scriptstyle (j_1)_*} \\ \pi_1(X_0, x_0) & \xrightarrow[(j_2)_*]{} & \pi_1(X, x_0) \end{array}$$

is a pushout of $((i_1)_*, (i_2)_*)$. Hence, $\pi_1(X, x_0)$ is uniquely determined up to isomorphism.

Proof. The given inclusions provide the following diagram of inclusions.

$$\begin{array}{ccc} w(X_0, x_0) & \longrightarrow & w(X_1, x_0) \\ \downarrow & & \downarrow \\ w(X_0, x_0) & \longrightarrow & w(X, x_0) \end{array}$$

The corresponding diagram of the fundamental group is commutative. If G is any group and

$$\begin{array}{ccc} \pi_1(X_0, x_0) & \xrightarrow{(i_1)_*} & \pi_1(X_1, x_0) \\ {\scriptstyle (i_2)_*}\downarrow & & \downarrow{\scriptstyle u_1} \\ \pi_1(X_0, x_0) & \xrightarrow[u_2]{} & G \end{array}$$

is a commutative diagram then we have to show that there exists exactly one homomorphism $h: \pi_1(X, x_0) \to G$ with $h \circ (j_1)_* = u_1$ and $h \circ (j_2)_* = u_2$. First of all we define a function $g: w(X, x_0) \to G$ and show that $g(w)$ depends only on the homotopy class rel $\{0, 1\}$ of w. Hence g defines a map $h: \pi_1(X, x_0) \to G$. Let $w \in w(X, x_0)$. If $w \in w(X_\nu, x_0)$, $\nu = 1, 2$, then we define $g(w) = u_\nu([w])$, where $[w]$ is the class of w in $\pi_1(X_\nu, x_0)$. The value $g(w)$ is uniquely determined because if $w \in w(X_1, x_0)$ and $w \in w(X_2, x_0)$ then $w \in w(X_0, x_0)$ and therefore $u_1([w]) = u_2([w])$. Moreover, the value $g(w)$ depends only on the homotopy class of w in $\pi_1(X_\nu, x_0)$ if $w \in w(X_\nu, x_0)$. Now, let the path w be not completely in some X_ν. Then there exist numbers $0 = \tau_0 < \tau_1 < \cdots < \tau_n = 1$ with $w([\tau_{i-1}, \tau_i]) \subset X_\nu$, $\nu = 1$ or 2 (lemma of Lebesgue). Let $w_i: I \to X$ be defined by $w_i(t) = w((1-t)\tau_{i-1} + t\tau_i)$, $i = 1, \dots, n$. For $i = 1, \dots, n-1$ let further $r_i: I \to X$ be a path with $r_i(0) = x_0$ and $r_i(1) = w(\tau_i)$ such that $r_i(t) \in X_\nu$ for all t if $w(\tau_i) \in X_\nu$, $\nu = 1$ or 2. For $i \in \{0, 1\}$ we take r_i to be the constant path $x_0 \in X_0$, compare Figure 2.9.

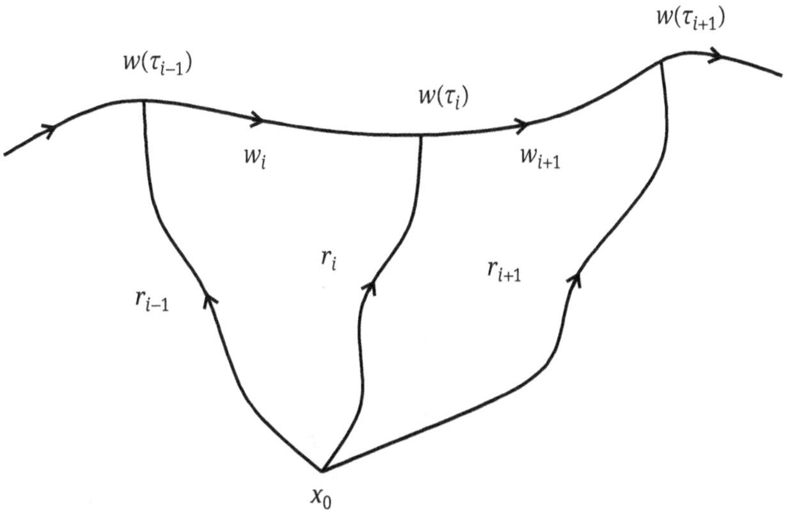

Figure 2.9: Porting a path.

Especially r_i is completely in X_0 if $w(\tau_i) \in X_0$. Now we define $g(w)$ by

$$g(w) = g(r_0 w_1 r_1^{-1}) \cdot g(r_1 w_2 r_2^{-1}) \cdots g(r_{n-1} w_n r_n^{-1}),$$

where $g(r_i w_{i+1} r_{i+1}^{-1}) = u_\nu([r_i w_{i+1} r_{i+1}^{-1}])$. We have to show that

(a) $g(w)$ is independent of the choice of the paths r_i;

(b) $g(w)$ is independent of the chosen partition $0 = \tau_0 < \tau_1 < \cdots < \tau_n = 1$; and

(c) If $w' \in w(X, x_0)$ and $w' \simeq w \text{ rel } \{0,1\}$ then $g(w') = g(w)$.

For (a) let $s_j : I \to X$ with $s_j(0) = x_0$, $s_j(1) = w(\tau_j)$ be another path for $j \in \{1, \dots, n-1\}$. Then

$$\begin{aligned} g(r_{j-1} w_j s_j^{-1}) g(s_j w_{j+1} r_{j+1}^{-1}) &= g(r_{j-1} w_j r_j^{-1} r_j s_j^{-1}) g(s_j r_j^{-1} r_j w_{j+1} r_{j+1}^{-1}) \\ &= g(r_{j-1} w_j r_j^{-1}) g(r_j s_j^{-1}) g(s_j r_j^{-1}) g(r_j w_{j+1} r_{j+1}^{-1}) \\ &= g(r_{j-1} w_j r_j^{-1}) g(r_j w_{j+1} r_{j+1}^{-1}). \end{aligned}$$

Here we have used the fact that $g(u)$ for $u \in w(X_\nu, x_\nu)$ depends only on the homotopy class of u rel $\{0,1\}$. Hence $g(w)$ does not change if we replace r_j by s_j, $j = 1, \dots, n-1$.

Assertion (b) can be seen as follows: Any two suitable partitions of I have a common refinement. Hence, it is enough to show: By the addition of a point τ to the partition $0 = \tau_0 < \cdots < \tau_n = 1$ we do not change the value $g(w)$. Suppose that $\tau \in (\tau_{j-1}, \tau_j)$ and $r : I \to X$ a path with $r(0) = x_0$, $r(1) = w(\tau)$ such that the $r(t)$ are in the same X_ν as $w(\tau)$ for all $t \in I$; compare Figure 2.10.

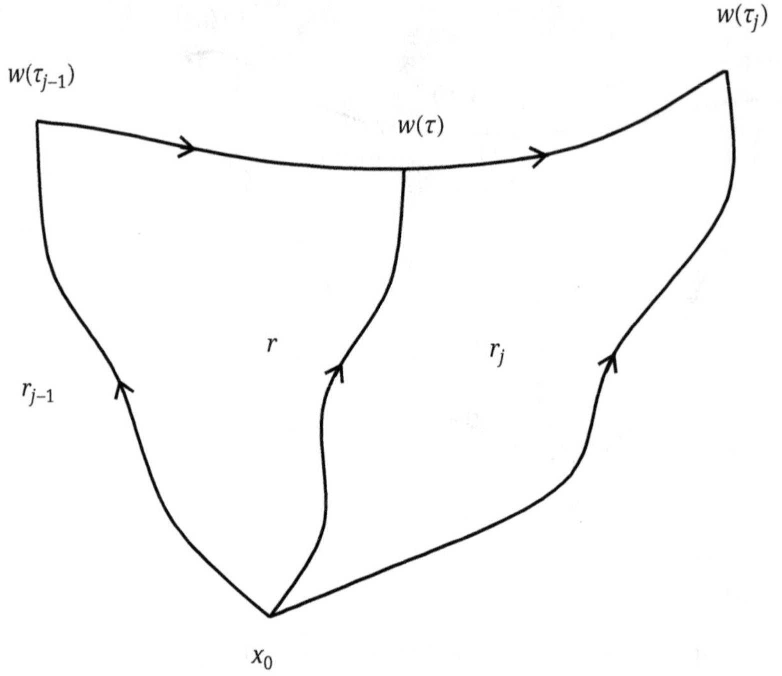

Figure 2.10: Independence of the decomposition.

Then in the definition of $g(w)$, only the value $g(r_{j-1}w_jw_j^{-1})$ could change. We have $w_j \simeq w_j^1 w_j^2$ rel $\{0,1\}$, where $w_j^1(t) = w((1-t)\tau_{j-1} + t\tau)$ and $w_j^2(t) = w((1-t)\tau + t\tau_j)$ for all $t \in I$.

Since $g(w)$ depends for all $u \in w(X_\nu, x_0)$ only on the homotopy class rel $\{0,1\}$ in $w(X_\nu, x_0)$, we have

$$\begin{aligned} g(r_{j-1}w_jr_j^{-1}) &= g(r_{j-1}w_j^1w_j^2r_j^{-1}) \\ &= g(r_{j-1}w_j^1r^{-1}rw_j^2r_j^{-1}) \\ &= g(r_{j-1}w_j^1r^{-1})g(rw_j^2r_j^{-1}), \end{aligned}$$

and the value of $g(w)$ is unchanged. Hence, $g(w)$ is independent of the partition.

For (c) let $w' \in [w]$ and $H: I \times I \to X$, a homotopy rel $\{0,1\}$ from w to w'. We may partition I in $0 = t_0 < t_1 < \cdots < t_n = 1$ and $0 = \tau_0 < \tau_1 < \cdots < \tau_n$ such that for each $(s,r) \in \{1,\ldots,n\} \times \{1,\ldots,k\}$ the image of

$$Q_{s,r} = \{(x,y) \in I \times I \mid t_{s-1} \le x \le t_j, \tau_{r-1} \le y \le \tau_r\}$$

under H is completely contained in some X_ν, $\nu = 1$ or 2; see Figure 2.11.

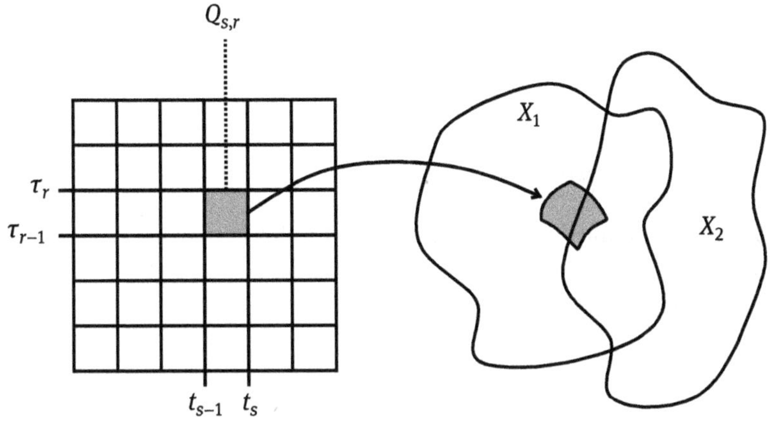

Figure 2.11: Images of $Q_{s,r}$.

Let $w_l = H|_{I\times\{\tau_l\}}$ for all $l \in \{0,\dots,k\}$. We show that $g(w_{l-1}) = g(w_l)$ for all $l \in \{1,\dots,k\}$. Then $g(w) = g(w')$ because $w_0 = w$ and $w_k = w'$. We connect x_0 with each point $H(t_s,\tau_r)$ by a path $u_{s,r}$ which is completely in those X_ν which contain $H(t_s,\tau_r)$. Thereby we choose $u_{s,r}$ as the constant path x_0 if $s = 0$ or $s = n$. Furthermore, we define paths w_r^i for $i \in \{1,\dots,n\}$, $r \in \{0,\dots,k\}$ and v_i^s for $s \in \{0,\dots,n\}$, $i \in \{1,\dots,k\}$ by $w_r^i(t) = H((1-t)t_{i-1} + t_i, \tau_r)$ and $v_r^i(t) = H(t_i, (1-t)\tau_{r-1} + t\tau_r)$. These paths w_r^i and v_r^i connect $H(t_{i-1},\tau_r)$ with $H(t_i,\tau_r)$ and $H(t_i,\tau_{r-1})$ with $H(t_i,\tau_r)$ as shown in Figure 2.12.

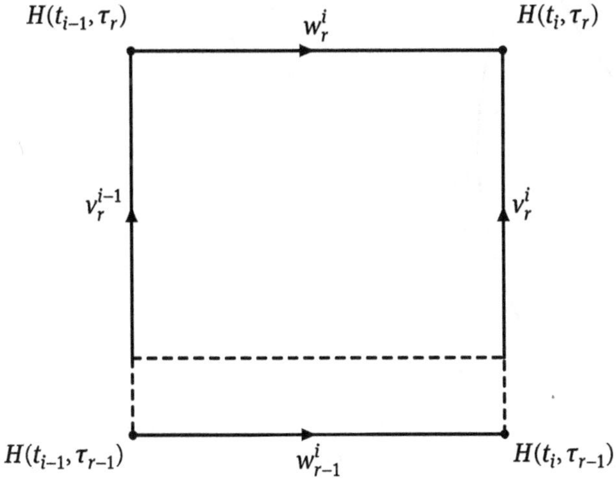

Figure 2.12: Connections via paths.

We have $w^i_{r-1} \simeq (v^{i-1}_r w^i_r (v^i_r)^{-1})$ rel $\{0,1\}$ via

$$F(t,s) = \begin{cases} v^{i-1}_r(4st) & \text{if } t \in [0, \frac{1}{4}],\\ H((2-4t)t_{i-1} + (4t-1)t_i, (1-s)\tau_{r-1} + s\tau_r) & \text{if } t \in [\frac{1}{4}, \frac{1}{2}],\\ v^i_r(s(2-2t)) & \text{if } t \in [\frac{1}{2}, 1]. \end{cases}$$

Since $H(Q(i,r))$ is contained in some X_ν, we get with this homotopy rel $\{0,1\}$ that $g(u_{i-1,r-1} w^i_{r-1} u^{-1}_{i,r-1}) = g(u_{i-1,r-1}((v^{i-1}_r w^i_r)(v^i_r)^{-1}) u^{-1}_{i,r-1})$. From (a) we now get

$$\begin{aligned} g(w_{r-1}) &= \prod_{i=1}^{n} g(u_{i-1,r-1} w^i_{r-1} u^{-1}_{i,r-1})\\ &= \prod_{i=1}^{n} g((u_{i-1,r-1} v^{i-1}_r) w^i_r ((v^i_r)^{-1} u^{-1}_{i,r-1}))\\ &= \prod_{i=1}^{n} g(u_{i-1,r} w^i_r u^{-1}_{i,r})\\ &= g(w_r). \end{aligned}$$

The independence of the choice of paths is shown in Figure 2.13.

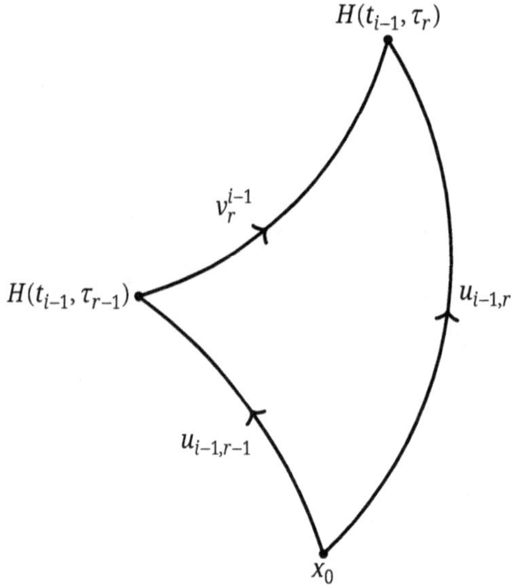

Figure 2.13: The independence of the choice of paths.

Therefore $g(w)$ depends only on the homotopy class rel $\{0,1\}$ of w. Hence, g induces a map $h: \pi_1(X, x_0) \to G$ by $h([w]) = g(w)$. By construction, h is a homeomorphism and $h \circ (j_1)_* = u_1$, $h \circ (j_2)_* = u_2$. We now show the uniqueness of h: Let $h': \pi_1(X, x_0) \to G$ be a homomorphism with $h' \circ (j_1)_* = u_1$ and $h' \circ (j_2)_* = u_2$.

Let $w \in w(x, x_0)$. Then, from the considerations above, there exist paths $w_1, \dots, w_s$ with $w_i \in w(X_\nu, x_0)$, $\nu = 1$ or 2, such that $w \simeq w_1 \cdots w_s \operatorname{rel} \{0,1\}$. Then $h'([w]) = h'([w_1]) \cdots h'([w_j]) = h([w_1]) \cdots h([w_j]) = h(w)$, because $h'([w_i]) = u_\nu([w_i]) = h([w_i])$ for $w_i \in w(X_\nu, x_0)$ for $\nu = 1$ or 2. □

We repeat some notations and properties. Let $A \subset X$ and $i \colon \hookrightarrow X$ be the inclusion. Then A is a retract of X, if there exists a retraction map $r \colon X \to A$ with $r \circ i = \mathrm{id}_A$. If also $i \circ r \simeq \mathrm{id}_X$ then A is a deformation retract. Moreover, A is a *strong deformation retract* of X if there exists a retraction map $r \colon X \to A$ with $i \circ r \simeq \mathrm{id}_x \operatorname{rel} A$.

Theorem 2.2.17. *Let X be path connected, $f \colon S^{n-1} \to X$ a map and $Y = X \cup_f e^n$, $n \geq 2$, $y_0 \in Y$. Then*

$$\pi_1(Y, y_0) \cong \pi_1(X, f(1)) / \langle\langle f_*(\pi_1(S^{n-1}, 1)) \rangle\rangle.$$

Proof. We show the theorem for $n = 2$, the general case is analogous (see Corollary 2.2.19). Let $\mathrm{pr} \colon X \sqcup D^2 \to Y$ be the natural quotient map. We choose $X_1 = Y \setminus \{\mathrm{pr}(0)\}$, $X_2 = \mathrm{pr}(e^2)$, $e^2 = \mathring{D}^2$, $X_2 \cong e^2$, $X_0 = X_1 \cap X_2 \cong e^2 \setminus \{0\}$. Then X_1 has $\mathrm{pr}(X)$ as a strong deformation retract. To see this we give the retraction: $r \colon X \sqcup (D^2 \setminus \{0\}) \to X \sqcup S^1$ via

$$r(x) = \begin{cases} x & \text{if } x \in X \text{ or } x \in S^1, \\ \frac{x}{\|x\|} & \text{if } x \in D^2 \setminus \{0\}, \end{cases}$$

and a homotopy $F(X \sqcup (D^2 \setminus \{0\})) \times I \to X \sqcup (D^2 \setminus \{0\})$ via

$$F(x,t) = \begin{cases} x & \text{if } x \in X, \\ (1-t)x + t\frac{x}{\|x\|} & \text{if } x \in D^2 \setminus \{0\}. \end{cases}$$

With this we define $\bar{r} \colon X_1 \to \mathrm{pr}(X)$ via $\bar{r}(\mathrm{pr}(x)) = \mathrm{pr}(rx)$ and $\bar{F} \colon X_1 \times I \to X_1$ via $\bar{F}(\mathrm{pr}(x), t) = \mathrm{pr}(F(x,t))$; $\bar{r}$ and $\bar{F}$ are well defined and continuous by definition of the quotient topology. Hence $\mathrm{pr}(X)$ becomes a strong deformation retract of X_1. Now, $\pi_1(X_2) = \pi_1(e^2) = \{1\}$. By the theorem of Seifert and van Kampen, and the respective example in Section 1.5, we get $\pi_1(Y, y_0) \cong \pi_1(X_1, y_0) / \langle\langle (i_1)_*(\pi_1(X_0, x_0)) \rangle\rangle$ with $y_0 \in X_0$. We want to describe the factor group in detail. For this purpose, we consider the following commutative diagram of continuous maps:

$$\begin{array}{ccccc} e^2 \setminus \{0\} & \xrightarrow{\mathrm{pr}|_{e^2 \setminus \{0\}}} & X_0 & \xrightarrow{i_1} & X_1 \\ \Big\downarrow {\scriptstyle x \mapsto \frac{x}{\|x\|}} & & & & \Big\downarrow {\scriptstyle \bar{r}} \\ S^1 & \xrightarrow[f]{} & X & \xrightarrow[\mathrm{pr}|_X]{} & \mathrm{pr}(X) \end{array}$$

This gives a commutative homomorphism of (fundamental) groups where we do not explicitly write down the basis points:

$$\begin{array}{ccccc}
\pi_1(e^2 \setminus \{0\}) & \xrightarrow{\cong} & \pi_1(X_0) & \xrightarrow{(i_1)_*} & \pi_1(X_1) \\
\downarrow & & & & \downarrow\cong \\
\pi_1(S^1) & \xrightarrow[f_*]{} & \pi_1(X) & \xrightarrow[\cong]{} & \pi_1(\mathrm{pr}(X))
\end{array}$$

□

From this we get $\pi_1(X_1)/\langle\langle(i_1)_*(\pi_1(X_0))\rangle\rangle \cong \pi_1(X)/\langle\langle f_*(\pi_1(S^{n-1}))\rangle\rangle$.

Corollary 2.2.18. *For each $n \in \mathbb{N}$ there exists a space X_n with $\pi_1(X_n) \cong \mathbb{Z}/n\mathbb{Z} = \mathbb{Z}_n$.*

Proof. Define $f_n: S^1 \to S^1$, $f_n(z) = z^n$ and $X_n = S^1 \cup_{f_n} e^2$, $x_0 = f_n(1)$. Then $\pi_1(X_n, x_0) = \pi_1(S^1, 1)/(f_n)_*(\pi_1(S^1, 1))$. Now $(f_n)_*(\pi_1(S^1, 1)) \cong n\pi_1(S^1, 1) \cong n\mathbb{Z}$, hence $\pi_1(X_n, x_0) \cong \mathbb{Z}/n\mathbb{Z}$. □

Corollary 2.2.19. *Let X be a path connected space and $f: S^{n-1} \to X$ a map, $n \geq 3$. Then $\pi_1(X \cup_f e^n, x_0) \cong \pi_1(X, x_0)$.*

Proof. We only need to apply $\pi_1(D^n \setminus \{0\}, x_0) \cong \pi_1(S^{n-1}, x_0)$ for $n \geq 3$ (see Theorem 2.1.13). □

2.3 Universal coverings, edge path groups and applications

We now continue with the theory of covering spaces and CW complexes. To have the earlier results all fully available we assume that the basis spaces are path connected topological spaces and all maps are continuous (if not stated otherwise). The covering spaces $\tilde{X}$ are path connected and locally path connected (if not stated otherwise).

Definition 2.3.1.
1. A *homomorphism between two coverings* $p_1: \tilde{X}_1 \to X$ and $p_2: \tilde{X}_2 \to X$ is a map $f: \tilde{X}_1 \to \tilde{X}_2$ such that $p_1 = p_2 \circ f$, that is, the diagram

$$\begin{array}{ccc}
\tilde{X}_1 & \xrightarrow{f} & \tilde{X}_2 \\
& {\scriptstyle p_1}\searrow \quad \swarrow{\scriptstyle p_2} & \\
& X &
\end{array}$$

   is commutative.
2. If f is a homeomorphism, then f is an *isomorphism*.
3. If $\tilde{X}_1 = \tilde{X}_2 = \tilde{X}$, $p_1 = p_2 = p$ and f an isomorphism, then f is an *automorphism* or *deck transformation*.

Let $A(\tilde{X}, p)$ be the group of all deck transformations. If $\tilde{x}_0 \in p^{-1}(x_0)$ then also $f(\tilde{x}_0) \in p^{-1}(x_0)$ for all $f \in \mathrm{Aut}(\tilde{X}, p)$.

Theorem 2.3.2. 1. *If f, g are homomorphisms between $p_1: \tilde{X}_1 \to X$ and $p_2: \tilde{X}_2 \to X$ such that $f(\tilde{x}) = g(\tilde{x})$ for some $\tilde{x} \in \tilde{X}_1$ then $f = g$.*
2. *The group $A(\tilde{X}, p)$ acts without fixed points on $\tilde{X}$, that is, if $f \in A(\tilde{X}, p)$ with $f(\tilde{x}) = \tilde{x}$ for some $\tilde{x} \in \tilde{X}$, then $f = 1_{\tilde{X}}$.*
3. *Let $p_1: \tilde{X}_1 \to X$ and $p_2: \tilde{X}_2 \to X$ be covering maps and $\tilde{x}_i \in \tilde{X}_i$ with $p_1(\tilde{x}_1) = p_2(\tilde{x}_2)$.*
   (a) *A homomorphism $f: \tilde{X}_1 \to \tilde{X}_2$ with $f(\tilde{x}_1) = \tilde{x}_2$ exists if and only if $(p_1)_*(\pi_1(\tilde{X}_1, \tilde{x}_1))$ is a subgroup of $(p_2)_*(\pi_1(\tilde{X}_2, \tilde{x}_2))$.*
   (b) *There exists an isomorphism $f: \tilde{X}_1 \to \tilde{X}_2$ with $f(\tilde{x}_1) = \tilde{x}_2$ if and only if $(p_1)_*(\pi_1(\tilde{X}_1, \tilde{x}_1)) = (p_2)_*(\pi_1(\tilde{X}_2, \tilde{x}_2))$.*
4. *Let $p: \tilde{X} \to X$ be a covering map, $x_0 \in X$ and $\tilde{x}_1, \tilde{x}_2 \in p^{-1}(x_0)$. Then there exists a deck transformation f with $f(\tilde{x}_1) = f(\tilde{x}_2)$ if and only if $p_*(\pi_1(\tilde{X}_1, \tilde{x}_1)) = p_*(\pi_1(\tilde{X}_2, \tilde{x}_2))$.*
5. *Two coverings $p_1: \tilde{X}_1 \to X$ and $\tilde{p}_2: \tilde{X}_2 \to X$ of X are isomorphic if and only if the subgroups $(p_1)_*(\pi_1(\tilde{X}_1, \tilde{x}_1))$ and $(p_2)_*(\pi_1(\tilde{X}_2, \tilde{x}_2))$ of $\pi_1(X, x_0)$ are conjugate for some points $\tilde{x}_1 \in \tilde{X}_1$ and $\tilde{x}_2 \in \tilde{X}_2$ with $p_1(\tilde{x}_1) = p_2(\tilde{x}_2)$.*

Proof. 1. This follows from Theorem 2.2.6.
2. Since we may consider f (and g) as a lift of p_1 this statement follows from the first one.
3. (a) If there is such an f, then $(p_1)_* = (p_2)_* \circ f_*$ and then $(p_1)_*(\pi_1(\tilde{X}_1, \tilde{x}_1)) \subset (p_2)_*(\pi_1(\tilde{X}_2, \tilde{x}_2))$. On the other side, we define $f: \tilde{X}_1 \to \tilde{X}_2$ as follows: For $\tilde{x} \in \tilde{X}_1$ we choose a path w in $\tilde{X}_1$ from $\tilde{x}_1$ to $\tilde{x}$. We consider the path $p_1 \circ w$ in X from $p_1(\tilde{x}_1)$ to $p_2(\tilde{x})$. Let $\tilde{w}$ be the unique lift from $p_1 \circ w$ to a path $\tilde{w}$ in $\tilde{X}_2$ with start point $\tilde{x}_2$ (see Theorem 2.2.7). We then define $f: \tilde{X}_1 \to \tilde{X}_2$ via $f(\tilde{x}) = \tilde{w}(1)$. We have to show that this definition is independent of the choice of w. For this purpose, let v be another path in $\tilde{X}_1$ from $\tilde{x}_1$ to $\tilde{x}$. By our precondition there exists a loop $\tilde{v}$ in $\tilde{X}_2$ with start point $\tilde{x}_2$ such that $(p_1)_*([vw^{-1}]) = (p_2)_*([\tilde{v}])$. Then $p_1 \circ v \simeq (p_2 \circ \tilde{v})(p_1 \circ w)$ rel $\{0, 1\}$ in X and the result follows from Theorems 2.2.6 and 2.2.10.
   (b) The above definition provides a function $f: \tilde{X}_1 \to \tilde{X}_2$ with $p_1 = p_2 \circ f$ and $f(\tilde{x}_1) = \tilde{x}_2$. Note that f is continuous by the covering property of p_1 and p_2 and because $\tilde{X}_1$ is locally path connected. The proof is exactly the same as that for Theorem 2.2.15.
4. This follows directly from statement 3.(b).
5. If f is an isomorphism then the statement follows from part 3. The converse follows from Theorem 2.1.13 and part 3, because without loss of generality we may assume equality. □

Lemma 2.3.3. *Let $p_1: \tilde{X}_1 \to X$ and $p_2: \tilde{X}_2 \to X$ be covering maps and $f: \tilde{X}_1 \to \tilde{X}_2$ a homomorphism. Then $f: \tilde{X}_1 \to \tilde{X}_2$ is a covering map.*

Proof. We show that f is surjective: Let $\tilde{x}_1 \in \tilde{X}_1$ and set $\tilde{x}_2 = f(\tilde{x}_1)$. Let $y \in \tilde{X}$ be arbitrary, and $\tilde{w}_2$ be a path from $\tilde{x}_2$ to y. We may lift the path $p_2 \circ \tilde{w}_2$ uniquely to a path $\tilde{w}_1$ in $\tilde{X}_1$ with $\tilde{\omega}_1(0) = \tilde{x}_1$ (see Theorem 2.2.7). From $p_2 \circ f \circ \tilde{w}_1 = p_1 \circ \tilde{w}_1 = p_2 \circ \tilde{w}_2$ and $f \circ \tilde{w}_1(0) = \tilde{w}_2(0)$ we get $\tilde{w}_2 = f \circ \tilde{w}_1$ and $y = f \circ \tilde{w}_1(1)$. We now have to find elementary neighborhoods for the points of $\tilde{X}_2$. Let $y \in \tilde{X}_2$. Let U_1 and U_2 be elementary neighborhoods of $p_2(y)$ related to p_1 and p_2, respectively. Then the path connected component of $U_1 \cap U_2$ which contains $p_2(y)$ is an elementary component U of $p_2(y)$ for both coverings. The component of $p_2^{-1}(U)$ which contains y is then an elementary neighborhood of y for $f: \tilde{X}_1 \to \tilde{X}_2$. □

Remark 2.3.4. Let G be a group and $U < G$. The *normalizer* $N_G(U)$ of U in G is the subgroup of G, which contains all elements $g \in G$ with $g^{-1}Ug = U$.

We have $U \subset N_G(U)$ and $U \lhd N_G(U)$. The group $N_G(U)$ is the biggest subgroup of G which contains U as a normal subgroup. If $U \lhd G$, then $N_G(U) = G$. If $p: \tilde{X} \to X$ is a covering map and $\tilde{x}_0 \in \tilde{X}$, $\tilde{x}_0 \in p^{-1}(x_0)$, $x_0 \in X$, then we denote with $N(p_*(\pi_1(\tilde{X}, \tilde{x}_0)))$ the normalizator of $p_*(\pi_1(\tilde{X}, \tilde{x}_0))$ in $\pi_1(X, x_0)$.

Theorem 2.3.5. *Let $p: \tilde{X} \to X$ be a covering map. The group $A(\tilde{X}, p)$ of deck transformations is isomorphic to the factor group $N(p_*(\pi_1(\tilde{X}, \tilde{x}_0)))/p_*(\pi_1(\tilde{X}, \tilde{x}_0))$.*

Proof. We define an allocation $\psi: A(\tilde{X}, p) \to \pi_1(X, p(\tilde{x}_0)) \bmod p_*(\pi_1(\tilde{X}, \tilde{x}_0))$ as follows: If $f \in A(\tilde{X}, p)$ choose a path $\tilde{a}$ in $\tilde{X}$ from $f(\tilde{x}_0)$ to $\tilde{x}_0$. Let $a \in \pi_1(X, x_0)$ with $x_0 = p(\tilde{x}_0)$ be the homotopy class represented by the path $p \circ \tilde{a}$ and $\psi(f)$ the residue class mod $p_*(\pi_1(\tilde{X}, \tilde{x}_0))$ of a. The allocation ψ is certainly defined. By Theorem 2.3.2 there exists a deck transformation f with $f(\tilde{x}_0) = \tilde{a}(1)$ if and only if $p_*(\pi_1(\tilde{X}, \tilde{x}_0)) = p_*(\pi_1(\tilde{X}, \tilde{a}(1))) = a^{-1}p_*(\pi_1(\tilde{X}, \tilde{x}_0))a$, that is, if a is in the normalizer $N(p_*(\pi_1(\tilde{X}, \tilde{x}_0)))$ in $\pi_1(X, p(\tilde{x}_0))$. It follows $\operatorname{im}(\psi) = N(p_*(\pi_1(\tilde{X}, \tilde{x}_0)))/p_*(\pi_1(\tilde{X}, \tilde{x}_0))$. Let also $g \in A(\tilde{X}, p)$ and $\tilde{b}$ a path in $\tilde{X}$ from $g(\tilde{x}_0)$ to $\tilde{x}_0$. Then $\tilde{c} = f(\tilde{b})\tilde{a}$ is a path from $(f \circ g)(\tilde{x}_0)$ to $\tilde{x}_0$ such that $p \circ \tilde{c}$ represents the element $\psi(f \circ g)$. From $p \circ \tilde{c} = \underbrace{p \circ f(\tilde{b}) p \circ \tilde{a}}_{\substack{\text{Definition of } \tilde{c} \text{ and} \\ \text{the product of paths}}} = (p \circ \tilde{b})(p \circ \tilde{a})$, $p \circ g = p$, we get $\psi(f \circ g) = \psi(f)\psi(g)$, hence ψ is a homomorphism: $p_*(\pi_1(\tilde{X}, \tilde{x}_0)) = p_*(\pi_1(\tilde{X}, f(\tilde{b})\tilde{a}(1))) = [p \circ \tilde{b}]^{-1}[p \circ \tilde{a}]^{-1}p_*(\pi_1(\tilde{X}, \tilde{x}_0))[p \circ \tilde{a}][p \circ \tilde{b}] = (ab)^{-1}p_*(\pi_1(\tilde{X}, \tilde{x}_0))[p \circ \tilde{a}][p \circ \tilde{b}] = (ab)^{-1}p_*(\pi_1(\tilde{X}, \tilde{x}_0))ab$ because $[p \circ \tilde{b}] = p \circ f(\tilde{b})$.

It remains to be shown that ψ is injective. To this end, $\psi(f) = 1$ implies $a \in p_*(\pi_1(\tilde{X}, \tilde{x}_0))$, that is $p \circ \tilde{a} \simeq p \circ \tilde{a}'$, where $\tilde{a}': (I, \partial I) \to (\tilde{X}, \tilde{x}_0)$ is a loop. From Theorem 2.1.8 we get $\tilde{a}(1) = \tilde{a}'(1)$, that is, $f(\tilde{x}_0) = \tilde{x}_0$. Hence, f is the identity by Theorem 2.3.2. □

Definition 2.3.6. 1. A covering $p: \tilde{X} \to X$ is called *regular* if $p_*(\pi_1(\tilde{X}, \tilde{x}_0))$ is a normal subgroup of $\pi_1(X, p(\tilde{x}_0))$.
2. A covering $p: \tilde{X} \to X$ is called *universal covering* if $\pi_1(\tilde{X}_0, \tilde{x}_0) = \{1\}$.

By Theorem 2.3.2 two universal coverings are isomorphic.

Lemma 2.3.7. *A covering $p: \tilde{X} \to X$ is regular if and only if two paths in $\tilde{X}$ which cover the same path in X, are both closed or both not closed.*

Proof. Let $w: I \to X$ be a closed path, $\tilde{w}_1, \tilde{w}_2: I \to \tilde{X}$, may cover w and let $\tilde{w}_1$ be closed. Let $\tilde{a}: I \to \tilde{X}$ with $\tilde{a}(0) = \tilde{w}_1(0)$, $\tilde{a}(1) = \tilde{w}_2(0)$; further let $a = p \circ \tilde{a}$. Then $\tilde{w}_2$ if and only if the lift of awa^{-1} which starts with $\tilde{w}_1(0)$ is closed; compare Figure 2.14.

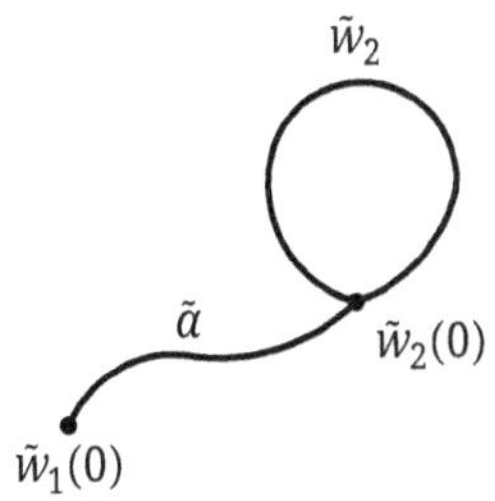

Figure 2.14: Connecting $\tilde{w}_1(0)$ and $\tilde{w}_2(0)$.

The latter is equivalent to $[awa^{-1}] = [a][w][a]^{-1} \in p_*(\pi_1(\tilde{X}, \tilde{w}_1(0)))$ and this is equivalent to $[a]p_*([\tilde{w}_1])[a]^{-1} \in p_*(\pi_1(\tilde{X}, \tilde{w}_1(0)))$. Hence we have obtained the result by going through both directions. □

Theorem 2.3.8. 1. *If $p: \tilde{X} \to X$ is a regular covering then*

$$A(\tilde{X}, p) \cong \pi_1(X, x_0)/p_*(\pi_1(\tilde{X}, \tilde{x}_0))$$

for points $x_0 \in X$, $\tilde{x}_0 \in \tilde{X}$ with $p(x_0) = x_0$.
2. *If $p: \tilde{X} \to X$ is a universal covering of X then $A(\tilde{X}, p) \cong \pi_1(X, x_0)$.*

Proof. The result follows directly from Theorems 2.1.13, 2.3.2, 2.3.5 and Lemma 2.3.7. □

Remark 2.3.9. The isomorphism between $A(\tilde{X}, p)$ and $\pi_1(X, x_0)/p_*(\pi_1(\tilde{X}, \tilde{x}_0))$ can be described for regular coverings as follows: Let $w: I \to X$, $w(0) = w(1) = x_0$. Let $\tilde{w}$ be the lifted path with $\tilde{w}(0) = \tilde{x}_0$. There exists exactly one deck transformation g with $g(\tilde{x}_0) = \tilde{w}(1)$. The map $[w] \mapsto g$ defines a homomorphism from $\pi_1(X, x_0)$ onto $A(\tilde{X}, p)$ which has just $p_*(\pi_1(\tilde{X}, \tilde{x}_0))$ as the kernel. We have $g = 1$ if and only if $[\tilde{w}] \in \pi_1(\tilde{X}, \tilde{x}_0)$ if and only if $[w] \in p_*(\pi_1(\tilde{X}, \tilde{x}_0))$. Note that the isomorphism is not canonical, since there is an arbitrariness in the choice of $\tilde{x}_0$. As another choice we have a conjugation of $\pi_1(X, x_0)$ behind it.

Definition 2.3.10. Let Y be a topological space and G be a group of homeomorphisms of Y. Then G is said to act *properly discontinuous* on Y if for each $y \in Y$ there exists a neighborhood U such that $U \cap gU = \emptyset$ for $g \in G, g \neq 1$.

We remark that this definition is a bit different from those in Subsection 3.3.1 where we, for instance, allow fixed points. However, in our situation where we consider deck transformations the definitions are equivalent.

Theorem 2.3.11. 1. *Let $p: \tilde{X} \to X$ be a covering map. Then $A(\tilde{X}, p)$ acts properly discontinuous on $\tilde{X}$.*
2. *Let Y be a (path connected and locally path connected) topological space and G be a group which acts properly discontinuous on Y. The canonical projection $p: Y \to Y/G$ from Y onto the quotient space Y/G (with respect to the equivalence relation $y_1 \sim y_2$ if and only if there is a $g \in G$ with $g(y_1) = y_2$) is a regular covering with G the group of deck transformations. In particular, the sequence*

$$\{1\} \to \pi_1(Y) \to \pi_1(Y/G) \to G \to \{1\}$$

is exact.

Proof. 1. For $\tilde{x}_0 \in \tilde{X}$ take an elementary neighborhood U of $p(\tilde{x}_0) = x_0$. Then $p^{-1}(U) = \bigcup_{\tilde{x}\in p^{-1}(x_0)} \tilde{U}_{\tilde{x}}$, where $\tilde{U}_{\tilde{x}}$ is a neighborhood of $\tilde{x}$ and $U_{\tilde{x}_1} \cap U_{\tilde{x}_2} = \emptyset$ for $\tilde{x}_1, \tilde{x}_2 \in p^{-1}(x_0)$, $\tilde{x}_1 \neq \tilde{x}_2$, especially $g(\tilde{U}_{\tilde{x}_0}) \cap \tilde{U}_{\tilde{x}_0} = \emptyset$ for each deck transformation $g \neq 1$. By definition, $A(\tilde{X}, p)$ acts properly discontinuous.

2. We have to find an elementary neighborhood for $x \in Y/G$. Let $y \in Y$ with $p(y) = x$. There exists a neighborhood U of y such that $gU \cap U = \emptyset$ for all $g \neq 1$. Note that U contains a path connected open neighborhood V of y because Y is locally path connected. We would like to show that $W = p(V)$ is an elementary neighborhood of x. Since p is an open map, W is open in Y/G, and p maps V injectively and continuously onto W, in fact homeomorphically because p is open. This is the case because there are no two elements $y_1, y_2 \in V$ with $y_1 = g(y_2)$ for some $g \in G$, $g \neq 1$.
Let V' be a component of $p^{-1}(W)$. Then there exists a $g \in G$ with $g(V) = V'$. Then p maps V' also homeomorphically onto W because V is homeomorphically mapped onto V by g^{-1}. Hence, W is an elementary neighborhood of x, and $p: Y \to Y/G$ is a covering map.
Each $g \in G$ is a deck transformation by construction. Conversely, let $f: Y \to Y$ be a deck transformation and $f(y) = y'$. Then $p(y) = p(y')$ and hence $g(y) = y'$ for some $g \in G$. Therefore, $f \circ g^{-1}$ is a deck transformation with y' as fixed point. This gives $f \circ g^{-1} = 1$ and $f = g$. Hence, $G = A(Y, p)$. Furthermore, it follows from $p^{-1}(y) = Gy$ that the covering is regular because two paths which cover a path from Y/G can be mapped by a homeomorphism from G to each other, and hence they are both open or both closed (see Lemma 2.3.7). □

Theorem 2.3.12. *Let X be a space which has a universal covering $q: Y \to X$. Then there exists for each conjugacy class of subgroups of $\pi_1(X, x_0)$ a covering map $p: \tilde{X} \to X$ such that $p_*(\pi_1(\tilde{X}, \tilde{x}_0))$, $\tilde{x}_0 \in p^{-1}(x_0)$, belongs to the conjugacy class.*

Proof. Let $\psi: \pi_1(X, x_0) \to A(Y, q)$ be the isomorphism between the fundamental group and the group of the deck transformations as in Remark 2.3.9 whereby we have chosen a point $y_0 \in q^{-1}(x_0)$. Let $G \subset \pi_1(X, x_0)$ be a subgroup which belongs to the given conjugacy class. Then G acts on Y via $g(y) = \psi(g)(y)$, $g \in G$, $y \in Y$. Let $\tilde{X} = Y/G$ and $r: Y \to \tilde{X}$ be the projection $p: Y \to \tilde{X}$, and let $p(r(y)) = q(y)$, that is,

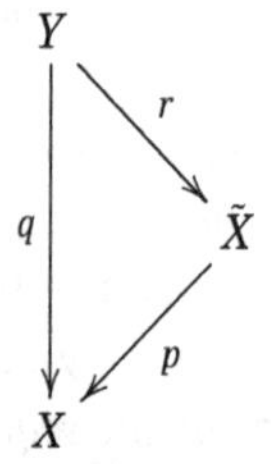

commutes. The map p is continuous and surjective: $r(q^{-1}(A)) = p^{-1}(A)$, $A \subset X$ open, which implies that p is continuous because q is continuous and r is open.

Let $U \subset X$ be an elementary neighborhood with respect to $q: Y \to X$ and $\tilde{U}$ a path component of $q^{-1}(U)$. Then $q^{-1}(U)$ is the union of the disjoint sets $f(\tilde{U})$ with $f \in A(Y, q)$, and as a consequence $p^{-1}(U)$ is the union of the disjoint sets $r(f'(\tilde{U}))$ where f' goes through a system of coset representatives of $A(Y, q)/\psi(G)$. Each of these sets is mapped homeomorphically onto U by p. Therefore $p: \tilde{X} \to X$ is a covering map.

It remains to show that $p_*(\pi_1(\tilde{X}, r(y_0)))$ belongs to the conjugacy class. A loop $w: (I, \partial I) \to (X, x_0)$ lifts up with respect to $p: \tilde{X} \to X$ to a loop if and only if one of the following conditions holds for the lift $w': (I, 0) \to (Y, y_0)$ of w with respect to $q: Y \to X$:

- $r(y_0) = rw'(1)$;
- $y_0 \in G(w'(1))$;
- $g(y_0) = w'(1)$ for some $g \in G$; or
- $\psi[w] = g$.

Therefore $p_*(\pi_1(\tilde{X}, r(y_0))) = G$. □

Remark 2.3.13. Theorem 2.3.12 shows that the construction of coverings for subgroups of a fundamental group succeeds if a universal covering exists. Hence it is important to look at conditions for the existence of a universal covering. Let $p: \tilde{X} \to X$ be a universal covering and $x \in X$, $\tilde{x} \in p^{-1}(x)$, U an elementary neighborhood of x, $i: U \to X$ the inclusion and V the component of $p^{-1}(U)$ which contains $\tilde{x}$. Then we have the following commutative diagram:

$$\begin{array}{ccc} \pi_1(V, \tilde{x}) & \longrightarrow & \pi_1(\tilde{X}, \tilde{x}) \\ {\scriptstyle p|_V}\downarrow & & \downarrow \\ \pi_1(U, x) & \xrightarrow[i_*]{} & \pi_1(X, x) \end{array}$$

Since $(p|_V)_*$ is an isomorphism we get $\operatorname{im}(i_*) = \{1\}$. Hence, for each $x \in X$ there exists a neighborhood U such that the homomorphism $i_*: \pi_1(U, x) \to \pi_1(X, x)$ is trivial. Then X is called *semi locally simply connected.*

Theorem 2.3.14. *Let X be connected, locally path connected and semi locally simply connected. Then there exists for each conjugacy class of subgroups of $\pi_1(X, x_0)$ a covering map $p: \tilde{X} \to X$ such that for $\tilde{x} \in p^{-1}(x_0)$ the subgroup $p_*(\pi_1(\tilde{X}, \tilde{x}))$ belongs to the conjugacy class.*

Proof. By Theorem 2.3.12 it is enough to construct a universal covering $p: \tilde{X} \to X$.

1. Construction of $\tilde{X}$.
   Take $x_0 \in X$. Let $\tilde{X}$ be the set of the homotopy classes of the paths starting at x_0, whereby at the homotopies the termini remain fixed. The covering map $p: \tilde{X} \to X$ assigns to each path class the end point of the paths from the class. Now we give $\tilde{X}$ a topological structure such that $\tilde{X}$ is simply connected and p is a covering map.

2. Topology on $\tilde{X}$.
   Note that X has a basis of open sets U with the following properties:
   U is path connected, and $i_*: \pi_1(U) \to \pi_1(X)$ is trivial ($i: U \hookrightarrow X$). Hence, each loop in U is contractible in X, in other words: If two paths in U have the same termini then they are homotopic with a homotopy which fixes the termini. These sets U with this properties are called *basic sets*. Let a be a path with $a(0) = x_0$ and U a basic set which contains $a(1)$. Then let $([a], U)$ be the set of all classes of paths of the form aa' where a' is a path in U with start point $a(1)$. We get the same set if we replace a by another path of the class of a. As a basis for the topology of $\tilde{X}$ we take the sets $([a], U)$ where the a goes through the paths in X with $a(0) = x_0$ and the U goes through the basic sets with $a(1) \in U$.
   To show that we have a basis of a topology it is enough to show that for $z \in ([a], U) \cap ([b], V)$ a basic set exists and a path c with $x(0) = x_0$, $c(1) \in W$ and $z \in ([c], W) \subset ([a], U) \cap ([b], V)$. To this end, choose c arbitrarily in the homotopy class z and take an open path connected set W with $p(z) = c(1) \in W \subset U \cap V$. Then W is a basic set.
3. Verification of the covering property.
   (a) Let $[a] \in \tilde{X}$ and U be a basic set which contains $a(1)$. Then $p|_{([a],U)}$ is a unique map from $([a], U)$ onto U.
   (b) Let U be a basic set and $x \in U$. Then $p^{-1}(U) = \bigcup_\lambda ([a_\lambda], U)$, where $[a_\lambda]$ goes through the classes of paths from x_0 to x. The sets $([x_\lambda], U)$ are point wise disjoint.

   Assertion (b) shows that p is continuous, because the $([a], U)$ are open and form a basis for the topology of $\tilde{X}$. Hence, $p|_{([a],U)}$ is a unique continuous map from $([a], U)$ onto U. Since each open subset of $([a], U)$ is a union of open sets of the form $([b], V)$ with open $V \subset U$, then $p|_{([a],U)}$ is an open map, and hence, a homeomorphism onto U. We get as the image a union of open sets $V \subset U$.
   Since the sets $([a], U)$ from (b) are pairwise disjoint, U satisfies for $p: \tilde{X} \to X$ the conditions we need for elementary neighborhoods.
4. The space $\tilde{X}$ is path connected.
   Let $\tilde{x}_0 \in \tilde{X}$ the class of the constant path at x_0; let $z \in \tilde{X}$ be represented by the path $f: I \to X, f(0) = x_0$. Define $f_s: I \to X$ by $f_s(t) = f(st)$. Then $f_1 = f$ and f_0 is the constant path at $\tilde{x}_0$. Let z_s be the class of f_s. The map $s \mapsto z_s$ is a path in $\tilde{X}$ from $\tilde{x}_0$ to z. For this purpose, we still have to show the continuity, that is, for $s_0 \in I$ and a basic set U, which contains $f_{s_0}(1)$, we have to find a real number $\delta > 0$ such that $[f_s] \in ([f_{s_0}], U)$ for $|s - s_0| < \delta$. But this only means $f_s(1) = f(s) \in U$ since we get a path homotopy to f_s by going through f_s and proceeding to $f(s)$ or going back.
   The steps 1. to 4. show that $p: \tilde{X} \to X$ is a covering map, however, $\tilde{X}$ is not necessarily locally path connected at this stage.
5. The space $\tilde{X}$ is simply connected.
   Let $w: (I, \partial I) \to (X, x_0)$ be a loop. Define $w_s(t) = w(st)$ and $\tilde{w}: I \to \tilde{X}$, $\tilde{w}(s) = [w_s]$. The path $\tilde{w}$ is the lift of w with $\tilde{w}(0) = \tilde{x}_0$ because $p(\tilde{w}(s)) = p[w_s] = w_s(1) = w(s)$, and $\tilde{w}(1) = [w_1]$. We have $[w] \in p_*(\pi_1(\tilde{X}, \tilde{x}_0))$ if and only if $\tilde{w}(1) = \tilde{x}_0$ if and

only if w is nullhomotopic, hence $p_*(\pi_1(\tilde{X}, \tilde{x}_0)) = \{1\}$. Since p_* is injective we get $\pi_1(\tilde{X}, \tilde{x}_0) = \{1\}$. □

Theorem 2.3.15. *Each finite CW complex (X, ζ) is locally contractible, that is, each point has a contractible neighborhood.*

Proof. We show by induction on the number of cells in ζ that $x \in X$ has a neighborhood U which is contractible to $\{x\}$. If ζ has only one cell, then X is a point and there is nothing to show. Let $X = X_0 \cup e^q$, where (X_0, ζ_0) is a subcomplex of X for which the statement holds. For each $x \in e^q$ we see that e^q itself is a neighborhood of x which is contractible to $\{x\}$. Let $x \in X_0$, and let V be a neighborhood of V which is open in X_0 and contractible in V to $\{x\}$. Let $F: D^q \to X$ be a characteristic map of e^q. The set $F^{-1}(V)$ is an open subset of S^{q-1}. The set $W = \{\tau \in D^q \mid \tau \neq 0 \text{ and } \frac{\tau}{\|\tau\|} \in F^{-1}(V)\}$ is an open subset of D^q which contains $F^{-1}(V)$ as a deformation retract. The set $U = V \cup F(W)$ is a neighborhood of x in X which contains V as a deformation retract and, hence, U is contractible to $\{x\}$. □

Corollary 2.3.16. *Each finite connected CW complex has a universal covering.*

Example 2.3.17. From the Riemann mapping theorem we directly see that the upper half plane is a universal covering of the oriented compact surface group F_g, $g \geq 2$.

Theorem 2.3.18. *Let (X, ζ) be a finite CW complex. Let $p: \tilde{X} \to X$ be a covering map. Then $\tilde{X}$ has a cell partition such that $\tilde{X}$ is a (not necessarily finite) CW complex with* $\dim(\tilde{X}) = \dim(X)$.

The cells of $\tilde{X}$ are path components of $\mathrm{pr}^{-1}(e)$ where e goes through the cells of $\tilde{X}$. The projection pr *maps the cells of $\tilde{X}$ homeomorphically onto the cells of X.*

Proof. Let $e \subset X$ be a cell. Since e is locally path connected, the same holds for $\mathrm{pr}^{-1}(e)$. Hence $\mathrm{pr}^{-1}(e)$ is the topological sum of its path components, and for each such component $\tilde{e}$ we see that $\mathrm{pr}|_{\tilde{e}}: \tilde{e} \to e$ is a covering map. But e is simply connected, hence, $\tilde{e}$ is homeomorphic to e and hence a cell. All these cells form a cell partition of $\tilde{X}$ which has the properties C and W (see Remark 2.1.29). If we lift the characteristic map of e to $\tilde{X}$ (which is possible by Theorem 2.2.15) then we get a characteristic map of the lifted cell $\tilde{e}$. □

Corollary 2.3.19. 1. *Subgroups of free groups are free.*
2. *If F is a free group of finite rank n and $H \subset F$ a subgroup of index k, then H is free of rank $k(n-1)+1$.*

Proof. 1. We only give the proof for a free group of rank $n < \infty$ because we worked mostly with finite CW complexes. The general case is completely analogous.
Let F be a free group of rank n and H a subgroup of F. Let Y be the bouquet of n circles obtained by gluing together a collection of n circles along a single point. We have $\pi_1(Y, y_0) = F$. There exists a covering $p: \tilde{Y} \to Y$ with $p_*(\pi_1(\tilde{Y}, \tilde{y}_0)) \cong H \cong \pi_1(\tilde{Y}, \tilde{y}_0)$. Thus $\tilde{Y}$ is a graph by Theorem 2.3.18, and hence $\pi_1(\tilde{Y}, \tilde{y}_0)$ is a free group.

2. Let n be the rank of F and $[G : H] = k < \infty$. The complex Y as in the first statement has one 0-cell Y_0 and n 1-cells, $\tilde{Y}$ consists of k 0-cells and kn 1-cells. This can be seen as follows. Let $\tilde{y}_0, \tilde{y}_1, \dots$ be all points in $p^{-1}(y_0)$ and for $i = 0, 1, \dots$ let $\tilde{w}_i$ be a path in $\tilde{Y}$ from $\tilde{y}_0$ to $\tilde{y}_i$. Then $a_i = [p \circ \tilde{w}_i] \in \pi_1(Y, y_0)$, and the sets $a_0 p_*(\pi_1(\tilde{X}, \tilde{x}_0))$, $a_1 p_*(\pi_1(\tilde{X}, \tilde{x}_1)), \dots$ in $\pi_1(Y, y_0)$ are exactly the pairwise disjoint residue classes of $p_*(\pi_1(\tilde{Y}, \tilde{y}_0))$ in $\pi_1(Y, y_0)$. Since $H \cong \pi_1(\tilde{Y}, \tilde{y}_0)$ we see that H has the rank $kn + 1 - k = k(n-1) + 1$. □

In an analogous manner we may show the following theorem.

Theorem 2.3.20 (Theorem of Kurosh; see Section 1.4). *Each subgroup $U \subset G * H$ is a free product $U = F * U_1 * \cdots$, with F a free group, and each U_i is in $G * H$ conjugate to a subgroup of G or H.*

Sketch of Proof. We choose pointed spaces (X, x_0) and (Y, y_0) as sketched in Figure 2.15 such that $G = \pi_1(X, x_0)$ and $H = \pi_1(Y, y_0)$.

Figure 2.15: Pointed spaces.

We choose (Z, z_0) as sketched in Figure 2.16.

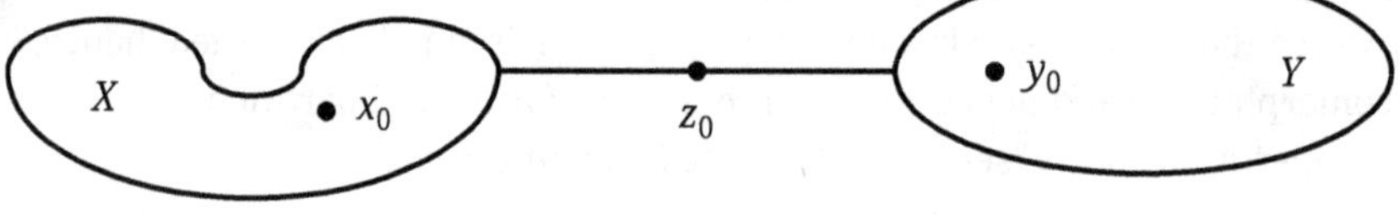

Figure 2.16: Connecting pointed spaces.

We then have $\pi_1(Z, z_0) = \pi_1(X, x_0) * \pi_1(Y, y_0)$ by Theorem 2.2.16. □

We now want to complete this section by giving covering space arguments to prove more results in infinite group theory. We consider (finite) CW complexes and presentations of their fundamental groups. Many of the results go through for infinite CW complexes, just using conditions C and W. From now on, a CW complex (X, ζ) shall satisfy the additional condition EF: Each 2-cell has a characteristic map $F: D^2 \to X$ such that S^1 is divided in consecutive arcs $I_j = \{e^{2\pi i t} \mid t_j \le t \le t_{j+1}\}$, $0 = t_0 < t_1 < \cdots < t_n = 1$, and $F|_{I_j}$ is a characteristic map for a 1-cell.

Definition 2.3.21. Let (X, ζ) be a (finite) CW complex with property EF. An *edge* σ is a directed one-dimensional cell from ζ, σ^{-1} denotes the inverse edge. An *edge path* is a word $\sigma_1\sigma_2\cdots\sigma_m$ of edges whereby σ_{i+1} has the end point of σ_i as its start point. Two edge paths are called *elementary equivalent* if one goes out of the other by adding or canceling edge paths of the form $\sigma\sigma^{-1}$ or by replacing a part of an edge path w by an edge part w' where ww'^{-1} is the image of the boundary of a 2-cell from ζ, compare Figure 2.17.

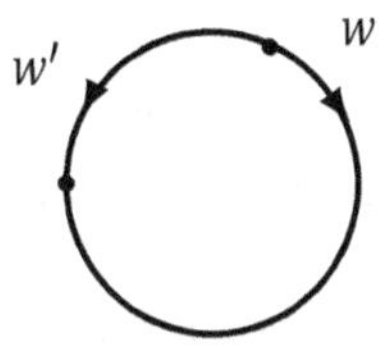

Figure 2.17: Image of the 2-cell.

Two edge paths w and w' are called *equivalent* if they can be connected by a (finite) chain of edge paths in which two adjacent members are elementary equivalent: $w = w_1, w_2, \ldots, w_k = w'$ where w_i is elementary equivalent to w_{i+1}. The product of two edge paths is given by going through one after the other (if possible).

Now let x_0 be a 0-cell. From the set of all equivalence classes of those edge paths which start and end at x_0 together with the above product we get a group, the edge path group, denoted by $\pi(X, \zeta; x_0)$. It depends only on the 2-skeleton of X^2.

Lemma 2.3.22. *A connected one-dimensional CW complex is a tree if each closed edge path is equivalent to the trivial path. This is equivalent to:*
1. *There is no simply closed edge path.*
2. *The group of the edge paths is trivial.*
3. *Any edge path can be transformed by successive canceling of parts of the form $\sigma\sigma^{-1}$ into a simple path which connects start point and end point or into the trivial path.*
4. *Two simple paths with the same termini are equal.*

Calculation 2.3.23 (Calculation of the edge path group). Let $K = (X, \zeta)$ be a finite CW complex with property EF. Let X be connected and x_0 be a 0-cell. Then there exists in X^1 a tree B as subcomplex which contains all 0-cells (a spanning tree). For each cell from $X^1 \setminus B$ we choose a direction; we denote this system of edges by C. Each edge $\sigma \in C$ determines exactly one simply closed edge path which contains σ as the only edge from C: We go from x_0 through B to the start point of σ, go over σ and then we go from the end point of σ through B back to x_0. The set of equivalence classes of these edge paths is denoted by $\mathcal{S}$ (recall that all equivalent edge paths start and end at x_0). We write $s(\sigma)$ for the equivalence class of $\sigma \in C$. We have the following.
1. The set $\mathcal{S}$ is a generating system of $\pi(X, \zeta; x_0)$.

Let e^2 be a 2-cell from ζ. Because of the condition EF the boundary of e^2 is an edge path $w = \sigma_1^{\varepsilon_1}\sigma_2^{\varepsilon_2}\cdots\sigma_n^{\varepsilon_n}$, $\varepsilon_i = \pm 1$. Let $i_1, i_2, \ldots, i_n$ be the indices for $\sigma_{i_j} \in C$. Then $s(\sigma_{i_1})^{\varepsilon_{i_1}}\cdots s(\sigma_{i_n})^{\varepsilon_{i_n}}$ is a relation.
This we can see as follows. Take an edge path v from x_0 to the start point of $\sigma_1^{\varepsilon_1}$ (which is the end point of $\sigma_n^{\varepsilon_n}$). Then vwv^{-1} is an edge path which starts in x_0 and is equivalent to vv^{-1}. Let $\mathcal{R}$ be the system of relations which we obtain in such matter.
2. The set $\mathcal{R}$ is a system of defining relations for $\pi(X,\zeta;x_0)$.
3. Hence, we have the representation $\pi(X,\zeta;x_0) = \langle \mathcal{S} \mid \mathcal{R}\rangle$.

Corollary 2.3.24. 1. *The edge path group of a connected graph K, i. e. a one-dimensional connected CW complex, is free. If α_0 and α_1 are the numbers of vertices and edges, respectively, then K is free of rank $\alpha_1 - \alpha_0 + 1$ (because $\alpha_0 - 1$ edges are in the spanning tree B).*
2. *The edge path group of the CW complex of the oriented closed surface F_g, $g \geq 1$, has the presentation $\langle a_1, b_1, \ldots, a_g, b_g \mid \prod_{i=1}^{g}[a_i, b_i] = 1\rangle$.*
The edge path group of the CW complex of the non-oriented closed surfaces N_g, $g \geq 2$, has the presentation $\langle a_1, \ldots, a_g \mid a_1^2\cdots a_g^2 = 1\rangle$.

Theorem 2.3.25. *For each finite connected CW complex with property EF we have $\pi(X,\zeta,x_0) \cong \pi_1(X,x_0)$.*

Proof. For each edge σ of X let $\varphi: I \to X$ be the characteristic map which maps 0 to the start point and 1 to the end point of σ. We show that $\psi: \pi(X,\zeta;x_0) \to \pi_1(X,x_0)$, $\psi([\sigma_1\cdots\sigma_n]) = [\psi(\sigma_1)\cdots\psi(\sigma_n)]$, is an isomorphism. Since equivalent edge paths are homotopic, ψ is uniquely defined and a homomorphism.
1. The map ψ is injective.
By Theorem 2.1.43 (theorem of the existence of a cellular approximation) for a path $\gamma: (I,\partial I) \to (X,x_0)$ there exists a path $\gamma': (I,\partial I) \to (X,x_0)$ with $\gamma \simeq \gamma'$ and $\gamma'(I) \subset X^1$. By property EF γ' is homotopic to an edge path.
2. The map φ is injective.
Let U be a subgroup of the edge path group $\pi(X,\zeta;x_0)$. Analogously as for the case of the fundamental group we may construct a covering space of X whose edge path group is isomorphic to U. Also we have the analogous results for the group of deck transformations. For more details see pp. 29–37 in [281].
Let $\tilde{\tilde{p}}: \tilde{\tilde{X}} \to X$ be the covering for the trivial subgroup of $\pi(X,\zeta;x_0)$, and let $\tilde{p}: \tilde{X} \to X$ be the universal covering of X. Since $\pi(\tilde{\tilde{X}})$ is trivial, also $\pi_1(\tilde{\tilde{X}})$ is trivial because φ is surjective.
Therefore $\tilde{\tilde{p}}: \tilde{\tilde{X}} \to X$ and $\tilde{p}: \tilde{X} \to X$ are isomorphic coverings. Especially the group of deck transformations, which are $\pi(X,\zeta;x_0)$ and $\pi_1(X,x_0)$, respectively, are isomorphic in a canonical manner as described. Hence, φ is injective. □

Theorem 2.3.26. *Let $G = \langle \mathcal{S} \mid \mathcal{R} \rangle$ be a finitely presented group. Then there exists a finite connected CW complex (X, ζ) with property EF and $x_0 \in X$ such that $G = \pi(X, \zeta; x_0) = \pi_1(X, x_0)$.*

Proof. Let $|\mathcal{S}| = k \geq 1$. Let X^1 be the bouquet of k circles S_s^1, $s \in \mathcal{S}$, obtained by gluing together the k circles S_s^1, $s \in \mathcal{S}$ along a single common point x_0. This, as in the case of the free group, is a 1-dimensional CW complex with one 0-cell and k 1-cells $e_s^1 = S_s^1 \setminus \{x_0\}$.

Let $f_0\colon (I, \partial I) \to (S_s^1, x_0)$ be the characteristic map of e_j^1. For each relation $R(s) = R = s_1^{\varepsilon_1} \cdots s_n^{\varepsilon_n}$, $R \in \mathcal{R}$, define $\varphi_R\colon (I, \partial I) \to (X^1, x_0)$ as follows: In the interval $\frac{i}{n} \leq t \leq \frac{i+1}{n}$, $i = 0, 1, \ldots, n-1$, let

$$\varphi_R(t) = \begin{cases} f_{s_{i+1}}(n(t - \frac{i}{n})) & \text{if } \varepsilon_i = +1, \\ f_{s_{i+1}}(n(\frac{i+1}{n} - t)) & \text{if } \varepsilon_i = -1. \end{cases}$$

Now $X = X^1 \cup (\bigcup_{R \in \mathcal{R}} e_r^2)$ where e_R^2 is attached via φ_R is the desired CW complex. □

We point out that a CW complex with the above property is called an *Eilenberg–MacLane space* of type K$(G, 1)$ where the 1 refers to the fact that G is isomorphic to the first homotopy group of the complex. More generally, any connected CW complex X is called an Eilenberg–MacLane space of type K(G, n) if $\pi_n(X) \cong G$. These spaces have rich applications in singular cohomology and higher homotopy theory.

Remark 2.3.27. This shows again that the fundamental group of a graph is free, and, using coverings, that each subgroup U of a free group is free.

We proceed with some more examples and group theoretical applications of CW complexes in the next sections.

2.3.1 Surface groups

At first we consider the special case of surface groups. Let

$$F_g = e^0 \cup e_1^1 \cup \cdots \cup e_g^1 \cup e_1'^1 \cup \cdots \cup e_g'^1 \cup e^2$$

be the oriented surface of genus $g \geq 2$. We know that F_g is homeomorphic to a ball with g attached handles; see Figure 2.18.

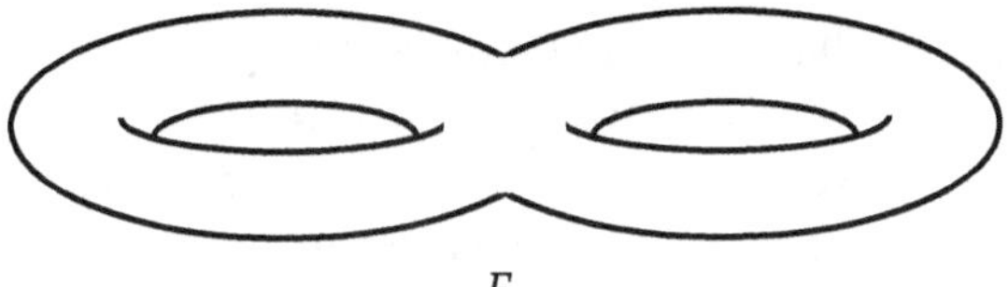

Figure 2.18: Surface of genus 2.

Let U be a subgroup of $\pi_1(F_g)$ of index k. Let $p: \tilde{X} \to F_g$ be a covering map with $\pi_1(\tilde{X}) = U$. There are k preimages of e^0, $2gk$ preimages of the $e_i^1, e_i'^1$ and k preimages of e^2; compare Figure 2.19.

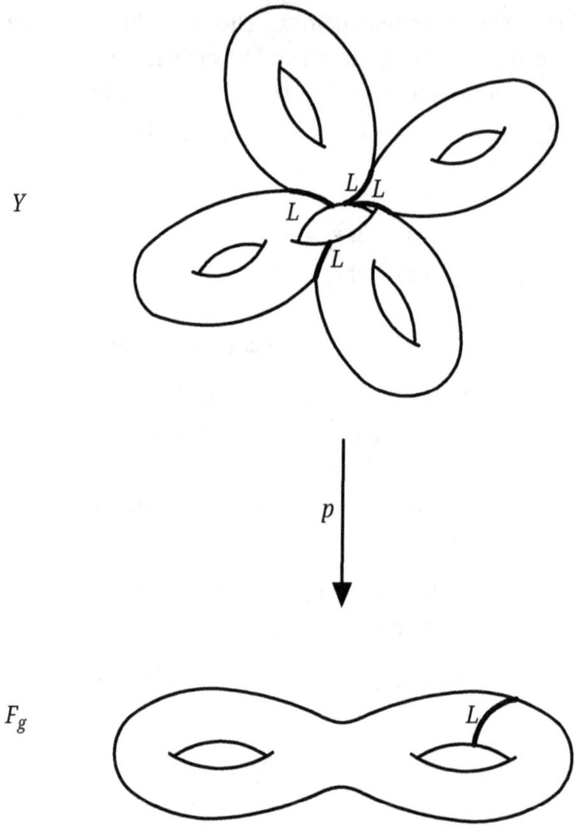

Figure 2.19: Cutting along a loop.

By cutting along the loop L and identifying them we see that Y is a covering space of F_g with fundamental group U, in fact Y is homeomorphic to $\tilde{X}$ and hence, homeomorphic to a ball with $k(g-1)+1$ attached handles. So $\tilde{X}$ is a closed surface and U is a surface group.

We remark that a subgroup U of $\pi_1(F_g)$ of infinite index is free; see [281].

An alternative proof is as follows. Let $G = \pi_1(F_g)$, $g \geq 2$. By the Riemannian mapping theorem the upper half plane $\mathfrak{H} = \{z \in \mathbb{C} \mid \operatorname{Im}(z) > 0\}$ is a universal covering of F_g. The group G acts properly discontinuous on $\mathfrak{H}$ with a fundamental domain $\mathfrak{F}_g \subset \mathfrak{H}$. Let H be a subgroup of G of finite index k. Then H has a simply connected fundamental domain in $\mathfrak{H}$ consisting of k copies $\mathfrak{F}_g$. Respective identification of the boundary arcs via boundary substitutions leads to an oriented compact surface of genus $k(g-1)+1$.

If K is a subgroup of G of infinite index then K cannot have a fundamental domain which is totally contained in $\mathfrak{H}$. Hence, a fundamental domain for K must have free boundary components in $\mathbb{R} \cup \{\infty\}$, that is, each fundamental domain for K has infinite hyperbolic area. Therefore K is free.

The situation is a bit different for compact non-orientable surfaces N_g of genus $g \geq 2$. If $p: \tilde{X} \to N_g$ is a covering map then $\tilde{X}$ is a compact surface, orientable or non-orientable. Let U be a subgroup of N_g of finite index k then U is isomorphic to the fundamental group of a non-orientable surface of genus $k(g-2)+2$ or the fundamental group of an orientable surface of genus $\frac{k}{2}(g-2)+1$. If $G = \pi_1(N_g)$ then G contains a characteristic subgroup of index 2 which is isomorphic to $\pi_1(F_{g-1})$. Again, a subgroup $\pi_1(N_g)$ of infinite index is free. More details for N_g can be found in [281].

In this connection we mention the solution of the *Nielsen realization problem*, given in the following version:

Theorem 2.3.28. *Let G be a torsion-free group which contains an orientable surface group of genus $g \geq 2$ and of finite index as a subgroup. Then G is a surface group.*

A wonderful proof together with a description of the history of some single steps can be found in [280].

2.3.2 Covering space proofs of Theorems 1.2.52 and 1.3.23

We now prove Theorem 1.2.52.

Theorem. *Let $F = \langle y_1, \ldots, y_n \mid \rangle$ be a free group of rank $n < \infty$ with basis $Y = \{y_1, \ldots, y_n\}$. Then the fixed point subgroup of any automorphism is of finite rank.*

Proof. We give a proof that can be found in [61]. Let $\alpha \in \operatorname{Aut}(F)$ and $H = \operatorname{Fix}(\alpha)$. Let X be the coset graph of H in F, that is, the graph whose vertices are the right cosets of H in F and the edges are of the form u—v, $u = Hg$, $v = Hgx_i$, $g \in F$, $x_i \in Y \cup Y^{-1}$. Then $H \cong \pi_1(X)$. We assign names for the vertices of X as follows. Let p be a path from the basis point corresponding to the coset H to a given vertex. Then in traversing p one reads off a word w in the basis Y of F. The name assigned to the end point of p is the reduced form of $\alpha(w^{-1})w$. Since in traversing a closed path starting at the basis point one reads off an element of H, the name is independent of the path chosen. To show that $\pi_1(X)$ is finitely generated one assigns a direction to each geometric edge of X. For this purpose, consider an edge $u \overset{x}{—} v$ with label $x \in Y \cup Y^{-1}$, where, say, $u = Hg$ and $v = Hgx$ for $g \in F$. Now, if u has some name z then it follows that v has name $\alpha(x)^{-1}zx$. Then the geometric edge (formally consisting of the inverse pair of edges which have u and v as end points) is directed from u to v if the displayed generator x is canceled when $\alpha(x)^{-1}zx$ is reduced to normal form otherwise from v to u. If the word z is long in comparison with $\alpha(x)$ then the edge will be directed towards u. Since there are only finitely many words of the form $\alpha(x), x \in Y \cup Y^{-1}$, this shows that with finitely many exceptions, there is, at each vertex, at most one outwardly directed edge. Now a graph in which the geometric edges can be directed in such a way that at every vertex there is at most one outwardly directed edge that contains at most one circuit. If follows that if infinitely many edges are deleted from X then the result is a union of trees. Hence $\pi_1(X)$ is finitely generated. □

We now prove Theorem 1.3.23.

Theorem. *If G is a non-trivial normal subgroup of the free group F_n on $X = \{x_1, \ldots, x_n\}$, $n < \infty$. Then G has finite index in F_n if and only if it is finitely generated.*

Proof. If G has finite index then the coset graph Γ of G in F is finite and, hence, G is finitely generated. Conversely, let G be finitely generated. Then G has certainly finite index if $n \leq 1$. Now, let $n \geq 2$.

Let $Q = \{Gu \mid$ there is a $v \in F_n, v \neq 1$, with $uv \in G$ and uv reduced$\}$. Let $K(G,X)$ be the subgraph of the coset graph Γ which is induced by Q. Let Gu be a right coset and $a_1 \cdots a_k$ reduced with $k \geq 1$, $a_i \in X \cup X^{-1}$. There exist $a, b \in X \cup X^{-1}$ with $a_k = a \neq b$. Since $G \neq \{1\}$ there is a shortest non-empty reduced word $w \in G$. This word cannot be of the form cvc^{-1} with $c \in X \cup X^{-1}$ because $cvc^{-1} \in G$ implies $v \in G$ since G is normal. If w neither begins with a^{-1} nor ends in a, then $a_1 \cdots a_k w a_k^{-1} \cdots a_1^{-1}$ is reduced and describes a path from G to G in $K(G,X)$. As a consequence $Gu \in Q$. Thus we may assume that w either begins with a^{-1} or ends in a. It cannot be both as we pointed out above. By symmetry let w end in a. Then the first letter of w is some $c \neq a^{-1}$. By interchanging the role of b and b^{-1} if necessary we may assume that $b \neq c \neq a^{-1}$. As a consequence $uv = a_1 \cdots a_k b^{-1} w b a_k^{-1} \cdots a_1^{-1}$ is a path from G to G in $K(G,X)$, and therefore $Gu \in Q$. But this means altogether that $K(G,X)$ is the complete coset graph Γ. Therefore the index of G in F_n is finite. □

2.3.3 Group amalgam description of the fundamental groups of certain CW complexes

– Let $G = G_1 * G_2$ be a finitely generated free product and U be a subgroup of G. Then $U = F * (*_{i \in I} U_i)$ where F is free and the U_i are conjugate to subgroups of G_1 or G_2. Let K_1' and K_2' be disjoint CW complexes, $x_1 \in K_1'^0$, $x_2 \in K_2'^0$ the basis points. Let $K = K_1' \cup K_2' \cup \{x_0, e_1^1, e_2^1\}$, where e_i^1 has the points x_0, x_i as termini; see Figure 2.20.

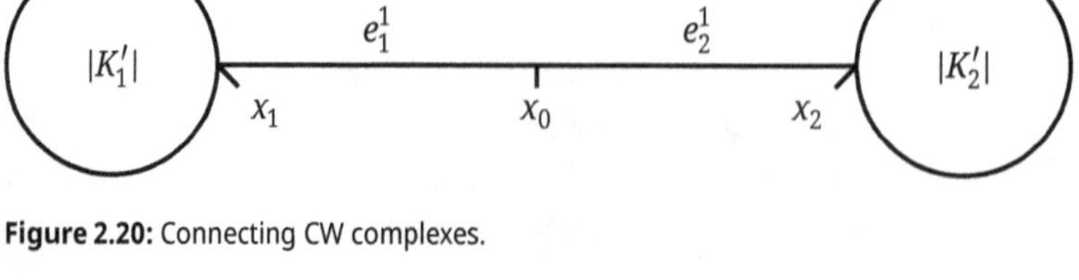

Figure 2.20: Connecting CW complexes.

Then $\pi_1(K, x_0) = \pi_1(K_1, x_0) * \pi_1(K_2, x_0)$. Now use the covering space argument.

- Let $K_0 = (X_0, \zeta_0)$ and $K = (X, \zeta)$ be finite CW complexes. The pair K_0 is called a *subcomplex* of K if $X_0 \subset X$ is a subspace and $\zeta_0 \subset \zeta$ is a subset. Let K be a finite connected CW complex with property EF, K_1, K_2 and $K_1 \cap K_2$ connected subcomplexes such that $K = K_1 \cup K_2$. Choose a basis point $x_0 \in (K_1 \cap K_2)^0$. Let $\varphi^{(i)}: K_1 \cap K_2 \to K_i$ be embeddings. If $\varphi_*^{(1)}$ and $\varphi_*^{(2)}$ are injective, then $\pi_1(K, x_0) = \pi_1(K_1, x_0) *_{\pi_1(K_1 \cap K_2, x_0)} \pi_1(K_2, x_0)$ by Theorem 2.2.16 of Seifert and van Kampen.
- *Gluing complexes together.*
  Let $K = (X, \zeta)$ be a finite connected CW complex with property EF. Let L_1, L_2 be connected subcomplexes with $L_1 \cap L_2 = \emptyset$. Let $L_i(X_i, \zeta_i)$ and $\varphi: L_1 \to L_2$ be a complex isomorphism, that is, φ is injective and cellular ($\varphi(X_1^q) \subset X_2^q$ for all $q \geq 0$). The basis point x_0 shall not be in $L_1 \cup L_2$. Let X_1 and $\varphi(X_1) = X_2$ be 0-cells of L_1 and L_2, respectively. Let K/φ be the complex, which comes from K by identifying L_1 and L_2. Let σ_1 and σ_2 be paths from x_0 to x_1 and x_1 to x_2, respectively. Then we get a presentation of $\pi_1(K/\varphi, x_0)$ from $\pi_1(K, x_0)$ by adding a generator t and relations $[\sigma_1 \lambda \sigma_1^{-1}] = [\tau \sigma_2 \varphi(\lambda) \sigma_2^{-1} \tau^{-1}]$, $t = [\tau]$, where λ runs through paths from a generating system of $\pi_1(L_1, x_1)$, τ denotes the path $\sigma_1^{-1}\sigma_2$ and t its homotopy class; compare Figure 2.21.

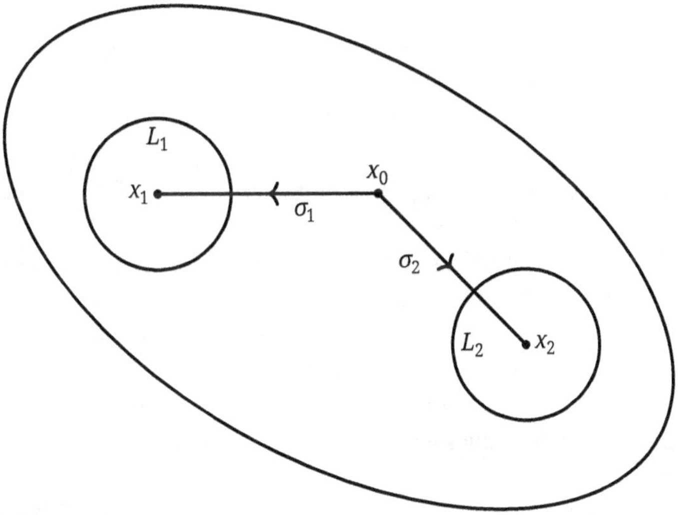

Figure 2.21: Identifying subcomplexes.

2.3.4 Topological proof of the Reidemeister–Schreier theorem

Theorem 2.3.29 (Reidemeister–Schreier). *Let $G = \langle S \mid R\rangle$ be finitely presented and $U < G$. Furthermore, let $U = \{a_i \mid i \in I\}$ be a system of representatives of the cosets Ua. Let $\bar{g}$ be the coset representative of g. Let $a_0 = 1$; furthermore, every initial subword of $a_i \in U$ is also in U, and let $Ua_i \neq Ua_j$ if $i \neq j$. Then U is generated by the non-trivial elements of the form $as(\overline{as})^{-1}$ and has defining relations aRa^{-1}, written in $as(\overline{as})^{-1}$.*

Proof. We consider the CW complex X with $\pi(X, \zeta; x_0) \cong \langle S \mid R\rangle$ (S and R are not necessarily finite, the construction is similar if S or R is infinite). Let $U < G$, $p: \tilde{X} \to X$ be the corresponding covering and $\tilde{x}_0$ be a basis point of $\tilde{X}$. We choose a spanning tree $\tilde{B}$ in the 1-skeleton of $\tilde{X}$. Then there exists a path with start point $\tilde{x}_0$ to each 0-cell $\tilde{x} \in \tilde{X}$ in $\tilde{B}$, which represents a reduced word in the generators S. This word represents a coset which also represents $\tilde{X}$, and is uniquely determined by the coset. If we omit the last letter in the word then we get a path with the same properties, which goes to the previous 0-cell and whose word represents the respective coset. By induction we see that the choice of the spanning tree $\tilde{B}$ in $\tilde{X}$ gives a set of representatives for the cosets, in which each initial subword also represents a coset. Conversely each system of coset representatives which satisfies the above Schreier condition represents a spanning tree in $\tilde{X}$. We now find a generating system for U in $\tilde{X}$ as follows. If $\tilde{\sigma} \in \tilde{X}^1 \setminus \tilde{B}$ we go from $\tilde{x}_0$ through $\tilde{B}$ to the start point $\tilde{\sigma}$, go over $\tilde{\sigma}$ itself and then through $\tilde{B}$ back to $\tilde{x}_0$. So we get a generating system for U. To get the defining relations we go through the 1-cells of the boundaries of the 2-cells and note in which order and direction we pass these 1-cells. □

2.3.5 Covering space interpretation of residual finiteness

We recall that a group is residually finite if for all $g \in G\setminus\{1\}$ there exists a subgroup $H_g \subset G$ of finite index with $g \notin H_g$. Put differently, $\bigcap_{[G:H]<\infty} H = \{1\}$. Any finitely generated subgroup of the general linear group $\mathrm{GL}(n, R)$ is residually finite for a commutative ring with unit R (Theorem 1.1.18 of Malcev). Hence we see that free groups and surface groups are residually finite. In particular fundamental groups of compact hyperbolic manifolds are residually finite. It is an open question, if there are hyperbolic groups that are not residually finite.

Within the scope of this section we would like to point out a topological characterization of residually finite fundamental groups of CW complexes that can be found in [235].

Remark 2.3.30. Let X be a CW complex and $G = \pi_1(X)$ be its fundamental group. Then G is residually finite if and only if for all compact subsets $K \subset \tilde{X}$ of the universal covering $P: \tilde{X} \to X$ there exists a finite covering $p: Y \to X$ such that $p^{-1}(P(K)) \to Y$ is an embedding.

This theorem may also be applied to check if an immersion can be lifted to an embedding in a finite covering.

2.3.6 Subgroup separability

In the following we present the observations by Aab and Rosenberger in [1].

Definition 2.3.31. A group G is *subgroup separable* (or LERF) if for any finitely generated subgroup H of G and any element $g \in G \setminus H$ a subgroup H^* of finite index in G exists such that

(a) $H < H^*$; and
(b) $g \notin H^*$.

The following four statements are direct consequences of the definition:

1. A group G is subgroup separable if and only if for any finitely generated subgroup H of G and any element $g \in G \setminus H$ there is a finite group K and a homomorphism $\varphi: H \to K$ such that $\varphi(g) \notin K$.
2. If K is a subgroup of the subgroup separable group G then K is also subgroup separable.
3. If the subgroup G of K is subgroup separable and the index of G in K is finite then K is subgroup separable.
4. Let G be a subgroup separable group, H a finitely generated subgroup of G and $g_1, \dots, g_n \in G \setminus H$, $n \in \mathbb{N}$, then there exists a subgroup H_n^* of finite index in G so that $H < H_n^*$ and $g_1, \dots, g_n \notin H_n^*$.

Definition 2.3.32. A group G is *strongly subgroup separable* if the following holds:

(a) If K is a finitely generated subgroup of G and $h, g \in G$ such that $K \cap h\langle g \rangle = \emptyset$ then there exists a subgroup L of finite index in G such that $K < L$ and $L \cap h\langle g \rangle = \emptyset$.
(b) If K is a finitely generated subgroup of G and $g \in G$ such that $K \cap \langle g \rangle = \{1\}$, then there exists an $r \in \mathbb{N}$ such that for any $s \in \mathbb{N}$ there is a normal subgroup Q_s of finite index in G with $Q_s \cap \langle g \rangle = KQ_s \cap \langle g \rangle = \langle g^{rs} \rangle$.

A strongly subgroup separable group G is subgroup separable. To see this apply condition (a) with $g = 1$.

Definition 2.3.33. A group G is *strict subgroup separable* if the following holds: If $x, g_1, \dots, g_n \in G \setminus \{1\}$, where x is an element of infinite order, and $K_1, \dots, K_n$ are finitely generated subgroups of G, $m, n \in \mathbb{N}$, then there exists a $k \in \mathbb{N}$ such that for any $l \in \mathbb{N}$ there is a normal subgroup $N = N(l)$ of finite index in G, such that the following five conditions hold:

(a) $N \cap \langle x \rangle = \langle x^{kl} \rangle$;
(b) If $K_j \cap \langle x \rangle = \{1\}$ then $K_j N \cap \langle x \rangle = \langle x^{kl} \rangle$;

(c) If $K_j \cap \langle x \rangle \neq \{1\}$ then $K_jN \cap \langle x \rangle = K_j \cap \langle x \rangle$;
(d) If $g_i \notin \langle x \rangle$ then $g_i \notin N\langle x \rangle$; and
(e) If $g_i\langle x \rangle \cap K_j = \emptyset$ then $g_i\langle x \rangle \cap K_jN = \emptyset$.

Remark 2.3.34. A strict subgroup separable group is subgroup separable.

Proof. Define $K_1 = H$, $n = 1$. If $g \in G \setminus H$ is of infinite order then define $x = g$, $m = 1$, $g_1 \in G \setminus \{1\}$ and $l \geq 2$ arbitrary; then $\varphi\colon G \to G/N$, $\varphi(g) \notin \varphi(K_1)$ because of (b) and (c). If $g \in G \setminus H$ is of finite order and H has finite index in G, then $H^* = H$ is the group for Definition 2.3.31. If $g \in G \setminus H$ is of finite order and H has infinite index in G, then there exists an element y of infinite order with $\langle y \rangle \cap H = \{1\}$. If $gy^p \in g\langle y \rangle \cap H$ then define $g_1 = g, x = y^{p+1}$; if $gy^p \notin g\langle y \rangle \cap H$ for all $p \in \mathbb{Z}$ then define $g_1 = g, x = y, m = 1$, arbitrary. Then $\varphi\colon G \to G/N$, $\varphi(g) \notin \varphi(K_1)$ because of (e). □

Lemma 2.3.35. *If G is strong subgroup separable then G is strict subgroup separable.*

Lemma 2.3.35 is a straightforward consequence of the proof of the "Lemma of Free Groups" by Brunner, Burns and Solitar [39].

Example 2.3.36. Free groups, see [39], and Fuchsian groups, see [204] and [169], are strong (hence strict) subgroup separable groups.

Theorem 2.3.37. *Let A, B be any strict subgroup separable groups, then any amalgamated free product $G = A *_C B$ with $C \cong \mathbb{Z}$ is subgroup separable.*

To prove this theorem we apply Tretkoff's covering space arguments [259]. Tretkoff gave a covering space argument of the result by Brunner, Burns and Solitar. The proof is given as a sequence of seven lemmas. Only the proof of the sixth lemma is different from Tretkoff's proof because there Tretkoff used the fact that free groups are potent; see the next definition.

Definition 2.3.38. A group G is *potent* with respect to a non-trivial element $g \in G$ provided: if n is any positive integer dividing the order of g, or any positive integer if g has infinite order, then there is a finite quotient Q of G in which g is mapped to an element of order n.

The property of potency for the factors is not essential. Hence we here briefly describe the main covering constructions and give a detailed proof of Tretkoff's sixth lemma whereas we do not give detailed proofs of the other lemmas which we may take over from [259]. We leave the proofs of the other lemmas as exercises. We give here Theorem 2.3.37 as an example of how useful covering space arguments are to prove group theoretic properties. The maps $\varphi\colon C \hookrightarrow A$ and $\psi\colon C \hookrightarrow B$ will be the inclusions which define the amalgamated free product $G = A *_C B$. We may assume that A and B are finitely generated. Let X, Y, U be two-dimensional CW complexes with property EF and with basis points x_0, y_0, u_0 and $\pi_1(X) = A, \pi_1(Y) = B, \pi_1(U) = C$ where we omit the basis point to simplify notations and require the mappings to preserve the basis points. Furthermore, let $\Phi: U \to X$ and $\Psi: U \to Y$ be mappings such that $\varphi = \Phi_*: C = \pi_1(U) \to \pi_1(X) = A$ and $\psi = \Psi_*: C = \pi_1(U) \to \pi_1(Y) = B$ are the induced mappings. Let W be the quotient space of $X \cup Y \cup ([0,1] \times U)$ with identifications: $\varphi(u) = (0,u)$ and $\psi(u) = (1,u)$ for each $u \in U$. Choose $w_0 = (\frac{1}{2}, u_0)$ as the basis point of W. Then $\pi_1(W) = A *_C B$. We will view U as the subspace $\{\frac{1}{2}\} \times U$ of W, as depicted in Figure 2.22, and use the following notations:

- $\bar{E} = [0,1] \times U$, $E = (0,1) \times U$;
- $\bar{E}_A = [0, \frac{1}{2}] \times U$, $\bar{Z}_A = X \cup \bar{E}_A$; and
- $\bar{E}_B = [\frac{1}{2}, 1] \times U$, $\bar{Z}_B = Y \cup \bar{E}_B$.

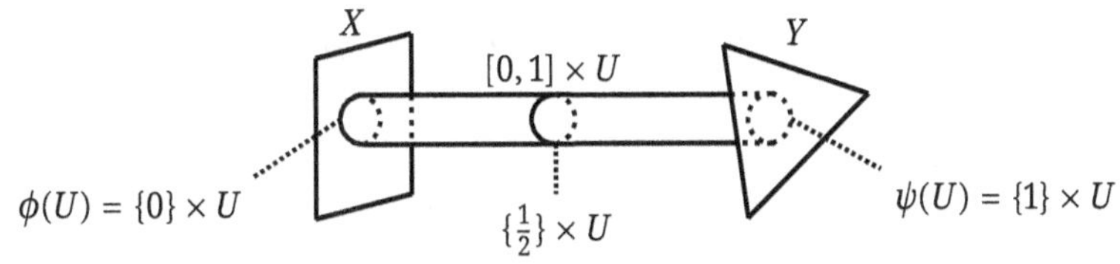

Figure 2.22: Subspaces of W.

The map $p: W(H) \to W$ should be the covering map such that $p_*(\pi_1(W(H))) = H$ for the given subgroup $H < G$. Choose a basis point $\tilde{w}_0 \in p^{-1}(w_0) \subset W(H)$ and let $\tilde{\delta}$ be the lift of the element $g_0 \in G \setminus H$, starting at $\tilde{w}_0$. Then $\tilde{\delta}$ does not end in $\tilde{w}_0$. Recall that we may consider g_0 as a path in W (see Theorem 2.3.26). Divide $W(H)$ into the path components over X, Y and E: $p^{-1}(X) = \dot{\bigcup}_{i \in I} \tilde{X}_i$, $p^{-1}(Y) = \dot{\bigcup}_{j \in J} \tilde{Y}_j$, $p^{-1}(E) = \dot{\bigcup}_{k \in K(i,j)} \tilde{E}_{ijk}$, $i \in I, j \in J$. For $U \subset E$ let $p^{-1}(U) = \dot{\bigcup}_{k \in K(i,j)} \tilde{U}_{ijk}$. Now we present $W(H)$ through a bipartite graph $\Gamma(H)$ with vertices $\{\tilde{x}_i\}$, $i \in I$, and $\{\tilde{y}_j\}$, $j \in J$, and edges $\{\tilde{e}_{ijk}\}$, $k \in K(i,j)$ where $\tilde{e}_{ijk}$ starts at $\iota(\tilde{e}_{ijk}) = \tilde{x}_i$ and ends at $\tau(\tilde{e}_{ijk}) = \tilde{y}_j$. We parametrize $\tilde{e}_{ijk}$ through $\tilde{e}_{ijk}: [0,1] \to \tilde{e}_{ijk}$ and define the midpoint $\tilde{m}_{ijk} = \tilde{e}_{ijk}(\frac{1}{2})$. Define a mapping $f_H: W(H) \to \Gamma(H)$ as follows: $f_H(\tilde{X}_i) = \{\tilde{x}_i\}$, $f_H(\tilde{Y}_j) = \{\tilde{y}_j\}$ and $f_H(\tilde{E}_{ijk}) = \{\tilde{e}_{ijk}\}$, where $f_H(\tau, \tilde{u}) = \varepsilon_{ijk}(\tau)$, $0 < \tau < 1$, and $\tilde{u} \in \tilde{U}_{ijk}$. The map f_H is continuous and surjective, and we get the following.

Lemma 2.3.39. *The group $\pi_1(\Gamma(H))$ is free of finite rank.*

For the next lemma we need a mild example of a graph of groups. We give a short overview of graph of groups in Subsection 2.3.7. Here we mention it for the brief description of Tretkoff's second lemma. We now define a graph of groups $(\Gamma(H), \mathcal{H})$ as follows: $\mathcal{H}(\tilde{x}_i) = \pi_1(\tilde{X}_i)$, $\mathcal{H}(\tilde{y}_j) = \pi_1(\tilde{Y}_j)$, $\mathcal{H}(\tilde{e}_{ijk}) = \pi_1(\tilde{E}_{ijk}) = \pi_1(\tilde{U}_{ijk})$, and as homomorphisms $\mathcal{H}(\tilde{e}_{ijk}) \hookrightarrow \mathcal{H}(\tilde{x}_i)$ and $\mathcal{H}(\tilde{e}_{ijk}) \hookrightarrow \mathcal{H}(\tilde{y}_j)$. We take the induced mappings of the inclusions $\tilde{U}_{ijk} \hookrightarrow \tilde{X}_i$ and $\tilde{U}_{ijk} \hookrightarrow \tilde{Y}_j$. With this definition the fundamental group of $(\Gamma(H), \mathcal{H}) = \pi_1(\Gamma(H), \mathcal{H})$ is isomorphic to H. The group H is finitely generated, therefore there exist a finite number of subspaces $\tilde{X}_1, \ldots, \tilde{X}_m, \tilde{Y}_1, \ldots, \tilde{Y}_n$ and $\tilde{E}_{ijk}$ such that the lifts $\tilde{S}_1, \ldots, \tilde{S}_t$ of the generators of H and $\tilde{\delta}$ are in the union of these subspaces. We define $\Delta(H) = f_H(\tilde{S})$; $(\Delta(H), \mathcal{H})$ is a subgraph of $(\Gamma(H), \mathcal{H})$.

Lemma 2.3.40. *The group* $\pi_1(\Delta(H), \mathcal{H})$ *is isomorphic to* H*, that is,* $\pi_1(\Delta(H), \mathcal{H}) \cong H$.

Lemma 2.3.41. *Each vertex group* $\mathcal{H}(\tilde{v})$ *of* $(\Delta(H), \mathcal{H})$ *is finitely generated.*

To shorten the description of Tretkoff's argument a bit we consider $\mathcal{H}(\tilde{v})$ and $\mathcal{H}(\tilde{e})$ as subgroups of G and elements of C as elements of A and B. Let $\tilde{v}$ be any vertex of $\Delta(H)$, $\tilde{e}_1, \dots, \tilde{e}_r$ be the edges of $\Delta(H)$ which have $\tilde{v}$ as a vertex and $\tilde{m}_1, \dots, \tilde{m}_r$ be the midpoints of the edges $\tilde{e}_1, \dots, \tilde{e}_r$. Let $\tilde{Z} = \tilde{Z}(\tilde{v}) \subset W(H)$ be the inverse image under f_H of $\tilde{v}$ and the closed halves of the edges $\tilde{e}_1, \dots, \tilde{e}_r$ which contain $\tilde{v}$. Let $Z = Z(v) = p(\tilde{Z}) \subset W$, and we call $\tilde{U}_j = f_H^{-1}(\tilde{m}_j) \subset \tilde{Z}$ the *distinguished faces* of $\tilde{Z}$. We choose basis points $\tilde{u}_j \in \tilde{U}_j \cap p^{-1}(w_0)$, $j = 1, \dots, r$, and let $\tilde{u}_1$ be the basis point of $\tilde{Z}$, and let $c \in C$ be a generator of C. Then there are paths $\tilde{a}_j$ in $\tilde{Z}$ which start at $\tilde{u}_1$ and end at $\tilde{u}_j$ with $a_j = [p(\tilde{a}_j)] \in \mathcal{G}(v)$ where $\mathcal{G}(v) = \pi_1(X)$ or $\pi_1(Y)$, for the homotopy classes and $a_1 = [p(\tilde{a}_1)] = 1$ such that the following holds.

Lemma 2.3.42. *There are* $\mu_1, \dots, \mu_r$ *with* $\mu_r \in \mathbb{N}$ *or* ∞ *such that:*
1. *The midpoint* $\tilde{m}_j$ *is the image under* f_H *of the end point* $\tilde{a}(1)$*;*
2. $a_j^{-1} c^{\mu_j} a_j \in \mathcal{H}(\tilde{v})$ *if* $\mu_j < \infty$*;*
3. $a_j^{-1} c^{\nu} a_j \notin \mathcal{H}(\tilde{v})$ *if* $0 < |\nu| < \mu_j$*; and*
4. $a_j^{-1} c^{\nu} a_k \notin \mathcal{H}(\tilde{v})$ *for all integers* ν *when* $j \neq k$.

Lemma 2.3.43. *There is a subgroup* $\mathcal{H}^*(\tilde{v})$ *of finite index in* $\mathcal{G}(v)$ *where* $\mathcal{G}(v) = \pi_1(X)$ *or* $\pi_1(Y)$*, and there is a sequence of positive integers* $\nu_1, \dots, \nu_r$ *such that:*
1. $\mathcal{H}(\tilde{v}) < \mathcal{H}^*(\tilde{v}) < \mathcal{G}(v)$*;*
2. $a_j^{-1} c^{\nu_j} a_j \in \mathcal{H}^*(\tilde{v})$*;*
3. $a_j^{-1} c^{\nu} a_j \notin \mathcal{H}^*(\tilde{v})$, $0 < |\nu| < \nu_j$*; and*
4. $a_j^{-1} c^{\nu} a_k \notin \mathcal{H}^*(\tilde{v})$, $j \neq k$, $0 \leq \nu \leq \max(\nu_j, \nu_k)$.

Now we discuss the lemma which is different from Tretkoff's original lemma. Let $p^*: \tilde{Z}^* \to Z$ be a covering such that $p^*_*(\pi_1(\tilde{Z}^*)) = \mathcal{H}^*(\tilde{v})$. Using Lemma 2.3.43 we get a corresponding sequence $\tilde{U}_1^*, \dots, \tilde{U}_r^*$ of path components of $(p^*)^{-1}(U) \cap \tilde{Z}^*$ to the distinguished faces $\tilde{U}_1, \dots, \tilde{U}_r$. The path components of $((p^*)^{-1}(U) \cap \tilde{Z}^*) \setminus (\tilde{U}_1^* \cup \cdots \cup \tilde{U}_r^*)$ will be called *free faces*. Furthermore, there exists a covering $\dot{p}: \tilde{Z} \to \tilde{Z}^*$ such that the diagram

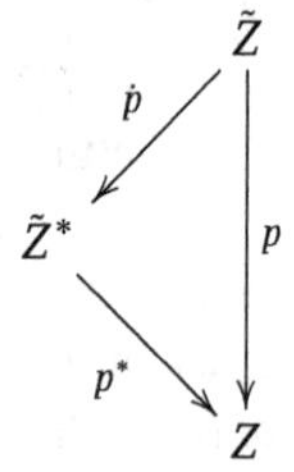

of covering maps is commutative. For short, we denote subsets of $\tilde{Z}(\tilde{x}_i)$ and $\tilde{Z}(\tilde{y}_j)$ which are joined together by $\tilde{e}_{ijk}$, by $\tilde{U}_{ijk}$ and $\tilde{U}_{j,k}$, respectively. Analogously we get $\tilde{U}^*_{i,k}$ and $\tilde{U}^*_{j,k}$. We define $\tilde{U}^*_{ijk} = [0,1] \times \tilde{U}^*_{i,k} = [0,1] \times \tilde{U}^*_{j,k}$ and $\tilde{S}^*$ to be the union of all $\tilde{Z}^*(\tilde{v})$, $\tilde{v}$ a vertex of $\Delta(H)$ and all $\tilde{U}^*_{ijk}$.

With the adequate identifications we get the commutative diagram

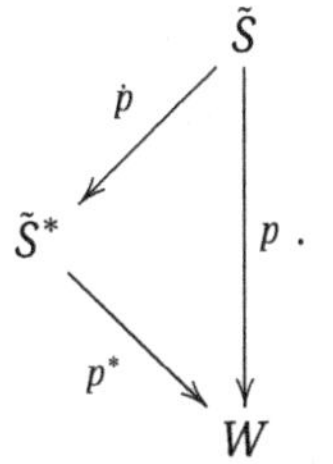

We have $H = p_*(\pi_1(\tilde{S})) \subset (p^*)_*(\pi_1(\tilde{S}^*))$, and $(p^*)^{-1}(w)$, $w \in W$, is finite. If $p^*: \tilde{S}^* \to W$ would be a covering then the subgroup $H^* = (p^*)_*(\pi_1(\tilde{S}^*))$ would be a finite index subgroup of G and would solve the problem. But p^* is a covering if and only if each $\tilde{Z}^*$ contains no free face. So it remains to consider the case that $\tilde{Z}^*$ contains a free face. Let $\tilde{V}_1^*$ be a free face, for instance $\tilde{V}^* = \tilde{Z}^*(\tilde{x}_i)$. Note that $p^*: \tilde{Z}^*(\tilde{x}_i) \to Z(x_i) = A$ is a finite-sheeted covering, therefore $\tilde{V}_1^* \subset (p^*)^{-1}(\varphi(U))$ is a finite-sheeted covering space of U. The number $\nu \in \mathbb{N}$ of sheets is the index of $(p^*)_*(\pi_1(\tilde{V}_1^*))$ in $\pi_1(U) = C$. Because B is subgroup separable there exists a subgroup K_1 of B which contains the subgroup $\langle c^\nu \rangle$ such that $c, c^2, \ldots, c^{\nu-1} \notin K_1$. Then there exists a covering $f: \tilde{Z}^*(K_1) \to Z(K_1)$ such that $f_*(\pi_1(\tilde{Z}^*(K_1))) = K_1$. Because $c, c^2, \ldots, c^{\nu-1} \notin K_1$ and $c^\nu \in K_1$ each path component $\tilde{U}^*$ over $U \subset Z(K_1)$ is a ν-sheeted covering of U and there exist $\mu_1 = \frac{[B:K_1]}{\nu}$ different path components over U. Hence $\pi_1(U) = C \cong \mathbb{Z}$ is an Abelian group and therefore any covering of U is regular, and so all ν-sheeted covering spaces are homeomorphic. Take μ_1 copies of $\tilde{S}^*$ and identify each free face corresponding to $\tilde{V}_1^*$ with a path component over U in $\tilde{Z}^*(K_1)$, denote the resulting space by $\tilde{S}_1$ and combine the projections f and p^* to $p_1^*: S_1^* \to W$. If there is a second free face $\tilde{V}_2^*$ of $\tilde{S}$, then we have μ_1 copies of $\tilde{V}_2$ in $\tilde{S}_1^*$. Repeat the above construction for V_2^* and take μ_1 copies of $\tilde{Z}^*(K_2)$ and $\mu_1\mu_2$ copies of $\tilde{S}_1^*$. Identifying as above yields a space $\tilde{S}_2^*$ and a projection $p_2^*: \tilde{S}_2^* \to W$. $\tilde{S}^*$ contains only a finite number of free faces, so we get a space $W(H^*) = \tilde{S}_n^*$ and a projection $p^* = p_n^*: W(H^*) \to W$, which is now a finite-sheeted covering of W (we only identified a finite number of copies of $\tilde{S}^*$). Defining $H^* = (p^*)_*(\pi_1(W(H^*)))$ gives the following lemma.

Lemma 2.3.44. *There is a subgroup H^* of finite index in G that contains H.*

Finally, from the construction, we get the following.

Lemma 2.3.45. *We have $g_0 \notin H^*$.*

This finishes the proof of Theorem 2.3.37.

Remark 2.3.46. One could expect that Theorem 2.3.37 holds also if A and B are just subgroup separable. However, this is not the case. Let F_n be the free group of rank n. The direct product $F_n \oplus \mathbb{Z}$ is subgroup separable but $(F_2 \oplus \mathbb{Z}) *_{\mathbb{Z}} (F_3 \oplus \mathbb{Z})$ is not subgroup separable; see [169]. We mention that there are quite a lot activities in research and the literature concerning the subgroup separability of groups. For instance, subgroup separability of HNN extensions are considered in [205, 160] and [231]. As a further example we mention the recent result by H. Wilton [273], who showed that finitely generated fully residually free groups are subgroup separable.

2.3.7 Graphs of groups

We finish this section with a brief overview on graphs of groups.

Definition 2.3.47. A *graph of groups* is a pair $(\mathcal{G}, X)$ where X is a connected graph and $\mathcal{G}$ consists of a family $(G_v \mid v \in V(X))$ of *vertex groups* and a family $(G_\sigma \mid \sigma \in E(X))$ of *edge groups* where the following holds:
1. For any edge σ we have $G_{\sigma^{-1}} = G_\sigma$.
2. For any edge σ there are homomorphisms $\kappa_\sigma\colon G_\kappa \to G_{s(\sigma)}$ and $\lambda_\sigma\colon G_\sigma \to G_{t(\sigma)}$ such that $\lambda_\sigma = \kappa_{\sigma^{-1}}$. Here $s(\sigma)$ is the start point and $t(\sigma)$ is the end point of σ. A *path* in $(\mathcal{G}, X)$ is a sequence $(g_0, \sigma_1, g_1, \dots, \sigma_r, g_r)$ where $g_i \in G_{v_i}$ and $(v_0, \sigma_1, v_1, \dots, \sigma_r, v_r)$ is a path in X.
3. The equivalence relation $\simeq$ of homotopy equivalence of paths in $(\mathcal{G}, X)$ is induced by elementary homotopy equivalences $(\sigma, \lambda_\sigma(h), \sigma^{-1}, (\kappa_\sigma(h))^{-1}) \simeq (1)$ where $1 \in G_{s(\sigma)}$ and $(g, \sigma, 1, \sigma^{-1}, g') \simeq (gg')$. If v is a vertex of X then the homotopy classes of closed paths at v form a group under the operation of concatenation of representatives—with the natural rule $(\dots, g)(g', \dots) = (\dots, gg', \dots)$. This group is denoted by $\pi_1(\mathcal{G}, X, v)$ and called the *fundamental group* of $(\mathcal{G}, X)$.

Obviously, if all the vertex groups are trivial, then $\pi_1(\mathcal{G}, X, v)$ is isomorphic to the usual fundamental group $\pi_1(X, v)$, and the map which deletes the group entries in any path in $(\mathcal{G}, X)$ is always an epimorphism from $\pi_1(\mathcal{G}, X, v)$ to $\pi_1(X, v)$.

Theorem 2.3.48. *Let $(\mathcal{G}, X)$ be a graph of groups and v_0 a vertex of X. Let T be a spanning tree of X and let $E^+(X)$ be an orientation of X, that is, a subset of $E(X)$ containing exactly one member of each pair of inverse edges of X. Then $\pi_1(\mathcal{G}, X, v_0)$ is the group obtained from the free product $P = F * (*_{v \in V(X)} G_v)$ where F is the free group on a set $\{t_\sigma \mid \sigma \in E^+(X)\}$ in one-to-one-correspondence with $E^+(X)$ by adding the relations $t_\sigma \lambda_\sigma(h) t_\sigma^{-1} = \kappa_\sigma(h)$ for $\sigma \in E^+(X)$ and $h \in G_\sigma$, $t_\sigma = 1$ for $\sigma \in E(T) \cap E^+(X)$.*

Proof. Write $\pi_1(\mathcal{G}, X, T)$ for the quotient of the free product P by the given relations. Then there is an obvious map $p\colon \pi_1(\mathcal{G}, X, v_0) \to \pi_1(\mathcal{G}, X, T)$ given by $(g_0, \sigma_1^{\varepsilon_1}, g_1, \dots, \sigma_n^{\varepsilon_n}, g_n) \mapsto g_0 t_{\sigma_1}^{\varepsilon_1} g_1 \cdots t_{\sigma_n}^{\varepsilon_n} g_n$, where $\varepsilon_i = \pm 1$ and $\sigma_i \in E^+(X)$, $1 \le i \le n$. Conversely consider the map $f: P \to \pi_1(\mathcal{G}, X, v_0)$ given by $x \mapsto (1, \sigma_1, 1, \ldots, \sigma_r, x, \sigma_r^{-1}, 1, \ldots, \sigma_1^{-1}, 1)$, $t_\sigma \mapsto (1, \sigma_1, 1, \ldots, \sigma_r, 1, \sigma, 1, \tau_1, 1, \ldots, \tau_s, 1)$ where in the first case $x \in G_v$ and $(\sigma_1, \ldots, \sigma_r)$ is the unique reduced path in T from the basis point v_0 to v, and in the second case the edge sequences $(\sigma_1, \ldots, \sigma_r)$ and $(\tau_1, \ldots, \tau_s)$ define the reduced paths from v_0 to $s(\sigma)$ and from $t(\sigma)$ back to v_0. Since this map is compatible with the relations factored out, f induced a homomorphism $\tilde{f}: \pi_1(\mathcal{G}, X, T) \to \pi_1(\mathcal{G}, X, v_0)$, which is inverse to the map p. □

Corollary 2.3.49. *The natural map $\kappa_v: G_v \to \pi_1(\mathcal{G}, X, T)$ is an embedding.*

Examples 2.3.50. 1. Let X consist of the graph $v \overset{\sigma}{—} w$ and $\mathcal{G} = \{(G_v, G_w), G_\sigma\}$. The only possible choice for T is $T = X$, and so $\pi_1(\mathcal{G}, X, v)$ is the free product $G_v * G_w * F(t_\sigma)$ modulo the relations $t_\sigma = 1$, $t_\sigma \kappa_\sigma(h) t_\sigma^{-1} = \lambda_\sigma(h)$ for $h \in G_\sigma$. Then $\pi_1(\mathcal{G}, X, v)$ is (isomorphic to) the free product of the two vertex groups amalgamating the two copies of the edge group.
2. Let X consist of a single vertex and a single loop at that vertex as in Figure 2.23. Then $\pi_1(\mathcal{G}, X, v)$ is just the HNN extension with the vertex group as basis group and the two copies of the edge group conjugate by the generator corresponding to the loop.

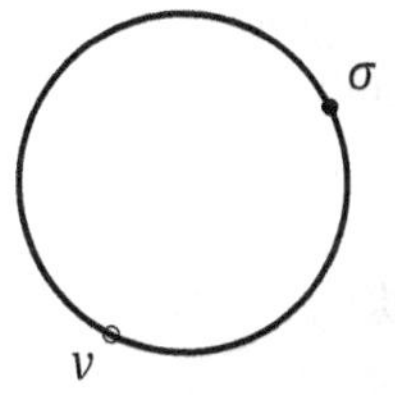

Figure 2.23: Graph with a single vertex and a single loop.

In the following we want to say that a group H *acts without inversion* on a graph X if for all $g \in G, \sigma \in E(X), g(\sigma) \neq \sigma^{-1}$. Our last theorem in this subsection is obvious from our discussion of covering maps.

Theorem 2.3.51. *Let the group G act without inversion on a tree X, and let $Y = X/G$ be the quotient graph. For any vertex v and edge σ of Y let $\tilde{v}$ and $\tilde{\sigma}$ denote a vertex and an edge of X lying over v and σ, respectively. Then G is (isomorphic to) the fundamental group of the graph of groups $(\mathcal{G}, Y)$ whose vertex groups and edge groups, respectively, are stabilizers of the form $G_v = \mathrm{Stab}(\tilde{v})$ and $G_\sigma = \mathrm{Stab}(\tilde{\sigma})$.*

This has the following consequence. We say that G *acts freely* on a graph X if every vertex of X has trivial stabilizer.

Corollary 2.3.52. *A group is free if and only if it acts freely without inversion on a tree.*

If a group acts freely on a tree then so does any subgroup. Hence, we have the following corollary.

Corollary 2.3.53. *A subgroup of a free group is free.*

3 Hyperbolic groups

3.1 Introduction: the theory of hyperbolic groups

A main principle in the study of geometric objects is the analysis of their isometry groups. Gromov [118] to a certain extent followed the reversed principle making groups geometric objects and studying the algebraic properties of the groups by means of geometric properties of their Cayley graphs. In particular, he introduced hyperbolic groups, that is, groups whose Cayley graphs have geometric properties of hyperbolic geometry. This finally led to *geometric group theory*. Gromov's initial idea was two-fold. Firstly, he wanted to generalize the theory of small cancellation groups. Secondly, he wanted to extend the techniques applied to the study of the fundamental groups of hyperbolic manifolds—such as surface groups—using geometric properties of the manifolds.

We recall the following definitions for groups.

Definition 3.1.1. Let Γ be a group. Then Γ is said to be *of finite type* if Γ is finitely generated and Γ is called *finitely presentable* if Γ is of finite type and represented by finitely many defining relations.

In the following we need a special kind of generating systems for groups of finite type. Let Γ be a group of finite type and $S \subset \Gamma$ be a generating system for Γ. Then S is called a *valid* generating system if it has the following two properties:

(a) Let 1 denote the neutral element of Γ, then $1 \notin S$.

(b) The set S is a *symmetric generating system*, that is, if $\gamma \in S$ then also $\gamma^{-1} \in S$.

In the following, the pair (Γ, S) denotes a group Γ together with a valid generating system S. Given such a pair we define a metric on Γ with respect to S in the following way.

Given a pair (Γ, S) as above. Then define $l_S\colon \Gamma \to [0,\infty)$ as follows: If $\gamma \in \Gamma$, then $l_S(\gamma) = 1$ if $\gamma = 1$ and if $\gamma \neq 1$ then let $l_S(\gamma)$ be the minimal length of a word that is completely constructed of elements from S that represent γ. This length is also called *S-length*. We now define the desired metric $d_S\colon \Gamma \times \Gamma \to [0,\infty)$ via

$$d_S(\gamma_1,\gamma_2) = l_S(\gamma_1^{-1}\gamma_2)$$

and check that d_S is indeed a metric:

1. The equivalence $l_S(\gamma) = 0$ if and only if $\gamma = 1$ implies the equivalence $d_S(\gamma_1,\gamma_2) = 0$ if and only $\gamma_1 = \gamma_2$.
2. We have $d_S(\gamma_1,\gamma_2) = l_S(\gamma_1^{-1}\gamma_2) = l_S(\gamma_2^{-1}\gamma_1) = d_S(\gamma_2,\gamma_1)$, because S is symmetric.
3. We have $d_S(\gamma_1,\gamma_2) \leq d_S(\gamma_1,\beta) + d_S(\beta,\gamma_2)$ for all $\gamma_1,\gamma_2,\beta \in \Gamma$. This is because of $\gamma_1^{-1}\gamma_2 = \gamma_1^{-1}\beta\beta^{-1}\gamma_2$.

Remarks 3.1.2. 1. The metric structure on (Γ, S) depends on the choice of S. Say $\Gamma = \mathbb{Z}$ and $S = \{\pm 1\}$, then $d_S(0,1) = 1$, and if $S' = \{\pm 2, \pm 3\}$, then

$$d_{S'}(0,1) = 2.$$

https://doi.org/10.1515/9783111340043-003

2. The metric structure on (Γ, S) is induced by the natural metric structure of the Cayley graph with respect to (Γ, S): The vertices are elements of Γ, and two vertices γ_1 and γ_2 are connected by an edge if and only if there exists a $\sigma \in S$ with $\gamma_1\sigma = \gamma_2$. If we parametrize in such a way that any edge of the Cayley graph has length 1, then the metric of (Γ, S) is induced from that of the Cayley graph of (Γ, S). Here we extend the metric for the Cayley graph in the usual way for all pairs of points of edges by transforming any edge to an interval of length 1. In this manner the Cayley graph becomes a geodesic metric space. We always consider the Cayley graph in this way, which should not lead to misunderstandings. Any closed path represents a relation. If we insert a 2-cell for any closed path in the Cayley graph then we obtain a simply connected 2-dimensional complex, the *Cayley complex*.

The construction of the Cayley graph depends on the choice of S as well as on the metric on (Γ, S). We would like to have an equivalence relation that permits to connect the different metric spaces for Γ if we alter S.

Definition 3.1.3. Let (X, d) and (X', d') be metric spaces. Then (X, d) and (X', d') are *quasi isometric*, if there are functions $f: X \to X'$ and $g: X' \to X$ together with constants $\lambda > 0$ and $C \geq 0$, such that
(a) $d'(f(x), f(y)) \leq \lambda d(x, y) + C$ for all $x, y \in X$;
(b) $d(g(x'), g(y')) \leq \lambda d'(x', y') + C$ for all $x', y' \in X'$;
(c) $d(g(f(x)), x) \leq C$ for all $x \in X$; and
(d) $d'(f(g(x')), x') \leq C$ for all $x' \in X'$.

We interpret this situation from high distance, that is, C can hardly be distinguished from 0:
- Conditions (a) and (b) suggest that f and g are Lipschitz.
- Conditions (c) and (d) suggest that f and g are inverse isometries.

Theorem 3.1.4. *Quasi isometry is an equivalence relation in the class of metric spaces.*

Proof. Of course quasi isometry is reflexive and symmetric. We show transitivity. Let (X, d) and (X', d') as well as (X', d') and (X'', d'') be quasi isometric. Thus we have functions $X \underset{g}{\overset{f}{\rightleftarrows}} X'$, $X' \underset{g'}{\overset{f'}{\rightleftarrows}} X''$ and constants λ, C and λ', C', respectively, such that the conditions (a)–(d) are satisfied. We look for functions $X \underset{g''}{\overset{f''}{\rightleftarrows}} X''$ and constants λ'', C'', such that conditions (a)–(d) are satisfied again. Set $f'' = f' \circ f$ and $g'' = g \circ g'$, $\lambda'' = \lambda\lambda'$ and $C'' = 2C + 2C' + \lambda' C + \lambda C'$. We check the conditions step by step:
(a) Let $x, y \in X$. Then

$$\begin{aligned} d''(f''(x), f''(y)) &= d''(f'(f(x)), f'(f(y))) \\ &\leq \lambda' d'(f(x), f(y)) + C' \end{aligned}$$

$$\leq \lambda'[\lambda d(x,y) + C] + C' = \lambda'\lambda d(x,y) + C' + \lambda' C$$
$$\leq \lambda'' d(x,y) + C''.$$

(b) This is analogous.
(c) Let $x \in X$. Then (according to our assumption)

$$d'(g' \circ f'(f(x)), f(x)) \leq C'$$

and hence because of (b)

$$d(g(g' \circ f'(f(x))), g(f(x))) \leq \lambda C' + C,$$

which gives

$$d(g'' \circ f''(x), x) \leq d(g'' \circ f''(x), g(f(x))) + d(g(f(x)), x)$$
$$\leq (\lambda C' + C) + C = \lambda C' + 2C \leq C''.$$

(d) This is analogous. □

Theorem 3.1.5. *Let Γ be a group of finite type with finite valid generating systems S and S', that is, S and S' are symmetric and do not contain 1. Then the metric spaces (Γ, S) and (Γ, S') are quasi isometric.*

Proof. We look for suitable f, g, λ and C. Take $f = \mathrm{id}_{(\Gamma,S)}$, $g = \mathrm{id}_{(\Gamma,S')}$, $C = 0$ and $\lambda = \max(\{l_{s'}(\gamma) \mid \gamma \in S\} \cup \{l_s(\gamma') \mid \gamma' \in S'\})$.

We verify condition (a): Let $x, y \in (\Gamma, S)$. Then

$$d_{S'}(f(x), f(y)) = l_{S'}(f(x)^{-1}f(y)) = l_{S'}(x^{-1}y).$$

Our definition of λ permits $l_{S'}(x^{-1}y) \leq \lambda l_S(x^{-1}y)$ because, if we write $x^{-1}y$ as a product (of elements of S) with length k, then we can surely write $x^{-1}y$ as a product (of elements of S') of length $\leq \lambda k$. Hence

$$d_{S'}(f(x), f(y)) \leq \lambda l_S(x^{-1}y) = \lambda d_S(x,y) + C.$$

The proof of (b) is analogous and that of (c) and (d) is obvious because f and g are inverses for each other. □

We observe: The quasi isometry class of the metric spaces for (Γ, S), S finite, is an invariant of the group Γ of finite type and does not depend on the (finite) generating set S.

We ask: *Is this invariant suitable in order to study group theoretical properties of Γ and to what extent does quasi isometry preserve group theoretic properties?*

Remark 3.1.6. Two finite groups Γ_1, Γ_2 are quasi isometric (more precisely, the metric spaces for (Γ_1, S_1) and (Γ_2, S_2), S_1, S_2 finite, are quasi isometric for C large enough).

In general we call two groups Γ_1 and Γ_2 of finite type *quasi isometric*, if the metric spaces for (Γ_1, S_1), S_1 a finite generating set for Γ_1, and (Γ_2, S_2), S_2 a finite generating set for Γ_2, are quasi isometric. We can generalize the above observation.

Definition 3.1.7. Two groups Γ_1 and Γ_2 are called *commensurable* if there exist subgroups $\Gamma_i' < \Gamma_i$, $i = 1, 2$, of finite index with $\Gamma_1' \cong \Gamma_2'$.

We note that two finite groups are commensurable (as, say, $\{1\}$ is of finite index in both of them).

Remark 3.1.8. More generally we can show that two commensurable groups of finite type are quasi isometric. The other direction is not true. There are examples of two groups of finite type that are quasi isometric but not commensurable. There are interesting interconnections: Say Γ is a group of finite type that is quasi isometric to $\mathbb{Z}^n$, $n \in \mathbb{N}$, then Γ must contain a subgroup Γ' of finite index with $\Gamma' \cong \mathbb{Z}^n$. Or let Γ be of finite type and quasi isometric to a free group, then Γ has a free subgroup of finite index.

Aiming at the notion of hyperbolic groups we will first deal with the concept of a hyperbolic metric space. For this we will fix the following notation: If (X, d) is a metric space, then we usually write $|x - y|$ instead of $d(x, y)$; $x, y \in X$.

Definition 3.1.9. 1. Let $x_0, x_1 \in X$ with $a = |x_1 - x_0|$. A *geodesic segment* in X starting at x_0 and ending in x_1 is an isometry $g: [0, a] \to X$ with $g(0) = x_0$ and $g(a) = x_1$. We say that X is a *geodesic space* if for all $x_0, x_1 \in X$ there is a geodesic segment in X starting at x_0 and ending at x_1.
2. A *geodesic triangle* in X with $x, y, z \in X$ as vertices is the union of three geodesic segments with (pairwise) x, y and z as end points.

Note that the definition explicitly allows degenerated triangles, for instance take $y = z$ and the geodesic segments from x to y and x to z are different. A first example of a geodesic space is the Cayley graph for a group of finite type with finite generating system S. If the Cayley graph is not a tree, then it contains a circle (or embedded loop). Hence there is more than one geodesic segment allowed between the same pair of points.

We fix the following notation: Let $x_0, x_1 \in X$ for a geodesic space X. Although several geodesic segments in X with end points x_0 and start points x_1 are allowed, we denote by $[x_0, x_1]$ a given geodesic segment with x_0 and x_1 as start and end points, respectively.

Definition 3.1.10. Let $\delta \geq 0$. We say that a geodesic space X satisfies the *Rips condition* for the constant δ if for every geodesic triangle $[x, y] \cup [y, z] \cup [z, x]$ in X and for every $u \in [x, y]$ the following holds: $d(u, [y, z] \cup [z, x]) \leq \delta$; see Figure 3.1.

We call a geodesic space X *hyperbolic* if it satisfies the Rips condition for a constant $\delta \geq 0$.

Theorem 3.1.11. *Let X_1 and X_2 be geodesic spaces that are quasi isometric. If X_1 is hyperbolic then also X_2 is hyperbolic.*

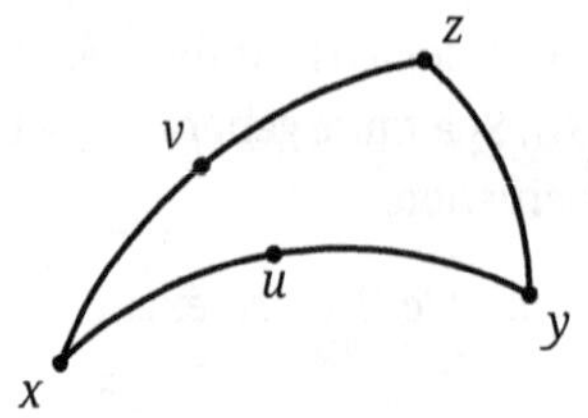

Figure 3.1: Geodesic triangle.

We give the proof in Theorem 3.7.15. Hence quasi isometries respect hyperbolicity.

Definition 3.1.12. Let Γ be a group of finite type. The group Γ is called *hyperbolic* if there is a finite generating system S such that the metric space for (Γ, S)—or the Cayley graph for (Γ, S)—is a hyperbolic space.

According to Theorem 3.1.11 the definition of a hyperbolic group is independent of the choice of S.

Remark 3.1.13. Although the definition of a hyperbolic group is inspired by the definition of hyperbolic manifolds, it goes beyond this geometric motivation just by its algebraic structure. The fundamental group of compact hyperbolic manifolds are hyperbolic groups. These fundamental groups are torsion-free—the universal cover is contractible. On the other hand finite groups are always hyperbolic—just choose as the Rips constant δ the maximal distance in the Cayley graph. Hence hyperbolic groups are not necessarily torsion-free. However, hyperbolic groups have at most a finite number of conjugacy classes of torsion elements (cf. Subsection 3.4.2). The theory of hyperbolic groups is suitable for Burnside problems: Determine the infinite torsion groups of finite type, see [178].

3.2 Hyperbolic and metric spaces

3.2.1 The Gromov product in hyperbolic metric spaces and $\mathbb{R}$-trees

We begin with an equivalent definition of a hyperbolic space. Let again (X, d) be a metric space. As before we write $|x - y|$ instead of $d(x, y)$.

Definition 3.2.1. Let $x \in X$. The *Gromov product* of two points $y, z \in X$ with basis x is

$$(y \mid z)_x := \frac{1}{2}(|y - x| + |z - x| - |z - y|).$$

If no misunderstandings occur, that is, if x is known and fixed, we write $(y|z)$ instead of $(y \mid z)_x$, in particular as an abbreviation in technical proofs.

Remark 3.2.2. We have $0 \leq (y \mid z)_x \leq \min\{|y - x|, |z - x|\}$.

Proof. The triangle inequality yields $|y - z| \leq |y - x| + |x - z|$. Thus

$$(y \mid z)_x = \frac{1}{2}(|y - x| + |x - z| - |y - z|) \geq 0.$$

By symmetry, it suffices for the upper bound to show that $(y \mid z)_x \leq |y - x|$. Suppose $(y \mid z)_x > |y - x|$. Then we have $\frac{1}{2}(|y - x| + |x - z| - |y - z|) > |y - x|$ and hence $|y - x| + |x - z| - |y - z| > 2|y - x|$. But then

$$0 > |y - x| + |y - z| - |x - z| = 2(x \mid z)_y,$$

which is a contradiction. Hence we have $(y \mid z)_x \leq |y - x|$. □

Remark 3.2.3 (Geometric interpretation of the Gromov product). We consider a metric tree, that is, a connected graph without circles where every edge is equipped with a metric isometric to $[0, \bar{a}] \subset \mathbb{R}$, where $\bar{a}$ is the length of the given edge. Let T be such a tree and $x, y, z \in T$. In the non-degenerate case we have the situation depicted in Figure 3.2.

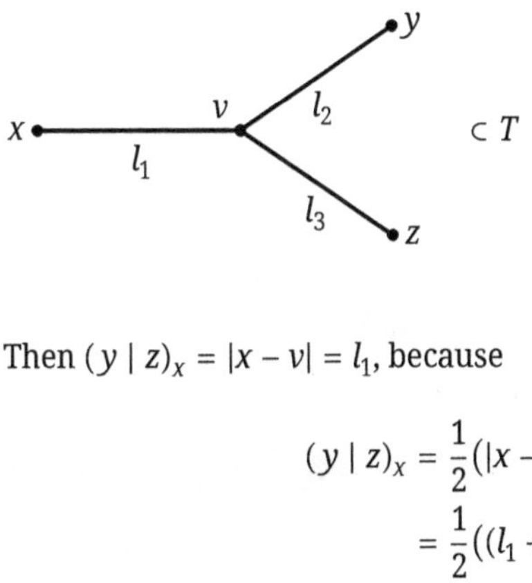

Figure 3.2: Metric tree.

Then $(y \mid z)_x = |x - v| = l_1$, because

$$\begin{aligned}(y \mid z)_x &= \frac{1}{2}(|x - y| + |x - z| - |y - z|) \\ &= \frac{1}{2}((l_1 + l_2) + (l_1 + l_3) - (l_2 + l_3)) = l_1.\end{aligned}$$

Definition 3.2.4. Let $\delta \in [0, \infty)$. We say that X is *δ-hyperbolic* if

$$(x \mid z)_w \geq \min\{(x \mid y)_w, (y \mid z)_w\} - \delta$$

for all $w, x, y, z \in X$. If the value of δ is known or of no importance we simply say that X is hyperbolic.

Remark 3.2.5. We claim that X is δ-hyperbolic if and only if for all $w, x, y, z \in X$ we have

$$|x - z| + |y - w| \leq \max\{|x - y| + |z - w|, |x - w| + |z - y|\} + 2\delta.$$

Proof. If X is δ-hyperbolic and $w, x, y, z \in X$, then

$$(x \mid z)_w \geq \min\{(x \mid y)_w, (y \mid z)_w\} - \delta.$$

Suppose $(x \mid z)_w \geq (x \mid y)_w - \delta$. Then we have

$$\begin{aligned} &\frac{1}{2}(|x-w| + |z-w| - |x-z|) \geq \frac{1}{2}(|x-w| + |y-w| - |x-y|) - \delta \\ \Leftrightarrow \quad & |z-w| - |x-z| + 2\delta \geq |y-w| - |x-y| \\ \Leftrightarrow \quad & |x-z| + |y-w| \leq |x-y| + |z-w| + 2\delta. \end{aligned}$$

The case $(x \mid z)_w \geq (y \mid z)_w - \delta$ follows analogously and we finally have

$$|x-z| + |y-w| \leq \max\{|x-y| + |z-w|, |x-w| + |z-y|\} + 2\delta.$$

For the other direction we reverse our argumentation. □

Remark 3.2.6 (Geometric interpretation of the notion δ-hyperbolic). We consider the notion in terms of geodesic quadrilaterals with vertices w, x, y, z as depicted in Figure 3.3.

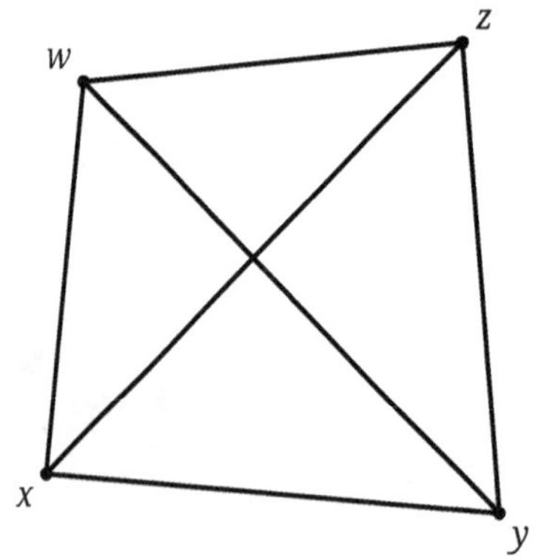

Figure 3.3: Geodesic quadrilateral.

We consider the following three sums of lengths:
- The sum of the lengths of the diagonals =: l_1.
- The sum of the lengths of the ‘roof’ and ‘basement’ =: l_2.
- The sum of the lengths of the ‘sides’ =: l_3.

If X is hyperbolic, then the maximum $\{l_1, l_2, l_3\}$ cannot exceed the maximum of the two other l_i by more than 2δ.

Remarks 3.2.7. 1. The space $\mathbb{R}^n$ (with respect to the Euclidean metric) is not hyperbolic for $n \geq 2$ because big quadrilaterals have diagonals of lengths $\sqrt{2}\times$side length.

2. Metric trees are 0-hyperbolic.
   To see this, let w, x, y, z be points such that no three of them lie on a geodesic path. Then we may have the situation depicted in Figure 3.4.

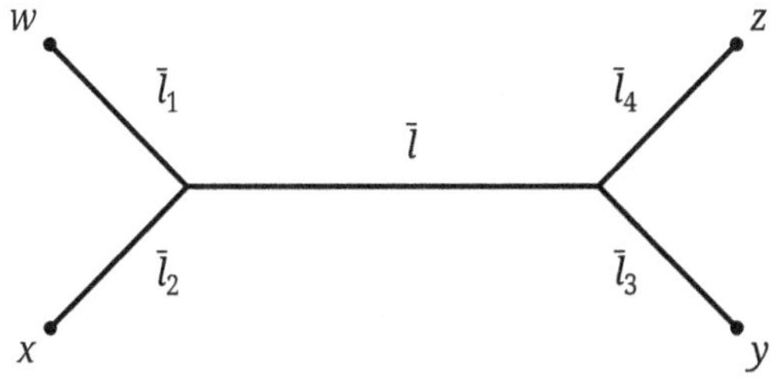

Figure 3.4: Points on a geodesic path.

   The three sums of lengths in the geodesic quadrilateral are

$$(\bar{l}_1 + \bar{l} + \bar{l}_3) + (\bar{l}_2 + \bar{l} + \bar{l}_4) \quad \text{(diagonals)},$$
$$(\bar{l}_1 + \bar{l} + \bar{l}_4) + (\bar{l}_2 + \bar{l} + \bar{l}_3) \quad \text{(roof and basement)},$$
$$(\bar{l}_1 + \bar{l}_2) + (\bar{l}_3 + \bar{l}_4) \quad \text{(sides)}.$$

   The first two sums are equal, the latter is the smallest. Hence, this quadrilateral satisfies the condition of being 0-hyperbolic. (In general in a 0-hyperbolic space the two biggest sums must be equal.) Similarly we treat the other possibilities for w, x, y and z.

Theorem 3.2.8. *Let F be a 0-hyperbolic and finite metric space. Then F can be isometrically embedded in a metric tree.*

Proof. We fix $w \in F$ as basis point. For all $x \in F$ let $|x|$ abbreviate the distance $|x - w|$ in the given metric. For every $x \in F \backslash \{w\}$ let I_x denote the interval $[0, |x|]$ in $\mathbb{R}$. We proceed in three steps.

1. Let $\tilde{T} = \bigcup_{x \in R(x)} I_x$. On $\tilde{T}$ we define an equivalence relation $\sim$ as follows:

$$t \sim t' \quad \text{if and only if} \quad t = t' \quad \text{and} \quad t \le (x \mid x')_w,$$

   where $t \in I_x$ and $t' \in I_{x'}$ ($t = t'$ is suitable because of $I_x, I_{x'} \subset$ R). The motivation for $t \le (x \mid x')_w$ lies in the picture for the geometric interpretation of the Gromov product.
   We prove that $\sim$ is an equivalence relation:
   - Reflexivity: $(x \mid x)_w = |x|$, because

$$(x \mid x)_w = \frac{1}{2}(|x - w| + |x - w| - |x - x|) = \frac{1}{2}(|x| + |x| - 0).$$

   - Symmetry: follows immediately as the Gromov product is symmetric.

- Transitivity: Let $t \sim t'$ and $t' \sim t''$. Then $t = t'$ and $t' = t''$ yield $t = t' = t''$, hence $t = t''$.
  According to $t \leq (x \mid x')_w$ we have $t \leq (x \mid x')_w$ as well as $t \leq (x' \mid x'')_w$. As F is 0-hyperbolic, we have $(x \mid x'')_w \geq \min\{(x \mid x')_w, (x' \mid x'')_w\}$. It follows that $t \leq (x \mid x'')_w$, hence $t \sim t''$.

We set $T := \tilde{T}/\sim$.

2. We would like to define a metric on T. To this end, we define a map

$$\tilde{d}\colon \tilde{T} \times \tilde{T} \to \mathbb{R}_+ = [0, \infty), \quad (t, t') \mapsto t + t' - 2\min\{t, t', (x \mid x')_w\}.$$

Of course, $\tilde{d}$ takes values in $\mathbb{R}_+$. Right from the definition we see that $\tilde{d}$ is symmetric. Furthermore, we have $\tilde{d}(t, t'') \leq \tilde{d}(t, t') + \tilde{d}(t', t'')$.
To see this, we have to show that

$$\begin{aligned} t + t'' - 2\min\{t, t'', (x \mid x'')_w\} \leq{} & t + t' - 2\min\{t, t', (x \mid x')_w\} \\ & + t' + t'' - 2\min\{t', t'', (x' \mid x'')_w\}. \end{aligned}$$

Hence we have to show alternatively that

$$\min\{t, t', (x \mid x')_w\} + \min\{t', t'', (x' \mid x'')_w\} \leq t' + \min\{t, t'', (x \mid x'')_w\}.$$

Because of $\min\{t, t', (x \mid x')_w\} \leq t'$ and $\min\{t', t'', (x' \mid x'')_w\} \leq t'$ it suffices to show

$$\min\{t, t', (x \mid x')_w\} \leq \min\{t, t'', (x \mid x'')_w\} \tag{3.1}$$

or

$$\min\{t', t'', (x' \mid x'')_w\} \leq \min\{t, t'', (x' \mid x'')_w\}. \tag{3.2}$$

If $\min\{t, t'', (x \mid x'')_w\} = t''$, then (3.2) holds; if $\min\{t, t'', (x \mid x'')_w\} = t$, then (3.1) holds.
Now suppose that (3.1) does not hold and $\min\{t, t'', (x \mid x'')_w\} = (x \mid x'')_w$. As F is 0-hyperbolic, it follows that $(x \mid x'')_w \geq \min\{(x \mid x')_w, (x' \mid x'')_w\}$.
The assumption that (3.1) does not hold implies that $(x \mid x'')_w \leq (x \mid x')_w$. Hence we have $(x \mid x'')_w \geq (x' \mid x'')_w$ and thus (3.2) holds. Hence the triangle inequality holds for $\tilde{d}$.
We claim that $\tilde{d}$ respects $\sim$.
This can be seen as follows: Let $t \sim \bar{t}$ and $t' \sim \bar{t}'$. We have to show $\tilde{d}(t, t') = \tilde{d}(t, \bar{t}')$. Note that $t \sim \bar{t}$ yields $t = \bar{t}$ and $t \leq (x \mid \bar{x})_w$, and analogously $t' \sim \bar{t}'$ yields $t' = \bar{t}'$ and $t' \leq (x' \mid \bar{x}')_w$. For $\tilde{d}(t, t') = \tilde{d}(\bar{t}, \bar{t}')$ we have to show that

$$t + t' - 2\min\{t, t', (x \mid x')_w\} = \bar{t} + \bar{t}' - 2\min\{\bar{t}, \bar{t}', (\bar{x} \mid \bar{x}')_w\}$$

hence $\min\{t, t', (x \mid x')_w\} = \min\{\bar{t}, \bar{t}', (\bar{x} \mid \bar{x}')_w\}$; hence

$$\min\{t, t', (x \mid x')_w\} = \min\{t, t', (\bar{x} \mid \bar{x}')_w\} \tag{3.3}$$

because $t = \bar{t}$, $t' = \bar{t}'$. We consider two cases.

- Let $\min\{t, t', (x \mid x')_w\} = t$ (the case $\min(t, t', (x \mid x')_w) = t'$ is analogous). We have to show that $t \leq (\bar{x} \mid \bar{x}')_w$. F being 0-hyperbolic yields

  $$(\bar{x} \mid \bar{x}')_w \geq \min\{(\bar{x} \mid x)_w, (x \mid \bar{x}')_w\}$$

  and

  $$(x \mid \bar{x}')_w \geq \min\{(x \mid x')_w, (x' \mid \bar{x}')_w\}.$$

  Thus $(\bar{x} \mid \bar{x}')_w \geq \min\{(\bar{x} \mid x)_w, (x \mid x')_w, (x' \mid \bar{x}')_w\} \geq \min_w\{t, (x \mid x')_w, t'\} = t$. Hence, if $\min\{t, t', (x \mid x')_w\} = t$ then $\min\{t, t', (\bar{x} \mid \bar{x}')_w\} = t$. Analogously we treat the cases where t or t' equals the values on the right hand side of (3.3).
- Now let $(x \mid x')_w = \min\{t, t', (x \mid x')_w\}$. Again we have

  $$(\bar{x} \mid \bar{x}')_w \geq \min\{t, (x \mid x')_w, t'\} = (x \mid x')_w.$$

  As F is 0-hyperbolic, it follows that

  $$(x, x')_w \geq \min\{(x \mid \bar{x})_w, (\bar{x}, \bar{x}')_w, (\bar{x}', x')_w\} \geq \min\{t, (\bar{x}, \bar{x}')_w, t'\}$$

  which shows (3.3) for this case. The case $(\bar{x} \mid \bar{x}')_w = \min\{t, t', (\bar{x}, \bar{x}')_w\}$ is analogous. Hence $\tilde{d}$ respects the equivalence relation $\sim$. Finally we have to show that $\tilde{d}(t, t') = 0$ if and only if $t \sim t'$: If $t \sim t'$, then $t = t'$ and $t \leq (x \mid x')_w$, hence

  $$\tilde{d}(t, t') = t + t' - 2\min\{t, t', (x \mid x')_w\} = 0.$$

  On the other hand, if $t = t'$, but $t > (x \mid x')_w$ then

  $$t + t' - 2\min\{t, t', (x \mid x')_w\} > 0$$

  hence $\tilde{d}(t, t') > 0$. If $t \neq t'$ then we always have $\tilde{d}(t, t') > 0$. Hence $\tilde{d}$ induces a metric d on T.

3. We show that T is a tree:
   - $\tilde{T}$ is a finite disjoint union of intervals as F is finite.
   - If I_x and $I_{x'}$ are two of these disjoint intervals in $\tilde{T}$, then the images of I_x and $I_{x'}$ give a (possibly degenerate) tripod in T, compare Figure 3.5.
   This we obtain by identifying of the common subinterval $[0, (x|x')_w]$ in I_x and $I_{x'}$. Finally F is embedded into T in a natural way:
     - w is mapped onto the equivalence class of 0 in an I_x.
     - If $X \in F \setminus \{w\}$, then x is mapped onto the equivalence class of $|x|$ in $[0, |x|] = I_x$.

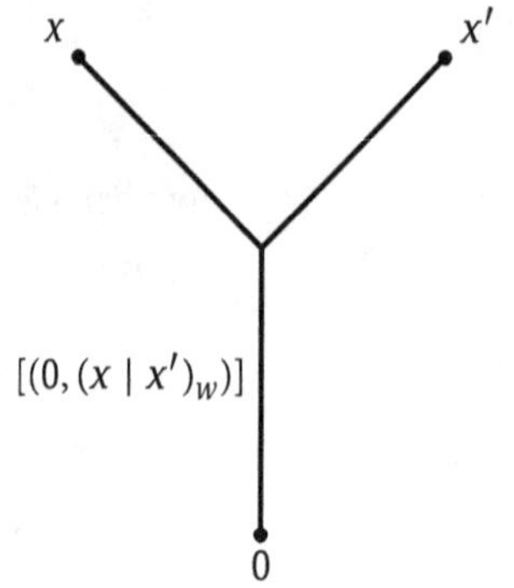

Figure 3.5: Tripod.

The fact that this gives an isometry follows from

$$|x| + |x'| - 2\min\{|x|, |x'|, (x|x')_w\} = |x - x'|.$$

□

Theorem 3.2.8 can be extended as we only need the finiteness of F in order to obtain a 1-dimensional simplicial complex quickly. We would like to make this precise.

Definition 3.2.9. A metric space T is called a *real tree*, or an $\mathbb{R}$-tree, if the following two conditions hold:

(a) Two (different) points in T are the end points of a uniquely determined geodesic segment in T.

(b) If two geodesic segments meet in exactly one common end point, then their union is a geodesic segment.

Remark 3.2.10. Hence Theorem 3.2.8 also holds for infinite F as long as T is a real tree. Thus, if F is a 0-hyperbolic geodesic metric space, then F can be isometrically embedded into a real tree. Of course a real tree is a 0-hyperbolic geodesic metric space. Hence a geodesic metric space is 0-hyperbolic if and only if it can be embedded isometrically in a real tree (see [209, Proposition 2.18]).

Remark 3.2.11. In Corollary 2.3.52 we showed that a group G is free if and only if it acts freely without inversions on a tree. At this stage we should mention Rips' theorem (see for instance [107]). A finitely generated group acting freely without inversions on $\mathbb{R}$-trees is a free product of free Abelian groups and surface groups. Certainly the respective group actions on $\mathbb{R}$-trees are analogously defined as for trees. Rips' theorem does not hold in general for infinitely generated groups. Zastrow [277] constructed an infinitely generated group which acts freely on an $\mathbb{R}$-tree but which is not a free product of surface groups and Abelian groups.

3.2.2 Approximation by trees

In the following we consider δ-hyperbolic metric spaces for a $\delta \geq 0$. We will show that these can be approximated by trees although Theorem 3.2.8 does not hold in general.

Definition 3.2.12. Let X, Y be metric spaces where X is a subspace of Y. Then X is called a *subray* with origin x if $x \in X$ and there exists an isometric embedding $(X, x) \to (\mathbb{R}_+, 0)$. (As before we write $|x|$ instead of $|x - w|$, if w is the chosen basis point.)

Theorem 3.2.13 (Approximation Theorem). *Let F be a δ-hyperbolic metric space with basis point w and let $k \in \mathbb{N}$.*
1. *If $|F| \leq 2^k + 2$, then there is a (pointed) metric tree T and a map $\Phi\colon F \to T$, such that*
   - *Φ respects the distance to the basis points, that is, $|\Phi(x)| = |x|$ for all $x \in F$.*
   - *$|y - x| - 2k\delta \leq |\Phi(y) - \Phi(x)| \leq |y - x|$ for all $x, y \in F$.*
2. *Suppose there are subrays F_i with origins w_i in F, where $i = 1, \ldots, n$ and $n \leq 2^k$, such that $F = \bigcup_{i=1}^n F_i$. Set $c = \max_{1 \leq i \leq n} |w_i|$. Then there is a (pointed) real tree T and a map $\Phi\colon F \to T$, such that*
   - *$|\Phi(x)| = |x|$ for all $x \in F$.*
   - *$|y - x| - 2(k+1)\delta - c \leq |\Phi(y) - \Phi(x)| \leq |y - x|$ for all $x, y \in F$.*

The proof of Theorem 3.2.13 is based on two lemmas that we will discuss now.

Lemma 3.2.14 (First version). *Let $|F| \leq 2^k + 2$, $L \in \mathbb{N}$ and $x_1, \ldots, x_L$ be a sequence of points in F. Then*

$$(x_1 \mid x_L)_w \geq \min_{2 \leq i \leq L} (x_{i-1} \mid x_i)_w - k\delta.$$

(Note that $k\delta$ depends on $|F|$ but not on L.)

Proof. If one $x_j = w$ then

$$\begin{aligned}(x_j, x_{j\pm1})_w &= \frac{1}{2}(|x_j - w| + |x_{j\pm1} - w| - |x_j - x_{j\pm1}|) \\ &= \frac{1}{2}(|w - w| + |x_{j\pm1} - w| - |w - x_{j\pm1}|) = 0\end{aligned}$$

and we have nothing to show. For $L \leq 2$ the lemma is trivial and for $L = 3$ it holds as F is δ-hyperbolic.

Now let $x_j \neq w$ for $j = 1, \ldots, L$ and $L \geq 4$. For $n \in \mathbb{N}$ with $n \leq k$ and $L \leq 2^n + 1$ we show inductively

$$(x_1 \mid x_L)_w \geq \min_{2 \leq i \leq L} (x_{i-1} \mid x_i)_w - n\delta.$$

If $n = 2$, then $L = 4$ or 5. Let $L = 4$. We have to show that

$$(x_1 \mid x_4)_w \geq \min\{(x_1 \mid x_2), (x_2 \mid x_3), (x_3, x_4)\} - 2\delta.$$

Now

$$\begin{aligned}(x_1 \mid x_4)_w &\geq \min\{(x_1 \mid x_3), (x_3 \mid x_4)\} - \delta \\ &\geq \min\{\min\{(x_1 \mid x_2), (x_2 \mid x_3)\} - \delta, (x_3 \mid x_4)\} - \delta \\ &\geq \min\{(x_1 \mid x_2), (x_2 \mid x_3), (x_3 \mid x_4)\} - 2\delta.\end{aligned}$$

Let $L = 5$. We have to show that

$$(x_1 \mid x_5)_w \geq \min\{(x_1 \mid x_2), (x_2 \mid x_3), (x_3 \mid x_4), (x_4, x_5)\} - 2\delta.$$

We have

$$\begin{aligned}(x_1 \mid x_5)_w &\geq \min\{(x_1 \mid x_3), (x_3 \mid x_5)\} - \delta \\ &\geq \min\{\min\{(x_1 \mid x_2), (x_2 \mid x_3)\} - \delta, \min\{(x_3 \mid x_4), (x_4 \mid x_5)\} - \delta\} \\ &\geq \min\{(x_1 \mid x_2), (x_2 \mid x_3), (x_3 \mid x_4), (x_4 \mid x_5)\} - 2\delta.\end{aligned}$$

We now do the induction: Let m be the largest integer $\leq \frac{L}{2}$. Then

$$2 \leq m \leq 2^{n-1} + 1 \quad \text{and} \quad 2 \leq L - m \leq 2^{n-1} + 1.$$

The induction hypothesis yields

$$(x_1 \mid x_m) \geq \min_{2 \leq j \leq m} \{(x_{j-1} \mid x_j)\} - (n-1)\delta$$

as well as

$$(x_m \mid x_L) \geq \min_{m+1 \leq j \leq L} \{(x_{j-1} \mid x_j)\} - (n-1)\delta.$$

Hence,

$$\begin{aligned}(x_1 \mid x_L) &\geq \min\{(x_1 \mid x_m), (x_m \mid x_L)\} - \delta \\ &\geq \min_{2 \leq j \leq L} \{(x_{j-1} \mid x_j)\} - (n-1)\delta - \delta \\ &= \min_{2 \leq j \leq L} \{(x_{j-1} \mid x_j)\} - \delta n.\end{aligned}$$

This shows the lemma for $L \leq 2^k + 1$. Now let $L > 2^k + 1$. Because of $|F| \leq 2^k + 2$ and $x_j \neq w$ for all j there exists $p, q \in \{1, \ldots, L\}$ with $p < q$ and $x_p = x_q$. We have

$$\min_{2 \leq j \leq L} \{(x_{j-1} \mid x_j)\} \leq \min_{\substack{2 \leq j \leq L \\ j \neq p+1, \ldots, q}} \{(x_{j-1} \mid x_j), (x_p \mid x_{q+1})\},$$

that is, we remove the subsequence $\{x_{p+1}, \ldots, x_q\}$ from $\{x_1, \ldots, x_L\}$ (note $x_p = x_q$). We proceed accordingly until we obtain a subsequence of length $\leq 2^k + 1$ so that we can apply the argument above. □

Lemma 3.2.15 (Second version). *Assume that the second condition of Theorem* 3.2.13 *holds for F. Let $L \in \mathbb{N}$ and $x_1, \dots, x_L$ be a sequence of points in F. Then*

$$(x_1 \mid x_L)_w \geq \min_{2 \leq j \leq L}\{(x_{j-1} \mid x_j)_w\} - (k+1)\delta - 2c.$$

Proof. Let $x, y \in F$. Then $(x \mid y)_w \leq \min\{|x|, |y|\}$ (see Remark 3.2.2). We consider two claims.

1. For $i \in \{1, \dots, n\}$ we have $(x \mid y)_{w_i} - c \leq (x \mid y)_w \leq (x \mid y)_{\omega_i} + c$.

   Proof. The inequalities $|x - w| \leq |x - w_i| + |w_i - w|$ and $|y - w| \leq |y - w_i| + |w_i - w|$ follow from the triangle inequality. Hence

   $$|x| \leq |x - w_i| + |w_i| \quad \text{and} \quad |y| \leq |y - w_i| + |w_i|$$

   and it follows that

   $$\frac{1}{2}(|x| + |y| - |x - y|) \leq \frac{1}{2}(|x - w_i| + |y - w_i| - |x - c|) + c.$$

   Hence $(x \mid y)_w \leq (x \mid y)_{w_i} + c$. For the lower bound of $(x \mid y)_w$ we use $|x - w_i| \leq |x - w| + |w - w_i|$, $|y - w_i| \leq |y - w| + |w - w_i|$ and proceed as before. □

2. Let $x, y \in F_i$ for an $i \in \{1, \dots, n\}$. Then $(x \mid y)_w \geq \min\{|x|, |y|\} - 2c$.
   This can be seen as follows: As F_i is a subray with origin w_i, there is an isometric embedding $(F_i, w_i) \to (\mathbb{R}_+, 0)$, and hence

   $$(x \mid y)_{w_i} = \min\{|x - w_i|, |y - w_i|\}$$

   because there are only degenerated tripods in $\mathbb{R}_+$. The triangle inequality yields

   $$|x - w| \leq |x - w_i| + |w_i - w| \leq |x - w_i| + c$$

   hence $|x| - c \leq |x - w_i|$. Analogously $|y| - c \leq |y - w_i|$. Hence

   $$(x \mid y)_{w_i} \geq \min\{|x| - c, |y| - c\} = \min\{|x|, |y|\} - c.$$

   Finally, the first claim yields

   $$\begin{aligned}(x \mid y)_w &\geq (x \mid y)_{w_i} - c \\ &\geq (\min\{|x|, |y|\} - c) - c \\ &= \min\{|x|, |y|\} - 2c.\end{aligned}$$

   Now let $p, q \in \{1, \dots, L\}$ with $p < q$ and $x_p, x_q \in F_i$ for an i. Then

   $$(x_p \mid x_q)_{w_i} = \min\{|x_p - w_i|, |x_q - w_i|\}$$

and from the above

$$(x_p \mid x_q)_w \geq \min\{|x_p|, |x_q|\} - 2c.$$

But then we have (note $|x_p| \geq (x_p \mid x_{p+1})$, $|x_q| \geq (x_{q-1} \mid x_q)$)

$$\begin{aligned}(x_p \mid x_q)_w &\geq \min\{(x_p, x_{p+1}), (x_{q-1}, x_q)\} - 2c \\ &\geq \min_{p+1 \leq j \leq q}\{(x_{j-1} \mid x_j)\} - 2c.\end{aligned} \tag{3.4}$$

Strategy: Starting with the above sequence $\{x_1, \ldots, x_L\}$ we look for pairs x_p, x_q as before. Having found them we construct (according to equation (3.4)) the subsequence $\{x_1, \ldots, x_p, x_q, \ldots, x_L\}$. We proceed as long as it is possible. Finally we obtain a sequence of the form $\{y_1, y_2, \ldots, y_M\}$ with $y_1 = x_1, y_M = x_L$ and, if $y_{p'}$ and $y_{q'}$ are in the same F_i, then p' and q' are neighbors and unique. This implies $M \leq 2n \leq 2 \cdot 2^k = 2^{k+1}$. The first version (Lemma 3.2.14) yields

$$(x_1 \mid x_L)_w \geq \min_{2 \leq j \leq M}\{(y_{j-1} \mid y_j)\} - (k+1)\delta$$

and with equation (3.4):

$$(x_1 \mid x_L)_w \geq \min_{2 \leq j \leq L}\{(x_{j-1} \mid x_j)\} - (k+1)\delta - 2c. \qquad \square$$

Lemma 3.2.16. *We define maps $F \times F \to \mathbb{R}_+$ via*
- $(x \mid y)' = \sup\{\min_{2 \leq j \leq L}(x_{j-1} \mid x_j)\}$, *where we take the supremum over all sequences of the form $x = x_1, x_2, \ldots, x_L = y$; and*
- $|x - y|' = |x| + |y| - 2(x \mid y)'$.

We write $\delta' = (k+1)\delta + 2c$.

Then:
1. *The map $(x, y) \mapsto |x - y|'$ is a pseudo metric on F.*
2. *We have $|x - y| - 2\delta' \leq |x - y|' \leq |x - y|$ for all $x, y \in F$.*
3. *We have $|x|' = |x|$ for all $x \in F$ (where $|x|' = |x - w|'$ with basis point w in F).*
4. *This pseudo metric is 0-hyperbolic in the following sense:*

$$(x \mid z)' \geq \min\{(x \mid y)', (y \mid z)'\} \quad \textit{for all } x, y, z \in F.$$

Proof. 1. Lemma 3.2.14 implies that $(x \mid y)'$ is bounded, hence $(x \mid y)' \in \mathbb{R}_+$. We have $|x - y|' \in \mathbb{R}_+$ under the assumption that the triangle inequality holds for $|\cdot|'$ (which we will prove independently from $|x - y|' \in \mathbb{R}_+$) because

$$|x - y|' \leq |x - x|' + |x - y|'$$

hence $|x - x|' \geq 0$ and

$$0 \leq |x - x|' \leq |x - y|' + |y - x|' = 2|x - y|',$$

hence $|x - y|' \geq 0$.

We now show the triangle inequality. Fix $x, y, z \in F$. We would like to show

$$|x - z|' \leq |x - y|' + |y - z|',$$

that is,

$$|x| + |z| - 2(x \mid z)' \leq |x| + |y| - 2(x \mid y)' + |y| + |z| - 2(y \mid z)'$$

and hence

$$(x \mid y)' + (y \mid z)' \leq |y| + (x \mid z)'.$$

Without loss of generality let $(x \mid y)' \leq (y \mid z)'$. Given $\varepsilon > 0$ we choose sequences $x = u_1, u_2, \ldots, u_L = y$ and $y = v_0, v_1, \ldots, v_M = z$, such that

$$(x \mid y)' \leq \min_{2 \leq j \leq L} \{(u_{j-1} \mid u_j)\} + \varepsilon,$$
$$(x \mid y)' \leq \min_{2 \leq j \leq M} \{(v_{j-1} \mid v_j)\} + \varepsilon.$$

Let $z_1, \ldots, z_{L+M}$ denote the sequence $u_1, u_2, \ldots, u_L = v_0, v_1, \ldots, v_M$. Then $z_1 = u_1 = x$, $z_{L+M} = v_M = z$, and $z_1, \ldots, z_{L+M}$ is a chain connecting x with z. Hence

$$\begin{aligned} (x \mid z)' &\geq \min_{2 \leq j \leq L+M} (z_{j-1} \mid z_j) \\ &\geq \min\{(x \mid y)' - \varepsilon, (y \mid z)' - \varepsilon\} \\ &= \min\{(x \mid y)', (y \mid z)'\} - \varepsilon \\ &= (x \mid y)' - \varepsilon \end{aligned}$$

(because of $(x \mid y)' \leq (y \mid z)'$). As $\varepsilon > 0$ is arbitrary it follows that

$$(x \mid z)' \geq (x \mid y)'. \tag{3.5}$$

We also have

$$|y| \geq (y \mid v_1) = (v_0 \mid v_1) \geq \min_{1 \leq j \leq M} \{(v_{j-1} \mid v_j)\} \geq (y \mid z)' - \varepsilon.$$

Again, as $\varepsilon > 0$ is arbitrary, it follows that

$$|y| \geq (y \mid z)'. \tag{3.6}$$

From equations (3.5) and (3.6) follows the triangle inequality for $|\cdot|'$. As $|\cdot|'$ is clearly symmetric it remains to show (for $|\cdot|'$ being a pseudo metric) $|x-x|' = 0$ for all $x \in F$. This holds as $(x \mid y)' \geq (x \mid y)$ for all $x, y \in F$. Hence $|x|+|y|-2(x \mid y) \geq |x|+|y|-2(x \mid y)'$ and $|x-y| \geq |x-y|'$ for all $x, y \in F$. In particular $0 = |x-x| \geq |x-x'| \geq 0$, hence $|x-x'| = 0$. This shows that $|\cdot|'$ is a pseudo metric.

2. We know that $|x-y|' \leq |x-y|$ for all $x, y \in F$. For the lower bound of $|x-y|'$ we use Lemma 3.2.15. We have

$$(x \mid y)_w \geq (x \mid y)'_w - (k+1)\delta - 2c$$

and hence

$$-(x \mid y)'_w \geq -(x \mid y)_w - (k+1)\delta - 2c.$$

Thus

$$\begin{aligned} |x-y|' &= |x| + |y| - 2(x \mid y)'_w \\ &\geq |x| + |y| - 2(x \mid y)_w - 2(k+1)\delta - 4c \\ &= |x-y| - 2\delta'. \end{aligned}$$

3. Let $x \in F$. We consider the chain $w = x_0, x_1, \ldots, x_L = x$, which connects w with x. Because of

$$(x_0 \mid x_1)_w = (w \mid x_1)_w = \frac{1}{2}(|w-w| + |x-w| - |w-x|) = 0$$

we have $\min_{2 \leq j \leq L}\{(x_{j-1} \mid x_j)\} = 0$. Hence $(w \mid x)'_w$ is the supremum of a set that only contains the 0. Hence

$$|x|' = |x-w|' = |x| + |w| - 2(x \mid w)' = |x|.$$

4. This has already been shown in the proof of the triangle inequality for $|\cdot|'$. □

We are now ready to give the proof of Theorem 3.2.13.

Proof of Theorem 3.2.13. The pseudo metric $|\cdot|'$ defines an equivalence relation on F via

$$x \sim y \quad \text{if and only if} \quad |x-y|' = 0.$$

Note that the triangle inequality yields the facts that $x \sim y$ and that $y \sim z$ implies $x \sim z$. Set $F' = F/{\sim}$. Then F' is a metric space because $|\cdot|'$ becomes a metric $|\cdot|''$ on F'. According to Lemma 3.2.16 F' is 0-hyperbolic. Theorem 3.2.8 together with the remarks on real trees gives an isometric embedding,

$$\varphi\colon (F', |\cdot|'') \to T,$$

where T denotes a real tree. Let Φ be the composition

$$F \overset{\text{quotient}}{\longrightarrow} F' \overset{\varphi}{\longrightarrow} T.$$

The composition Φ has the desired properties according to Lemmas 3.2.14, 3.2.15 and 3.2.16. □

Remark 3.2.17. If $F = \{x, y, z\}$ is a metric space then F is 0-hyperbolic.

Proof. A point, say z, is chosen as a basis point. Hence $(y \mid z)_z = 0$ and $(x \mid z)_x = 0$. Furthermore, $(x \mid y)_z \geq 0$, $(x \mid x)_z = |x - z|$, $(y \mid y)_z = |y - z|$. We have to show $(a \mid b)_x \geq \min\{(a \mid c)_x, (b \mid c)_z\}$ for all $a, b, c \in \{x, y, z\}$. This is clear if a, b or c equals z. Now let neither a, nor b or c be equal to z. This yields two cases.

1. Let $a = b$, say $a = b = x$. If $c = x$ there is nothing to show. Now let $c = y$. Then

$$(x \mid x)_x = |x - z| \geq \frac{1}{2}(|x - z| + |y - z| - |x - y|) = (x \mid y)_z$$

according to the triangle inequality $|y - z| \leq |x - z| + |x - y|$.

2. Let $a \neq b$, say $a = x$, $b = y$. Without loss of generality $c = x$. Then $(x \mid y)_x \geq (x \mid y)_x$, and there is nothing to show.

Theorems 3.2.8 and 3.2.13 now give: If x, y, z are three points in a metric space, then there is an isometric embedding

$$\Phi\colon \{x, y, z\} \to T_\Delta,$$

where T_Δ is a (possibly degenerated) tripod; compare Figure 3.6.

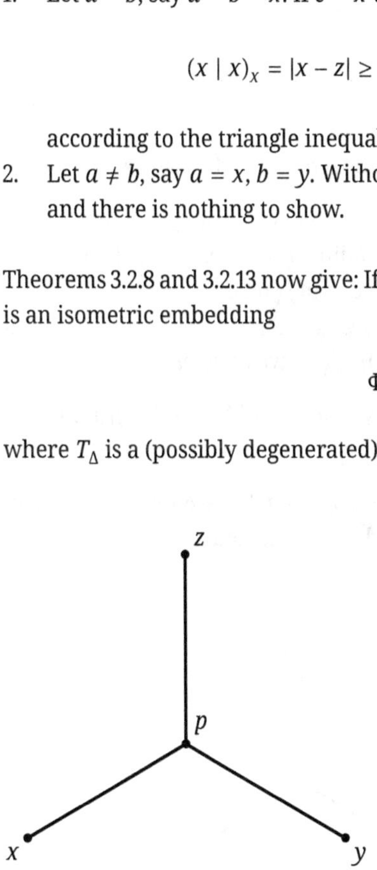

Figure 3.6: Isometric embedding.

This follows from the following: Let $l_1 = |x - y|$, $l_2 = |x - z|$ and $l_3 = |y - z|$. We look for a tripod as in Figure 3.7 with $k_1 + k_2 = l_1$, $k_1 + k_3 = l_2$ and $k_2 + k_3 = l_3$. Because of

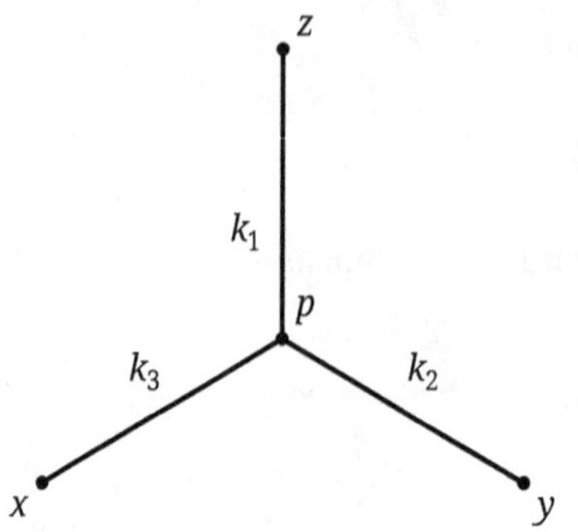

Figure 3.7: Tripod with described lengths.

$$\det\begin{pmatrix} 1 & 1 & 0 \\ 1 & 0 & 1 \\ 0 & 1 & 1 \end{pmatrix} = -2 \neq 0$$

such a tripod exists. □

Recall that $[x,y]$ denotes a geodesic segment connecting x and y (see Section 3.1).

Definition 3.2.18. Let $\Delta = [x,y] \cup [y,z] \cup [z,x]$ be a geodesic triangle in the metric space X. Let $f_\Delta\colon \{x,y,z\} \to T_\Delta$ be an isometric embedding as above. As $[x,y]$, $[y,z]$ and $[z,x]$ are geodesic segments, there is a unique extension of f_Δ on Δ, also denoted f_Δ, such that $f_\Delta|_{[x,y]}, f_\Delta|_{[y,z]}$ and $f_\Delta|_{[z,x]}$ are isometric embeddings. Let $\delta \geq 0$. We say that the triangle Δ is *δ-thin* (or *δ-slim*), if for all $u, v \in \Delta$ with $f_\Delta(u) = f_\Delta(v)$ we have $|u - v| \leq \delta$.

Remark 3.2.19. We claim that Δ is δ-thin if and only if $|u - v| \leq |f_\Delta(u) - f_\Delta(v)| + \delta$ for all $u, v \in \Delta$.

Proof. If Δ is δ-thin then we obviously have $|u - v| \leq |f_\Delta(u) - f_\Delta(v)| + \delta$ for all $u, v \in \Delta$. In the other direction we typically have the situation depicted in Figure 3.8.

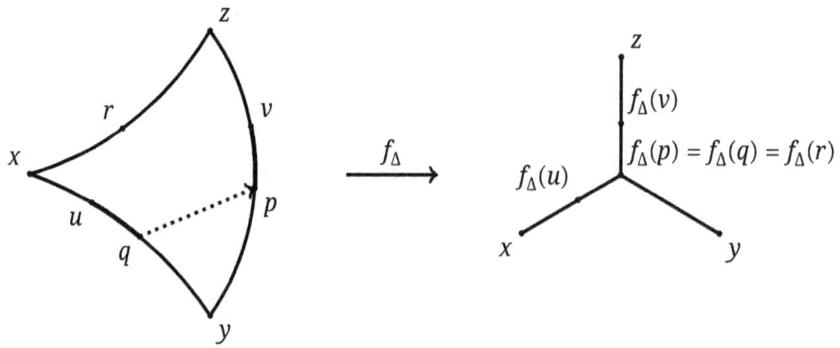

Figure 3.8: Mapping of a triangle onto a tripod.

This yields

$$\begin{aligned}|u-v| &\le |u-q| + |q-p| + |p-v| \\ &\le |f_\Delta(u) - f_\Delta(q)| + |f_\Delta(p) - f_\Delta(v)| + \delta \\ &\le |f_\Delta(u) - f_\Delta(v)| + \delta,\end{aligned}$$

where we have used $f_\Delta(p) = f_\Delta(q)$ in the last inequality. □

Lemma 3.2.20. *Let $\Delta = [x,y] \cup [y,z] \cup [z,x]$ be a geodesic triangle in the metric space (X,d). Then*
1. *$(y \mid z)_x \le d(x,[y,z])$; and*
2. *If in addition Δ is δ-thin, we also have $d(x,[y,z]) \le (y \mid z)_x + \delta$.*

Proof. 1. Let $f_\Delta : \Delta \to T_\Delta$ be as above. Let v be the point of valency 3; see Figure 3.9.

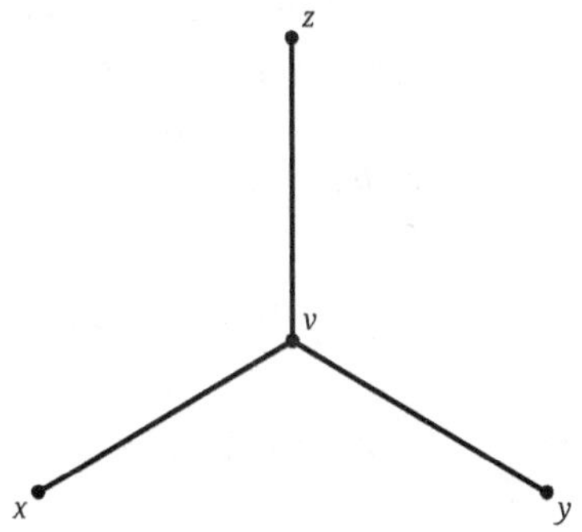

Figure 3.9: Point with valency 3.

Then there are uniquely determined points $p \in [y,z]$, $q \in [z,x]$ and $r \in [x,y]$ that are mapped onto v under f_Δ. Because of the local isometry of f_Δ we have

$$(y \mid z)_x = \frac{1}{2}(|y-x| + |z-x| - |z-y|) = |q-x| = |r-x|; \tag{3.7}$$

compare Figure 3.10.

As $[y,z]$ is compact there is a $w \in [y,z]$ with $|w-x| = d(x,[y,z])$. Then there is a $w' \in [x,y] \cup [z,x]$ with $f_\Delta(w') = f_\Delta(w)$. Without loss of generality let $w' \in [z,x]$. Equation (3.7) yields $(y \mid z)_x \le |w'-x|$. Furthermore, we have $|w'-x| = |z-x| - |z-w'|$; hence

$$(y \mid z)_x \le |w'-x| = |z-x| - |z-w'| = |z-x| - |z-w| \le |x-w| = d(x,[y,z]),$$

where the second equality follows from the isometric embedding and the last inequality follows from the triangle inequality.

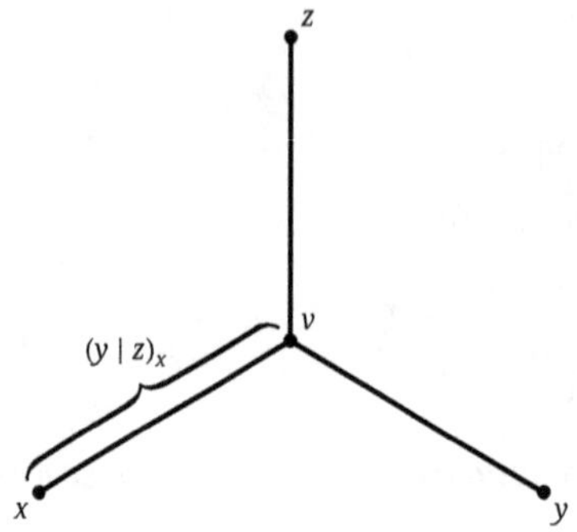

Figure 3.10: Gromov product as length of a tripod.

2. If Δ is in addition δ-thin, then $|q - p| \leq \delta$, hence

$$d(x, [y, z]) \leq |x - q| + |q - p| \leq (y \mid z)_x + \delta. \qquad \square$$

Definition 3.2.21. Let $\Delta = [x, y] \cup [y, z] \cup [z, x]$ and $f_\Delta \colon \Delta \to T_\Delta$ be as above. Furthermore, let V be the point in T_Δ with valency 3, we call it the *tripod point*. Then there exist uniquely determined points $p \in [y, z]$, $q \in [z, x]$ and $r \in [x, y]$ that are mapped onto V under f_Δ. The set $\{p, q, r\}$ is called *descriptive triple* of Δ. The *height* or the *measure* of Δ is the diameter of $\{p, q, r\}$, denoted as $\operatorname{diam}(p, q, r)$, that is,

$$\sup\{|p - q|, |q - r|, |r - p|\}.$$

The *minimal height* or the *minimal measure* of Δ is

$$\min\{\operatorname{diam}(\{u, v, w\}) \mid u \in [y, z], v \in [z, x], w \in [x, y]\},$$

that is,

$$\min\{\sup\{|u - v|, |v - w|, |w - u|\} \mid u \in [y, z], v \in [z, x], w \in [x, y]\}.$$

Lemma 3.2.22. *Let $\Delta = [x_1, x_2] \cup [x_2, x_3] \cup [x_3, x_1]$ be a geodesic triangle of minimal measure δ and height δ'. Then $\delta \leq \delta' \leq 4\delta$.*

Proof. The inequality $\delta \leq \delta'$ follows from the definition. We show $\delta' \leq 4\delta$. Let $p_1 \in [x_2, x_3]$, $p_2 \in [x_3, x_1]$ and $p_3 \in [x_1, x_2]$ be the descriptive triple of Δ. Let $g_1 \colon [0, |x_2 - x_3|] \to [x_2, x_3] \subset X$, $g_2 \colon [0, |x_3 - x_1|] \to [x_3, x_1] \subset X$ and $g_3 \colon [0, |x_1 - x_2|] \to [x_1, x_2] \subset X$ be the isometries whose images are $[x_2, x_3]$, $[x_3, x_1]$ and $[x_1, x_2]$, respectively. Then

$$(t_1, t_2, t_3) \mapsto \sup\{|g_1(t_1) - g_2(t_2)|, |g_2(t_2) - g_3(t_3)|, |g_3(t_3) - g_1(t_1)|\}$$

is a continuous map in $\mathbb{R}_+$, defined on a compact and fixed parallelepiped. There also exists a $(\bar{t}_1, \bar{t}_2, \bar{t}_3)$ that attains the minimum. Set $q_1 = g_1(\bar{t}_1) \in [x_2, x_3]$, $q_2 = g_2(\bar{t}_2) \in [x_3, x_1]$ and $q_3 = g_3(\bar{t}_3) \in [x_1, x_2]$. Then δ is the diameter of $\{q_1, q_2, q_3\}$; see Figure 3.11.

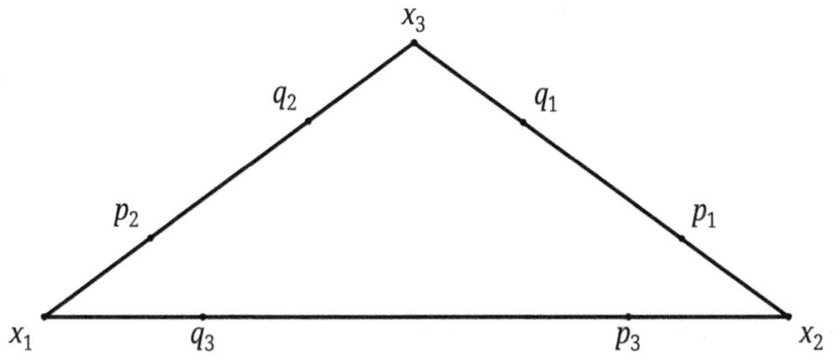

Figure 3.11: δ as the diameter of $\{q_1, q_2, q_3\}$.

Let $\binom{1\,2\,3}{i\,j\,k}$ be a permutation of $\{1, 2, 3\}$.

Set $a_i = |x_j - x_k|$, $b_{i,k} = |p_i - x_k|$ and $c_{i,k} = |q_i - x_k|$. Then

$$b_{i,k} = b_{j,k} = (x_i \mid x_j)_{x_k} = \frac{1}{2}(a_i + a_j - a_k) \quad \text{and} \quad b_{i,j} + b_{i,k} = a_i = c_{i,j} + c_{i,k}.$$

We have $|q_i - q_j| \le \delta$. Because of $|q_i - x_k| \le |q_i - q_j| + |q_j - x_k|$ and $|q_j - x_k| \le |q_i - q_j| + |q_i - x_k|$ we get $||q_i - x_k| - |q_j - x_k|| \le |q_i - q_j| \le \delta$. Because of this we have $|c_{i,k} - c_{j,k}| \le \delta$. The above equations yield

$$\begin{aligned} 2b_{i,k} &= a_i + a_j - a_k = c_{i,j} + c_{i,k} + b_{j,i} + b_{j,k} - c_{k,i} - c_{k,j} \quad \text{or} \\ 2b_{i,k} &= c_{i,j} + c_{i,k} + b_{k,i} + b_{i,k} - c_{k,i} - c_{k,j}, \quad \text{hence} \\ 0 &= c_{i,j} + c_{i,k} + b_{k,i} - b_{i,k} - c_{k,i} - c_{k,j} \quad \text{and} \end{aligned}$$

$$c_{i,k} - b_{i,k} - c_{k,i} + b_{k,i} = c_{k,j} - c_{i,j}.$$

Finally, we have

$$|c_{i,k} - b_{i,k} - c_{k,i} + b_{k,i}| = |c_{k,j} - c_{i,j}| \le \delta. \tag{3.8}$$

For every cyclic permutation of $\{1, 2, 3\}$ set $d_i = c_{i,j} - b_{i,j}$. Because of $b_{i,j} + b_{i,k} = c_{i,j} + c_{i,k}$ we obtain $d_i = c_{i,j} - b_{i,j} = -(c_{i,k} - b_{i,k})$, hence by (3.8) we have $|-d_i - d_k| \le \delta$, $|-d_j - d_i| \le \delta$ and $|-d_k - d_j| \le \delta$.

We now have

$$2|d_i| = |d_i + d_j + d_i + d_k - d_j - d_k| \le |d_i + d_j| + |d_i + d_k| + |d_j + d_k| \le 3\delta.$$

Hence $|d_i| \le \frac{3}{2}\delta$. Now

$$\begin{aligned} |d_i| &= |c_{i,j} - b_{i,j}| \\ &= \big||q_i - x_j| - |p_i - x_j|\big| \\ &= |q_i - p_i| \end{aligned}$$

for all i. We then have

$$|p_j - p_k| \leq |p_j - q_j| + |q_j - q_k| + |q_k - p_k| \leq \frac{3}{2}\delta + \delta + \frac{3}{2}\delta = 4\delta$$

and hence $\delta' \leq 4\delta$. □

3.2.3 Equivalent hyperbolicity properties for geodesic metric spaces

In this subsection we further characterize the hyperbolicity of a geodesic metric space.

Theorem 3.2.23. *Let X be a geodesic metric space. We consider the following properties of X, where $\delta \in [0, \infty)$.*
(P1, δ): *X is δ-hyperbolic.*
(P2, δ): *Any geodesic triangle in X is δ-thin.*
(P3, δ): *X satisfies the Rips condition of the constant δ, that is, for every geodesic triangle $\Delta = [x,y] \cup [y,z] \cup [z,x]$ we have $d(u, [y,z] \cup [z,x]) \leq \delta$ for all $u \in [x,y]$.*
(P4, δ): *δ is an upper bound for the heights of geodesic triangles in X.*
(P5, δ): *δ is an upper bound for the minimal measure of geodesic triangles in X.*

These properties are equivalent in the following sense:

For every ordered pair $i,j \in \{1,\dots,5\}$ there is a constant $c_{i,j}$ with $1 \leq c_{i,j} \leq 4$, such that the following holds: If X satisfies the property (P_i, δ), then X also satisfies the property $(P_j, c_{i,j}\delta)$.

Proof. The implications $(P2,\delta) \Rightarrow (P3,\delta)$, $(P2,\delta) \Rightarrow (P4,\delta)$ and $(P4,\delta) \Rightarrow (P5,\delta)$ follow from the definitions; $(P5,\delta) \Rightarrow (P4,4\delta)$ follows from Lemma 3.2.22. We show the implications
1. $(P1,\delta) \Rightarrow (P2,4\delta)$,
2. $(P2,\delta) \Rightarrow (P1,2\delta)$,
3. $(P3,\delta) \Rightarrow (P2,4\delta)$, and
4. $(P4,\delta) \Rightarrow (P2,2\delta)$.

1. Suppose $(P1,\delta)$ holds. Let $\Delta = [x,y] \cup [y,z] \cup [z,x]$ be a geodesic triangle with the associated map $f_\Delta: \Delta \to T_\Delta$. Let u and v be different points in Δ with $f_\Delta(u) = f_\Delta(v)$; see Figure 3.12. We show $|u - v| \leq 4\delta$, which shows the first implication. Without loss of generality let $u \in [x,y]$ and $v \in [z,x]$.
   For $t = |x - u|$ we have

$$t = |f_\Delta(x) - f_\Delta(u)| = |f_\Delta(x) - f_\Delta(v)| \leq (y \mid z)_x$$

   and $(u \mid y)_x = (v \mid z)_x = t$, as $u \in [x,y]$ and $v \in [z,x]$. As X is δ-hyperbolic, it follows that

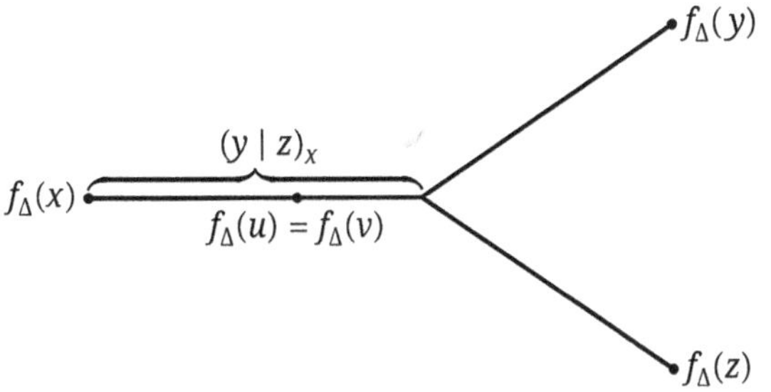

Figure 3.12: $P(1, \delta)$ implies $P(2, 4\delta)$.

$$\begin{aligned}(u \mid v)_x &\geq \min\{(u \mid y)_x, (y \mid v)_x\} - \delta \\ &\geq \min\{(u \mid y)_x, \min\{(y \mid z)_x, (z \mid v)_x\} - \delta\} - \delta \\ &\geq \min\{(u \mid y)_x, (y \mid z)_x, (z \mid v)_x\} - 2\delta \\ &= t - 2\delta.\end{aligned}$$

By definition

$$(u \mid v)_x = \frac{1}{2}(|u - x| + |v - x| - |u - v|) = t - \frac{1}{2}|u - v|$$

and hence $t - \frac{1}{2}|u - v| \geq t - 2\delta$, that is, $|u - v| \leq 4\delta$.

2. Suppose $(P2, \delta)$ holds. We show $(P1, 2\delta)$. Let $x_0, x_1, x_2, x_3 \in X$. We consider the six geodesic segments $[x_0, x_1]$, $[x_0, x_2]$, $[x_0, x_3]$, $[x_1, x_2]$, $[x_1, x_3]$ and $[x_2, x_3]$. We consider x_0 as basis point (and omit this in our notation). We show

$$(x_1 \mid x_2) \geq \min\{(x_1 \mid x_3), (x_3 \mid x_2)\} - 2\delta.$$

Set $t := \min\{(x_1 \mid x_3), (x_3 \mid x_2)\}$. If $t \leq (x_1 \mid x_2)$, there is nothing to show. Now let $t > (x_1, x_2)$. Because of $t \leq (x_1 \mid x_3)_{x_0}$ and $t \leq (x_3 \mid x_2)_{x_0}$ the segments $[x_0, x_1]$, $[x_0, x_2]$ and $[x_0, x_3]$ have length larger than t. Hence there is a uniquely determined x_i' in $[x_0, x_i]$ with $|x_i' - x_0| = t$ for $i = 1, 2, 3$.
For $\Delta_j = [x_0, x_j] \cup [x_j, x_3] \cup [x_0, x_3]$ for $j = 1, 2$ the geodesic triangle is determined by $\{x_0, x_j, x_3\}$. Let $f_{\Delta_j} =: f_j\colon \Delta_j \to T_j := T_{\Delta_j}, j = 1, 2$, be the associated maps of the tripods. Because of

$$|x_j' - x_0| = |x_3' - x_0| = t \leq (x_j \mid x_3) \quad \text{for } j = 1, 2$$

we have $f_j(x_j') = f_j(x_3')$ for $j = 1, 2$. As $(P2, \delta)$ holds and $f_j(x_j') = f_j(x_3')$ for $j = 1, 2$, we must have $|x_j' - x_3'| \leq \delta$ for $j = 1, 2$. Hence

$$|x_1' - x_2'| \leq |x_1' - x_3'| + |x_3' - x_2'| \leq 2\delta.$$

Let $\Delta = [x_0, x_1] \cup [x_1, x_2] \cup [x_0, x_2]$ and $f_\Delta: \Delta \to T_\Delta$ be the associated map of the tripod. The condition $t > (x_1 \mid x_2)$ implies that there exists a $y_j \in [x_1, x_2]$ for $j = 1, 2$ with $f_\Delta(x_j') = f_\Delta(y_j)$ and $|x_j' - x_1| = |y_j - x_1|$.
Further, $|x_j' - y_j| \leq \delta$ for $j = 1, 2$, because Δ is δ-thin. Hence

$$|y_1 - y_2| \leq |y_1 - x_1'| + |x_1' - x_2'| + |x_2' - y_2| \leq |x_1' - x_2'| + 2\delta.$$

Thus, we have

$$\begin{aligned}
2\delta \geq |x_1' - x_2'| &\geq |y_1 - y_2| - 2\delta \\
&= (|x_2 - x_1| - |y_1 - x_1| - |y_2 - x_2|) - 2\delta \\
&= |x_2 - x_1| - |x_1' - x_1| - |x_2' - x_2| - 2\delta \\
&= |x_2 - x_1| - (|x_1 - x_0| - |x_1' - x_0|) - (|x_2 - x_0| - |x_2' - x_0|) - 2\delta \\
&= (|x_1' - x_0| + |x_2' - x_0|) - (|x_2 - x_0| + |x_1 - x_0| - |x_2 - x_1|) - 2\delta \\
&= 2t - 2(x_1 \mid x_2)_{x_0} - 2\delta,
\end{aligned}$$

hence $(x_1 \mid x_2)_{x_0} \geq t - 2\delta$.

3. We show $(P3, \delta) \Rightarrow (P2, 4\delta)$. Suppose $(P3, \delta)$, that is, the Rips condition for δ holds but $(P2, 4\delta)$ does not hold. Hence there is a geodesic triangle, say Δ, in X, that is not (4δ)-thin. Let $\Delta = [x, y] \cup [y, z] \cup [z, x]$, then without loss of generality there exist points $u \in [x, y]$ and $v \in [z, x]$ with
   - $|u - x| = |v - x| \leq (y \mid z)_x$ (inequality follows from continuity); and
   - $|u - v| > 4\delta$.

   (The first property implies $f_\Delta(u) = f_\Delta(v)$, where $f_\Delta: \Delta \to T_\Delta$ is the associated map on the tripod.) We show that this Δ does not satisfy the Rips condition for δ. Note that

$$d(v, [x, y]) = \min\{d(v, [x, u]), d(v, [u, y])\}.$$

By Lemma 3.2.20 we have

$$d(v, |x, u|) \geq (x \mid u)_v \quad \text{and} \quad d(v, [u, y]) \geq (u \mid y)_v.$$

Hence $d(v, [x, y]) \geq \min\{(x \mid u)_v, (u \mid y)_v\}$. Because of $|u - x| = |v - x|$ we have

$$2(x \mid u)_v = |u - v| + |x - v| - |u - x| = |u - v|.$$

Thus, because of $(y \mid x)_v \geq 0$, we also have

$$\begin{aligned}
2(u \mid y)_v &= |u - v| + |y - v| - |u - y| \\
&= |u - v| + |y - v| - (|y - x| - |u - x|) \\
&= |u - v| + (|y - v| + |u - x| - |y - x|) \\
&= |u - v| + (|y - v| + |v - x| - |y - x|)
\end{aligned}$$

$$= |u - v| + 2(y \mid x)_v$$
$$\geq |u - v|.$$

Hence

$$d(v, [x,y]) \geq \frac{1}{2}|u - v| > \frac{1}{2}(4\delta) = 2\delta.$$

Because of $d(v, [x,y]) > 2\delta$ we have $d(v, x) > 2\delta$; hence there is a unique $p \in [x, v]$ with $|p - v| = \delta$. Because of $d(v, [x,y]) > 2\delta$ and $d(p, v) = \delta$ we must have $d(p, [x,y]) > \delta$. Next, we consider $d(x, [y, z]) \leq d(p, [y, z]) + |p - x|$, that is,

$$d(p, [y, z]) \geq d(x, [y, z]) - |p - x|.$$

Lemma 3.2.20 yields $d(x, [y, z]) \geq (y \mid z)_x$, hence

$$d(p, [y, z]) \geq (y \mid z)_x - |p - x|.$$

Because of $(y \mid z)_x > |v - x|$ we have

$$d(p, [y, z]) > |v - x| - |p - x| = |p - v| = \delta.$$

It follows that $d(p, [x,y] \cup [y, z]) > \delta$. Hence the Rips condition for δ is not satisfied. This is a contradiction to the above assumption, hence the third implication holds.

4. We would like to show $(P4, \delta) \Rightarrow (P2, 2\delta)$, that is, we assume that every geodesic triangle in X is bounded by δ and conclude that every geodesic triangle in X is (2δ)-thin. At first, we claim that, if $(P4, \delta)$ implies that $(P2, 2\delta)$ holds for all $\delta > 0$, then $(P4, 0) \Rightarrow (P2, 0)$. This can be seen as follows: If the height of every geodesic triangle is 0 then every $\delta > 0$ gives an upper bound for the height of a geodesic triangle. Now assume the implication $(P4, \delta) \Rightarrow (P2, 2\delta)$ holds. If Δ is a geodesic triangle and $f_\Delta\colon \Delta \to T_\Delta$ the associated map on the tripod T_Δ, then this implies $|u - v| < 2\delta$ if $f_\Delta(u) = f_\Delta(v)$. As this holds for all $\delta > 0$, it follows that $|u - v| = 0$, hence $f_\Delta(u) = f_\Delta(v)$ implies $|u - v| = 0$, that is, X satisfies $(P2, 0)$, hence it is also 0-thin.
We would now like to show the fourth implication under the assumption $\delta > 0$. To this end, let $\Delta = [x, y] \cup [y, z] \cup [z, x]$ be a geodesic triangle in X and $f_\Delta\colon \Delta \to T_\Delta$ be the associated map on the tripod T_Δ. Let $p_I \in [x, y]$ and $q_I \in [z, x]$ be the unique points in the geodesic segments, that get mapped on the tripod T_Δ under f_Δ. Because of $(P4, \delta)$ the height of Δ is bounded by δ, hence $|p_I - q_I| \leq \delta$. Similarly set $t_I = |p_I - x| = |q_I - x| = (y \mid z)_x$. For every $s \in [0, t_I]$ let $p_s \in [x, p_I]$ and $q_s \in [q_I, x]$ the uniquely determined points (in their respective geodesic segments) with $|p_s - x| = s = |q_s - x|$. In order to show $(P2, 2\delta)$ for X, it suffices to show $|p_s - q_s| < 2\delta$ for all $s \in [0, t_I]$.
We prove the desired result by means of a construction of a finite amount of sequences.

Construction of the sequences

We construct sequences $(p_i)_{i\in\mathbb{N}} \subset [x, p_I]$ and $(q_i)_{i\in\mathbb{N}} \subset [q_I, x]$ recursively as follows: Let $p_1 = p_I$, $q_1 = q_I$. If p_n and q_n are defined, then we define p_{n+1}, q_{n+1} in the following way: Let Δ_n be the geodesic triangle $\Delta_n = [x, p_n] \cup [p_n, q_n] \cup [q_n, x]$ where $[x, p_n] \subset [x, p_I]$, $[q_n, x] \subset [q_I, x]$. Let $f_{\Delta_n} =: f_n : \Delta_n \to T_{\Delta_n}$ be the associated map from Δ_n on the tripod. The points $p_{n+1} \in [x, p_n]$ and $q_{n+1} \in [q_n, x]$ are determined because they are mapped onto the tripod point of T_{Δ_n}. As δ is an upper bound for the height of geodesic triangles in X, it follows that $|p_i - q_i| \le \delta$ for all i. The map $f_i : \Delta_i \to T_{\Delta_i}$ yields

$$|p_i - p_{i+1}| + |q_i - q_{i+1}| = |p_i - q_i|$$

for all i and hence

$$|p_i - p_{i+1}| \le \delta \quad \text{and} \quad |q_i - q_{i+1}| \le \delta \tag{3.9}$$

for all i. The sequences (p_i) and (q_i), $i \in \mathrm{N}$, satisfy a monotony criterion in $[x, p_I]$ and $[q_I, x]$, respectively, because of

$$|x - p_{i+1}| \le |x - p_i| \quad \text{and} \quad |x - q_{i+1}| \le |x - q_i|$$

for all i. As $[x, p_I]$ and $[q_I, x]$ are closed, (p_i) and (q_i) converge, say

$$p_i \to p_\infty \in [x, p_I] \quad \text{and} \quad q_i \to q_\infty \in [q_I, x].$$

We claim: $p_\infty = q_\infty$. This can be seen as follows: Let $\varepsilon > 0$ be arbitrary. Then there exists an N with

- $|p_\infty - p_n| < \frac{\varepsilon}{4}$ and $|q_\infty - q_n| < \frac{\varepsilon}{4}$ for all $n \ge N$.
- $|p_m - p_n| < \frac{\varepsilon}{4}$ and $|q_m - q_n| < \frac{\varepsilon}{4}$ for all $m, n \ge N$.
  (We used the Cauchy criterion for convergence.)

From $|p_{N+1} - p_N| + |q_{N+1} - q_N| = |p_N - q_N|$ follows $|p_N - q_N| < \frac{\varepsilon}{2}$. Hence

$$|p_\infty - q_\infty| \le |p_\infty - p_N| + |p_N - q_N| + |q_N - q_\infty| < \frac{\varepsilon}{4} + \frac{\varepsilon}{2} + \frac{\varepsilon}{4} = \varepsilon.$$

As $\varepsilon > 0$ is arbitrary, it follows that $|p_\infty - q_\infty| = 0$, hence $p_\infty = q_\infty$.
Set $t_I' = |p_\infty - x| = |q_\infty - x|$. Suppose $p_s \in [p_\infty, p_I]$ and $q_s \in [q_I, q_\infty]$ are identified by f_Δ. Then equation (3.9) implies that there exists an i with

$$|p_i - p_s| = |q_i - q_s| \le \frac{1}{2}\delta.$$

Hence

$$|p_s - q_s| \le |p_s - p_i| + |p_i - q_i| + |q_i - q_s| \le \frac{1}{2}\delta + \delta + \frac{1}{2}\delta = 2\delta.$$

We now conclude the proof of the fourth implication: If $t_I' \leq \delta$, then there is nothing to show (note that p_I and q_I are mapped on the tripod under f_Δ). Now let $t_I' > \delta$. Set $t_{II} = t_I' - \frac{1}{2}\delta$. Let $p_{II} \in [x, p_\infty]$ and $q_{II} \in [q_\infty, x]$ be the points with distance t_{II} from x. Then the above remark yields

$$|p_{II} - q_{II}| \leq |p_{II} - p_\infty| + |p_\infty - q_\infty| + |q_\infty - q_{II}| = \frac{1}{2}\delta + 0 + \frac{1}{2}\delta = \delta.$$

Hence p_{II} and q_{II} can be chosen as new start points for a second pair of sequences (constructed as above) in $[x, p_{II}]$ and $[q_{II}, x]$, respectively. This process terminates after finitely many steps because of $\delta > 0$. □

Definition 3.2.24. Let X be a geodesic metric space. Let $d \geq 0$ and $g\colon [0, d] \to X$ be a geodesic segment, say $[x_0, x_1]$. The *natural parameterization* of $[x_0, x_1]$ is the map $P\colon [0, 1] \to X$, such that $P(t) = x_t \in [x_0, x_1]$ and $|x_t - x_0| = td$. Let $[x_0, x_1]$ and $[y_0, y_1]$ be naturally parametrized geodesic segments in X. If there exists a $\delta \geq 0$, such that, for all naturally parametrized pairs of geodesic segments, $[x_0, x_1]$, $[y_0, y_1]$, the following holds:

$$|x_t - y_t| \leq (1 - t)|x_0 - y_0| + t|x_1 - y_1| + \delta$$

then we call X *δ-convex*. If $\delta = 0$, then X is just called *convex*.

Examples 3.2.25. 1. The unit sphere $S^2 \subset \mathbb{R}^3$ is π-convex, because the geodesics are contained in half circles of great circles (if S^2 carries the usual metric) and such half circles have maximum length π.

2. The space $\mathbb{R}^2$ is convex.

   Proof. Let $[x_0, x_1]$ and $[y_0, y_1]$ be geodesic segments in $\mathbb{R}^2$. The natural parametrization for $[x_0, x_1]$ and $[y_0, y_1]$, respectively, is $P_x(t) = x_0 + t(x_1 - x_0)$ or $P_y(t) = y_0 + t(y_1 - y_0)$, respectively. Hence

$$\begin{aligned} |x_t - y_t| &= \big|x_0 + t(x_1 - x_0) - y_0 - t(y_1 - y_0)\big| \\ &= \big|(1 - t)(x_0 - y_0) + t(x_1 - y_1)\big| \\ &\leq (1 - t)|x_0 - y_0| + t|x_1 - y_1|. \end{aligned}$$ □

3. A tripod, see Figure 3.13, is convex. This can easily be seen by considering cases.

Theorem 3.2.26. *Let X be a geodesic space wherein any geodesic triangle is δ-thin. Then X is 2δ-convex.*

Proof. Let $[x_0, x_1]$ and $[y_0, y_1]$ be two naturally parametrized geodesic segments in X. We consider two cases.

- Special case: $x_0 = y_0$.
  Let $[x_1, y_1]$ be a geodesic that connects x_1 and y_1. Set $\Delta = [x_0, x_1] \cup [y_0, y_1] \cup [x_1, y_1]$ and let $f_\Delta =: f\colon \Delta \to T_\Delta$ be the associated map on the tripod. We have

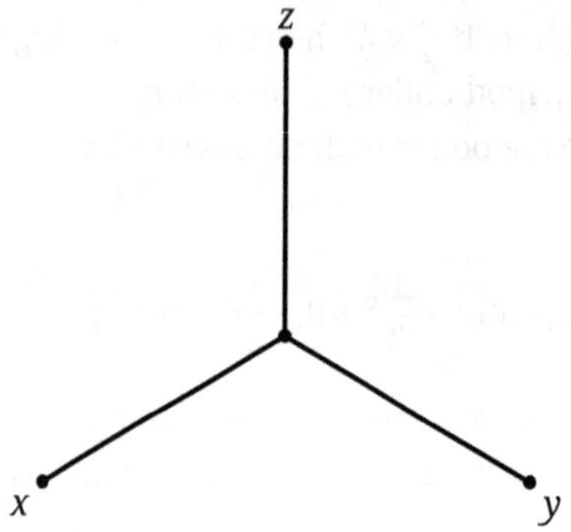

Figure 3.13: A tripod is convex.

$$|x_t - y_t| \leq |f(x_t) - f(y_t)| + \delta$$

because Δ is δ-thin. As T_Δ is convex we have

$$|f(x_t) - f(y_t)| \leq (1-t)|f(x_0) - f(y_0)| + t|f(x_1) - f(y_1)|.$$

Note that the image under f of a naturally parametrized geodesic segment is again naturally parametrized. Hence

$$\begin{aligned}|x_t - y_t| &\leq |f(x_t) - f(y_t)| + \delta \\ &\leq (1-t)|f(x_0) - f(y_0)| + t|f(x_1) - f(y_1)| + \delta \\ &= t|f(x_1) - f(y_1)| + \delta \\ &= t|x_1 - y_1| + \delta,\end{aligned}$$

where we use $x_0 = y_0$ in the second equality.

– General case.
Let $[z_0, z_1]$ be a geodesic segment that is naturally parametrized with $x_0 = z_0$ and $y_1 = z_1$. The triangle inequality and the special case yield

$$\begin{aligned}|x_t - y_t| &\leq |x_t - z_t| + |z_t - y_t| \\ &\leq (t|x_1 - z_1| + \delta) + ((1-t)|z_0 - y_0| + \delta) \\ &= t|x_1 - y_1| + (1-t)|x_0 - y_0| + 2\delta.\end{aligned}$$

Hence X is 2δ-convex. □

Remark 3.2.27. If $x_1 = y_1$, then an argument as the one above implies that

$$|x_t - y_t| \leq (1-t)|x_0 - y_0| + \delta.$$

Remark 3.2.28. $\mathbb{R}^2$ is convex, but it is not hyperbolic. Hence the other direction of Theorem 3.2.26 does not hold.

3.2.4 The upper half plane as a hyperbolic metric space

At the end of this section we would like to apply Theorem 3.2.23 in order to show that the upper half plane $\mathfrak{H} = \{z = x + iy \in \mathbb{C} \mid y > 0\}$ is a proper hyperbolic metric space together with the metric given by the differential

$$\frac{|dz|}{y}, \quad z = x + iy, \quad y > 0.$$

Here we call a metric space *proper* if all its closed balls are compact. The following discussion can be read in detail in [127]. In order to simplify notations we denote the upper half plane $\mathfrak{H}$ just by H in the following.

At first, H together with the given metric is a proper metric space. We denote the distance between two points $P, Q \in H$ by $d_H(P, Q)$. It is determined by minimalizing all lengths $\int dz_H$ over all piecewise smooth paths γ in H from P to Q. For this geometry the geodesics are vertical paths (with respect to the x-axis) and circles and arcs of circles orthogonal to the x-axis; see Figure 3.14.

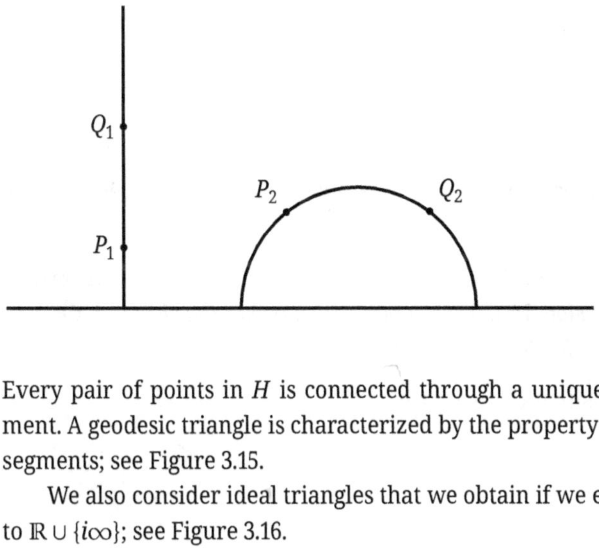

Figure 3.14: Geodesics in H.

Every pair of points in H is connected through a uniquely determined geodesic segment. A geodesic triangle is characterized by the property that all its sides are geodesic segments; see Figure 3.15.

We also consider ideal triangles that we obtain if we extend the geodesic segments to $\mathbb{R} \cup \{i\infty\}$; see Figure 3.16.

The orientation preserving isometries are exactly those linear fractional transformations

$$z \mapsto \frac{az + b}{cz + d}, \quad a, b, c, d \in \mathbb{R}, \quad ad - bc = 1, \quad z \in H.$$

These form a group that is isomorphic to $\mathrm{PSL}(2, \mathbb{R}) \cong \mathrm{SL}(2, R)/\{\pm E_2\}$. An important property of all ideal triangles having all end points on $\mathbb{R} \cup \{i\infty\}$ is that they are equivalent under the group of orientation preserving isometries of H. The reason for this is that all

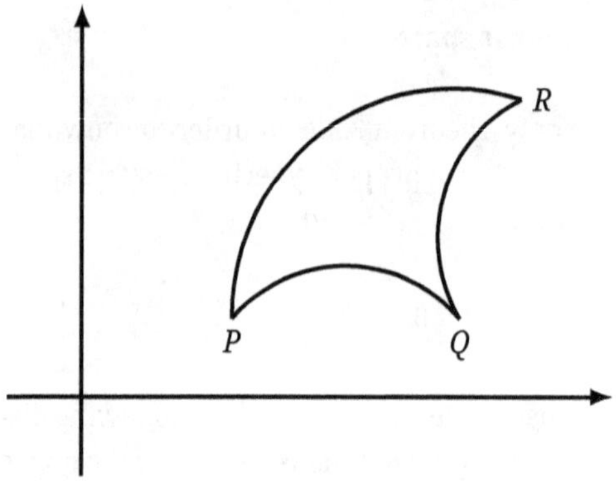

Figure 3.15: Geodesic triangle.

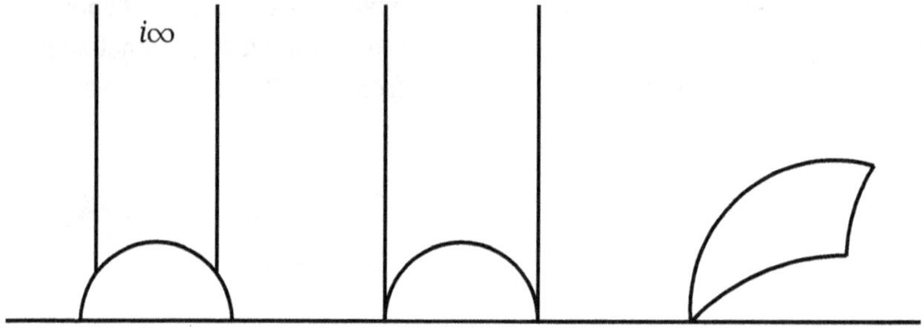

Figure 3.16: Ideal triangles.

linear fractional transformations as above act transitively on the set of three pairwise different points on $\mathbb{R} \cup \{i\infty\}$.

Theorem 3.2.29. *There is a $\delta > 0$ such that for all geodesic triangles in H with vertices A, B, C and all points P on the side AB there is a point Q on one of the sides AC, CB with $d_H(P,Q) \leq \delta$. The optimal value for δ is $\ln(1+\sqrt{2})$.*

For a visualization of the situation, see Figure 3.17.

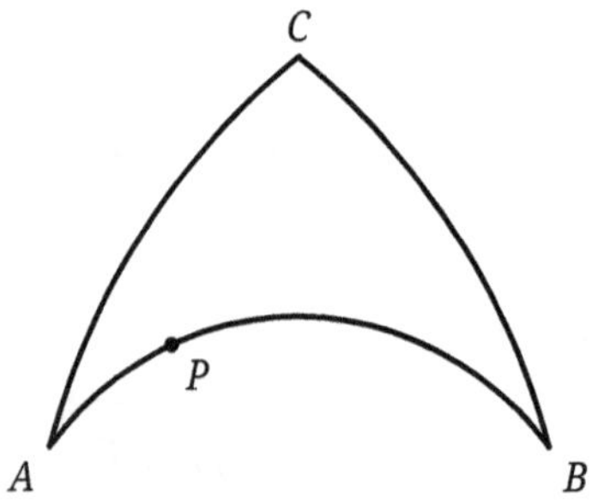

Figure 3.17: Geodesic triangle in H.

Proof. When moving C away from P to $i\infty$ along the geodesic determined by A and B, respectively, the distance from P to the opposite sides becomes at most larger. Analogously, we can move A and B in the direction of the real axis such that we can assume that the triangle ABC is ideal with $A, B, C \in \mathbb{R} \cup \{i\infty\}$.

As all such ideal triangles are equivalent we can assume that $A = (0,0)$, $B = (2,0)$ and $C = i\infty$ such that P lies on the circles with center $(1,0)$ and radius 1. It suffices to show that in Figure 3.18 $d_H(P,Q) \geq d_H(P_1,Q_1)$, where $P = (1,1)$ and PQ, P_1Q_1 are geodesic segments. The map $z \mapsto kz$ induces an isometry of H, where $k \in \mathbb{R}_+^*$. As PQ and P_1Q_1 lie on circles with center $(0,0)$, there exists a uniquely determined k such that $z \mapsto kz$ maps the segment P_1Q_1 on the arc P_2Q, where $P_2 \in PQ$.

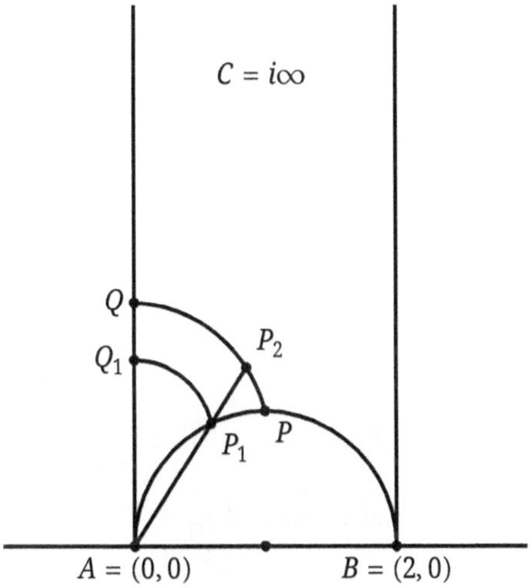

Figure 3.18: Calculation of δ.

Hence $d_H(P_1,Q_1) = d_H(P_2,Q) \leq d_H(P,Q)$. Thus all geodesic triangles in H are δ-thin with $\delta = d_H(P,Q)$. If we parametrize the arc PQ on the circle with radius $\sqrt{2}$ and center $(0,0)$ by $x = \sqrt{2}\cos(\Theta)$, $y = \sqrt{2}\sin(\Theta)$, $\frac{\pi}{4} \leq \Theta \leq \frac{\pi}{2}$, we get

$$\int_{PQ} \frac{|dz|}{y} = \int_{\frac{\pi}{4}}^{\frac{\pi}{2}} \frac{d\Theta}{\sin(\Theta)} = \frac{1}{2}\ln \frac{1-\cos(\Theta)}{1+\cos(\Theta)}\bigg|_{\frac{\pi}{4}}^{\frac{\pi}{2}} = \ln(1+\sqrt{2}). \qquad \square$$

3.3 Hyperbolic geodesic metric spaces and hyperbolic groups

The aim of this section is to prove that commensurable groups are quasi isometric.

3.3.1 Properly discontinuous actions of groups and quasi isometries

Definition 3.3.1. 1. A metric space is called *proper* if all its closed balls are compact.
2. A group Γ acts *properly discontinuous* on a metric space X if for all compact subsets $K \subset X$ there are only finitely many $\gamma \in \Gamma$ with $\gamma(K) \cap K \neq \emptyset$.

We usually write, if no misunderstandings can occur, γx instead of $\gamma(x)$.

Remark 3.3.2. A geodesic metric space X is proper if and only if X is locally compact and complete, see Chapters 9 and 11 in [48] or [261] for details.

Theorem 3.3.3. *Let X be a proper geodesic metric space and Γ be a subgroup of the isometry group of X that acts properly discontinuous on X. If the quotient space X/Γ is compact then Γ is of finite type and Γ, together with the word length metric (see Section 3.1) with respect to a valid finite generating system, is quasi isometric to X.*

Proof. The group Γ is a subgroup of the isometry group of X and hence Γ acts on X. Let $\pi: X \to X/\Gamma$ be the quotient map. We define a metric on X/Γ as follows: If d is the metric on X, then we define $\bar{d}$ on X/Γ via

$$\bar{d}(\bar{x},\bar{y}) := d(\pi^{-1}(\bar{x}), \pi^{-1}(\bar{y}))$$

for all $\bar{x},\bar{y} \in X/\Gamma$. First, if $\pi(x) = \bar{x}$, $\pi(y) = \bar{y}$, then

$$\pi^{-1}(\bar{x}) = \Gamma x = \{z \mid z = \gamma x \text{ for a } \gamma \in \Gamma\}$$

and accordingly $\pi^{-1}(\bar{y}) = \Gamma y$. Hence $\bar{d}(\bar{x},\bar{y}) = d(\Gamma x, \Gamma y)$. As Γ is a subgroup of the isometry group of X, we have $d(\gamma_1 x, \gamma_2 y) = d(x, \gamma_1^{-1}\gamma_2 y)$, hence $\bar{d}(\bar{x},\bar{y}) = d(x, \Gamma y)$. With this at hand, we can easily show that $\bar{d}$ is a metric:
- The inequality $\bar{d}(\bar{x},\bar{y}) \geq 0$ is obvious. If $\bar{x} = \bar{y}$, then there exists a $\gamma \in \Gamma$ with $x = \gamma y$, and thus we have $\bar{d}(\bar{x},\bar{y}) \leq d(x, \gamma y) = d(x,x) = 0$. If $\bar{x} \neq \bar{y}$, then $x \notin \Gamma y$. As Γ acts properly discontinuous and X is proper we have $\bar{d}(\bar{x},\bar{y}) > 0$.
- Of course $\bar{d}$ is symmetric.
- For the triangle inequality let $\bar{x},\bar{y},\bar{z} \in X/\Gamma$. As Γ acts properly discontinuous and X is proper we have $\gamma_1, \gamma_2 \in \Gamma$ with

$$\bar{d}(\bar{x},\bar{z}) = d(x, \gamma_1 z) \quad \text{and} \quad \bar{d}(\bar{z},\bar{y}) = d(z, \gamma_2 y).$$

As Γ is a subgroup of the isometry group of X it follows

$$\begin{aligned} \bar{d}(\bar{x},\bar{z}) + \bar{d}(\bar{z},\bar{y}) &= d(x, \gamma_1 z) + d(z, \gamma_2 y) \\ &= d(x, \gamma_1 z) + d(\gamma_1 z, \gamma_1\gamma_2 y) \\ &\geq d(x, \gamma_1\gamma_2 y) \geq d(x, \Gamma y) = \bar{d}(\bar{x},\bar{y}). \end{aligned}$$

As usual the quotient topology of X/Γ is equivalent to the topology induced by $\bar{d}$ (see [48]). Hence X/Γ is a metric space with $\bar{d}$ as metric. From now on we write again $|x-y|$ instead $d(x,y)$. Let R be the diameter of X/Γ. This exists because X/Γ is compact. We fix $x_0 \in X$ and define

$$B := \{x \in X \mid |x - x_0| \leq R\}, \quad S := \{\gamma \in \Gamma \mid \gamma \neq 1 \text{ and } \gamma B \cap B \neq \emptyset\}.$$

We note that B is compact because X is proper and S is finite because Γ acts properly discontinuous. We prove four properties associated with X, Γ, S and B.

- We have $X = \bigcup_{\gamma\in\Gamma} \gamma B$: Let $y \in X$. Set $\bar{x}_0 = \pi(x_0), \bar{y} = \pi(y)$. As the diameter of X/Γ equals R we have $\bar{d}(\bar{x}_0, \bar{y}) \leq R$, that is, $d(x_0, \gamma y) \leq R$ for a $\gamma \in \Gamma$. Hence $\gamma y \in B$, thus $y \in \gamma^{-1}(B)$.
- The set S is symmetric, that is, $S = S^{-1}$: If $\gamma \in \Gamma \backslash \{1\}$ with $\gamma B \cap B \neq \emptyset$, then there exist $x, y \in B$ with $\gamma(x) = y$ thus $x = \gamma^{-1}(y)$. Hence $B \cap \gamma^{-1}(B) \neq \emptyset$, and thus $\gamma^{-1} \in S$. We now write $\lambda := \sup_{\gamma\in S}\{|x_0 - \gamma(x_0)|\}$ and

$$r = \begin{cases} \infty & \text{if } \Gamma = S \cup \{1\}, \\ \inf\{d(B, \gamma B) \mid \gamma \in \Gamma\backslash(S \cup \{1\})\} & \text{if } \Gamma \neq S \cup \{1\}. \end{cases}$$

- We have $r > 0$: Let $S' = \{\gamma \in \Gamma \mid \gamma(2B) \cap (2B) \neq \emptyset\}$, where $2B := \{x \mid |x - x_0| \leq 2R\}$. Then S' is finite because $2B$ is compact and Γ acts properly discontinuous. This also yields $\{\gamma \in \Gamma \mid \gamma(B) \cap (2B) \neq \emptyset\} \subset S'$ and further $S \cup \{1\} \subset S'$. The claim now follows from the fact that the distance between disjoint compact subsets is strictly larger than 0.
- Suppose there are $x, y \in X$ and $\gamma \in \Gamma$ with $y \in B$, $x \in \gamma B$ and $|y - x| < r$. Then $\gamma \in S \cup \{1\}$: If $\gamma \notin S \cup \{1\}$, then $d(B, \gamma B) \geq r$. But $d(B, \gamma B) \leq d(y, x) < r$. Hence $\gamma \in S \cup \{1\}$.

The proof of the theorem now follows from the following three assertions:
(1) The set S generates Γ, and for every $\gamma \in \Gamma$ we have $d_S(1, \gamma) \leq \frac{1}{r}|x_0 - \gamma x_0| + 1$ (recall that $d_S(\gamma_1, \gamma_2) = l_S(\gamma_1^{-1}\gamma_2)$).
(2) We have $|x_0 - \gamma x_0| \leq \lambda d_S(1, \gamma)$ for all $\gamma \in \Gamma$.
(3) The orbit of x_0, that is $\Gamma x_0 \subset X$, is quasi isometric to X.

We first show the properties (1)–(3) and then how the theorem follows from them.
(1) The claim follows immediately if $r = \infty$, that is, $\Gamma = S \cup \{1\}$. Now let $\Gamma \neq S \cup \{1\}$. Let $\gamma \in \Gamma$ and $k \in \mathbb{N} \cup \{0\}$ be minimal with $|x_0 - \gamma x_0| < kr + R$. Let $[x_0, \gamma x_0]$ be the geodesic segment from x_0 to γx_0. We can choose a finite sequence $x_0, x_1, \ldots, x_{k+1} \in [x_0, \gamma x_0]$ with $|x_1 - x_0| \leq R$ and $|x_i - x_{i-1}| < r$ for $i = 2, \ldots, k+1$ and $x_{k+1} = \gamma x_0$. This is possible because $X = \bigcup_{\gamma\in\Gamma} \gamma B$. That is why there also exist $\gamma_i \in \Gamma$ with $x_i \in \gamma_i B$ for $i = 1, \ldots, k+1$. We choose $\gamma_1 = 1$ and $\gamma_{k+1} = \gamma$. Set $\sigma_i := \gamma_i^{-1}\gamma_{i+1}$ for $i = 1, \ldots, k$. We have $\sigma_1 \cdots \sigma_k = (1^{-1}\gamma_2)(\gamma_2^{-1}\gamma_3)\cdots(\gamma_k^{-1}\gamma_{k+1}) = \gamma_{k+1} = \gamma$, further $\gamma_i^{-1}(x_i) \in B$ for all i. Hence $\gamma_i^{-1}x_{i+1} = \gamma_i^{-1}\gamma_{i+1}\gamma_{i+1}^{-1}x_{i+1} = \sigma_i(\gamma_{i+1}^{-1}x_{i+1}) \in \sigma_i(B)$.

As γ_i is an isometry we have $|\gamma_i^{-1}x_i - \gamma_i^{-1}x_{i+1}| = |x_i - x_{i+1}| < r$ for all i. For $z = \gamma_i^{-1}x_{i+1}$, $x = \gamma_i^{-1}x_i$ we have $z \in \sigma_i(B)$ and $|z - x| < r$. The last remark above yields $\sigma_i \in S \cup \{1\}$ for all i. Also S generates the group Γ. With this, we have also shown that $d_S(1,\gamma) \leq k$, where k is minimal as above with $|x_0 - \gamma x_0| < kr + R$, hence $(k-1)r + R \leq |x_0 - \gamma x_0|$ hence $k \leq \frac{1}{r}|x_0 - \gamma x_0| + 1 - \frac{R}{r} \leq \frac{1}{r}|x_0 - \gamma x_0| + 1$. This is why

$$d_S(1,\gamma) \leq \frac{1}{r}|x_0 - \gamma x_0| + 1.$$

(2) Let $\gamma \in \Gamma$. We use induction over $n = d_S(1,\gamma)$. For $n = 1$ the assertion follows from the definition of λ. Suppose the assumption holds for all $\sigma \in \Gamma$ with $d_S(1,\sigma) \leq n-1$, $n \geq 2$. Let $d_S(1,\gamma) = n$. Then $\gamma = \sigma_1 \cdots \sigma_n$ with $\sigma_i \in S$ for all i. Hence

$$\begin{aligned}|x_0 - \gamma x_0| &= |x_0 - (\sigma_1 \cdots \sigma_n)x_0| \\ &\leq |x_0 - \sigma_1 x_0| + |\sigma_1 x_0 - \sigma_1(\sigma_2 \cdots \sigma_n)(x_0)| \\ &= |x_0 - \sigma_1 x_0| + |x_0 - (\sigma_2 \cdots \sigma_n)(x_0)| \\ &\leq \lambda + \lambda \underbrace{d_S(1, \sigma_2 \cdots \sigma_n)}_{\leq n-1} \\ &\leq \lambda + \lambda(n-1) \\ &= \lambda n = \lambda d_S(1,\gamma).\end{aligned}$$

(3) We look for maps $\Gamma x_0 \underset{g}{\overset{f}{\rightleftarrows}} X$ and constants $C \geq 0$ and $\Lambda > 0$ such that the four conditions for quasi isometries (see Definition 3.1.3) hold. Let f be the inclusion. We would like to define g. To this end we use $X = \bigcup_{\gamma \in \Gamma} \gamma(B)$. Let $g: X \to \Gamma x_0$ be a map with $g(x) = \gamma_x(x_0)$ and $x \in \gamma_x(B)$ (Note that usually there are many γ_x possible. All we do is to fix for every $x \in X$ a γ_x, such that $x \in \gamma_x(B)$). We may assume that $g(\gamma(x_0)) = \gamma(x_0)$ for $\gamma \in \Gamma$. The first and third conditions for quasi isometries hold for all $\Lambda \geq 1$. We assume $\Lambda = 1$ and $C = 2R$.
The second condition for quasi isometry: We show that $d(g(x), g(y)) \leq d(x,y) + 2R$ holds for all $x, y \in X$. We remark that $d(g(x), x) \leq R$ and $d(g(y), y) \leq R$ because the diameter of B is equal to R and the definition of g. That is why

$$\begin{aligned}d(g(x), g(y)) &\leq d(g(x), x) + d(x,y) + d(y, g(y)) \\ &\leq R + d(x,y) + R \\ &\leq d(x,y) + 2R.\end{aligned}$$

The fourth condition for quasi isometry, $d(f \circ g(x), x) \leq 2R$, holds because f is the inclusion:

$$d(f \circ g(x), x) = d(g(x), x) \leq R \leq 2R.$$

We now show how (1), (2) and (3) imply Theorem 3.3.3. According to (3) Γx_0 is quasi isometric to X. Thus it suffices to show that Γ is quasi isometric to Γx_0. This follows now from (1) and (2). To this end we have to construct maps $\Gamma \underset{g}{\overset{f}{\rightleftarrows}} \Gamma x_0$ and constants $C \geq 0$ and $\Lambda > 0$, such that the conditions for quasi isometry hold. To this end, let $f(\gamma) = \gamma x_0$, and for $x \in \Gamma x_0$ we fix $\gamma_x \in \Gamma$ with $\gamma_x(x_0) = x$ and set $g(x) = \gamma_x$. Let $\Lambda := \max\{\lambda, \frac{1}{r}\}$. For the definition of C we consider the stabilizer of x_0

$$\mathrm{Stab}(x_0) = \{\gamma \in \Gamma \mid \gamma x_0 = x_0\}.$$

As Γ acts properly discontinuous, $\mathrm{Stab}(x_0)$ is finite and hence its diameter (with respect to d_S) is finite.

We set $C = 1 + \mathrm{diam}(\mathrm{Stab}(x_0))$ and verify the conditions (a)–(d) for quasi isometries as in Definition 3.1.3.

(a) We show $|f(\gamma_1) - f(\gamma_2)| \leq \Lambda d_S(\gamma_1, \gamma_2) + C$.
We have

$$\begin{aligned} |f(\gamma_1) - f(\gamma_2)| &= |\gamma_1 x_0 - \gamma_2 x_0| = |x_0 - \gamma_1^{-1}\gamma_2 x_0| \\ &\leq \lambda d_S(1, \gamma_1^{-1}\gamma_2) \\ &= \lambda d_S(\gamma_1, \gamma_2) \leq \Lambda d_S(\gamma_1, \gamma_2) + C, \end{aligned}$$

where the inequality follows from (2).

(b) We show $d_S(g(\gamma_1 x_0), g(\gamma_2 x_0)) \leq \Lambda|\gamma_1 x_0 - \gamma_2 x_0| + C$.
Let in addition $g(\gamma_i x_0) = \gamma_i$. Hence $\gamma_i' x_0 = \gamma_i x_0$, thus $(\gamma_i')^{-1}\gamma_i x_0 = x_0$, that is,

$$(\gamma_i)^{-1}\gamma_i \in \mathrm{Stab}(x_0) \quad \text{for } i = 1, 2.$$

Set $\eta_i = (\gamma_i')^{-1}\gamma_i \in \mathrm{Stab}(x_0)$ for $i = 1, 2$. Then $\gamma_i = \gamma_i'\eta_i$ for $i = 1, 2$. Now we have

$$\begin{aligned} d_S(g(\gamma_1 x_0), g(\gamma_2 x_0)) &= d_S(\gamma_1', \gamma_2') = d_S(1, (\gamma_1')^{-1}\gamma_2') \\ &\leq \frac{1}{r} \cdot |x_0 - (\gamma_1')^{-1}\gamma_2'(x_0)| + 1 \\ &= \frac{1}{r}|\gamma_1' x_0 - \gamma_2' x_0| + 1 \\ &= \frac{1}{r}|\gamma_1\eta_1 x_0 - \gamma_2\eta_2 x_0| + 1 \\ &= \frac{1}{r}|\gamma_1 x_0 - \gamma_2 x_0| + 1 \\ &\leq \Lambda|\gamma_1 x_0 - \gamma_2 x_0| + C, \end{aligned}$$

where the first inequality follows from (1) and the last equality follows from $\eta_i \in \mathrm{Stab}(x_0)$.

(c) We show that $d_S(g \circ f(\gamma), \gamma) \leq C$. To this end, let $g \circ f(\gamma) = \gamma$.
Note that as in the second condition we have $\gamma = \gamma\eta$ for an $\eta \in \mathrm{Stab}(x_0)$. Hence

$$d_S(g \circ f(\gamma), \gamma) = d_S(\gamma', \gamma) = d_S(\gamma\eta, \gamma) = d_S(1, (\gamma\eta)^{-1}\gamma) = d_S(1, \eta^{-1}).$$

As $\mathrm{Stab}(x_0) < \Gamma$, it follows that $1, \eta^{-1} \in \mathrm{Stab}(x_0)$, hence

$$d_S(1, \eta^{-1}) \leq \mathrm{diam}(\mathrm{Stab}(x_0)) \leq C,$$

that is, $d_S(g \circ f(\gamma'), \gamma) \leq C$.

(d) This condition is obvious because $f \circ g = \mathrm{id}_{\Gamma_0}$.

This concludes the proof. □

Examples 3.3.4. Let $w_1, w_2 \in \mathbb{C}\backslash\{0\}$ with $\mathrm{im}(\frac{w_1}{w_1}) > 0$. Let $\Gamma = \langle \gamma_1, \gamma_2 \rangle$, where $\gamma_1 : z \mapsto z + w_1$ and $\gamma_2 : z \mapsto z + w_2$. Then it follows from Theorem 3.3.3 that

1. $\mathbb{C}$ is a proper geodesic metric space (with respect to the usual metric in $\mathbb{C}$);
2. Γ is a subgroup of the isometry group of $\mathbb{C}(= \mathbb{R}^2)$, that properly discontinuously acts on $\mathbb{C}$;
3. $\mathbb{C}/\Gamma$ is compact;
4. $\mathbb{C}$ is quasi isometric to Γ (with respect to the metric d_S for $S = \{\pm\gamma_1, \pm\gamma_2\}$); and
5. $\mathbb{C}$ is not hyperbolic.

Corollary 3.3.5. 1. *Let $\Gamma_1 < \Gamma_2$ be of finite index and let Γ_2 be of finite type. Then Γ_1 and Γ_2 are quasi isometric.*
2. *Let $1 \to \Delta \to \Gamma_1 \to \Gamma_2 \to 1$ be a short exact sequence of groups with Δ finite. Let Γ_2 be of finite type. Then Γ_1 and Γ_2 are quasi isometric.*

Proof. 1. Let S be a valid finite generating system of Γ_2 and let X be the Cayley graph of the pair (Γ_2, S). We collect and prove some properties of X and Γ_1.

- The graph X is proper: First observe that $L_n := \{\gamma \in \Gamma_2 \mid l_s(\gamma) \leq n\}$ is finite for all $n \in \mathbb{N}$. Set $X_n = L_n \cup E_n$, where the edge e belongs to E_n if and only if both end points of e belong to L_n. Then X_n is a finite set of vertices and edges as the degree of each vertex of X equals the order of S, that is, it is finite. Hence X_n is compact. Of course we have $X = \bigcup_{n\in\mathbb{N}} X_n$, and any closed ball in X is a closed subset of one of the X_n. Hence every closed ball is compact (see [48]) and hence X is proper.
- The group Γ_1 can now be considered as a subgroup of the isometry group of X: Γ_1 acts via isometries on Γ_2 via the multiplication in Γ_2 (more precisely, on the metric space on (Γ_2, S)) because for $\gamma_1, \gamma_2 \in \Gamma_2$ and $\sigma \in \Gamma_1$ we have

$$d_S(\sigma\gamma_1, \sigma\gamma_2) = l_S(\gamma_1^{-1}\sigma^{-1}\sigma\gamma_2) = l_S(\gamma_1^{-1}\gamma_2) = d_S(\gamma_1, \gamma_2).$$

We consider the vertices of the Cayley graph. They are labeled by elements of Γ_2. An edge connects γ_1 and γ_2 if and only if $d_S(\gamma_1, \gamma_2) = 1$. As Γ_1 acts via isometries on Γ_2, it follows that $d_S(\sigma\gamma_1, \sigma\gamma_2) = 1$ for all $\sigma \in \Gamma_1$, if $d_S(\gamma_1, \gamma_2) = 1$. Hence the end points of an edge are transformed via elements in Γ_1 into end points of an edge. Hence we can parametrize the edges of X (an edge has length 1) such that Γ_1 acts via isometries on the Cayley graph of X.

- The action of Γ_1 on X is proper: It suffices to show that there exist only finitely $\gamma \in \Gamma_1$ with $\gamma(X_n) \cap X_n \neq \emptyset$ for a fixed $n \in \mathbb{N}$. We fix n. Let $\gamma \in \Gamma_1$ with $l_s(\gamma) > 2n$. If $\eta \in \Gamma_2$ and $l_s(\eta) \leq n$, then $l_s(\eta^{-1}\gamma) > n$. If $l_s(\eta^{-1}\gamma) \leq n$, then $\gamma = \eta(\eta^{-1}\gamma)$ would be a product of two elements with S-length smaller than or equal to n and hence $l_S(\gamma) \leq 2n$. This shows: If $l_g(\gamma) > 2n$, then $\gamma(X_n) \cap X_n = 0$. The claim follows as the number of elements in Γ_1 with S-length smaller than or equal to $2n$ is finite.

As now Γ_1 is of finite index in Γ_2, the quotient space Γ_2/Γ_1 is finite. As taking this quotient will never increase the valencies of the vertices we conclude that X/Γ_1 is compact. Hence the conditions of Theorem 3.3.3 for X and $\Gamma = \Gamma_1$ hold. Hence Γ_1 and Γ_2 are quasi isometric as X and Γ_2 are quasi isometric by Definition 3.1.3. This shows the first statement of Corollary 3.3.5.

2. As above let X be the Cayley graph of (Γ_2, S), where S is a valid finite generating system of Γ_2. We already know that X is proper. The action of Γ_1 on X is induced by the canonical homomorphism $\Gamma_1 \to \Gamma_2 \cong \Gamma_1/\Delta$. Hence the action is through isometries as Γ_2 acts via isometries on X (as above). As Γ_2 acts transitively on itself and the canonical homomorphism $\Gamma_1 \to \Gamma_2$ is surjective, Γ_1 acts transitively on the vertices of X. Hence X/Γ_1 is compact because Δ is finite. It remains to show that Γ_1 acts properly on X. Firstly, according to the first assertion Γ_2 acts properly on X. Thus, if C is compact in X and $L = \{\gamma \in \Gamma_2 \mid \gamma(C) \cap C \neq \emptyset\}$, then L is finite. Let $\varphi\colon \Gamma_1 \to \Gamma_2$ be the canonical homomorphism. If $\varphi^{-1}(L)$ is finite, then Γ_1 acts properly on X. But $\varphi^{-1}(L)$ is finite as $|\Delta| < \infty$. Hence Γ_1 and Γ_2 are quasi isometric according to Theorem 3.3.3. □

Remark 3.3.6. The valencies of the vertices in the proof of Corollary 3.3.5.1 can of course decrease taking the quotient. In addition, new vertices of valency 1 can develop. This happens with centers of edges $[\gamma_1, \gamma_2]$ if there exists a $\gamma \in \Gamma_2$ with $\gamma\gamma_1 = \gamma_2$ and $\gamma\gamma_2 = \gamma_1$ $(\gamma_1, \gamma_2 \in \Gamma_1)$. For example, the action of $\mathbb{Z}/2\mathbb{Z}$ on its Cayley graph changes the orientation of its one edge. The quotient space is depicted in Figure 3.19.

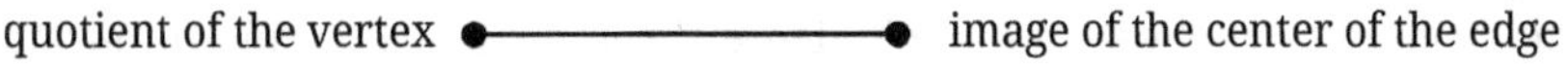
quotient of the vertex
image of the center of the edge

Figure 3.19: Quotient space.

Corollary 3.3.7. *Let $\widetilde{M}$ be a complete Riemannian manifold and let Γ be a subgroup of* $\mathrm{Isom}(\widetilde{M})$ *that acts properly on $\widetilde{M}$. Let $\widetilde{M}/\Gamma$ be compact. Then $\widetilde{M}$ and Γ are quasi isometric.*

Corollary 3.3.8. *Let M be a compact, connected Riemannian manifold with fundamental group Γ. If $\widetilde{M}$ is the universal covering of M, then Γ is quasi isometric to $\widetilde{M}$.*

Proof. The group Γ is isomorphic to the group of deck transformations of $\widetilde{M}$. The latter can be seen as a Riemannian manifold through lifting the metric under the projection map. Hence Γ acts via isometries on $\widetilde{M}$. This action is proper and M is proper because it is locally isometric to M and M is compact. □

3.3.2 Hyperbolic groups and proper actions on some hyperbolic space

Theorem 3.3.9. *Let Γ be a group. Then the following are equivalent:*

(1) *The group Γ is of finite type and hyperbolic.*

(2) *There is a proper hyperbolic geodesic metric space X on which Γ acts properly through isometries such that the diameter of X/Γ is finite.*

Proof.

- We assume that (1) holds. We can choose the Cayley graph of Γ for X. The group Γ being hyperbolic implies by definition that X is hyperbolic. Hence (2) holds.
- We assume that (2) holds. Theorem 3.3.3 implies that Γ is quasi isometric to the hyperbolic space X. Hence the Cayley graph of (Γ, S), S a valid generating system of Γ, is quasi isometric to X. Hence Γ is hyperbolic (which follows from Theorem 3.7.15 where we will show that Y is hyperbolic if X is hyperbolic and X, Y are quasi isometric spaces). □

Now we can put Corollary 3.3.5 differently.

Corollary 3.3.10. 1. *Let Γ_1 be a subgroup of the hyperbolic group Γ_2 with finite index. Then Γ_1 is also hyperbolic.*

2. *Let $1 \to \Delta \to \Gamma_1 \to \Gamma_2 \to 1$ be a short exact sequence of groups with Δ finite and Γ_2 hyperbolic. Then Γ_1 is also hyperbolic.*

We now describe in part Corollary 3.3.10 in a bit different manner. If X is a geodesic metric space then a subset A is *quasiconvex* if there is a constant $\epsilon > 0$ such that for any geodesic $[a, b]$ with endpoints $a, b \in A$ then $[a, b]$ is within an ϵ-neighborhood of A. A subgroup of a finitely generated group is quasiconvex if the vertices in the subgroup form a quasiconvex set in the Cayley graph. Hence Theorem 3.3.9 can be interpreted in the following way.

Theorem 3.3.11. *A quasiconvex subgroup of a hyperbolic group is hyperbolic.*

Thinking of the Cayley graph of a free group as a tree it is clear that a finitely generated subgroup of a finitely generated free group is quasiconvex. Further an infinitely generated subgroup of infinite rank is not hyperbolic and hence not quasiconvex. Therefore we have the following Corollary.

Corollary 3.3.12. *A subgroup of a finitely generated free group is quasiconvex if and only if it is finitely generated.*

We will conclude this section with some comments on the existence and construction of models for n-dimensional hyperbolic geodesic metric spaces. Our aim is to gain a little insight into hyperbolic geometry.

For $n = 2$ we have seen in Subsection 3.2.4 the model of the upper half plane. This model is equivalent to the standard model: The map $z \mapsto \frac{i-z}{i+z}$ maps the points 0, 1, ∞ on 1, i, -1, that is, it maps the real axis on the unit sphere and the point i on 0, and the upper half plane on the interior of the unit circle. The standard model can be extended. Let $n \geq 2$. An n-dimensional model for a hyperbolic n-space can be introduced in the following way: We consider the $x = (x_1, \ldots, x_{n+1}) \in \mathbb{R}^{n+1}$ with

$$x_1^2 + \cdots + x_n^2 - x_{n+1}^2 = -1 \quad \text{and} \quad x_{n+1} > 0.$$

This set can be seen as the point set P of a hyperboloid in $\mathbb{R}^{n+1}$. We imagine this looking at the level set $x_{n+1} = C$ for $C \geq 1$, that is, the intersection of P with the n-dimensional hyperplane $x_{n+1} = C$. For $C = 1$ this set only consists of the point $(0, 0, \ldots, 0, 1)$. If $C > 1$, then the equation $x_1^2 + \cdots + x_n^2 = C^2 - 1$ is the equation of an $(n-1)$-dimensional sphere with center at $(0, \ldots, 0, C)$ in the hyperplane $x_{n+1} = C$. Hence P is obtained by 'stacking' concentric $(n-1)$-spheres on top of each other; see Figure 3.20.

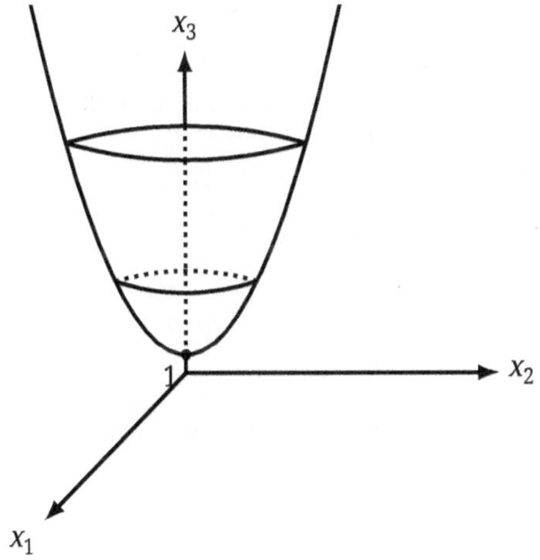

Figure 3.20: Hyperboloid.

We get a better imagination of this set if we consider the intersection with the (x_1, x_{n+1})-plane. This intersection is exactly the upper part of the hyperbola $x_1^2 - x_{n+1}^2 = -1$; see Figure 3.21.

Alternatively one can consider P as the graph of the function

$$f(x_1, \ldots, x_n) = \sqrt{1 + \sum_{i=1}^{n} x_i^2}.$$

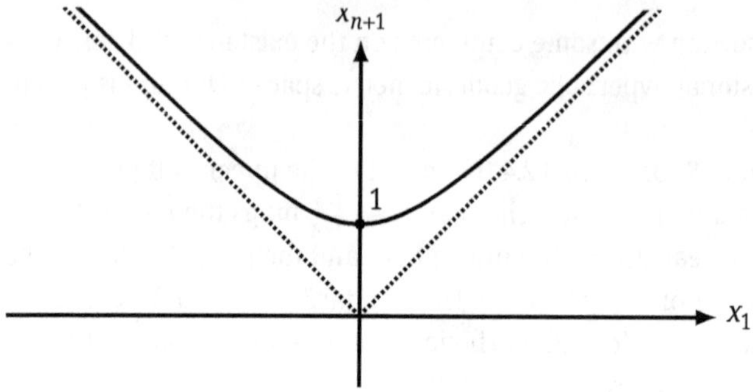

Figure 3.21: Intersection with the (x_n, x_{n+1})-plane.

This way, P becomes diffeomorphic to $\mathbb{R}^n$ via

$$(x_1, \dots, x_n) \mapsto (x_1, \dots, x_n, f(x_1, \dots, x_n)).$$

This point of view (of Riemannian differential geometry) is useful for the definition of a Riemannian metric on P.

As usual we now write X instead of P. If $p \in X$ then the tangent space $T_p(X)$ for X at p of all vectors $v \in \mathbb{R}^{n+1}$ with $v = g'(0)$, where $g(t)$ is a differentiable curve in X with $g(0) = p$. We note that $T_p(X)$ is an n-dimensional vector space over $\mathbb{R}$. In order to construct the desired Riemannian metric we have to define a positive-definite symmetric bilinear form on $T_p(X)$ for all $p \in X$. If we have such a bilinear form (that behaves smoothly as a function of p), we can define an arc length on X and hence a distance on X. We consider the symmetric bilinear form of the $\mathbb{R}^{n+1}$, given by

$$\langle x, y \rangle = x_1 y_1 + \dots + x_n y_n - x_{n+1} y_{n+1}.$$

We have $p \in X$ if and only if $\langle p, p \rangle = -1$.

Remark 3.3.13. If $p \in X$, then $v \in T_p(X)$ if and only if $\langle p, v \rangle = 0$.

Proof. Let $g(t) = (x_1(t), \dots, x_{n+1}(t))$ be a smooth curve in X with $g(0) = p$. Because of $g(t) \in X$ for all t we have $\langle g(t), g(t) \rangle = -1$ for all t. Hence

$$(x_1(t))^2 + \dots + (x_n(t))^2 - (x_{n+1}(t))^2 = -1.$$

Differentiation with respect to t yields

$$2x_1(t)x_1'(t) + \dots + 2x_n(t)x_n'(t) - 2x_{n+1}(t)x_{n+1}'(t) = 0.$$

If we set $t = 0$, we get $\langle p, g'(0) \rangle = 0$. Hence if $v \in T_p(x)$, then $\langle p, v \rangle = 0$. As $T_p(x)$ is an n-dimensional subspace of $\mathbb{R}^{n+1}$, we now have

$$T_p(X) = \{v \in \mathbb{R}^{n+1} \mid \langle v, p \rangle = 0\}$$

by a dimension argument. As $\langle\ ,\ \rangle$ is of the type $(-1, n)$, $\langle\ ,\ \rangle$ is a positive-definite symmetric bilinear form on $T_p(X)$ (theorem of Sylvester). Hence we obtain the Riemannian metric on X. □

We now consider the stereographic projection of

$$\{(x_1, \ldots, x_{n+1}) \in \mathbb{R}^{n+1} \mid x_{n+1} > 0\}$$

on the $\mathbb{R}^n$, given by

$$\pi(x) = \frac{(x_1, \ldots, x_n)}{1 + x_{n+1}}.$$

Figure 3.22 shows a sketch for $\mathbb{R}^3$.

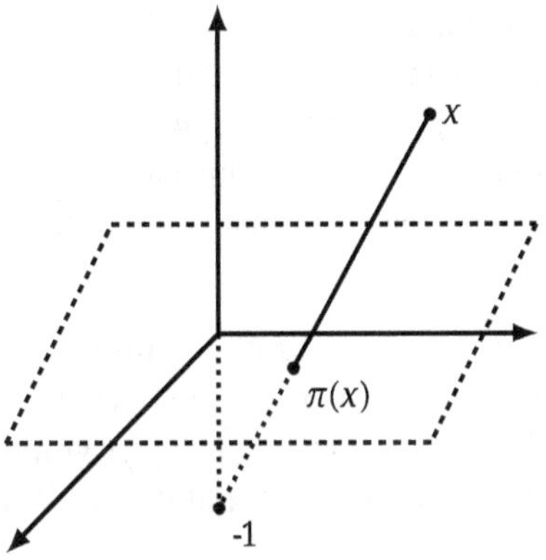

Figure 3.22: Stereographic projection.

The map $\pi|_X$ maps X on the open unit ball in $\mathbb{R}^n$, denoted by Y. Indeed $\pi|_X : X \to Y$ is a diffeomorphism. With this we can impose a metric on Y such that $\pi|_X$ is an isometry.

With this metric Y becomes the Poincaré model of a hyperbolic n-space. In fact, Y is a proper hyperbolic geodesic metric space (in the sense of our theory).

3.4 First properties of hyperbolic groups

In this section we prove in particular the following.

Theorem 3.4.1 (Rips). *Let Γ be a hyperbolic group. Then there is a contractible, locally finite simplicial complex P of finite dimension on which the action of Γ is simplicial, faithful and properly discontinuous and P/Γ is compact.*

We have the following implications:
1. Γ is finitely presented;
2. Γ only has a finite number of conjugacy classes of elements of finite order; and
3. If Γ is torsion-free, then it has finite cohomological dimension.

As before Γ will always denote a group of finite type and S denotes a valid generating system of Γ, that is, $S = S^{-1}$ and $1 \notin S$. Again, $d = d_s$ denotes the word metric of Section 3.1. Hence

$$d(\gamma\gamma_1, \gamma\gamma_2) = l_S(\gamma_1^{-1}\gamma^{-1}\gamma\gamma_2) = l_S(\gamma_1^{-1}\gamma_2) = d(\gamma_1, \gamma_2),$$

for all $\gamma, \gamma_1, \gamma_2 \in \Gamma$.

3.4.1 The Rips complex for a hyperbolic group

We start with the Rips construction of a simplicial complex. For the necessary topological background we refer to [247, 37, 48].

Definition 3.4.2. Let Γ and S be as above and $n \in \mathbb{N}$. A simplicial complex $P_n(\Gamma, S)$ whose vertices consist of elements of Γ and whose k-simplices, for $k \geq 1$, consist of the (unordered) $(k+1)$-tuples of the form $(\gamma_0, \gamma_1, \ldots, \gamma_k)$, where $\gamma_i \in \Gamma$ for all $i, \gamma_i \neq \gamma_j$ for $i \neq j$ and $d(\gamma_i, \gamma_j) \leq n$ for all $i, j \in \{0, 1, \ldots, k\}$ is called a *Rips complex*. We consider $P_n(\Gamma, S)$ together with the weak topology induced by the simplices (see [247]), that is, $A \subset P_n(\Gamma, S)$ is closed if and only if $A \cap T$ is closed in S for every closed simplex T in $P_n(\Gamma, S)$.

Remark 3.4.3. The Rips complex $P_n(\Gamma, S)$ is to some extent a generalization of the Cayley graph $\phi(\Gamma, S)$ for (Γ, S): The k-skeleton of $P_n(\Gamma, S)$, which we denote by $P_n(\Gamma, S)^{(k)}$, consists of all l-simplices in $P_n(\Gamma, S)$ with $0 \leq l \leq k$. In particular, $P_1(\Gamma, S)^{(1)}$ is the Cayley graph for (Γ, S). More generally, let $S_n = \{\gamma \in \Gamma \backslash \{1\} \mid d(\gamma, 1) \leq n\}$. Then $S_1 = S$ and $P_n(\Gamma, S)^{(1)}$ is the Cayley graph for (Γ, S_n).

Example 3.4.4. As an example Figure 3.23 shows $P_2(\mathbb{Z}, \{\pm 1\})^{(1)}$.

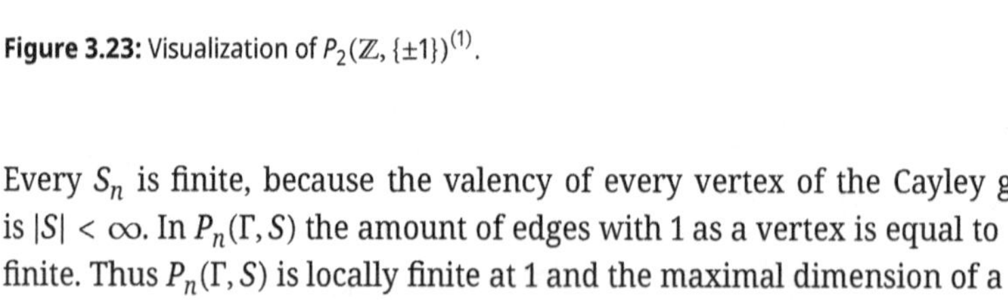

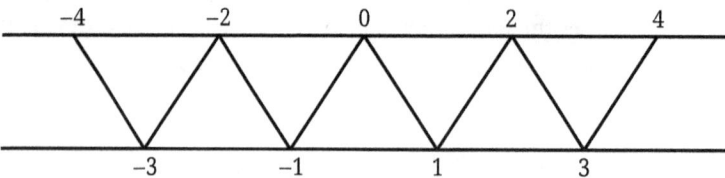

Figure 3.23: Visualization of $P_2(\mathbb{Z}, \{\pm 1\})^{(1)}$.

Every S_n is finite, because the valency of every vertex of the Cayley graph for (Γ, S) is $|S| < \infty$. In $P_n(\Gamma, S)$ the amount of edges with 1 as a vertex is equal to $|S_n|$, hence it is finite. Thus $P_n(\Gamma, S)$ is locally finite at 1 and the maximal dimension of a simplex with 1 as a vertex is at most $|S_n|$. The group Γ acts on the vertices of $P_n(\Gamma, S)$ as a permutation group by Cayley's theorem. And (as in the case of the Cayley graph for (Γ, S)) this action extends to $P_n(\Gamma, S)$, because the permutations preserve distances. It follows that $P_n(\Gamma, S)$

is locally finite at every vertex, hence $P_n(\Gamma, S)$ locally finite and $\dim(P_n(\Gamma, S)) \leq |S_n|$. Additionally, the operation (as in Cayley's theorem) on Γ is faithful. Finally, the permutations of the vertices induce a simplicial operation of Γ on $P_n(\Gamma, S)$.

Theorem 3.4.5. *The action of Γ on $P_n(\Gamma, S)$ has the following properties:*

(a) *The action is faithful, that is, Γ acts as a group of permutations on $P_n(\Gamma, S)$;*
(b) *The stabilizer of each simplex is finite;*
(c) *The action is proper;*
(d) *$P_n(\Gamma, S)/\Gamma$ is compact; and*
(e) *If Γ is torsion-free, then the action is free, that is, if $\gamma \in \Gamma$ fixes a simplex of $P_n(\Gamma, S)$ then $\gamma = 1$.*

Proof. Property (a) follows from the above discussion. Properties (b), (c) and (d) follow as in Corollary 3.3.5. It remains to show (e). First $\gamma \in \Gamma \backslash \{1\}$ fixes no edge of $P_n(\Gamma, S)$ because $\gamma\sigma = \sigma$ implies $\gamma = 1$. In addition, $\gamma \neq 1$ cannot fix a simplex of $P_n(\Gamma, S)$, because according to (b) the stabilizer of each simplex is finite, that is, a finite subgroup of Γ. But as Γ is torsion-free, it has no non-trivial finite subgroup. Hence the action is free. □

Example 3.4.6. Let $\Gamma = \mathbb{Z}/3\mathbb{Z} = \{1, \gamma, \gamma^2\}$ with $S = \{\gamma, \gamma^2\}$. Then $P_n(\Gamma, S) = P_1(\Gamma, S)$ for all $n \geq 1$ and $P_1(\Gamma, S)$ is a simple closed 2-simplex; compare Figure 3.24.

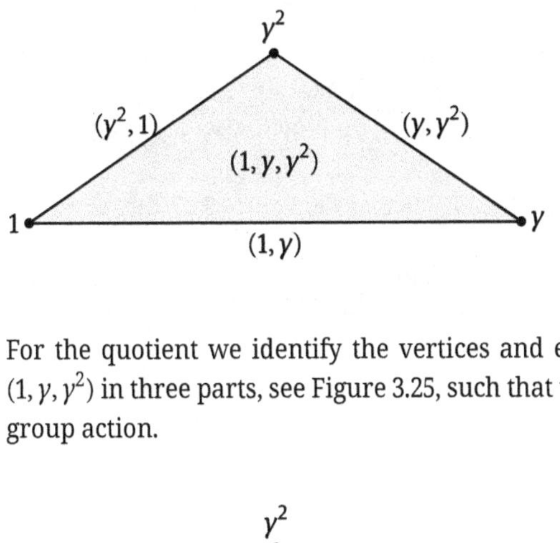

Figure 3.24: Visualization of $P_1(\mathbb{Z}/3\mathbb{Z}, \{\gamma, \gamma^2\})$.

For the quotient we identify the vertices and edges. For the 2-simplex we subdivide $(1, \gamma, \gamma^2)$ in three parts, see Figure 3.25, such that the inner vertices are identified via the group action.

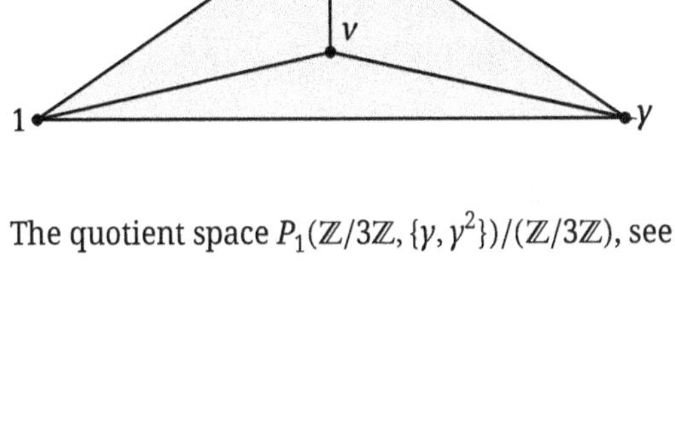

Figure 3.25: Barycentric subdivision of $P_1(\mathbb{Z}/3\mathbb{Z}, \{\gamma, \gamma^2\})$.

The quotient space $P_1(\mathbb{Z}/3\mathbb{Z}, \{\gamma, \gamma^2\})/(\mathbb{Z}/3\mathbb{Z})$, see Figure 3.26, is a CW complex.

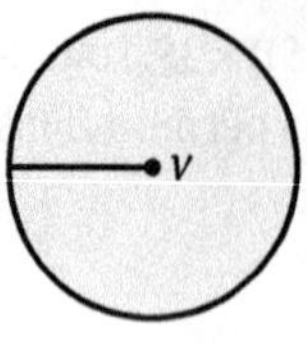

Figure 3.26: Quotient space.

Remarks 3.4.7. 1. Of course $P_n(\Gamma, S)/\Gamma$ is a finite-dimensional CW complex in general.
2. With the barycentric subdivision (see [247] and [48]) we can always obtain a simplicial quotient space. In general, one can show the following: If $P_n''(\Gamma, S)$ is the second barycentric subdivision of $P_n(\Gamma, S)$, then $P_n''(\Gamma, S)/\Gamma$ is a simplicial complex.

Lemma 3.4.8. *Let the Cayley graph of (Γ, S) be δ-hyperbolic. We fix an $n \geq 4\delta + 2$ and a vertex $\gamma_0 \in P_n(\Gamma, S)$. If γ is a vertex in $P_n(\Gamma, S)$ with $d(\gamma_0, \gamma) > [\frac{n}{2}]$, then there is a vertex $\gamma' \in P_n(\Gamma, S)$ with the following properties:*
(a) *$d(\gamma_0, \gamma') = d(\gamma_0, \gamma) - d(\gamma', \gamma)$;*
(b) *$d(\gamma', \gamma) = [\frac{n}{2}]$; and*
(c) *For every vertex $\gamma'' \in P_n(\Gamma, S)$ we have*

$$d(\gamma', \gamma'') \leq \max\left\{\left[\frac{n}{2}\right] + d(\gamma_0, \gamma'') - d(\gamma_0, \gamma), d(\gamma, \gamma'') - \left[\frac{n}{2}\right]\right\} + 2\delta.$$

Proof. Let $\phi(\Gamma, S)$ be the Cayley graph for (Γ, S). We have $\phi(\Gamma, S) \subset P_n(\Gamma, S)^{(1)}$. Let $[\gamma_0, \gamma]$ a geodesic segment in $\varphi(\Gamma, S)$ from γ_0 to γ. If $d(\gamma_0, \gamma) > [\frac{n}{2}]$, then there is according to Theorem 3.2.23 a vertex $\gamma' \in [\gamma_0, \gamma]$ with $d(\gamma', \gamma) = [\frac{n}{2}]$. As γ_0, γ' and γ lie on a fixed geodesic segment, we have $d(\gamma_0, \gamma') + d(\gamma', \gamma) = d(\gamma_0, \gamma)$, hence $d(\gamma_0, \gamma') = d(\gamma_0, \gamma) - d(\gamma', \gamma)$. This shows (a) and (b).

For (c) we use the square version of δ-hyperbolicity as in Figure 3.27, that is,

$$\underbrace{d(\gamma_0, \gamma) + d(\gamma', \gamma'')}_{\text{lengths of the diagonals}} \leq \max\{\underbrace{d(\gamma_0, \gamma'') + d(\gamma', \gamma)}_{\text{side lengths}}, \underbrace{d(\gamma_0, \gamma') + d(\gamma, \gamma'')}_{\text{lengths above and below}}\} + 2\delta.$$

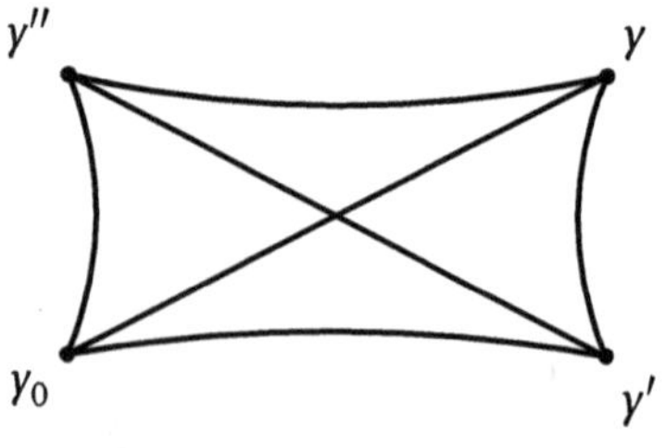

Figure 3.27: Square version.

Hence

$$d(\gamma', \gamma'') \leq \max\{d(\gamma_0, \gamma'') + d(\gamma', \gamma), d(\gamma_0, \gamma') + d(\gamma, \gamma'')\} + 2\delta - d(\gamma_0, \gamma)$$

and with that we get

$$d(\gamma',\gamma'') \le \max\{d(\gamma_0,\gamma'') + d(\gamma',\gamma) - d(\gamma_0,\gamma), d(\gamma_0,\gamma') + d(\gamma,\gamma'') - d(\gamma_0,\gamma)\} + 2\delta$$
$$= \max\left\{d(\gamma_0,\gamma'') + \left[\frac{n}{2}\right] - d(\gamma_0,\gamma), d(\gamma,\gamma'') - \left[\frac{n}{2}\right]\right\} + 2\delta$$

by (a) and (b). □

Theorem 3.4.9. *Let the Cayley graph of* (Γ, S) *be* δ*-hyperbolic. Let* K *be a finite simplicial complex with vertices* $\{p_0, p_1, \dots, p_k\}$*. Let* $n \in \mathbb{N}$ *be a natural number with* $n \ge 4\delta + 2$*, and* $f: K \to P_n(\Gamma, S)$ *be a simplicial map. Then there exists a homotopy* $h: K \times I \to P_n(\Gamma, S)$*, such that* $f \simeq f'$*, that is,* f *is homotopic to* f'*, where* $f': K \to P_n(\Gamma, S)$ *is simplicial and*

$$d(f'(p_0), f'(p_j)) \ge \frac{n}{2}$$

for all vertices $p_j \in K$.

Proof. Let p_i be a vertex of K with

$$d(f(p_0), f(p_i)) = \sup_{1\le j\le k}\{d(f(p_0), f(p_j))\}.$$

If $d(f(p_0), f(p_i)) \le [\frac{n}{2}]$, then we are done. Now let $d(f(p_0), f(p_i)) > [\frac{n}{2}]$. Set $\gamma_0 = f(p_0)$, $\gamma = f(p_i)$ for the application of Lemma 3.4.8. Then there is a vertex $\gamma' \in P_n(\Gamma, S)$ with $d(\gamma', f(p_i)) = [\frac{n}{2}]$ and the following properties:

(a) We have $d(f(p_0), \gamma') = d(f(p_0), f(p_i)) - [\frac{n}{2}]$.
(b) For every vertex $p_j \in K$ we have

$$d(\gamma', f(p_j)) \le \max\left\{\left[\frac{n}{2}\right] + d(f(p_0), f(p_j)) - d(f(p_0), f(p_i)), d(f(p_i), f(p_j)) - \left[\frac{n}{2}\right]\right\} + 2\delta$$
$$\le \max\left\{\left[\frac{n}{2}\right] + 2\delta, d(f(p_i), f(p_j)) - \left[\frac{n}{2}\right] + 2\delta\right\}$$

because $d(f(p_0), f(p_i)) \ge d(f(p_0), f(p_j))$.

Because of $n \ge 4\delta + 2$ we have $\frac{n}{2} \ge 2\delta + 1$, hence $[\frac{n}{2}] + 2\delta \le \frac{n}{2} + \frac{n}{2} = n$ and $[\frac{n}{2}] \ge 2\delta$. Thus $2\delta - [\frac{n}{2}] \le 0$.

Plugging into (b) yields

(c) $d(\gamma', f(p_j)) \le \max\{n, d(f(p_i), f(p_j))\}$ for every $p_j \in K$.

We know that two vertices in $P_n(\Gamma, S)$ are connected by an edge if and only if the distance between them is smaller than or equal to n. Hence by (c) we have that

(d) γ' are $f(p_i)$ are connected by an edge in $P_n(\Gamma, S)$, and,

(e) if $f(p_j)$ and $f(p_i)$ are connected by an edge, then also $f(p_j)$ and y' are connected by an edge $P_n(\Gamma, S)$.

Properties (d) and (e) allow us to define a homotopy $h_1: K \times I \to P_n(\Gamma, S)$ via the natural extension

$$h_1(p_j, t) = \begin{cases} f(p_j) & \text{if } f(p_j) \neq f(p_i), \\ ty' + (1-t)f(p_i) & \text{if } f(p_j) = f(p_i), \end{cases}$$

with $j \in \{0, 1, \ldots, k\}$. The natural extension is given by

$$h_1\left(\sum_{j=0}^{k} \lambda_j p_j, t\right) = \sum_{j=0}^{k} \lambda_j h_1(p_j, t),$$

where $0 \leq \lambda_j \leq 1$ for all j and $\sum_{j=0}^{k} \lambda_j = 1$. Set $f_1 = h_1|_{K\times\{1\}}$, then f_1 is simplicial. If $\sup_{1\leq j\leq k}\{d(f_1(p_0), f_1(p_j))\} \leq \frac{n}{2}$, then we are done. In the other case, we repeat the process. After finitely many steps we are done. □

Theorem 3.4.10. *Let the Cayley graph of* (Γ, S) *be* δ*-hyperbolic. If* $n \geq 4\delta + 2$*, then* $P_n(\Gamma, S)$ *is contractible.*

Proof. According to Theorem 3.4.9 any finite subcomplex of $P_n(\Gamma, S)$ can be homotopically transformed into a closed ball with center 1 and radius $[\frac{n}{2}]$. By definition of $P_n(\Gamma, S)$ the vertices of this ball are the vertices of a simplex. It follows that any map from S^m, $m \geq 0$, is nullhomotopic in $P_n(\Gamma, S)$. Hence we get for the homotopy group $\pi_m(P_n(\Gamma, S)) = 0$ for all $m \geq 0$. As $P_n(\Gamma, S)$ is a simplicial complex, $P_n(\Gamma, S)$ is contractible by Whitehead's theorem. □

We would now like to discuss how the Rips complex $P_n(\Gamma, S)$ for δ-hyperbolic Γ implies that

1. Γ is finitely presented;
2. Γ has only finitely many conjugacy classes of torsion elements; and
3. If Γ is torsion-free, then Γ has finite cohomological dimension.

In the following we assume that δ is a positive integer and that all geodesic triangles are δ-thin, that is, if $f_\Delta(u) = f_\Delta(v)$, then $|u - v| \leq \delta$ where f_Δ is a tripod-map ($\delta \in \mathbb{N}$ is no loss of generality, because if (Γ, S) is δ'-hyperbolic, then also $(\delta' + r)$ for all $r \in [0, \infty)$).

Definition 3.4.11. A path α is called *k-local geodesic*, if any subpath of length $\leq k$ is geodesic.

We use this notion in connection with Cayley graphs where the 'length of a path' is obvious.

Lemma 3.4.12. *Let $k = 4\delta$ and α be a k-local geodesic in $\phi(\Gamma, S)$ from 1 to γ. Let β be a geodesic in $\phi(\Gamma, S)$ from 1 to γ. Let $r \in \alpha$ and $s \in \beta$ be the vertices in $\phi(\Gamma, S)$ with distance 2δ of γ. Then $|r - s| \leq \delta$.*

Remark 3.4.13. As α is 4δ-local geodesic, we see that r lies on the subpath of α that consists of the last 4δ length units, and this is geodesic. Hence r is unique on α and exists ($\delta \in \mathbb{N}$ implies that r and s are vertices).

Proof of Lemma 3.4.12. We consider the subpath of $\phi(\Gamma, S)$ as pictured in Figure 3.28.

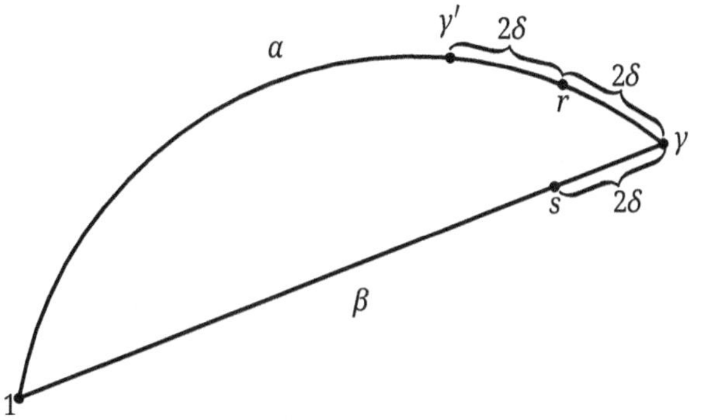

Figure 3.28: Subpath of $\phi(\Gamma, S)$.

If the length of α is smaller than or equal to $k = 4\delta$, then α and β are geodesics. In this case α and β are faces of a degenerated geodesic triangle (the third face is just the point γ). In this case $f_\Delta(r) = f_\Delta(s)$, hence $|r - s| \leq \delta$. Now suppose, that the length of α exceeds $k = 4\delta$. Let γ' be the vertex of α with distance $k = 4\delta$ from γ (see Figure 3.28). We have $|\gamma - \gamma'| = k = 4\delta$ because α is k-local geodesic. If on α the distance of 1 to γ' is smaller than or equal to 2δ, then the subpaths on α from 1 to γ' and from γ' to γ are geodesics, which we will denote by $[1, \gamma]_\alpha$ and $[\gamma, \gamma]_\alpha$. Then $[1, \gamma]_\alpha$, $[\gamma', \gamma]_\alpha$ and β give a geodesic triangle and f_Δ yields identifications of the subpaths $[\gamma', \gamma]_\alpha$, $[1, \gamma']_\alpha$ and β. Because the length of $[1, \gamma']_\alpha$ is at most 2δ, $[\gamma, r]_\alpha$ is identified at most with a part of $[1, \gamma]_\alpha$. Hence we have again $f_\Delta(r) = f_\Delta(s)$ and thus $|r - s| \leq \delta$; compare Figure 3.29.

Now let the distance between 1 and γ' on α exceed 2δ and let β' be a geodesic from 1 to γ as in Figure 3.30. If the length of β' is at most 2δ then β', $[\gamma', \gamma]_\alpha$ and β yield a geodesic triangle and with the same argument as above for $[1, \gamma]_\alpha$, $[\gamma', \gamma]_\alpha$ and β we get $|r - s| \leq \delta$. Hence we assume that the length of $\beta' > 2\delta$ and the length from 1 to γ' on α is also $> 2\delta$. Let $r' \in [1, \gamma]_\alpha$ and $s' \in \beta'$ with distance 2δ from γ' as in Figure 3.31.

Inductively (over the length of α) we can now assume $|r' - s'| \leq \delta$. Now $[r', r]_\alpha$ is geodesic, hence $|r' - r| = 4\delta$. This implies

$$4\delta = |r' - r| \leq |r' - s'| + |s' - r| \leq \delta + |s' - r|$$

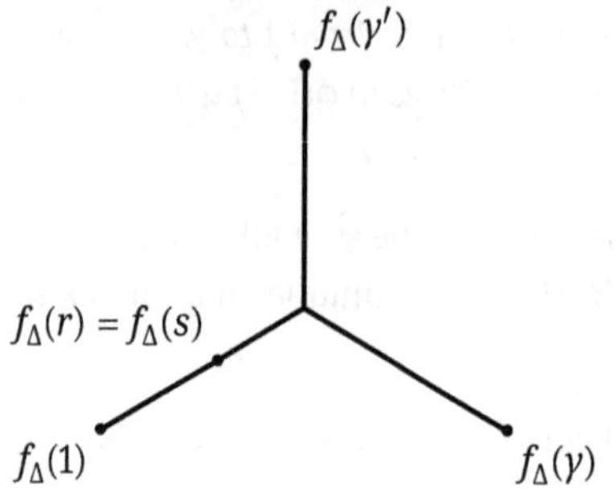

Figure 3.29: Visualization of $f_\Delta(r) = f_\Delta(s)$.

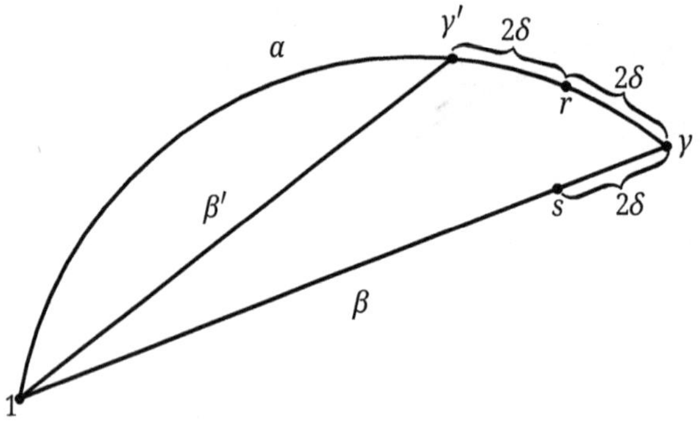

Figure 3.30: Distance exceeds 2δ.

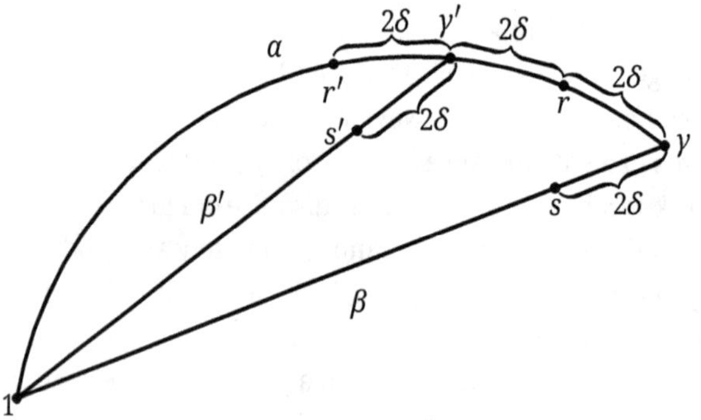

Figure 3.31: Induction steps.

hence $3\delta \le |s' - r|$. We now consider the geodesic triangle $\Delta = \beta' \cup [\gamma', \gamma]_\alpha \cup \beta$. Then $f_\Delta(s') = f_\Delta(r)$ or $f_\Delta(s) = f_\Delta(r)$. If we have $f_\Delta(s') = f_\Delta(r)$, then $|s' - r| \le \delta$, which is a contradiction. Hence $f_\Delta(s) = f_\Delta(r)$, that is, $|s - r| \le \delta$. □

Theorem 3.4.14. *Let k, α, β and δ be as in Lemma* 3.4.12. *Then α lies within the (closed) 3δ-neighborhood of β.*

Proof. Let $z \in \alpha$ be a point whose distance along α to 1 as well as to y is bigger than 2δ (if such a z does not exist we are done). Let $[1, z]$ and $[z, y]$ denote geodesics as in Figure 3.32.

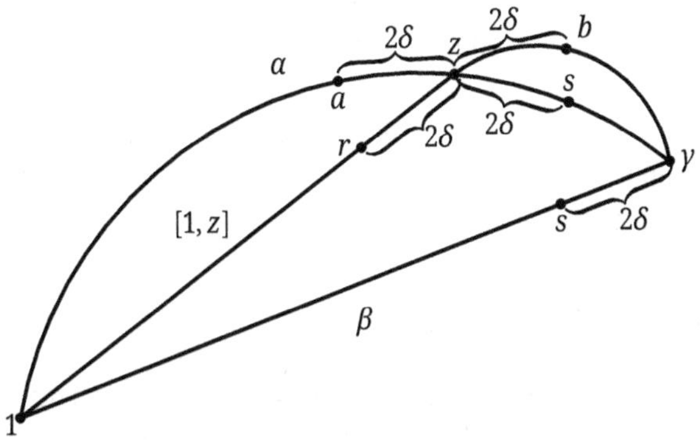

Figure 3.32: Constructing points.

Let $a, b \in \alpha$, $r \in [1, z]$ and $s \in [z, y]$ be points with distance 2δ from z (see Figure 3.32). According to Lemma 3.4.12 we have $|a - r| \le \delta$ and $|b - s| \le \delta$ (Lemma 3.4.12 does not show directly that $|b - s| \le \delta$, but left translation with y^{-1} is an isometry of $\phi(\Gamma, S)$ and Lemma 3.4.12 can now be applied to this situation). As α is a 4δ-local geodesic, $k = 4\delta$, we now have $|a - b| = 4\delta = k$, hence

$$\underbrace{|a-b|}_{=4\delta} \le \underbrace{|a-r|}_{\le\delta} + |r-s| + \underbrace{|s-b|}_{\le\delta}.$$

This yields $|r - s| \ge 2\delta$. As $[1, z] \cup [z, y] \cup \beta$ is a geodesic triangle and $|r - s| \ge 2\delta$, we have $d(r, \beta) \le \delta$ (as $\phi(\Gamma, S)$ is δ-hyperbolic), thus $d(z, \beta) \le |z - r| + d(r, \beta) \le 2\delta + \delta = 3\delta$. □

3.4.2 The Dehn algorithm and applications

Definition 3.4.15. We say that a group G *admits a Dehn algorithm*, if G has a finite presentation $\langle X \mid R \rangle$, such that the following holds: If $1 \neq w$ is a cyclically reduced word in $F(X)$ that is trivial in G then the length of w can be reduced by application of an element r that is cyclically reduced and conjugate to an element in $R^{\pm 1}$, that is, w contains a cyclically reduced subword that is also a subword of such an r and whose length is $> \frac{1}{2}|r|$, where $|r|$ denotes the free length of r. We say $\langle X \mid R \rangle$ is a Dehn algorithm for G.

Theorem 3.4.16. *Let (Γ, S) be a δ-hyperbolic group with $\delta \in \mathbb{N}$. Let R denote the words of $F(S)$, that are trivial in Γ and that have length $\le 8\delta$. Then $\langle S \mid R \rangle$ is a finite presentation for Γ, and $\langle S \mid R \rangle$ is a Dehn algorithm for Γ. In particular, the word problem in $\langle S \mid R \rangle$ is solvable.*

Proof. We have to show: If $w \in F(S)$ is cyclically reduced with w trivial in Γ then the length of w can be reduced with an element of R. Note that if $r \in R$, then so is r^{-1} and every cyclically reduced conjugate of r in R. Now let $w \in F(S)$ be cyclically reduced with w trivial in Γ. Let $k = 4\delta$. First let w, considered as a path in $\phi(\Gamma, S)$, be not k-local geodesic. Then w has a subpath λ of length $\leq k$ that is not geodesic. We can now assume that λ connects vertices of $\phi(\Gamma, S)$. Let β be a geodesic in $\phi(\Gamma, S)$ that connects the end points of λ. As the length of β is smaller than the length of λ, $\lambda\beta^{-1}$ is a closed path in $\phi(\Gamma, S)$ of length $< 2k = 8\delta$. Hence $\lambda\beta^{-1}$ can be considered as a conjugate of an element in R, whereby the elements of R can be considered as loops in $\phi(\Gamma, S)$ with basis point 1 and length $\leq 2k$. Hence in this case the length of w can be reduced by an element in R.

We now consider the case that w is k-local geodesic. As w represents a closed path with basis 1 in $\phi(\Gamma, S)$, it follows from Theorem 3.4.14, that w lies within a 3δ-neighborhood of 1. This implies that the length of w is smaller or equal to 3δ. If this was not the case, that is, if the length of w exceeds 3δ, then the initial path of w of length $3\delta + 1$ would be geodesic, hence it cannot be contained in a 3δ-neighborhood of 1, which yields a contradiction. Here we also have $w \in R$. The solvability of the word problem is obvious. □

Corollary 3.4.17. *The group (Γ, S) as above has only finitely many conjugacy classes of torsion elements.*

Proof. Let $\langle S \mid R \rangle$ be a Dehn algorithm for Γ. Let $\gamma \in \Gamma$ have finite order and let $\mathrm{conj}(\gamma)$ be the conjugacy class of γ in Γ. Let w be a word in $F(S)$ that has the shortest length among all words $v \in F(S)$ with $v \in \mathrm{conj}(\gamma)$. In particular, w is cyclically reduced. Let n be the order of w in Γ. Then $w^n = 1$ in Γ, hence there is a subword u in w^n and an $r \in R$ with $|u| > \frac{1}{2}|r|$, where $|\cdot|$ denotes the free length. The subword u cannot be a subword of w or a cyclic conjugate of w, because then w could be reduced in length contradicting the choice of w. Hence $|w| \leq \frac{1}{2}|r|$. This shows that the number of conjugacy classes of elements of finite order is bounded by the number of elements that have length at most $\max_{r \in R}\{\frac{1}{2}|r|\}$. □

As a direct consequence we get the following.

Corollary 3.4.18. *Every infinite hyperbolic group contains an element of infinite order.*

Remarks 3.4.19.
1. Our proof for Theorem 3.4.16 and Corollary 3.4.17 is more direct than that of Rips and uses only the Cayley graph (see [7]). Rips' proof uses the fact that an element of finite order, if it acts on a contractible Rips complex $P_n(\Gamma, S)$, has a fixed point.
2. The condition '$\delta \in \mathbb{N}$' in Theorem 3.4.16 and Corollary 3.4.17 is of course not necessary but is of a technical nature. Theorem 3.4.16 and Corollary 3.4.17 hold for every δ-hyperbolic group (Γ, S). If (Γ, S) is indeed δ-hyperbolic then is it also δ'-hyperbolic for $\delta' \geq \delta$.

3. Let (Γ, S) be a δ-hyperbolic group and $\langle S \mid R\rangle$ be a Dehn algorithm. Then $\langle S \mid R\rangle$ has a solvable conjugacy problem (see [118]). Using annular maps a proof can be given analogously to the respective proofs for small cancellation groups; see [177].
4. Let (Γ, S) be a torsion-free δ-hyperbolic group. Then (Γ, S) has a solvable isomorphism problem in the class of torsion-free hyperbolic groups (see [236]).
5. Let (Γ, S) be a δ-hyperbolic group. Then (Γ, S) has a solvable isomorphism problem in the class of hyperbolic groups (see [69]).

Theorem 3.4.16, together with Theorem 3.3.3 and Remark 3.2.27, yields the following.

Theorem 3.4.20. *Let (Γ, S) be a δ-hyperbolic group. Then the following hold:*
1. *The group (Γ, S) contains no copy of $\mathbb{Z} \oplus \mathbb{Z}$ as a subgroup.*
2. *If A is an Abelian subgroup of (Γ, S) and if A has an element of infinite order, then A is virtually infinite cyclic.*
3. *If (Γ, S) is torsion-free and A is an Abelian subgroup of (Γ, S), then A is cyclic or trivial. Especially, if g is an element of Γ of infinite order and not a proper power in Γ then $\langle g\rangle$ is malnormal in Γ.*
4. *Every non-trivial Abelian subgroup A of (Γ, S) is virtually cyclic.*

We leave the details as an exercise. The following is a direct consequence.

Corollary 3.4.21. *Let Γ be an infinite hyperbolic group.*
1. *Let $g \in \Gamma$ be an element of infinite order. Then $\langle g\rangle$ has finite index in its centralizer $C(g)$ in Γ, in particular $C(g)$ is virtually $\mathbb{Z}$. Especially, if Γ is torsion-free, then $C(g)$ is infinite cyclic.*
2. *If Γ is torsion-free then Γ is CT and CSA.*

We now look at the third statement of Theorem 3.4.1 (Rips).

Theorem 3.4.22. *Let (Γ, S) be a torsion-free δ-hyperbolic group. Then (Γ, S) has finite cohomological dimension.*

Proof. For the notions used in the following we refer to [38] and [247]. As Γ is torsion-free, the action of Γ on $P_n(\Gamma, S)$ is free. Now let $n \geq 4\delta+2$. Then $P_n(\Gamma, S)$ is contractible and hence Γ is isomorphic to the fundamental group $\pi_1(P_n(\Gamma, S)/\Gamma)$ (see Theorem 2.3.11). Furthermore, $P_n(\Gamma, S)/\Gamma$ is aspheric, that is, the higher homotopy groups are trivial and thus the cohomology groups of Γ are isomorphic to those of $P_n(\Gamma, S)/\Gamma$ (see [38, Chapter III]). As $P_n(\Gamma, S)/\Gamma$ is a finite-dimensional CW complex all the cohomology groups of dimension that exceed the dimension of $P_n(\Gamma, S)/\Gamma$ vanish. Hence Γ has finite cohomological dimension. □

At this point we would like to refer to two results of [201] that we state without proof.

Theorem 3.4.23. *Let (Γ, S) be a torsion-free δ-hyperbolic group. The Euler-characteristic and the homological dimension can be calculated by an algorithm.*

Theorem 3.4.24. *If* (Γ, S) *is* δ*-hyperbolic and* Γ *is virtually torsion-free then* $\operatorname{vcd}(\Gamma) \leq \dim(P_{4\delta+2}(\Gamma, S))$ *(where* $P_{4\delta+2}(\Gamma, S)$ *is the Rips complex whose dimension can be calculated by an algorithm).*

3.4.3 Linear isoperimetric inequalities

We now would like to study geometric properties that yield equivalent definitions of hyperbolic groups. In the following let G be a finitely presented group with a finite presentation $\langle S \mid R\rangle$. Let S be valid and every word in R be cyclically reduced. Furthermore, let R contain all inverses and all cyclic conjugates of elements of R. We consider G as a factor group of $F(S)$ under the canonical map $\pi: F(S) \to G$. Let N be the kernel of π hence $N = \langle\langle R\rangle\rangle$. Every element $w \in N$ can be written as

$$w = \prod_{i=1}^{k} v_i r_i v_i^{-1}$$

with reduced words $v_i \in F(S)$ and $r_i \in R$ for $i = 1, \ldots, k$. Here, the smallest possible k is called the *combinatorial area* $A(w)$ of w.

Definition 3.4.25. A group G as above has a *linear isoperimetric inequality*, if there exist a linear map f with $A(w) \leq f(|w|)$ for every $w \in N$, where $|\cdot|$ denotes the free length in $F(S)$.

Remark 3.4.26. The definition depends on the given generating system S. However, it is clear that the property of having a linear isoperimetric inequality is independent of the choice of a finite generating system S (see the proof of Theorem 3.1.5).

Theorem 3.4.27. *If* $\langle S \mid R\rangle$ *is a Dehn algorithm for* G *then* G *has a linear isoperimetric inequality.*

Proof. Let $\langle S \mid R\rangle$ be a Dehn algorithm for G, S a valid generating system and let R satisfy the above assumptions. As above, let N be the kernel of π. Let $w \in F(S)$ with $w \in N$. Then there exists an $r \in R$ with $r = r_1 r_2$, $|r_1| > |r_2|$ with $w = w_1 r_1 w_2$ in $F(S)$ and we have $w = w_1 r_2^{-1} w_2 w_2^{-1} r_2 r_1 w_2$ and $|w_1 r_2^{-1} w_2| < |w|$, $w_1 r_2^{-1} w_2 \in N$.

Now we can see inductively that w can be written as a product of $\leq 2|w|$ elements in $F(S)$ that are conjugates of words in R. If we set $f(x) = 2x$, we have $A(w) \leq f(|w|)$ with respect to the presentation $G = \langle S \mid R\rangle$. Thus G has a linear isoperimetric inequality. □

Corollary 3.4.28. *Let* (Γ, S) *be a* δ*-hyperbolic group. Then* Γ *has a linear isoperimetric inequality.*

Proof. This follows because Γ admits a Dehn algorithm. □

We will see that the converse also holds.

Theorem 3.4.29. *Let $G = \langle S \mid R\rangle$ be finitely presented. If G has a linear isoperimetric inequality then G is hyperbolic.*

The following proof is taken from [209] where we adopted our notation.

Proof. Let S be a valid finite generating system of G. In order to prove that G is hyperbolic we need to show that there is a $\delta \geq 0$ such that every geodesic triangle in $\Gamma(G,S)$ is δ-thin. Suppose, seeking a contradiction, that for any $L > 0$, there is a geodesic triangle in $\Gamma(G,S)$ that is not $2L$-thin.

For a given $L > 0$, let $\Delta_{xyz} = [x,y] \cup [y,z] \cup [z,x]$ be a geodesic triangle in $\Gamma(G,S)$ that is not $2L$-thin. Renaming x, y, z if necessary, we can assume that there is a point $p \in [x,y]$ such that $d_{\Gamma(G,S)}(p, [y,z] \cup [z,x]) > 2L$. Let r be the maximum of the lengths of words in R. Fix some $\varepsilon > r$ independently of L. Since we can take a large L arbitrarily, we can assume that $L > 4\varepsilon$. We cut off each vertex of Δ_{xyz} by a geodesic segment of length 4ε so that in the remaining part H, each side other than those contained in the original sides is a geodesic segment of length at least 4ε joining two of $[x,y]$, $[y,z]$, $[z,x]$. Since p is apart from the vertices of Δ_{xyz} at distance greater than $2L$, it is contained in the remaining part H. The remaining part H must be one of the following three types, compare Figure 3.33.
1. A hexagon obtained by deleting from Δ_{xyz} three triangles each containing a vertex.
2. A quadrilateral obtained by deleting from Δ_{xyz} one triangle containing a vertex and one quadrilateral containing a side.
3. A pentagon obtained by deleting from Δ_{xyz} two triangles sharing a part of a side, each of which contains a vertex, and one triangle containing the remaining vertex.

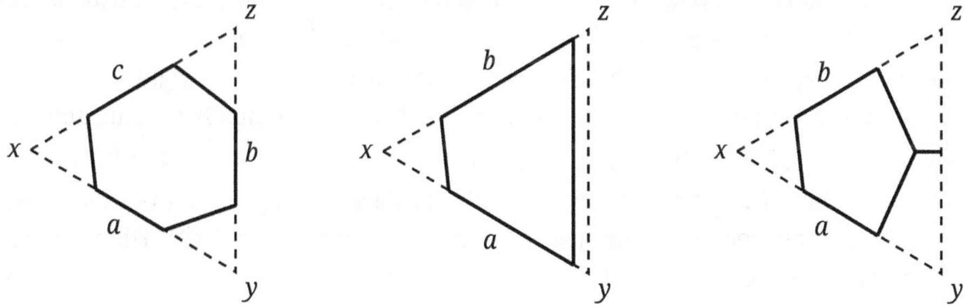

Figure 3.33: Possible cases for the remaining part H.

Since there are two or three sides of H along which Δ_{xyz} was not cut and which are contained in the sides of Δ_{xyz}, we call them a, b or a, b, c, respectively. Let $|a|$, $|b|$, $|c|$ be the lengths of a, b, c, respectively. We can assume that p lies on a. Since H forms a closed curve in $\Gamma(G,S)$, it corresponds to a conjugacy class of a word $w \in F(S)$ such that $p_S(w) = 1$. Since we assumed that G satisfies a linear isoperimetric inequality, there are constants A, B independent of w such that $A(w) \leq A|w| + B$.

We take a diagram, for instance a van Kampen diagram, D, of w such that each boundary of a 2-cell represents a conjugacy class of an element of R (see for instance [177] or [47]). Then we can construct a cellular map f from the 1-skeleton D^1 of D to $\Gamma(G,S)$ such that $f(\partial D) = H$ and each boundary of a 2-cell is mapped to a loop in $\Gamma(G,S)$ representing a conjugacy class of an element in R by taking each edge of D^1 to an edge of $\Gamma(G,S)$ representing the same element of S. We call a loop in $\Gamma(S,R)$ that is such an image of the boundary of a 2-cell by f, an *R-cycle*. The number of R-cycles in $f(D^1)$ is bounded by $A|w| + B$.

Now, we consider bounding the number of R-cycles in $f(D^1)$ from below. We take neighborhoods N_a, N_b, N_c (or N_a, N_b) of the sides a, b, c (or a, b unless H is a hexagon) in $f(D^1)$ as follows. First, for the side a, let $N_1(a)$ be the union of all R-cycles having edges contained in a. Since each R-cycle has length at most r, the neighborhood $N_a(a)$ contains at least $\frac{|a|}{r}$ R-cycles. Moreover, there is a unique arc on $\partial N_1(a)$ other than a connecting two sides of H, along which Δ_{xyz} was cut. Let a_1 denote this arc. Since the end points of a_1 are within distance $\frac{r}{2}$ from the end points of a, and a is a geodesic segment, we get $\text{length}(a_1) \geq |a| - r$ by the triangle inequality.

Similarly, let $N_2(a)$ be the union of all R-cycles having edges on a_1 and not contained in $N_1(a)$, that is, lying on the opposite side of a_1 from $N_1(a)$. Then $N_2(a)$ consists of at least $\frac{\text{length}(a_1)}{r} \geq \frac{|a|}{r-1}$ R-cycles. Let a_2 be the arc other than a_1 on $\partial N_2(a)$ connecting two sides of H, similarly to a_1 for $N_1(a)$. Then we have $\text{length}(a_2) \geq |a| - 2r$. Iterating this procedure $[\frac{\varepsilon}{r}]$ times, we set $N_a = \bigcup_{i=1}^{[\frac{\varepsilon}{r}]} N_i(a)$. We can easily see that, for $i \neq j$, the $N_i(a)$ and the $N_j(a)$ cannot contain common R-cycle. Hence, N_a contains at least $\frac{|a|[\frac{\varepsilon}{r}]}{r} - \frac{1}{2}([\frac{\varepsilon}{r}]([\frac{\varepsilon}{r}]+1))$ R-cycles. Moreover, since each point of N_a can be reached from a by connecting at most $[\frac{\varepsilon}{r}]$ arcs, each on an R-cycle, the set N_a is contained in the $(\varepsilon + r)$-neighborhood, hence 2ε-neighborhood of a. We define, by the same method, N_b, N_c when H is a hexagon (N_b otherwise). If $N_a \cap N_b \neq \emptyset$, then a and b can be connected by an arc with length $2\varepsilon + 2r < 4\varepsilon$. This contradicts our definition of H. Therefore, N_a, N_b, N_c are mutually disjoint, and the number of R-cycles contained in $N_a \cup N_b \cup N_c$ is at least $\frac{[\frac{\varepsilon}{r}](|a|+|b|+|c|)}{r} - \frac{3}{2}([\frac{\varepsilon}{r}]([\frac{\varepsilon}{r}] + 1))$. In the case when H is not a hexagon, we need to replace $|a| + |b| + |c|$ above by $|a| + |b|$. Let B_L be the intersection of the closed L-neighborhood of p and $f(D^1)$. Then, by our choice of p, we have $B_L \cap H \subset a$. Recall that ∂N_a contains an arc $a_{[\frac{\varepsilon}{r}]}$ connecting two sides of H. We can join $\partial B_L \cap a_{[\frac{\varepsilon}{r}]}$ to either of the two points of $\partial B_L \cap a$ by an arc with length less than or equal to 2ε, by tracking on the parts of R-cycles in ∂B_L. Let k be the subarc of $a_{[\frac{\varepsilon}{r}]}$ between two points of $\partial B_L \cap a_{[\frac{\varepsilon}{r}]}$ which are within distance 2ε from one of the two points of $\partial B_L \cap a$, respectively. As $B_L \cap a$ is a geodesic segment of length $2L$, we get $\text{length}(k) \geq 2L - 4\varepsilon$. Therefore, the number of R-cycles having edges on k and not contained in N_a is at least $\frac{2L-4\varepsilon}{r}$. Moreover, since the $2L$-neighborhood of p is disjoint from $b \cup c$ and $L > 4\varepsilon$, the R-cycles that we are counting are disjoint from $N_b \cup N_c$.

Thus, the number of R-cycles in $f(D^1)$ is at least

$$\frac{[\varepsilon/r](|a| + |b| + |c|)}{r} - \frac{3}{2}([\varepsilon/r]([\varepsilon/r] + 1)) + \frac{2L - 4\varepsilon}{r}.$$

If we fix ε such that $\frac{[\frac{\varepsilon}{r}]}{r} > A$, then the number of R-cycles in $f(D^1)$ can be bounded below by $A(|a| + |b| + |c|) + \frac{2L}{r} - C$ for a constant C independent of L. On the other hand, by the linear isoperimetric inequality, the number of R-cycles is bounded above by $A(|a|+|b|+|c|+12\varepsilon)+B$. We get a contradiction if we choose L to be sufficiently large. □

We can summarize our discussion as follows:

Theorem 3.4.30. *Let $G = \langle S \mid R\rangle$ be finitely presented. Then the following are equivalent:*
1. *The group G is hyperbolic.*
2. *The group G admits a Dehn algorithm.*
3. *The group G has a linear isoperimetric inequality.*

The next section is devoted to classical and recent examples of hyperbolic groups.

3.5 Examples of hyperbolic groups

We have already remarked that finite groups are obviously hyperbolic. As an example of a group that is not a hyperbolic group we just consider the Baumslag–Solitar groups $\mathrm{BS}(n,m) = \langle a, b \mid ba^n b^{-1} = a^m = 1\rangle$, $2 \le |m|$, $1 \le |n|$. Then the subgroup $\langle a^m, abab^{-1}\rangle$ is free Abelian of rank 2 and hence $\mathrm{BS}(n,m)$ is not hyperbolic.

In this section we will discuss more examples of hyperbolic groups. There has been a rich activity in the research on hyperbolic groups in recent years that led to new examples and insight into the structure of hyperbolic groups. We start with the discussion of some more basic examples and devote the following sections to new examples that have not been covered in the broader literature so far.

Examples 3.5.1. 1. Free groups of finite rank, and more generally, free products of finitely many hyperbolic groups are hyperbolic.
2. If A, B are hyperbolic and $A \cap B = H$ is a quasi convex subgroup, malnormal in either A or B, then the amalgamated product $A *_H B$ is hyperbolic. In particular if $H = \langle U\rangle = \langle V\rangle$ and U, V are elements of infinite order in A and B, respectively, and neither is a proper power then $A *_H B$ is hyperbolic. Since finitely generated subgroups of free groups are quasi convex it follows that if A and B are free groups and $A \cap B = H$ is a finitely generated subgroup malnormal in either A or B then $A *_A B$ is hyperbolic (see [27] and [156]).
3. A *separated HNN extension* is an HNN extension $K = \langle G, t \mid t^{-1}At = B\rangle$ where the associated subgroups have the property that $g^{-1}Ag \cap B = \{1\}$ for all $g \in G$. A separated HNN extension $K = \langle G, t \mid t^{-1}At = B\rangle$ of a hyperbolic group G is hyperbolic if the associated subgroups A and B are quasi convex in G and at least one is malnormal in G (see [156]).

   We now give a special application. In Subsection 1.6.3 we introduced the parafree groups

$$G_{i,j} = \langle a, b, t \mid a[b^i, a][b^j, t] = 1\rangle,$$

where $i, j \geq 1$. The $G_{i,j}$ can be expressed as an HNN extension

$$G_{i,j} = \langle a, b, t \mid t^{-1}a[b^i, a]b^j t = b^j\rangle.$$

The element $a[b^i, a]b^j$ is malnormal in $\langle a, b \mid \rangle$. Hence, we may apply the result in order to get that all $G_{i,j}$ are hyperbolic.

4. Let G be a non-Abelian, finitely generated fully residually free group with the property that every subgroup of infinite index is free. Then G is hyperbolic (see [82]).
5. Let $G = \langle a_1, \dots, a_n \mid R(a_1, \dots, a_n)^m = 1\rangle$, $n \geq 1$, $m \geq 2$ and $R(a_1, \dots, a_n)$ a nontrivial cyclically reduced word in the free group on $a_1, \dots, a_n$. Then G is hyperbolic. It follows from Newman's spelling theorem, Theorem 1.7.9, that G satisfies a Dehn algorithm.

3.5.1 Surface groups

Let $F = F(S)$ be free on the finite set S. Let $R \subset F$ be *symmetric*, that is, each $r \in R$ is cyclically reduced and with $r \in R$ all cyclic conjugates of r and r^{-1} are also in R. If r_1 and r_2 are different elements of R with $r_1 \equiv bc_1$ and $r_2 \equiv bc_2$, $b \in F$, then b is called a *piece* (with respect to R). Let $\lambda \in \mathbb{R}$, $\lambda > 0$ and $p, q \in \mathbb{N}$. We recall the *small cancellation conditions*:

- Condition $C'(\lambda)$: If $r \in R$ and $r \equiv bc$, where $b \in$ piece then $|b| < \lambda|r|$.
- Condition $T(q)$: Let $3 \leq h < q$ and $r_1, \dots, r_k \in R$ with $r_i r_{i+1} \neq 1$ for $i = 1, \dots, k$ whereby $r_{k+1} = r_1$. Then at least one of the products $r_1 r_2, \dots, r_{k-1} r_k, r_k r_1$ is reduced without shortening (with respect to $\equiv$).

Then we have the following.

Theorem 3.5.2. *Let $G = \langle S \mid R\rangle$ be finitely presented with R symmetric in $F(S)$. Let R satisfy $C'(\lambda)$ for $\lambda \leq \frac{1}{6}$ or $C'(\lambda)$ for $\lambda \leq \frac{1}{4}$ and $T(4)$. Then G is hyperbolic.*

Proof. From [177] we know that G admits a Dehn algorithm, and hence G is hyperbolic. □

Especially the orientable surface groups

$$G = \left\langle a_1, b_1, \dots, a_g, b_g \;\middle|\; \prod_{i=1}^{g} [a_i, b_i] = 1 \right\rangle$$

for $g \geq 2$ and the non-orientable surface groups

$$G = \langle a_1, \dots, a_g \mid a_1^2 \cdots a_g^2 = 1\rangle$$

for $g \geq 3$ are hyperbolic. This has the following consequence: Co-compact non-Euclidean crystallographic groups and therefore co-compact Fuchsian groups are hyperbolic. This follows from Corollary 3.3.10 because these groups contain hyperbolic surface groups.

3.5.2 Groups with sufficiently high-powered relators

We now characterize groups that become hyperbolic with sufficiently high-powered relators.

Definition 3.5.3. Let $G = \langle S \mid R \rangle$ be a finitely presented group. The pair (S, R) is called *admissible* if S and R are symmetric and if r_1, r_2 are two elements of R that are not inverse to each other, then neither r_1 nor r_2 is completely canceled in the product $r_1 r_2$.

Note that every finitely presented group has an admissible presentation. Given a finite presentation $\langle S \mid R \rangle$ with symmetric sets S and R, we can construct one by using Tietze transformations.

Theorem 3.5.4. *Let $G = \langle S \mid R \rangle$ such that the pair (S, R) is admissible. Then there is an $n \in \mathbb{N}$ such that $\langle S \mid \{r^n \mid r \in R\} \rangle$ is hyperbolic.*

Proof. We look at products of the form $r_1^m r_2^m$ where both r_1 and r_2 are elements of R and $m \in \mathbb{N}$. Since (S, R) is admissible and each $r \in R$ is cyclically reduced, no copy of r_1 or r_2 is canceled out completely in this product. It follows that a piece of r_1^m in the presentation $\langle S \mid \{r^m \mid r \in R\} \rangle$ cannot be longer than a piece of r_1 in $\langle S \mid R \rangle$. Choosing $m \geq 6$ we get a presentation that satisfies the small cancellation condition $C'(\frac{1}{6})$. Hence, $\langle S \mid \{r^m \mid r \in R\} \rangle$ is hyperbolic. □

We may apply this to strict Pride groups and generalized extended modular groups.

Definition 3.5.5. A *strict Pride group* is a finitely presented group $G = \langle S \mid R \rangle$ such that
(a) Every element of R involves at most two elements of S; and
(b) For every unordered pair $\{s_1, s_2\}$ there is at most one element $r \in R$ such that r involves exactly the elements s_1 and s_2.

Special cases are the *generalized triangle groups*

$$G = \langle a, b \mid a^p = b^q = R(a, b)^m = 1 \rangle,$$

where $p = 0$ or $p \geq 2$, $q = 0$ or $q \geq 2$, $m \geq 2$ and $R(a, b)$ a cyclically reduced word in the free product on a, b involving both a and b, as well as the *generalized tetrahedron groups*

$$G = \langle x, y, z \mid x^p = y^q = z^r = R_1(x, y)^n = R_2(y, z)^m = R_3(z, x)^s = 1 \rangle,$$

where $p = 0$ or $p \geq 2$, $q = 0$ or $q \geq 2$, $r = 0$ or $r \geq 2$, $2 \leq n, m, s$ and the respective $R_i(a, b)$, $i = 1, 2, 3$, a cyclically reduced word in the free product on a, b involving both a and b.

Definition 3.5.6. A *generalized extended modular group* is a group with a presentation

$$G = \langle x, y, z \mid x^p = y^q = z^r = R_1(x, y)^n = R_2(y, z)^m = (xyz)^s = 1 \rangle$$

with $p = 0$ or $p \geq 2$, $q = 0$ or $q \geq 2$, $r = 0$ or $r \geq q$, $2 \leq n, m, s$, and the respective $R_i(a, b)$, $i = 1, 2$ a cyclically reduced word in the free product on a, b involving both a and b.

Many of the strict Pride groups and the generalized extended modular groups appear in different contexts in group theory and low-dimensional topology.

Corollary 3.5.7. *Let $G = \langle S \mid R \rangle$ be a strict Pride group or a generalized extended modular group. Then there are (fairly small) natural numbers m_r, $r \in R$ such that $\langle S \mid \{r^{m_r} \mid r \in R\} \rangle$ is hyperbolic.*

For more details see [200] and [195]. We leave it as an exercise to find suitable m_r.

3.5.3 Baumslag–Solitar groups

We have seen in Example 3.5.1.5 that one-relator groups with torsion are hyperbolic. The case of torsion-free one-relator groups is still an open question: Gersten conjectured that these groups are hyperbolic if and only if they do not contain a copy of a Baumslag–Solitar group $\mathrm{BS}(1, m) = \langle x, y \mid yxy^{-1} = x^m \rangle$ where $m \neq 0$. This conjecture itself is still open but there are partial results.

Theorem 3.5.8 (See also Section 1.7). *Let G be a torsion-free one-relator group. Then the following are equivalent.*
1. *If $\langle x, y \rangle < G$ is a non-cyclic subgroup then there is a $k \in \mathbb{N}$ such that $\langle x^k, y^k \rangle$ is free of rank 2.*
2. *The group G does not contain a copy of a $\mathrm{BS}(1, m) = \langle x, y \mid yxy^{-1} = x^m \rangle$ where $m \neq 0$.*

However, commensurable Baumslag–Solitar groups $\mathrm{BS}(n, m)$ and $\mathrm{BS}(n', m')$ have quasi convex Cayley graphs which is of geometric interest.

In 1998 Farb and Mosher showed in [75] that the solvable Baumslag–Solitar groups $\mathrm{BS}(1, m)$ and $\mathrm{BS}(1, n)$ are commensurable if and only if there exists r, k and l such that $n = r^k$ and $m = r^l$.

In 2019 Casals-Ruiz, Kazachkov and Zakharov gave in [52] the following complete classification of commensurable Baumslag–Solitar groups. Here $\sim$ denotes commensurability relation.

Theorem 3.5.9. *Let $G_1 = BS(m_1, n_1)$ and $G_2 = BS(m_2, n_2)$ be two Baumslag–Solitar groups, where $1 \le |m_i| \le n_i$, $i = 1, 2$. Then the groups G_1 and G_2 are commensurable $\sim$ if and only if one of the following holds:*

(1) $|m_1| = |m_2| = 1$ *and* n_1, n_2 *are powers of the same integer, i. e.*

$$BS(1, n^{k_1}) \sim BS(1, n^{k_2}),$$

where $n, k_i \in \mathbb{N}$.

(2) $n_1 = n_2$ *and* $m_1 = \pm m_2$, *i. e.*

$$BS(m_1, n_1) \sim BS(\pm m_1, n_1).$$

(3) $|m_1| > 1$, $|m_2| > 1$, $m_1 | n_1$, $m_2 | n_2$ *and* $\frac{n_1}{|m_1|} = \frac{n_2}{|m_2|}$, *i. e.*

$$BS(\pm k, kn) \sim BS(\pm l, ln),$$

where $k, l, n \in \mathbb{N}$, $k, l > 1$.

We end this section with a characterization of hyperbolicity of groups of F-type. As a special case, we also look at cyclically pinched one-relator groups.

3.5.4 Groups of F-type

In the following, we consider in detail the groups of F-type that we have mentioned earlier in the book. This class of groups includes some known and important classes like Fuchsian groups of geometric rank ≥ 3, surface groups of genus ≥ 2, cyclically pinched one-relator groups and torus-knot groups, and prove algebraic and geometric properties of these groups.

We recall from Section 1.6 that a group G is of F-type, if it admits a presentation of the following form

$$G = \langle a_1, \dots, a_n \mid a_1^{e_1} = \dots = a_n^{e_n} = U(a_1, \dots, a_p)V(a_{p+1}, \dots, a_n) = 1 \rangle,$$

where $n \ge 2$, $e_i = 0$ or $e_i \ge 2$, for $i = 1, \dots, n$, $1 \le p \le n-1$, $U(a_1, \dots, a_p)$ is a cyclically reduced word in the free product on $a_1, \dots, a_p$ which is of infinite order and $V(a_{p+1}, \dots, a_n)$ is a cyclically reduced word in the product on $a_{p+1}, \dots, a_n$ which is of infinite order. With p understood we write U for $U(a_1, \dots, a_p)$ and V for $V(a_{p+1}, \dots, a_n)$.

Now, if $U = a_1^{\pm 1}$ then e_1 must equal zero since we assume that U has infinite order. In this case, G reduces to

$$G = \langle a_1, \dots, a_n \mid a_2^{e_2} = \dots = a_n^{e_n} = 1 \rangle$$

which is a free product of cyclic groups. Therefore if $p = 1$ (or $p = n - 1$) we restrict groups of F-type to those where $U = a_1^m$ (or $V = a_n^m$) with $m \geq 2$.

It follows then that in all cases the group G decomposes as a non-trivial free product with amalgamation

$$G = G_1 \underset{A}{*} G_2$$

with $G_1 \neq A \neq G_2$ where the factors are free products of the cyclics

$$G_1 = \langle a_1, \ldots, a_p \mid a_1^{e_1} = \cdots = a_p^{e_p} = 1 \rangle,$$
$$G_2 = \langle a_{p+1}, \ldots, a_n \mid a_{p+1}^{e_{p+1}} = \cdots = a_n^{e_n} = 1 \rangle,$$

and

$$A = \langle U^{-1} \rangle = \langle V \rangle.$$

We first mention the observation that a group of F-type is coherent. Observe that finitely generated subgroups of G_1 and G_2, respectively, are finitely related and every subgroup of A is finitely generated, in fact cyclic because A is cyclic. Hence, G is coherent by [151].

In what follows we make a further restriction. Suppose UV omits some generator. For instance, suppose that UV does not involve a_1. Then G is a free product $H_1 * H_2$, where

$$H_1 = \langle a_1 \mid a_1^{e_1} = 1 \rangle$$

and

$$H_2 = \langle a_2, \ldots, a_n \mid a_2^{e_2} = \cdots = a_n^{e_n} = UV = 1 \rangle.$$

This does not affect the validity of the upcoming results in the next subsections. Hence, we may assume that UV involves all the generators. One could assume that it would be convenient to have U and V of minimal length in their respective orbits under Nielsen transformations but unfortunately in doing so, we eventually lose the property that G_1 and G_2 are free products on the sets of the new generators, especially if $e_i \geq 2$ for some i. It is important for our purposes that the factors G_1 and G_2 are free products on exactly the given generators, respectively.

In the following subsections we consider essentially faithful representations in general and for groups of F-type and discuss many consequences of this. Then we classify the hyperbolic groups of F-type. The hyperbolic groups of F-type have a faithful representation into $\mathrm{PSL}(2, \mathbb{R})$ which gives interesting consequences for groups of F-type. Then we give some further algebraic properties and finally some quotients of groups of F-type.

Groups of F-type were originally introduced in [89] and [92]. A reasonable discussion can be found in [93].

3.5.4.1 Essential representations and algebraic consequences

Let G be a group. We say that a linear representation ρ over a field of characteristic 0 is an *essentially faithful representation* if ρ is finite-dimensional with torsion-free kernel. From Theorem 1.7.27 we already see that one relator groups with torsion and many torsion one-relator quotients of free products of cyclics admit an essentially faithful representation. In the following, we give some historic remarks on the notion of essentially faithful representations. The term was introduced in 1996 by B. Fine and G. Rosenberger (see [96]). Earlier, in 1985, they introduced the weaker concept of *essential representations*. Suppose G is a group with presentation

$$G = \langle a_1, \dots, a_n \mid a_1^{e_1} = \dots = a_n^{e_n} = R_1^{m_1} = \dots = R_k^{m_k} = 1 \rangle,$$

where $e_i = 0$ or $e_i \geq 2$ for $i = 1, \dots, n$, $m_j \geq 1$ for $j = 1, \dots, k$, and each R_j is a cyclically reduced word in the free product of the cyclic groups $\langle a_1 \rangle, \dots, \langle a_m \rangle$ of syllable length at least two.

A representation $\rho\colon G \to$ Linear Group over a field of characteristic zero is an essential representation if for each i the image $\rho(a_i)$ is of infinite order if $e_i = 0$ or of order e_i if $e_i \geq 2$, and for each j the image $\rho(R_j)$ has order m_j. This last term was used by B. Fine, J. Howie, R. Hidalgo, N. Kopteva, F. Levin, G. Rosenberger, R. Thomas, E. B. Vinberg and others in their work on generalized modular groups, generalized triangle groups, generalized tetrahedron groups, one-relator quotients of free products with amalgamation, groups of special NEC-type and Coxeter groups. In their paper [22], G. Baumslag, J. W. Morgan and P. Shalen described this phenomenon for generalized triangle groups as special representations.

We first give some general statements and show then that a group of F-type has an essentially faithful representation into $\mathrm{PSL}(2, \mathbb{C})$.

Proposition 3.5.10. *Let G be a finitely generated group. Then G admits an essentially faithful representation if and only if G is virtually torsion-free.*

Proof. Suppose G is finitely generated, and $\rho\colon G \to$ Linear Group is an essentially faithful representation. Since G is finitely generated $\rho(G)$ is a finitely generated linear group. From a result of Selberg, see [238], $\rho(G)$ is then virtually torsion-free. Let H be a torsion-free normal subgroup of $\rho(G)$ of finite index and let H^* be the pullback of H in G. H^* has finite index in G. If $g \neq 1$ has finite order then $\rho(g)$ has exactly the same order since ρ has torsion-free kernel. Therefore g cannot be in H^* since its image would then be an element of finite order in the torsion-free group H. Therefore H^* must be torsion-free and G is virtually torsion-free.

Conversely, suppose G is virtually torsion-free. Let H be a torsion-free subgroup of G of finite index. The intersection of the conjugates of H in G is a normal subgroup of finite index. Hence, G must contain a torsion-free normal subgroup H^* of finite index. Choose a faithful finite-dimensional representation ρ^* of the finite group G/H^*.

The composition of this with the natural homomorphism from G to G/H^* will give the desired representation. □

This has two immediate relations to the following.

Proposition 3.5.11. 1. *Let G be a finitely generated group. Then G is residually finite if and only if for each $g \in G$, $g \neq 1$, there exists a non-trivial linear representation ρ of G with $\rho(G) \neq \{1\}$.*
2. *Let G be a finitely generated group with a balanced representation*

$$G = \langle a_1, \ldots, a_n \mid R_1^{m_1} = \cdots = R_n^{m_n} = 1 \rangle,$$

where each R_i is a non-trivial cyclically reduced word in the free product on $\{a_1, \ldots, a_n\}$. If at least one $m_j \geq 2$, then G is non-trivial.

For Proposition 3.5.11.1 there is nothing to show. It is a consequence of the theorem of A. Malcev, see [187]. We just mention this for completeness. We now give a proof for Proposition 3.5.11.2.

Proof. Assume that $m_n \geq 2$. Consider the group

$$G^* = \langle a_1, \ldots, a_n \mid R_1^{m_1} = \cdots = R_{n-1}^{m_{n-1}} = 1 \rangle,$$

where $m_1, \ldots, m_{n-1} \geq 1$. Its Abelianization has torsion-free rank at least 1.

If we adjoin $R_n^{m_n}$ with $m_n \geq 2$ to the Abelianization, the resulting Abelian group is non-trivial. □

We now show that groups of F-type admit an essentially faithful representation into $\mathrm{PSL}(2, \mathbb{C})$. More concretely, we show the following.

Theorem 3.5.12. *Let G be a group of F-type. Then G has a representation $\rho: G \to \mathrm{PSL}(2, \mathbb{C})$ such that $\rho|_{G_1}$ and $\rho|_{G_2}$ are faithful. Further, if neither U nor V is a proper power then G has a faithful representation in* $\mathrm{PSL}(2, \mathbb{C})$.

Proof. Choose faithful representations

$$\rho_1: G_1 \to \mathrm{PSL}(2, \mathbb{C}) \quad \text{and} \quad \rho_2: G_2 \to \mathrm{PSL}(2, \mathbb{C})$$

such that

$$\rho_1(U^{-1}) = \pm \begin{pmatrix} t_1 & 0 \\ 0 & t_1^{-1} \end{pmatrix} \quad \text{and} \quad \rho_2(V) = \pm \begin{pmatrix} t_2 & 0 \\ 0 & t_2^{-1} \end{pmatrix},$$

where t_1 and t_2 are transcendental.

This may be done since U and V both have infinite order. If both G_1 and G_2 are cyclic, then we may choose $\rho(G)$ to be a cyclic group. Now, let G_1 or G_2 be non-cyclic. Then there exists an irreducible representation from G in $\mathrm{PSL}(2, \mathbb{C})$. We have that each of the n matrices have at least two degrees of freedom with the trace and determinant being specified.

Therefore, from the work of M. Culler and P. Shalen, see [68], or also from [124] and [224] in a different setting, the dimension of the character space, that is, the representation space modulo conjugation, of G in $\mathrm{PSL}(2,\mathbb{C})$ is at least $2n-1-3=2n-4$ which is positive for $n \geq 3$. Here, -1 represents a possible additional conjugation of G_1 or G_2 using the fundamental theorem of algebra. This, especially, implies that we may choose ρ_1 and ρ_2 such that $t_1 = t_2$. Now define $\rho: G \to \mathrm{PSL}(2,\mathbb{C})$ via $\rho_i = \rho|_{G_i}$ for $i = 1,2$.

This gives the desired representation of the theorem. Further if neither U nor V is a proper power, the above construction leads to the existence of a faithful representation of G because $\rho(g), g \in G_1 \setminus \langle U\rangle$, and $\rho(U)$ have no common fixed point, considered as a linear fractional transformation, analogously for G_2 and V. This gives the result that G has a faithful representation in $\mathrm{PSL}(2,\mathbb{C})$. □

We give the following consequence of Theorem 3.5.12.

Corollary 3.5.13. *Let G be a group of F-type. Then G admits an essentially faithful representation into* $\mathrm{PSL}(2,\mathbb{C})$.

We note that if both U and V are proper powers then there is no faithful representation in $\mathrm{PSL}(2,\mathbb{C})$. If $U = U_1^\alpha$, $\alpha \geq 2$, and $V = V_1^\beta$, $\beta \geq 2$, then $\rho(U_1)$ and $\rho(V_1)$ commute but U_1 and V_1 do not commute. If A and $B = A^k$, $|k| \geq 2$, are elements of infinite order in $\mathrm{PSL}(2,\mathbb{C})$, then they have the same fixed points, considered as linear fractional transformations, and hence commute.

We now describe some straightforward algebraic consequences which we get from the essentially faithful representation of a group of F-type into $\mathrm{PSL}(2,\mathbb{C})$.

Corollary 3.5.14. *Let G be a group of F-type. Then*
1. *G is virtually torsion-free.*
2. *If neither U nor V is a proper power then G is residually finite and thus Hopfian.*
3. *The following hold:*
   (i) *If $e_i \geq 2$ then a_i has order exactly e_i.*
   (ii) *Any element of finite order in G is conjugate to a power of some a_i.*
   (iii) *Any finite subgroup is cyclic and conjugate to a subgroup of some $\langle a_i\rangle$.*
   (iv) *Any Abelian subgroup is cyclic or free Abelian of rank 2.*

We remark that Corollary 3.5.14.1 follows directly from Corollary 3.5.13. Corollary 3.5.14.2 is a consequence of A. Malcev's Theorem [187] and Corollary 3.5.14.3 follows straightforward from the Nielsen cancellation method in free products with amalgamation.

The next result (Corollary 3.5.16) is concerned with the Tits alternative. We use for the Tits alternative some simple facts.

Remark 3.5.15. 1. A subgroup H of $\mathrm{PSL}(2,\mathbb{C})$ is *elementary* if the commutator of any two elements of infinite order has trace 2; or equivalently, G is elementary if any two elements of infinite order, regarded as linear fractional transformations, have at least one common fixed point. A non-elementary subgroup contains a free subgroup of rank two. Hence, if G is a group of F-type and $\rho: G \to \mathrm{PSL}(2,\mathbb{C})$ an essentially faithful representation such that $\rho(G)$ is non-elementary, then G has a free subgroup of rank 2.

2. If $G = \langle a_1, a_2 \mid a_1^p = a_2^q = 1\rangle$ with $2 \leq p, q$ and $p + q \geq 5$ then G contains a free subgroup of rank 2.

We now give the following reasoning. In many cases we get representations $\rho: G \to \mathrm{PSL}(2,\mathbb{C})$ with $\rho(G)$ non-elementary. Then we have Corollary 3.5.16. If each time $\rho(G)$ is elementary then we may apply Remark 3.5.15.2 or we may consider factor groups. For instance, let

$$G = \langle a, b, c \mid b^2 = c^2 = a^s(bc)^t = 1\rangle$$

with $s > 2$ or $t \geq 2$. Then G has the factor group

$$\overline{G} = \langle a, b, c \mid a^s = b^2 = c^2 = (bc)^t = 1\rangle$$

which certainly has a free subgroup of rank 2. In an analogous manner, we may consider the remaining groups.

Corollary 3.5.16. *Let G be a group of F-type. Then either G has a free subgroup of rank 2 or is solvable and isomorphic to a group with one of the following representations:*
(1) $H_1 = \langle a, b \mid a^2b^2 = 1\rangle$;
(2) $H_2 = \langle a, b, c \mid a^2 = b^2 = abc^2 = 1\rangle$; *and*
(3) $H_3 = \langle a, b, c, d \mid a^2 = b^2 = c^2 = d^2 = abcd = 1\rangle$.

From our previous result we get a close tie between the existence of non-Abelian free subgroups and SQ-universality. SQ-universality is related to the concept of large groups (in the sense of S. Pride).

A *large group* is a group with a finite index subgroup that maps onto the free group F_2 of rank 2. We have already seen that F_2 is SQ-universal. Now in [199], P. M. Neumann showed the following. If G is a subgroup of finite index in a group G, then G is SQ-universal if and only if H is SQ-universal. Hence, altogether, large groups are SQ-universal.

Theorem 3.5.17. *Let G be a group of F-type. If G is not solvable then G is large. In particular, a group of F-type is either SQ-universal or solvable.*

Proof. Suppose G is not solvable. We may assume without loss of generality that each $e_i \geq 2$. Let $\rho: G \to \mathrm{PSL}(2,\mathbb{C})$ be an essentially faithful representation, and let N be a normal torsion-free subgroup of finite index in $\rho(G)$. Let π be the canonical epimorphism from $\rho(G)$ onto the finite group $\rho(G)/N$. Now consider

$$X = \langle a_1, \ldots, a_n \mid a_1^{e_1} = \cdots = a_n^{e_n} = 1\rangle.$$

There is a canonical epimorphism $\varepsilon: X \to G$. Consider the sequence

$$X \xrightarrow{\varepsilon} G \xrightarrow{\rho} \rho(G) \xrightarrow{\pi} \rho(G)/N.$$

This yields an epimorphism $\psi: X \to \rho(G)/N$. Let $Y = \ker(\psi)$. Then Y is a normal subgroup of finite index j in X, and $Y \cap \langle a_i \rangle = \{1\}$ for $i = 1, \dots, n$. Then by the Kurosh theorem Y is a free group of finite rank r. The finitely generated free product of cyclic groups Y may be considered as a Fuchsian group of finite hyperbolic area. From the Riemann–Hurwitz formula we have

$$j\left(n - 1 - \left(\frac{1}{e_1} + \dots + \frac{1}{e_n}\right)\right) = r - 1.$$

Therefore

$$r = 1 - j\left(\left(\frac{1}{e_1} + \dots + \frac{1}{e_n}\right) - n + 1\right).$$

The group G is obtained from X by adjoining the additional relation $UV = 1$ and thus $G = X/K$ where K is the normal closure of UV in X. Since $K \subset Y$ the factor group Y/K may be regarded as a subgroup of some finite index j in G, and using the Reidemeister–Schreier method, Y/K can be defined on r generators and j relations. Then the deficiency of this presentation is given by

$$\begin{aligned} d = r - j &= 1 - j\left(\left(\frac{1}{e_1} + \dots + \frac{1}{e_n}\right) - n + 2\right) \\ &= 1 + j\left(n - 2 - \left(\frac{1}{e_1} + \dots + \frac{1}{e_n}\right)\right). \end{aligned}$$

If $n \geq 5$ or $n = 4$ and at least one $e_i \neq 2$, then $d \geq 2$. It follows then from [12] that G contains a subgroup of finite index which maps onto a free group of rank 2. Hence, G is large and therefore SQ-universal. Next suppose $n = 4$ and all $e_i = 2$. Then necessarily $p = 2$ and $U = (a_1 a_2)^s$, $V = (a_3 a_4)^t$ with $|s|, |t| \geq 1$. Since G is non-solvable, then without loss of generality we may assume that $s \geq 2$ and $t \geq 1$. Then G has as a factor group the free product

$$\overline{G} = \langle a_1, a_2, a_3 \mid a_1^2 = a_2^2 = (a_1 a_2)^s = a_3^2 = 1 \rangle.$$

The group $\overline{G}$ has as a normal subgroup of index 2 a group that is isomorphic to

$$\langle x, y, z \mid x^s = y^2 = z^2 = 1 \rangle.$$

Therefore, $\overline{G}$ and hence G also has a subgroup of finite index mapping onto a free group of rank 2. If $n = 3$ then the result follows by a similar case-by-case consideration, see [93, Chapter 8]. □

One of the most powerful techniques in the study of Fuchsian groups is the Riemann–Hurwitz formula relating the Euler characteristic of the whole group to that of a subgroup of finite index.

The concept of a rational Euler characteristic is extended to more general finitely presented groups. Further, these general rational Euler characteristics satisfy the Riemann–Hurwitz formula. For the general development of group homology and Euler characteristic we refer to [37].

Theorem 3.5.18. *Let G be a group of F-type. Then G has a rational Euler characteristic $\chi(G)$ given by*

$$\chi(G) = 2 + \sum_{i=1}^{n} a_i,$$

where $a_i = -1$ if $e_i = 0$ and $a_i = -1 + \frac{1}{e_i}$ if $e_i \geq 2$. If $|G : H| < \infty$ then $\chi(G)$ is defined and $\chi(H) = |G : H| \cdot \chi(G)$. In addition, G then is of finite homological type WFL, that is, G is virtually torsion-free and for every torsion free subgroup of finite index $\mathbb{Z}$ admits a finite free resolution over the group ring $\mathbb{Z}G$, and G has virtual cohomological dimension $\operatorname{vcd}(G) \leq 2$.

Proof. Since G is virtually finite we can apply the techniques of K. Brown, see [37], to get an Euler characteristic for G by the formula

$$\chi(G) = \chi(G_1) + \chi(G_2) - \chi(A).$$

Recall that $\chi(H)$ is defined if H is a free product of cyclics. Since A is infinite cyclic we have $\chi(A) = 0$. Therefore $\chi(G) = \chi(G_1) + \chi(G_2)$. G_1 and G_2 are free products of cyclic groups, so we can apply the computation rules for the free products, and we get

$$\chi(G) = 2 + \sum_{i=1}^{n} a_i,$$

where $a_i = -1$ if $e_i = 0$ and $a_i = -1 + \frac{1}{e_i}$ if $e_i \geq 2$. We note that the Euler characteristic of a group of F-type can be zero. In fact, for instance, $\chi(G) = 0$ if $n = 2$. Now, from Section 7.6 in [37], we get that G is of finite homological type WFL with $\operatorname{vcd}(G) \leq 2$. □

3.5.4.2 Additional algebraic results for groups of F-type

The first two results follow by a straightforward application of the Nielsen cancellation method in free products with amalgamation.

Theorem 3.5.19 (Freiheitssatz for groups of F-type). *Let G be a group of F-type. Suppose that UV involves all the generators. Then:*

1. *Any subset of $(n-2)$-many of the given generators generates a free product of cyclics of the obvious orders.*
2. *If both U and V are proper powers in the respective factors G_1 and G_2, then any subset of $(n-1)$-many of the given generators generates a free product of cyclics of the obvious orders.*

Theorem 3.5.20. *Let G be a group of F-type and let H be a non-cyclic two-generator subgroup of G. Then H is conjugate in G to a subgroup $\langle x,y\rangle$ satisfying one of the following conditions:*

(1) *$\langle x,y\rangle$ is a free product of cyclic groups.*

(2) *x^t is in $\langle U\rangle = \langle V\rangle$ for some natural number t and $y^{-1}x^ty$ is in $\langle a_1,\dots,a_p\rangle$ or $y^{-1}x^ty$ is in $\langle a_{p+1},\dots,a_n\rangle$.*

We could also prove Theorem 3.2 differently using [151].

If G is a group of F-type then $\langle U\rangle = \langle V\rangle$ is malnormal in G if neither U nor V is a proper power or is conjugate to a word of the form xy for elements x,y of order 2. Since non-cyclic two-generator subgroups of the free products of cyclic groups are free products of two cyclic groups, we get the following corollary.

Corollary 3.5.21. *Let G be a group of F-type. Suppose further that $\langle U\rangle = \langle V\rangle$ is malnormal in G. Then any two-generator subgroup of G is a free product of cyclics, and* $\operatorname{rank}(G) \ge 3$.

Indeed, the malnormality of $\langle U\rangle = \langle V\rangle$ in G is important. For instance, the following groups of F-type

$$\langle a_1, a_2, a_3, a_4 \mid a_1^2 = a_2^2 = a_3^2 = a_4^3 = a_1a_2a_3a_4 = 1\rangle$$

and

$$\langle a_1, a_2, a_3 \mid a_1^2 = a_2^3 = a_1a_2a_3^2 = 1\rangle$$

are two-generator groups.

We may extend Corollary 3.5.21 easily to the following.

Corollary 3.5.22. *Let G be a group of F-type and UV involve all the generators. Suppose further that $p \ge 3$, and $n - p \ge 3$, and that $\langle U\rangle = \langle V\rangle$ is malnormal in G. Then:*

1. *Any three-generator subgroup of G is a free product of cyclics.*
2. *Any four-generator subgroup of G is a free product of cyclics or a one-relator quotient of a free product of four cyclic groups.*

Proof. We apply the Nielsen cancellation method for free products with amalgamation as described in Theorem 1.5.19 and the remarks after it and, correspondingly, arguments like that given in [225].

1. Let $x,y,z \in G$. If $\langle x,y,z\rangle$ is already of rank two there is nothing to show. Hence, let $\operatorname{rank}(\langle x,y,z\rangle) = 3$. Without loss of generality, we may assume that one of the following two cases holds:
   a) $x,y,z \in \langle a_1,\dots,a_p\rangle = G_1$.
   b) $x,y \in \langle a_1,\dots,a_p\rangle = G_1$ with $z \notin G_2$.
   In case a) we are done. Now, let $x,y \in G_1$ and $z \notin G_2$. We consider the subgroup $\langle G_1, z\rangle$. We may assume that z has the normal form

$$z = \mu_1\mu_2\dots\mu_k,$$

$k \geq 1$ and $\mu_1, \mu_k \in G_2 \setminus \langle V \rangle$. But then $\langle y, x, z \rangle$ must be a free product of cyclics because $n - p \geq 3$.

2. Let $x, y, z, u \in G$. Assume that $H = \langle x, y, z, u \rangle$ is not a free product of cyclics. Then we must have $rk(H) = 4$. We use the Nielsen cancellation method together with some conjugation, if necessary. Since H is not a free product of four cyclic groups we may assume, by symmetry, that the following hold.
   (i) $x, y \in G_1$,
   (ii) a non-trivial power U^k, $k \geq 1$, of U is contained in $\langle x, y \rangle$,
   (iii) $\langle z, u \rangle \in G_2$, and
   (iv) $\langle U^k, z, u \rangle$ is a free product of two cyclic groups $\langle a \rangle$ and $\langle b \rangle$, that is $\langle U^k, z, u \rangle = \langle a \rangle * \langle b \rangle$, where $U^k = V^{-k}$ can be written as a word in z and u.

   Especially, $\langle V^{-k}, x, y, z, u \rangle$ is one-relator quotient of a free product of four cyclic groups. This gives automatically the desired result. □

We mention that we only need $n \geq 2$ and $n - p \geq 2$ to have the complete result in the case that $e_1 = \cdots = e_n = 0$.

If neither U nor V is a proper power, then from Theorem 3.5.12, G is both residually finite and Hopfian. This holds in general for groups of F-type. Recall that a group H is *conjugacy separable* if given any non-trivial $g, h \in G$ that are not conjugate then there exists a finite quotient H^* of H where images of g and h are still not conjugate. Conjugacy separability implies residual finiteness.

From work of R. B. J. F. Allenby in [6] we get that a group of F-type is conjugacy separable. Allenby actually proves the following.

Let

$$G = \langle a_1, \ldots, a_n \mid a_1^{e_1} = \cdots = a_n^{e_n} = (UV)^m = 1 \rangle,$$

where $n \geq 2$, $e_i = 0$ or $e_i \geq 2$, $1 \leq p \leq n-1$, $U = U(a_1, \ldots, a_p)$ is a cyclically reduced word of infinite order in the free product on $a_1, \ldots, a_p$, $V = V(a_{p+1}, \ldots, a_n)$ is a cyclically reduced word of infinite order in the free product on $a_{p+1}, \ldots, a_n$, and $m \geq 1$. Then G is conjugacy separable.

However, in Allenby's proof $m \geq 2$ is only necessary in the case where either U or V has finite order in the respective free product on the generators which they involve. In a group of F-type U and V are assumed to have infinite order so the result goes through. Hence, we have the following.

Theorem 3.5.23. *A group of F-type is conjugacy separable and, hence, residually finite and Hopfian.*

A detailed proof can be found in [93].

From work of M. Aab and G. Rosenberger, see [1], we can deduce that groups of F-type are subgroup separable.

Theorem 3.5.24. *A group of F-type is subgroup separable.*

C. Y. Tang, see [256], considered a class of groups which contains the groups of F-type.

Theorem 3.5.25. *A group of F-type has solvable generalized word problem, and hence, solvable word problem.*

From Theorem 4.6 in [185] we see that groups of F-type have solvable conjugacy problem. The applied technique also answers the *power conjugacy problem* for groups of F-type, that is, given two elements to determine if a power of one is conjugate to a power of the other. Hence, we have the following.

Theorem 3.5.26. *Groups of F-type have both solvable conjugacy problem and solvable power conjugacy problem.*

3.5.4.3 Hyperbolic groups of F-type

In the following we present the results by A. Juhász and G. Rosenberger, see [140], and consider the combinatorial curvature of a class of one-relator products which generalize groups of F-type. The main result is the following.

Theorem 3.5.27. *Let $G^{(1)} = *_{\alpha\in\mathcal{T}_1} G^{(1)}_\alpha$, $G^{(2)} = * \beta \in \mathcal{T}_2 G^{(2)}_\beta$, and let $G^{(0)} = G^{(1)} * G^{(2)}$. Let $U \in G^{(1)}$ and $V \in G^{(2)}$ be cyclically reduced words of infinite order in $G^{(0)}$. Assume that if $U \in G^{(1)}_\alpha$, then $G^{(1)}_\alpha$ is free and U is cyclically reduced in $G^{(1)}_\alpha$. Assume the analogue for V. Let G be the quotient of $G^{(0)}$ by the normal closure of UV in $G^{(0)}$. Then:*
1. *G has non-positive combinatorial curvature as a quotient of $G^{(0)}$.*
2. *G is hyperbolic in the sense of M. Gromov if and only if $G^{(i)}$ is hyperbolic for $i = 1, 2$, and the following property holds:*
   (†) *At least one of U or V is neither a proper power nor a product of two elements of order 2.*

Since groups of F-type are special cases of the groups mentioned in the theorem and cyclic groups are hyperbolic, we get the following corollary.

Corollary 3.5.28. *Groups of F-type are hyperbolic unless U is a proper power or a product of two elements of order 2 and V also is a proper power or a product of two elements of order 2. In the last case they have non-positive combinatorial curvature. In particular, they satisfy a quadratic isoperimetric inequality.*

We consider van Kampen diagrams. All the unexplained terms concerning them can be found in [35] and [177].

Remark 3.5.29. 1. Let M be a diagram over a free group or free product F. We shall denote by $|\partial M|$ the length of a cyclically reduced label of M over F.
2. Assume A is given by a presentation $\langle X \mid R\rangle$. Then we shall always assume that R is cyclically reduced. Also we shall assume that our van Kampen diagram contains a minimal number of regions for a simply connected given boundary label.
3. Recall from [118] that a presentation $\langle X \mid R\rangle$ has non-positive (negative) combinatorial curvature if every inner region D of every van Kampen diagram over R

is such that the excess $\kappa(D) = 2\pi + \sum_{i=1}^{n}(\theta_i - \pi)$ is non-positive (negative). Here, $\partial D = v_1 e_1 v_2 e_2 \cdots v_n e_n v_1$, v_i vertices on ∂D, e_i edges on ∂D, and θ_i are the inner angles of the polygon D at v_i. In general, one takes $\theta_i = \frac{2\pi}{d(v_i)}$ where $d(v_i)$ is the valency of v_i, that is, distribute the curvature equally among the regions containing v_i. However, from the point of view of the theory of groups with non-positive curvature, see [35] and the theory of hyperbolic groups, see [118], it is immaterial how the angles around v_i are distributed, as long as they sum up to 2π and $\theta_i - \frac{2\pi}{d(v_i)} < \frac{2\pi}{6}$. In the following we shall make use of this remark.

The following lemma is an immediate consequence of the construction of diagrams over free products. We omit its proof.

Lemma 3.5.30. *Let $A = \langle X \mid R\rangle$, $B = \langle Y \mid S\rangle$ be finitely generated, and let $G = A * B/D$, where D is the normal closure of $T \subset A * B$. Let $Q = R \cup S \cup T$. Thus $Q \subset F(X \cup Y)$. Assume that there are constants C and a such that the following isoperimetric inequalities hold for R-diagrams M, S-diagrams N and T-diagrams H respectively:*

$$\mathrm{Vol}(M) \le C|\partial M|^a, \quad \mathrm{Vol}(N) \le C|\partial N|^a \quad \textit{and} \quad \mathrm{Vol}(H) \le C|\partial H|^a,$$

where for a P-diagram Z, $\mathrm{Vol}(Z)$ is the number of regions of Z.

Then there is a constant $C' \ge C$ depending on Q such that for every word $W \in F(X \cup Y)$ which represents 1 in G there is a Q-diagram U with boundary label W such that $\mathrm{Vol}(U) \le C'|W|^a$.

We also mention the following corollary.

Corollary 3.5.31.
1. *If A and B are hyperbolic and the presentation of G as a quotient of the free product $A * B$ satisfies a linear isoperimetric inequality, then G is hyperbolic.*
2. *If A and B satisfy a polynomial isoperimetric inequality of degree k and G has non-positive combinatorial curvature as a quotient of $A * B$, then G satisfies a polynomial isoperimetric inequality of degree $\max(k, 2)$.*

Let R be the set of all the cyclically reduced cyclic conjugates of UV and $(UV)^{-1}$. We will describe the structure of the van Kampen diagrams for the groups given in Theorem 3.5.27. Let M be a reduced R-diagram, and let D be an inner region of M. Then D has a boundary cycle $v_0 \mu v_1 \nu v_0$ such that the label of μ is U and the label of ν is V. From now on, we shall not distinguish between a path in M and its label in $G^{(0)}$. Since the vertices v_0 and v_1 separate μ from ν we shall call them separating vertices, see Figure 3.34.

Theorem 3.5.32. *Separating vertices of D have valency at least 4.*

Proof. Assume v_0 has valency 3. Then there are regions D_1 and D_2 which contain v_0 on their boundary. Let x be the first letter on the edge common to D_1 and D_2 which contains v_0. If $x \in G^{(1)}$, then v_0 is a separating vertex of D_2, see Figure 3.35(a). Consequently a non-trivial tail of ν is a common edge of D and D_2. But then, since U is cyclically reduced and $U \neq U^{-1}$, D completely cancels D_2. This violates the assumption that M is reduced.

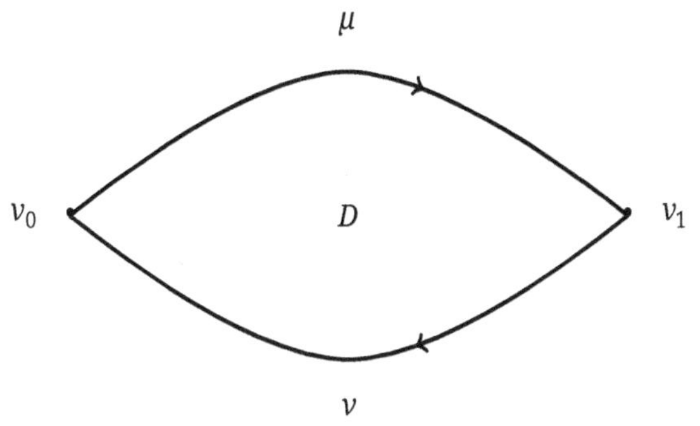

Figure 3.34: Separating vertices.

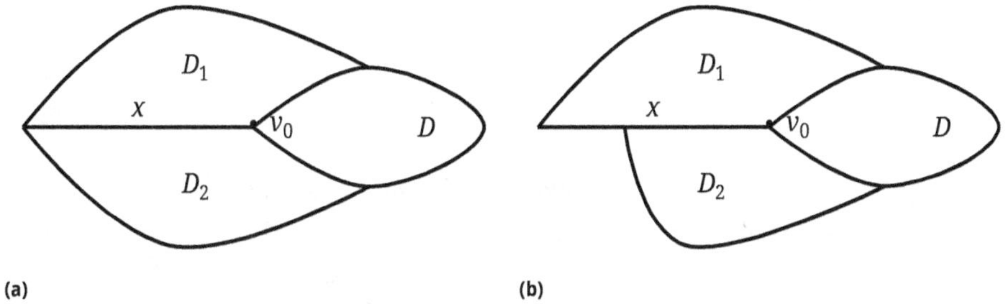

Figure 3.35: Valency 3.

Similarly, if $x \in G^{(2)}$, then v_0 is a separating vertex of D_1, by symmetry, see Figure 3.35(b), leading to the same contradiction. Thus v_0 has valency $\neq 3$.

Finally, assume v_0 has valency 2. Recall that U and V are cyclically reduced. Let first U and V have length at least 2. Then v_0 is a separating vertex of D_1, and D_1 cancels D completely as in Figure 3.36. Now let U or V have length 1. If, for instance, $U = x$ and v_0 has valency 2 then $x = x^{-1}$, that is, $U = U^{-1}$. This contradicts that U has infinite order. □

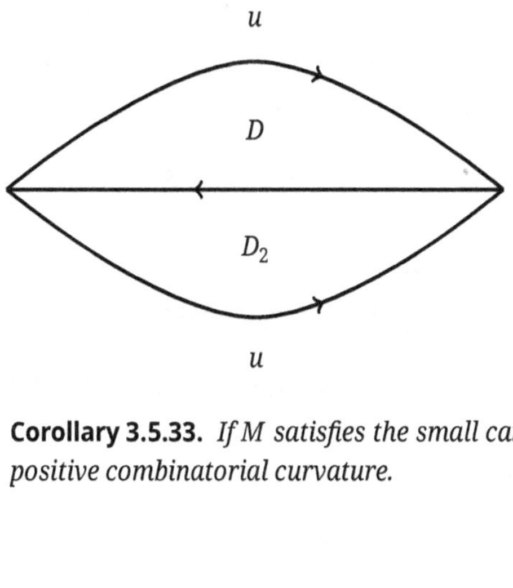

Figure 3.36: Valency 2.

Corollary 3.5.33. *If M satisfies the small cancellation condition $C(5)$, then M has a non-positive combinatorial curvature.*

The inner region D always has two vertices with valency ≥ 4 (namely v_0 and v_1).

Remark 3.5.34. Neither U nor V can be a piece, for then as we have seen, the diagram is not reduced. Consequently U and V are each the product of at least two pieces. Since by the theorem no region D_1 can be a neighbor of D with a common edge, see Figure 3.37, which contains v_0 as an inner vertex, D has at least 4 neighbors.

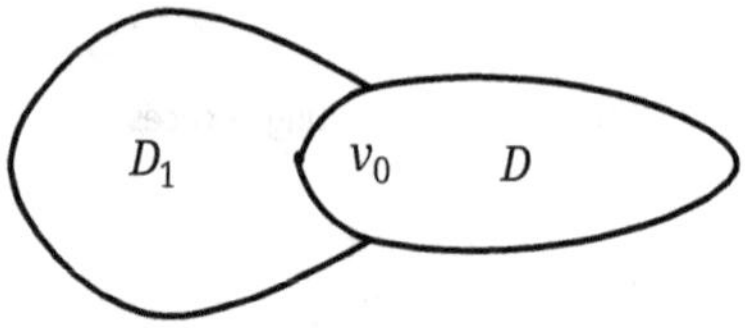

Figure 3.37: Neighbors of D.

This proves the following.

Corollary 3.5.35. 1. *M satisfies the small cancellation condition $C(4)$.*
2. *An inner region D has 4 neighbors if and only if its boundary cycle is subdivided into 4 edges in such a way that the label on the first two edges is U and the label on the second two edges is V. In particular, if such a region exists, then U and V are each the product of two pieces.*

Thus, in what follows, we shall study the situation when U is the product of two pieces. Denote by $\mathcal{D}_0$ the set of all the inner regions in M which have a boundary path with label in $\{U^{\pm 1}, V^{\pm 1}\}$ and have a decomposition to two pieces by a vertex which is not a separating vertex.

Denote the set of all the inner vertices of M which are not separating vertices and divide a boundary path of a region D with label $U^{\pm 1}$ or label $V^{\pm 1}$ to two parts by $\mathcal{V}_0$. For $i \geq 3$ denote by $\mathcal{V}_i$ the set of all the vertices of M not in $\mathcal{V}_0$ which have valency i. Denote by $\kappa(U)$ the contribution of all the vertices of an inner region D which are inner vertices of the boundary path μ with label U to the excess of D, and let $\kappa(V)$ be defined analogously. Then

$$\kappa(D) = 2\pi + \kappa(U) + \kappa(V) + (\theta_0 - \pi) + (\theta_t - \pi) = \kappa(U) + \kappa(V) + \theta_0 + \theta_t, \qquad (3.1)$$

where θ_0 and θ_t denote the angles of the two separating vertices v_0 and v_t, compare Figure 3.34.

Definition 3.5.36. 1. A *corner* consists of a pair (e, e^1) of edges which have the same initial vertex and are such that the path $e^{-1}e^1$ is a subpath of a boundary cycle of a region associated with the corner.
2. Given a vertex $u \in \mathcal{V}_0$, a *bad* corner at u is a corner (e, e^1) such that the path $e^{-1}e^1$ has label one of $U^{\pm 1}, V^{\pm 1}$. A *good* corner is a corner which is not bad.
3. Call two corners *adjacent* if they have an edge in common.

Theorem 3.5.37. *Let the edge pair (α^{-1}, β) define a bad corner at the vertex v of the region D. Then*
1. *The two corners at v which are adjacent to the given corner are good corners.*
2. *The vertex v is preceded or followed (in the boundary cycle of D) by a vertex of valency 4 which is a separating vertex of D.*

Proof. Let w_0 and w' be the separating vertices of F, see Figure 3.38. Then $w_1 \neq v$ since v is not a separating vertex by assumption and $w_0 \neq v_0$, otherwise either D_1 cancels D (if μ has label U) or U is not cyclically reduced (if μ_1 has label U^{-1}). □

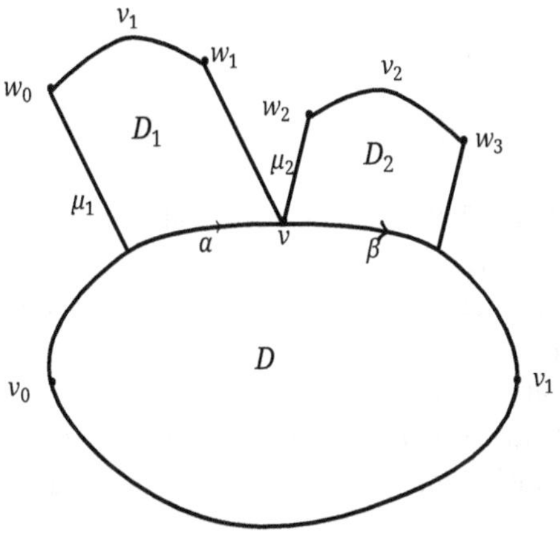

Figure 3.38: Separating vertices.

Now define the angles in M. If $v \in \mathcal{V}_i$, $i \geq 3$, then assign the value $\frac{2\pi}{i}$ to every angle having v as its vertex. If $v \in \mathcal{V}_0$ and $d = d(v) \geq 5$ then assign to each angle having v as its vertex the value $\frac{2\pi}{d}$. Let $\varepsilon = \frac{2\pi}{8m^2}$, $m = 2(|U| + |V|)$. If $v \in \mathcal{V}_0$ and $d(v) = 4$, then by Theorem 3.5.37 there are at most two regions which contain v on their boundary and belong to $\mathcal{D}_0$ (D_1 and D_3 in Figures 3.39(a) and (b)).

If D_1 and D_3 are in $\mathcal{D}_0$, then assign $\frac{2\pi}{4} - \varepsilon$ to γ_1 and γ_3 and assign $\frac{2\pi}{4} + \varepsilon$ to γ_2 and γ_4. If $D_1 \in \mathcal{D}_0$ and $D_3 \notin \mathcal{D}_0$, then assign $\frac{2\pi}{4} - \varepsilon$ to γ_1, $\frac{2\pi}{4} + \frac{\varepsilon}{2}$ to γ_2 and γ_4 and $\frac{2\pi}{4}$ to γ_3. Finally, if $d(v) = 3$, then by Theorem 3.5.37 exactly one of the regions containing v, say D_1, belongs to $\mathcal{D}_0$, see Figure 3.39(b). Assign $\frac{2\pi}{4} - \varepsilon$ to γ_1 and assign $\frac{2\pi}{3} + \frac{2\pi}{24} + \frac{\varepsilon}{2}$ to γ_2 and to γ_3.

Theorem 3.5.38. *Let D be an inner region of M. If D has a vertex $v \in \mathcal{V}_0$ on its boundary, then $\kappa(D) < 0$.*

Proof. Denote by a_i the number of inner vertices of μ (see Figure 3.34) in $\mathcal{V}_i$, $i \geq 3$, and let $a_4^* = \sum_{i\geq 4} a_i$. For $i \geq 5$, denote by b_i the number of inner vertices of μ in $\mathcal{V}_0$ with valency i. Define numbers b_4', b_4'', c_4 and c_3 as follows:
- b_4' is the number of inner vertices of μ in $\mathcal{V}_0$ which are of degree 4 and such that
  a) The corner associated with D is a good corner.

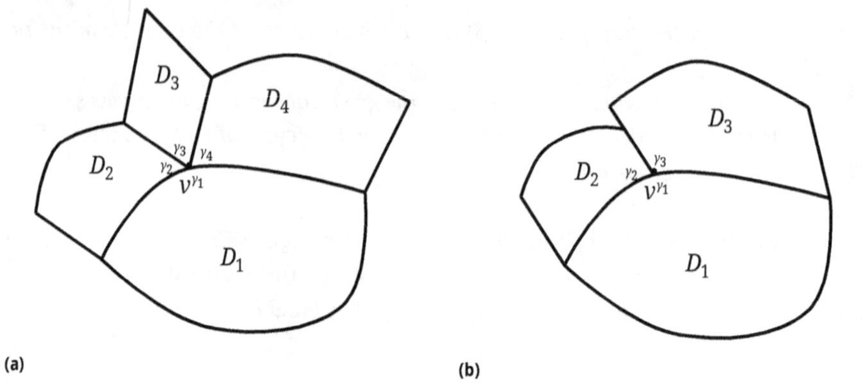

Figure 3.39: Two inner regions containing v.

b) Exactly one of the two corners adjacent to the corner associated with D is a bad corner.

- b_4'' is the number of inner vertices of μ in $\mathcal{V}_0$ which are of degree 4 and such that
  a) The corner associated with D is a good corner.
  b) Both the two corners adjacent to the corner associated with D are bad corners.
- c_4 is the number of inner vertices of μ in $\mathcal{V}_0$ which are of degree 4 and such that
  a) The corner associated with D is a good corner.
  b) Neither of the two corners adjacent to the corner associated with D are bad corners (so that the remaining corner which is 'opposite' to the corner associated with D, must be a bad corner).
- c_3 is the number of inner vertices of μ in $\mathcal{V}_0$ of degree 3 such that the corner associated with D is a good corner (in which case exactly one of the other two corners adjacent to the corner associated with D is a bad corner).

Assume $D \notin \mathcal{D}_0$ and evaluate $\kappa(U)$. Thus,

$$\begin{aligned}\kappa(U) &= \sum_{i\geq 3} a_i\left(\frac{2\pi}{i} - \pi\right) + \sum_{i\geq 5} b_i\left(\frac{2\pi}{i} - \pi\right) + b_4'\left(\frac{2\pi}{4} + \varepsilon - \pi\right)\\ &\quad + b_4''\left(\frac{2\pi}{4} + \frac{\varepsilon}{2} - \pi\right) + c_4\left(\frac{2\pi}{4} - \pi\right) + c_3\left(\frac{2\pi}{3} + \frac{2\pi}{24} - \pi + \frac{\varepsilon}{2}\right)\\ &\leq -\frac{\pi}{4}c_3 - \frac{\pi}{2}a_4^* - \frac{\pi}{2}(b_4' + b_4'') + \frac{\varepsilon}{2}(2b_4' + b_4' + c_3)\\ &\leq -\frac{\pi}{4}c_3 - \frac{\pi}{2}a_4^* - \frac{\pi}{2}(b_4' + b_4'') + \frac{\pi}{8m^2}(2m).\end{aligned}$$

Assume $c_3 > 0$. Then, by Theorem 3.5.37, $a_4^* \geq \frac{1}{2}c_3$, hence $-\frac{\pi}{2}(a_4^* + \frac{1}{2}c_3) \leq -\frac{\pi}{2} - \frac{\pi}{4}$, and

$$\kappa(U) \leq -\left(\frac{\pi}{4} + \frac{\pi}{2}\right) + \frac{\pi}{8m^2}(2m) = \frac{3}{4}\pi + \frac{\pi}{4m} < -\frac{3}{4}\pi + \frac{\pi}{4} = -\frac{\pi}{2}$$

as $m > 1$. Thus

$$\text{if } D \notin D_0 \text{ and } c_3 > 0, \quad \text{then } \kappa(U) < -\frac{\pi}{2}. \tag{3.2}$$

Assume $c_3 = 0$. Then $b_4' + b_4'' \neq 0$, hence, by Theorem 3.5.37, $a_4 > 0$. Consequently,

$$k(U) \leq -\frac{\pi}{2}a_4 - \frac{\pi}{2}(b' + b'') + \frac{\pi}{4m} \leq -\frac{\pi}{2} - \frac{\pi}{2} + \frac{\pi}{4m} < \frac{\pi}{2} - \frac{\pi}{2} + \frac{\pi}{4} < \frac{\pi}{2}.$$

$$\text{If } D \notin \mathcal{D}_0 \text{ and } c_3 = 0, \quad \text{then } \kappa(U) < -\frac{\pi}{2}. \tag{3.3}$$

Assume now that $D \in \mathcal{D}_0$. Then, by the choice of γ_1, $\kappa(U) = -\frac{\pi}{2} - \varepsilon$.

$$\text{If } D \in \mathcal{D}_0, \quad \text{then } \kappa(U) < -\frac{\pi}{2}. \tag{3.4}$$

Evaluate now $\kappa(D)$. By (3.1), (3.2), (3.3) and (3.4) we have

$$\kappa(D) = \kappa(U) + \kappa(V) + \frac{\pi}{2} + \frac{\pi}{2} < -\frac{\pi}{2} - \frac{\pi}{2} + \frac{\pi}{2} + \frac{\pi}{2} = 0$$

which yields the theorem. □

Corollary 3.5.39. *D has non-positive combinatorial curvature. D has zero curvature if and only if D has four neighbors and every boundary vertex is a separating vertex with valency* 4.

We now prove Theorem 3.5.27.

Proof of Theorem 3.5.27. We remark that we already have the first statement in Theorem 3.5.27 by the previous results. We now consider the relation (†). Let the notation be as in Figures 3.40(a) and (b). Then either v is a separating vertex for D_1 or v is a separating vertex for D_2.

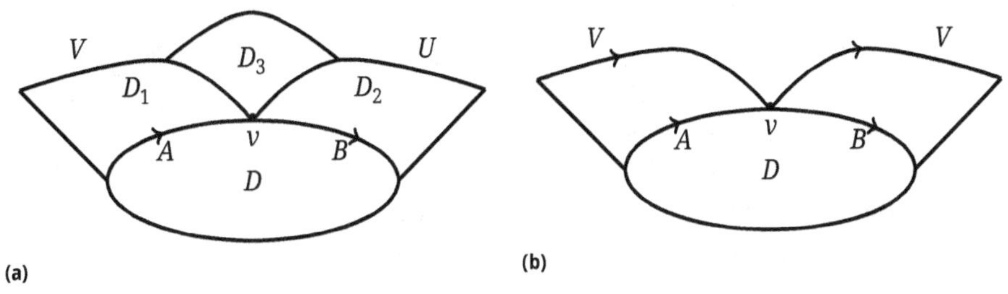

Figure 3.40: Separating vertex.

If v is a separating vertex for only one of D_1 and D_2, then it is a separating vertex for D_3 and hence for two neighboring regions of D (D_1 and D_3 or D_2 and D_3). But the boundary labels of such neighbors cancel each other, contradicting our assumption that M contains a minimal number of regions with the given boundary label. Consequently, v is a separating vertex for both D_1 and D_2, see Figure 3.40(b).

Let A be the label of the piece common to D and D_1, and let B be the label of the piece common to D and D_2. Then one of the following equations holds: $(AB)(AB) = U(AB)V$ (see Figure 3.41(a)) or $A = A^{-1}$ and $B = B^{-1}$, see Figure 3.41(b). In the first case $A = D^{\alpha}$ and $B = D^{\beta}$ and in the second case A and B have order 2 as required.

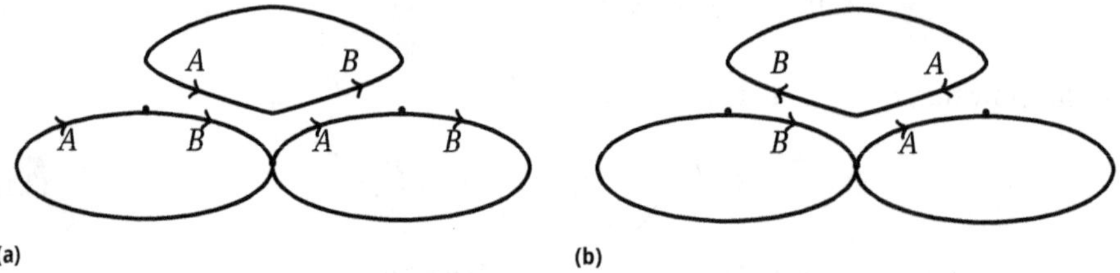

Figure 3.41: Common pieces.

Assume (†) holds. Then by Theorem 3.5.38 the combinatorial curvature at each inner region of a derived van Kampen diagram over R is strictly negative. Therefore G is hyperbolic.

Assume now that (†) does not hold. Then UV (or VU) has one of the following forms.

a) A^nB^m for $n, m \geq 2$.

b) $ABCD, A = A^{-1}, B = B^{-1}, C =^{-1}, D = D^{-1}$.

c) $ABC^n, n \geq 2, A = A^{-1}, B = B^{-1}$.

In case a) let $H = \langle A, B\rangle$. Then it follows from the fact that every van Kampen diagram over the symmetric closure of UV has a non-positive combinatorial curvature that

$$H \cong \langle a, b \mid a^n = b^{-m}\rangle.$$

Consequently, G is not hyperbolic.

A similar argument shows that the subgroup of G generated by A, B, C and D in case b) is isomorphic to $\langle a, b, c, d \mid a^2, b^2, c^2, d^2, abcd\rangle$ which contains a normal free Abelian subgroup generated by ab and bc.

Finally in case c) let $H = \langle B, BC^n\rangle$. Then, as above, $H \cong \langle a, b \mid a^2, (ab^n)^2\rangle$ in which $K = \langle b^n\rangle$ is an infinite cyclic normal subgroup such that $H/K = \mathbb{Z}_2 * \mathbb{Z}_n, n \geq 2$. Thus in all cases, G is not hyperbolic. □

Corollary 3.5.40. *Let G be a group of F-type. G is hyperbolic unless U is a proper power or a product of two elements of order 2 and V also is a proper power or a product of two elements of order 2.*

We just want to remark here that by the result of F. Dahmani and V. Guirardel, see [69], the isomorphism problem for a hyperbolic group of F-type is solvable in the class of hyperbolic groups.

It remains to answer the question if the isomorphism problem is solvable for a group of F-type in some class of groups. This is an open problem. There exists just a result in the special case of cyclically pinched one-relator groups, that is, in the case of groups of F-type with $e_1 = e_2 = \cdots = e_n = 0$. The isomorphism problem for a cyclically pinched one-relator group G is solvable in the class of one-relator groups, see [230].

Certainly, G has no faithful representation in $\mathrm{PSL}(2,\mathbb{R})$ if $U = XY$ with $X^2 = Y^2 = 1$ and $V = WZ$ with $W^2 = Z^2 = 1$ or $U = U_1^k$, $V = V_1^q$ with $k, q \geq 2$.

Concerning the question if there exists a faithful representation $\rho\colon G \to \mathrm{PSL}(2,\mathbb{R})$ we are left, up to symmetry, with the case $U = XY$ with $X^2 = Y^2 = 1$ and $V = V_1^q$ with $q \geq 2$. But in this case there does not exist such a ρ because

$$(\rho(X))^2 = (\rho(UY))^2 = (\rho(VY))^2 = 1$$

implies $\rho(V_1Y)^2 = 1$, but $(V_1Y)^2 \neq 1$ in G for any essential representation $\rho\colon G \to \mathrm{PSL}(2,\mathbb{R})$.

Our next aim is to characterize the hyperbolicity of a group of F-type by means of faithful representations in $\mathrm{PSL}(2,\mathbb{R})$.

Theorem 3.5.41. *Let G be a hyperbolic group of F-type. Then G has a faithful representation in* $\mathrm{PSL}(2,\mathbb{R})$.

We present the proof for the case that $e_i = 0$ for $i = 1,\dots,n$, U is not a proper power in $\langle a_1,\dots,a_p\rangle$, and V is not a proper power in $\langle a_{p+1},\dots,a_n\rangle$, that is, for the case of cyclically pinched one-relator groups. Details for the general case can be found in [84], it follows in an analogous manner. Hence, from now on let G be a cyclically pinched one-relator group. Now, for convenience, we write u instead of U and v instead of V.

Proof. If u or v is primitive then G is a free group, and the result holds certainly. Now, let u and v both be not primitive. We prove the existence of a faithful representation into $\mathrm{SL}(2,\mathbb{R})$. Since a hyperbolic cyclically pinched one-relator group as in the statement of the theorem is centerless this faithful representation can be extended to a faithful representation into $\mathrm{PSL}(2,\mathbb{R})$. We first embed F_1 into a free group $H_1 = \langle a, b\rangle$ of rank 2 and F_2 into a free group $H_2 = \langle c, d\rangle$ also of rank 2. It follows that u, v are both non-trivial and not primitive in H_1, H_2, respectively.

We now consider $H_1 = \langle a, b\rangle$. Choose $A, B \in \mathrm{SL}(2,\mathbb{R})$ with $\mathrm{tr}(A) = x > 2$ an algebraic number and $\mathrm{tr}(B) = y > 2$ also an algebraic number. Then $\mathrm{tr}(AB) = r$ and we will choose r later in a suitable manner.

The map $a \mapsto A$, $b \mapsto B$ defines a homomorphism $\phi_1\colon H_1 \to \mathrm{SL}(2,\mathbb{R})$. Let $U = \phi_1(u)$ be the image of u. From [95, 90] and also from [239] we get the following: the trace $\mathrm{tr}(U)$ is a nonconstant polynomial $f(r)$ in r with coefficients in $\mathbb{Z}[x,y]$; and without loss of generality we may assume that the highest coefficient is positive. Moreover all the coefficients are algebraic numbers.

We now argue analogously for $H_2 = \langle c, d\rangle$. Choose $C, D \in \mathrm{SL}(2,\mathbb{R})$ with $\mathrm{tr}(C) = z > 2$ an algebraic number and $\mathrm{tr}(D) = q > 2$ an algebraic number. We let $s = \mathrm{tr}(CD)$ and, just as for r, we will later choose s in a suitable manner. As before the map $c \mapsto C$, $d \mapsto D$ defines a homomorphism $\phi_2 : H_2 \to \mathrm{SL}(2,\mathbb{R})$. Let $V = \phi_2(v)$ be the image of v. As before $\mathrm{tr}(V)$ is a nonconstant polynomial $g(s)$ in s with coefficients in $\mathbb{Z}[z,q]$ and without loss of generality we may assume that the highest coefficient is positive. Moreover, all the coefficients are algebraic numbers. We make the following observations:
(1) We have $f(X) \to \infty$ if $X \to \infty$ and $g(X) \to \infty$ if $X \to \infty$.
(2) If we choose a sufficiently large transcendental number $t > 4$ then by the intermediate value theorem there exist an $r \in \mathbb{R}$ and an $s \in \mathbb{R}$ such that $f(r) = t = g(s)$.

The real numbers r, s have to be transcendental because the polynomials $f(X)$ and $g(X)$ have algebraic coefficients (if r was algebraic then $f(r)$ would also be algebraic). After a suitable conjugation of $\phi_1(H_1)$ and $\phi_2(H_2)$ in $\mathrm{SL}(2,\mathbb{R})$ we may assume that

$$U = \begin{pmatrix} t_1 & 0 \\ 0 & t_1^{-1} \end{pmatrix} = V, \quad t = t_1 + t_1^{-1}$$

with t_1 a real transcendental (recall that $t > 4$). We have the following facts:
(a) The group $\langle A, B\rangle$ is free of rank 2 because r is transcendental.
(b) The group $\langle C, D\rangle$ is free of rank 2 because s is transcendental.

Therefore ϕ_1 and ϕ_2 are monomorphisms and hence embeddings of the respective free groups into $\mathrm{SL}(2,\mathbb{R})$. The group F_1 is a subgroup of $H_1 = \langle a, b\rangle$ and F_2 is a subgroup of $H_2 = \langle c, d\rangle$. Hence

$$\phi_1|_{F_1} : F_1 \to \mathrm{SL}(2,\mathbb{R})$$

and $\phi_2|_{F_2} : F_2 \to \mathrm{SL}(2,\mathbb{R})$ are embeddings with

$$\phi_1|_{F_1}(u) = U = V = \phi_2|_{F_2}(v).$$

Recall that

$$U = \begin{pmatrix} t_1 & 0 \\ 0 & t_1^{-1} \end{pmatrix}.$$

Then the combination of $\phi_1|_{F_1}$ and $\phi_2|_{F_2}$ defines a homomorphism $\phi : G \to \mathrm{SL}(2,\mathbb{R})$ with $\phi|_{F_1} = \phi_1|_{F_1}$ and $\phi|_{F_2} = \phi_2|_{F_2}$. Now, u is not a proper power in F_1 and v is not a proper power in F_2. Hence exactly as in [239] and [228] we get from the homomorphism $\phi : G \to \mathrm{SL}(2,\mathbb{R})$ an injective homomorphism $\rho : G \to \mathrm{SL}(2,\mathbb{R})$. To see this let

$$\mathrm{tr}(G) = \{\mathrm{tr}(\phi(g)) \mid g \in G\}$$

and then choose a real transcendental number τ which is not algebraic over $K = \mathbb{Q}(\mathrm{tr}(G))$. Now define

$$T = \begin{pmatrix} \tau & 0 \\ 0 & \tau^{-1} \end{pmatrix}.$$

Define a homomorphism $\rho\colon G \to \mathrm{SL}(2,\mathbb{R})$ by $\rho(g) = T\phi(g)T^{-1}$ if $g \in F_1$ and $\rho(g) = \phi(g)$ if $g \in F_2$; since $\rho(u) = \phi(u) = U = V$ this does indeed give a homomorphism.

For $g \in G$ we write g_{ij} for the respective entry of $\rho(g)$. Then $\rho(g) = (g_{ij})$. We mention that $g_{12} \neq 0$ and $g_{21} \neq 0$ if $g \in F_1 \setminus \langle u \rangle$ or $g \in F_2 \setminus \langle v \rangle$ since F_1, F_2 are non-Abelian and u and v are not proper powers in F_1 and F_2, respectively. We now show that ρ is injective. Every element $g \in G$ is conjugate either to an element of F_1 or F_2 or to an element of the form $= x_1y_1 \cdots x_ky_k$ with $k \geq 1$ and $x_i \in F_1 \setminus \langle u \rangle$ and $y_i \in F_2 \setminus \langle v \rangle$ for $i = 1, \dots, k$. To prove that ρ is injective we may assume that $g = x_1y_1 \cdots x_ky_k$ as above. We now show that $\mathrm{tr}(\rho(g))$ is transcendental over K which proves that ρ is injective. For this purpose, we claim that g_{11} is a Laurent polynomial of degree $2k \geq 2$ over K, g_{22} is a Laurent polynomial of degree $\leq 2k$ over K and g_{12} and g_{21} are Laurent polynomials of degree $< 2k$ over K. We prove the claim by induction on k. If $k = 1$ then $g = xy$ with $x = x_1, y = y_1$. Recall

$$\rho(x) = T\phi(x)T^{-1} = \begin{pmatrix} \tau & 0 \\ 0 & \tau^{-1} \end{pmatrix} \begin{pmatrix} \alpha_{11} & \alpha_{12} \\ \alpha_{21} & \alpha_{22} \end{pmatrix} \begin{pmatrix} \tau^{-1} & 0 \\ 0 & \tau \end{pmatrix}$$

with $\alpha_{12} \neq 0 \neq \alpha_{21}$ because ϕ is injective on F_1. Then

$$\begin{aligned} g_{11} &= \alpha_{11}y_{11} + \alpha_{12}y_{12}\tau^2, \\ g_{12} &= \alpha_{11}y_{12} + \alpha_{12}y_{22}\tau^2, \\ g_{21} &= \alpha_{22}y_{11} + \alpha_{21}y_{11}\tau^{-2}, \\ g_{22} &= \alpha_{22}y_{11} + \alpha_{21}y_{21}\tau^{-2}. \end{aligned}$$

Note that possibly $y_{22} = 0$. This proves the claim for $k = 1$. Now let $k \geq 2$. We write $g = xy$ with $x = x_1y_1 \cdots x_{k-1}y_{k-1}$ and $y = x_ky_k$. By the inductive hypothesis and the case $k = 1$ already proved the claim holds for both x and y. Multiplication of $\rho(x)$ with $\rho(y)$ now proves the overall claim. Hence in particular $\mathrm{tr}(\rho(g))$ is transcendental over K, which proves that the homomorphism ρ is injective. □

Hence, we get the following.

Corollary 3.5.42. *Let G be a group of F-type. Then G is hyperbolic if and only if G has a faithful representation in* $\mathrm{PSL}(2,\mathbb{R})$.

This has now some additional algebraic consequences. We know that $\mathrm{PSL}(2,\mathbb{R})$ is commutative transitive. Hence, any hyperbolic group of F-type is commutative transitive.

In general, the class of CSA groups is a proven subclass of the class of commutative transitive groups, however they are equivalent in the presence of residual freeness. For hyperbolic groups of F-type we have the following.

Theorem 3.5.43. *Let G be a hyperbolic group of F-type. Assume that G is torsion-free or has only odd torsion, that is, e_i is odd if $e_i \geq 2$. Then G is CSA.*

Proof. We may assume that G is already a subgroup of $\mathrm{PSL}(2,\mathbb{R})$. Let A be a maximal Abelian subgroup of G. Since G is a subgroup of $\mathrm{PSL}(2,\mathbb{R})$ any two elements $a, b \in A$ have, considered as linear fractional transformations, the same fixed points.

Assume that G is not CSA. Then there are non-trivial elements $a, b \in A$ and $x \in G \setminus A$ with $xax^{-1} = b$. This means that x permutes non-trivially the fixed points of a, which are also the fixed points b. Hence, x must have order two. This gives a contradiction, and therefore G is CSA. □

An RG group H has the property that every Abelian subgroup is locally cyclic. Hence, it is often convenient to assume that every element of H is contained in a maximal cyclic subgroup of H. This last property is of course satisfied in hyperbolic groups. Certainly, an RG group is commutative transitive. Using Theorem 3.9.18 we easily get the following.

Theorem 3.5.44. *Let G be a hyperbolic group of F-type. Assume that G is torsion-free or has only odd torsion. Then G is an RG group.*

We present the following result and sketch its proof.

Theorem 3.5.45. *Let G be a group of F-type which is not a free product of cyclic groups. Assume that each subgroup of G of infinite index is a free product of cyclic groups. Then G is a co-compact planar discontinuous group.*

Sketch of proof. We first consider the case that G is not hyperbolic. We have to consider the situations for U and V which yield that G is not hyperbolic.

For these situations, we easily find subgroups of infinite index which are not free products of cyclic groups unless $G \cong \langle U, V \mid U^2 = V^2 \rangle$ or $G \cong \langle a, b, V \mid a^2 = b^2 = abV^2 = 1 \rangle$ or $G \cong \langle a, b, c, d \mid a^2 = b^2 = c^2 = d^2 = abcd = 1 \rangle$. But these three groups are co-compact Euclidean planar discontinuous groups.

Now let G be hyperbolic. We just consider G as a quotient of the cyclically pinched one-relator group where the a_i all have infinite order. Next we extend the arguments in [55]. Then we use the solution of Nielsen's realization problem by S. P. Kerckhoff in [152] (see also [280] for many details and applications). □

We end this section with the following conjecture.

Conjecture 3.5.46. *Let H be a finitely generated, non-elementary subgroup of* $\mathrm{PSL}(2,\mathbb{R})$. *Then H is finitely presented.*

3.5.4.4 One-relator amalgamated products and some algebraic consequences

In this subsection we give versions and extensions of some of the results in [102].

If $A, B \in \mathrm{PSL}(2, \mathbb{C})$ we say that the pair $\{A, B\}$ is *irreducible* if A, B, regarded as linear fractional transformations, have no common fixed point, that is, $\mathrm{tr}[A, B] \neq 2$.

Theorem 3.5.47 (Freiheitssatz). *Suppose $G = G_1 *_A G_2$ with $A = \langle a\rangle$ cyclic. Let $R \in G \setminus A$ be given in a reduced form $R = c_1 d_1 \cdots c_k d_k$ with $k \geq 1$ and $c_i \in G_i \setminus A$, $d_i \in G_2 \setminus A$ for $i = 1, \dots, k$. Assume that there exists a representation $\phi: G \to \mathrm{PSL}(2, \mathbb{C})$ such that $\phi|_{G_1}$ and $\phi|_{G_2}$ are faithful and the pairs $\{\phi(c_i), \phi(a)\}$ and $\{\phi(d_i), \phi(a)\}$ are irreducible for $i = 1, \dots, k$.*

Then the group $H = G/N(R^m)$, $m \geq 2$, admits a representation $\rho: H \to \mathrm{PSL}(2, \mathbb{C})$ such that $G_1 \to H \xrightarrow{\rho} \mathrm{PSL}(2, \mathbb{C})$ and $G_2 \to H \xrightarrow{\rho} \mathrm{PSL}(2, \mathbb{C})$ are faithful and $\rho(R)$ has order m. In particular, $G_1 \to H$ and $G_2 \to H$ are injective.

Proof. Let $\phi: G \to \mathrm{PSL}(2, \mathbb{C})$ be the given representation of G such that $\phi|_{G_1}$ and $\phi|_{G_2}$ are faithful and the pairs $\{\phi(c_i), \phi(a)\}$ and $\{\phi(d_i), \phi(a)\}$ are irreducible for $i = 1, \dots, k$.

We may assume that $\phi(a)$ has the form $\phi(a) = \pm\left(\begin{smallmatrix} s & 0 \\ 0 & s^{-1} \end{smallmatrix}\right)$ or $\phi(a) = \pm\left(\begin{smallmatrix} 1 & 1 \\ 0 & 1 \end{smallmatrix}\right)$.

Suppose first that $\phi(a) = \pm\left(\begin{smallmatrix} s & 0 \\ 0 & s^{-1} \end{smallmatrix}\right)$ and let $T = \pm\left(\begin{smallmatrix} t & 0 \\ 0 & t^{-1} \end{smallmatrix}\right)$ with t a variable whose value in $\mathbb{C}$ is to be determined.

Define

- $\rho(h_1) = \phi(h_1)$ for $h_1 \in G_1$; and
- $\rho(h_2) = T\phi(h_2)T^{-1}$ for $h_2 \in G_2$.

Since T commutes with $\phi(a)$ for any t, the map $\rho: H \to \mathrm{PSL}(2, \mathbb{C})$ will define a representation with the desired properties if there exists a value t such that $\rho(R)$ has order m.

Recall that a complex projective matrix B in $\mathrm{PSL}(2, \mathbb{C})$ will have finite order $m \geq 2$ if $\mathrm{tr}\, B = \pm 2\cos(\frac{\pi}{m})$. As in the statement of the theorem assume $R = c_1 d_1 \cdots c_k d_k$ with $k \geq 1$ and $c_i \in G_1 \setminus A$, $d_i \in G_2 \setminus A$ for $i = 1, \dots, k$, and assume that the pairs $\{\phi(c_i), \phi(a)\}$ and $\{\phi(d_i), \phi(a)\}$ are irreducible for $i = 1, \dots, k$. Define

$$f(t) = \mathrm{tr}(\phi(c_1)T\phi(d_1)T^{-1} \cdots \phi(c_k)T\phi(d_k)T^{-1}),$$

then $f(t)$ is a Laurent polynomial in t of degree $2k$ in both t and t^{-1}. The coefficients of t^{2k} and t^{-2k} are non-zero because the pairs $\{\phi(c_i), \phi(a)\}$ and $\{\phi(d_i), \phi(a)\}$ are irreducible.

Therefore by the fundamental theorem of algebra there exists a t_0 with $f(t_0) = 2\cos\frac{\pi}{m}$. With this choice of t_0 we have $\mathrm{tr}(\rho(R)) = 2\cos\frac{\pi}{m}$ and this $\rho(R)$ has order m. Therefore ρ is a representation with the desired properties. Now assume $\phi(a) = \pm\left(\begin{smallmatrix} 1 & 1 \\ 0 & 1 \end{smallmatrix}\right)$. In this case, define $T = \pm\left(\begin{smallmatrix} 1 & t \\ 0 & 1 \end{smallmatrix}\right)$ with t again a variable. Again T commutes with $\rho(a)$ and the proof goes through analogously as above giving the desired representation. □

Corollary 3.5.48. *Let $G = G_1 *_A G_2$ be a group of F-type. Assume further that $n \geq 4$, $2 \leq p \leq n-2$ and neither $U = U(a_1, \dots, a_p)$ nor $V = V(a_{p+1}, \dots, a_n)$ is a proper power in the free product on the generators they involve.*

Suppose that UV involves all the generators and let $R = c_1 d_1 \cdots c_k d_k$ *with* $k \geq 1$ *and* $c_i \in \langle a_1, \ldots, a_p \rangle \setminus \langle U \rangle$, $d_i \in \langle a_{p+1}, \ldots, a_n \rangle \setminus \langle V \rangle$ *for* $i = 1, \ldots, k$, *and let* $m \geq 2$. *Then the conclusion of Theorem 3.5.47 holds for* $H = G/N(R^m)$ *with*

$$G_1 = \langle a_1, \ldots, a_p \mid a_1^{e_1} = \cdots = a_p^{e_p} = 1 \rangle,$$
$$G_2 = \langle a_{p+1}, \ldots, a_n \mid a_{p+1}^{e_{p+1}} = \cdots = a_n^{e_n} = 1 \rangle,$$

and

$$A = \langle U^{-1} \rangle = \langle V \rangle.$$

Proof. By Theorem 3.5.12, the group G admits a faithful representation ϕ in $\mathrm{PSL}(2, \mathbb{C})$ such that the pairs $\{\phi(c_i), \phi(U)\}$ and $\{\phi(d_i), \phi(V)\}$ are irreducible for $i = 1, \ldots, k$. □

Corollary 3.5.49. *Let G be a group of F-type as in Corollary 3.5.48 and R be the relator as in Corollary 3.5.48. Suppose* $m \geq 8$. *Then* $H = G/N(R^m)$ *is virtually torsion-free.*

Proof. In G any element of finite order is conjugate to a power of a generator a_i. Since $m \geq 8$, from the torsion theorem for small cancellation products of D. Collins and F. Perraud, see [177] or [58], any element of finite order in G is conjugate to a power of a generator or a power of R.

Since $\rho(R)$ has exact order m and the representations of G_1 and G_2 are faithful, it follows that (the constructed) ρ is essentially faithful and therefore G is virtually torsion-free. □

Remark 3.5.50. Let G be a group of F-type with the presentation

$$G = G_1 \underset{A}{*} G_2$$

with $G_1 \neq A \neq G_2$,

$$G_1 = \langle a_1, \ldots, a_p \mid a_1^{e_1} = \cdots = a_p^{e_p} = 1 \rangle,$$
$$G_2 = \langle a_{p+1}, \ldots, a_n \mid a_{p+1}^{e_{p+1}} = \cdots = a_n^{e_n} = 1 \rangle,$$

$U = U(a_1, \ldots, a_p)$, $V = (a_{p+1}, \ldots, a_n)$, and $A = \langle U^{-1} \rangle = \langle V \rangle$. We call G a *special group of F-type* if UV involves all the generators, $n \geq 4$, $2 \leq p \leq n-2$, and neither U nor V is a proper power in the free product on the generators they involve.

Now let G be a special group of F-type and let $R \in G \setminus A$ be given in a reduced form $R = c_1 d_1 \cdots c_k d_k$ with $k \geq 1$ and $c_i \in G_1 \setminus A$, $d_i \in G_2 \setminus A$ for $i = 1, \ldots, k$. Then we may apply the theory of one-relator quotients $H = G/N(R^m)$ with $m \geq 2$ as described in [93] and the deficiency arguments in Section 3.5.4.1 analogously in this more general context. If we mirror the respective proofs there, then we easily get the following Theorem.

Theorem 3.5.51. *Let G be a special group of F-type, R be the relator as in Corollary 3.5.48, and $H = G/N(R^m)$, $m \geq 2$. Then the following hold.*

1. *For $i = 1, \dots, n$ let $a_i = 0$ if $e_i = 0$ and $a_i = \frac{1}{e_i}$, if $e_i \geq 2$. Then:*
   (i) *If $\sum_{i=1}^{n} a_i + \frac{1}{m} \leq n - 2$, then H has a subgroup of finite index mapping homomorphically on $\mathbb{Z}$. In particular, H is infinite.*
   (ii) *If $\sum_{i=1}^{n} a_i + \frac{1}{m} < n - 2$, then H has a subgroup of finite index mapping homomorphically onto a free group of rank 2. In particular, H is SQ-universal.*
2. *H is a non-trivial free product with amalgamation.*
3. *If $n \geq 5$ or $n = 4$ and at least one of the e_i is not equal to 2, then H has a free subgroup of rank 2.*

The details can be found in [102].

We have considered special groups G of F-type. We can make similar calculations for the hyperbolic groups of F-type which also have a faithful representation on $\mathrm{PSL}(2, \mathbb{C})$.

If now, for instance, $V = V_1^q$, $q \geq 2$, V_1 is not a proper power in $\langle a_{p+q}, \dots, a_n \rangle$, one has to be careful with the relator $R = c_1 d_1 \cdots c_k d_k$ for $k \geq 1$ where $c_i \in \langle a_1, \dots, a_p \rangle \setminus \langle U \rangle$ and $d_i \in \langle a_{p+1}, \dots, a_n \rangle \setminus \langle V \rangle$ for $i = 1, \dots, k$; G admits a faithful representation ϕ in $\mathrm{PSL}(2, \mathbb{C})$ but then a pair $\{\phi(d_i), \phi(V)\}$ is not irreducible if $d_i = V_1^t$ with $q \nmid t$. But everything goes through analogously if we consider only relators R of the form as above with $d_i \in \langle a_{p+1}, \dots, a_n \rangle \setminus \langle V_1 \rangle$.

3.6 Interlude: Marshall Hall's theorem and property S

We discuss some results by G. Baumslag, O. Bogopolski, B. Fine, A. Gaglione, G. Rosenberger and D. Spellman [18].

In the following we present a new elementary proof of Hall's theorem and then use this theorem to give a simple proof of property S for free groups. Recall that, if $\mathcal{P}$ is a group property, then a group G virtually satisfies $\mathcal{P}$ if G has a subgroup of finite index satisfying $\mathcal{P}$.

Theorem 3.6.1 (Marshall Hall). *Let H be a finitely generated subgroup of a free group F. Then H is virtually a free factor, that is, there is a subgroup G of finite index in F for which H is a free factor.*

The original proof of Hall used Schreier transversals (see Theorem 1.3.32). This proof has been translated into geometric language and a proof based on graph theory has been given. These graph-theoretic ideas were extended to torsion-free hyperbolic groups and Theorem 3.6.1 can be reobtained via a result of Stallings, Kapovich and Short (see [151]). Below we give a simple direct proof.

Proof. Without loss of generality we may assume that F is freely generated by a finite set X. Let (Γ_H, v) be a finite labeled graph with a distinguished vertex v representing H. This means that the following conditions are satisfied:
(1) The graph Γ_H is connected.
(2) The edges of this graph are labeled by letters from $X \cup X^{-1}$. If an edge e is labeled by x, the edge e^{-1} is labeled by x^{-1}.
(3) For any vertex of this graph, the edges going out of this vertex have different labels.
(4) The group H is generated by the labels of loops at v.

This graph can be obtained in the following way. Let $w_1, \dots, w_n$ be a Nielsen reduced set of generators for H. Draw a wedge of loops with a common vertex v. Divide each loop into edges and label them so that along the ith edge we can read the word w_i. Then, for any pair of these loops, we will identify their largest initial and terminal segments having the same labels.

A vertex u of Γ_H will be called *full* if the number of edges going out of u is maximal possible, that is $2n$. If each vertex of a subgroup graph is full, then the subgroup must be of finite index. Now we add some edges to the graph Γ_H to make each of the vertices full. The resulting graph will be denoted by $\tilde{\Gamma}_H$. Let $\tilde{\Gamma}'_H$ be a copy of the graph $\tilde{\Gamma}_H$ with the inverse edge labeling, that is, if $\gamma: \tilde{\Gamma}_H \to \Gamma_H$ is the isomorphism, then the label of any edge e of $\tilde{\Gamma}_H$ is inverse to the label of $\gamma(e)$. Finally, we identify the outer edges of $\tilde{\Gamma}_H$ with the corresponding edges of $\tilde{\Gamma}'_H$. We now obtain the labeled graph Δ_H which has each edge full. This graph (with distinguished vertex v) represents a subgroup G of F. The subgroup H is a free factor of G since Γ_H is a subgraph of Δ_H. Moreover, G is a subgroup of finite index in F since all the vertices of Δ_H are full. □

See Figure 3.42 for an illustration for $H = \langle ab^{-2}, ba^{-2} \rangle < F$ where F is free on a and b. We now use Hall's theorem to consider property S in free groups (the Greenberg–Stallings theorem).

Definition 3.6.2. A group G has *property S* if for any finitely generated commensurable subgroups A and B of G, $A \cap B$ has finite index in $\langle A, B \rangle$, the join in G of A and B.

Recall that in general two groups A and B are called (abstractly) commensurable if there are subgroups $H < A$ and $K < B$ such that H is isomorphic to K.

Clearly, any finite group or locally finite group satisfies property S, so in the infinite case, this property can be considered a fairly strong finiteness result. The genesis of this concept is the following theorem of Greenberg [117] and independently found by Stallings [244].

Theorem 3.6.3. *Free groups satisfy property S.*

In order to prove this using Hall's theorem, we need the following classical concept which can also be found in [143]. The conjugate of a subgroup H by an element $g \in G$ is denoted by H^g, that is $H^g = gHg^{-1}$.

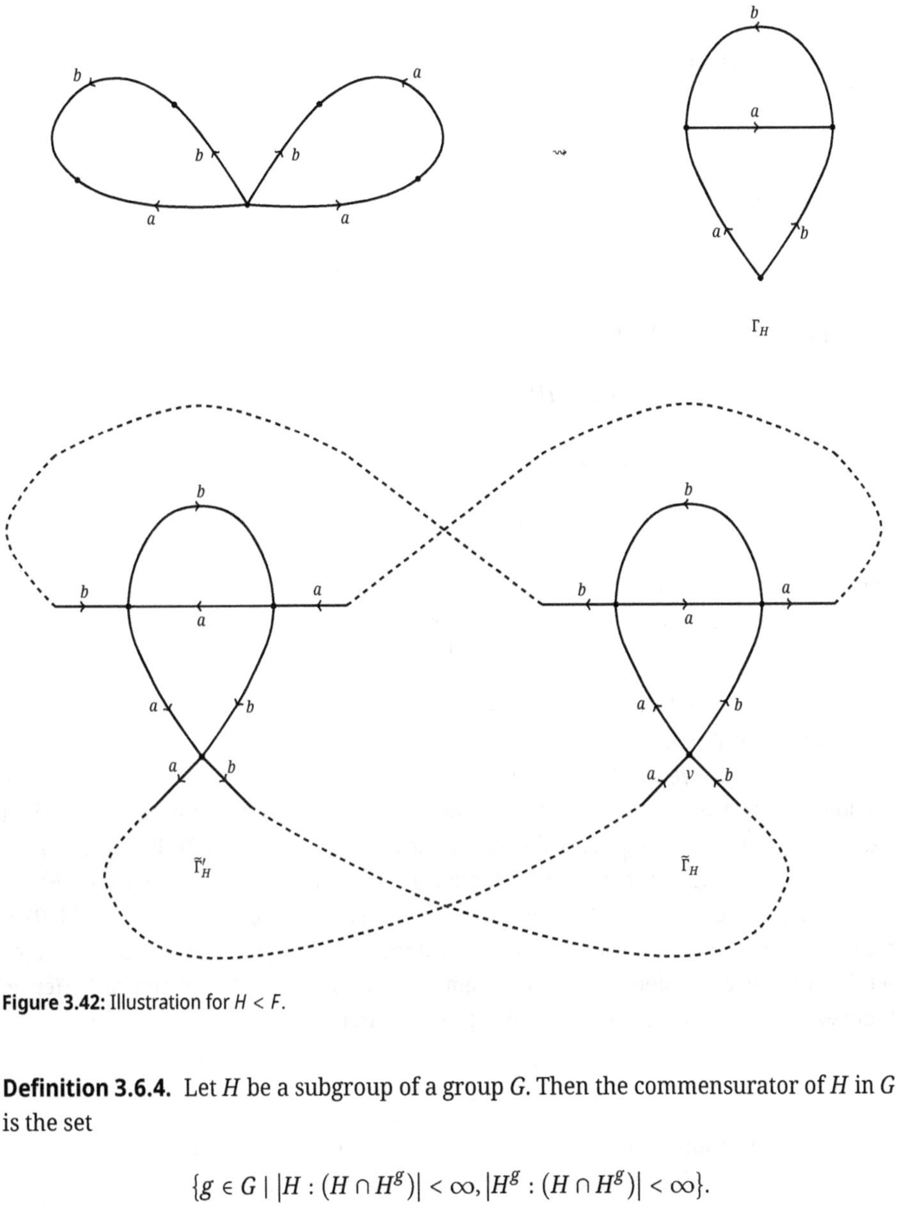

Figure 3.42: Illustration for $H < F$.

Definition 3.6.4. Let H be a subgroup of a group G. Then the commensurator of H in G is the set

$$\{g \in G \mid |H : (H \cap H^g)| < \infty, |H^g : (H \cap H^g)| < \infty\}.$$

We denote the commensurator by $\mathrm{comm}_G(H)$. In [144] the commensurator is called the *virtual normalizer*. The following lemma is not difficult to prove, though not entirely obvious, so we include the proof.

Lemma 3.6.5. *For any group G and subgroup H, the commensurator $\mathrm{comm}_G(H)$ is a subgroup. Further, if $|G : H| < \infty$, then $\mathrm{comm}_G(H) = G$.*

Proof. The second assertion is clear. We prove that $\mathrm{comm}_G(A)$ is a subgroup. We note that, if A, B, H are subgroups of G with $A \subset B$, then $|B : A| < \infty$ implies $|(B \cap H) : (A \cap H)| < \infty$. To see this, let $x, y \in B \cap H$. Then $x^{-1}y \notin A \cap H$ if and only if $x^{-1}y \notin A$. This says that two cosets $xB \cap H$ and $yB \cap H$ are distinct if and only if the cosets xA and yB are distinct. Then $|(B \cap H) : (A \cap H)| \leq |B : A|$. Let H be a subgroup of G and $x, y \in \mathrm{comm}_G(H)$. Then $|H : (H \cap H^x)| < \infty$ and $|H : (H \cap H^y)| < \infty$. Conjugating the first of these by y, we obtain

$$|H^y : (H^y \cap H^{xy})| < \infty.$$

Now apply the observation above with $H = H$, $B = H^y$ and $A = H^y \cap H^{xy}$. Then

$$|(H \cap H^y) : (H \cap H^y \cap H^{xy})| < \infty.$$

From $|H : (H \cap H^y)| < \infty$, it follows that

$$|H : (H \cap H^y \cap H^{xy})| < \infty,$$

and then we obtain $|H : H^{xy}| < \infty$. Therefore, $xy \in \mathrm{comm}_H(G)$. It is clear that $1 \in \mathrm{comm}_G(H)$, and if $x \in \mathrm{comm}_G(H)$, then so is x^{-1}. This completes the proof of the lemma. □

Proof of Theorem 3.6.3. Suppose that A and B are finitely generated commensurable subgroups of a free group F. Then $A \cap B$ has finite index in both A and B and is also finitely generated. We want to show that $A \cap B$ has finite index in the join $\langle A, B\rangle$. Since the join is also a free group, we may, without loss of generality, assume that the join $\langle A, B\rangle$ is all of F. Since $|A : (A \cap B)| < \infty$, we have $A \subset \mathrm{comm}_F(A \cap B)$. To see this, note that by Lemma 3.6.5 $\mathrm{comm}_A(A \cap B) = A$, but clearly $\mathrm{comm}_A(A \cap B) \subset \mathrm{comm}_F(A \cap B)$.

Similarly, $B \subset \mathrm{comm}_F(A \cap B)$. Since $\mathrm{comm}_F(A \cap B)$ is a subgroup, we have $\langle A, B\rangle \subset \mathrm{comm}_F(A \cap B)$, and therefore $F = \langle A, B\rangle = \mathrm{comm}_F(A \cap B)$ again by Lemma 3.6.5. Since $A \cap B$ is finitely generated by Hall's theorem, $A \cap B$ is virtually a free factor of F. Hence, there exists a subgroup H of finite index in F such that

$$H = (A \cap B) * K.$$

However, for any non-trivial $k \in K$, we have $(A \cap B) \cap (A \cap B)^k = \{1\}$. This follows directly from Section 1.4 on free products. Since $A \cap B$ is infinite, it follows that no non-trivial element of K can be in the commensurator. Since the commensurator is all of F, we must then have $K = \{1\}$ and therefore $H = A \cap B$ has finite index in the join. □

We now present some other results on finitely generated subgroups of free groups that can be proved in the same manner.

Theorem 3.6.6. *Let H be a finitely generated subgroup of a free group F. Then H has finite index if and only if for each $f \in F$, there exists an n such that $f^n \in H$.*

Proof. If $|F : H| = m < \infty$ and $f \in F$, then $1, f, f^2, \ldots, f^m$ cannot all be incongruent modulo H. Therefore, $f^k \in H$ for some k.

Conversely, suppose that, for each $f \in F$, we have $f^n \in H$ for some n. Since H is finitely generated, there exists a subgroup K of finite index in F such that $K = H * K_1$. However no non-trivial element of K_1 can have a power in H. Therefore, K_1 must be trivial and hence $H = K$ is of finite index. □

Corollary 3.6.7. *Let H be a finitely generated subgroup of a free group F. Then H has finite index if and only if $H \supset F(X^d)$ for some d, where $F(X^d)$ is the verbal subgroup of F generated by all dth powers.*

As a consequence of Theorem 3.6.6 and property S, we obtain the following, which seems to be difficult to prove directly.

Theorem 3.6.8. *Let A, B and H be finitely generated subgroups of a free group F with $H \subset A \cap B$. If each element of A and B to a sufficiently high power is in H, then each element of the join $\langle A, B\rangle$ to a sufficiently high power is in H.*

Proof. Since each element of A and B to a sufficiently high power is in H, it follows from Theorem 3.6.6 that H has finite index in both A and B. Then from property S; H has finite index in the join $\langle A, B\rangle$. It follows clearly that a sufficiently high power of each element in the join is in H. □

Theorem 3.6.9 (See Corollary 1.3.34). *A finitely generated non-trivial normal subgroup N of a free group F must be of finite index.*

Proof. Suppose that N is normal in F with $|F : N| < \infty$. Since N is finitely generated, there exists a subgroup K of finite index in F such that $K = N * K_1$. But then as before, $N \cap N^k = \{1\}$ for any non-trivial $k \in K_1$, contradicting the normality. Therefore, K_1 must be trivial and $N = K$ has finite index. □

We note that a version of this result was recently proved by Bridson and Howie [34].

Corollary 3.6.10. *Let H be a finitely generated subgroup of a free group F. Then H has finite index if and only if H contains a finitely generated non-trivial normal subgroup.*

Proof. If $|F : H| < \infty$, then the intersection of the conjugates of H is a finitely generated normal subgroup contained in H. Conversely, suppose that N is a finitely generated normal subgroup of F contained in H. Then from Theorem 3.6.9 N has finite index and therefore so does H. □

Corollary 3.6.10 provides a transparent proof of the residual finiteness of free groups.

Corollary 3.6.11. *A finitely generated free group F is residually finite.*

Proof. Let $g \in F$. From Hall's theorem, there exists a subgroup K of finite index in F such that $K = \langle g\rangle * K_1$. Since these are free groups, there is clearly a homomorphism ϕ

of K onto a finite group G with $\phi(g) \neq 1$. Let N be the kernel of ϕ so that $|K : N| < \infty$ and $g \notin N$. Since $|F : K| < \infty$, we have $|F : N| < \infty$. Take $N_1 = \cap_{h\in F} N^h$. Then $|F : N_1| < \infty$. Since $g \notin N$, it follows that $g \notin N_1$, completing the proof. □

The same argument can be extended to the stronger property of subgroup separability. Recall from Definition 2.3.31 that a group G is subgroup separable if given any finitely generated subgroup H of G and $g \notin H$, there exists a finite quotient $\bar{G}$ of G with $\bar{g} \notin \bar{H}$, where $\bar{g}$ and $\bar{H}$ are the respective images of g and H in $\bar{G}$. The Marshall Hall property was used originally to prove subgroup separability of free groups. A proof of this using Hall's theorem is an extension of the proof of the residual finiteness. Clearly, the subgroup separability implies residual finiteness. In the next theorem, we present another proof of Hall's result that free groups are subgroup separable.

Theorem 3.6.12. *Let F be a free group, H a finitely generated subgroup, and $\{g_1, \ldots, g_n\}$ a finite set of elements of $F \backslash H$. Then there exists a subgroup G of finite index in F, containing H and not containing $\{g_1, \ldots, g_n\}$.*

We remark that we have already seen Theorem 3.6.12 in a different description in Section 1.3 and we have proved it there with the Reidemeister–Schreier method.

Proof. Without loss of generality we may assume that F is finitely generated. It is sufficient to prove the theorem for $n = 1$. For arbitrary n, the result will then follow by induction. Indeed, we may assume that there exists a subgroup M of finite index in F that contains $\langle H, g_1, \ldots, g_{n-1}\rangle$ and does not contain g_n. By induction we can find the desired subgroup G in M.

Assume then that $n = 1$. Let $g = g_1$ and $H_1 = \langle H, g\rangle$. From Hall's theorem, there exists a subgroup L in F such that $H_1 * L$ has finite index in F. Hence, we may assume $F = H_1 * L$. Consider the retraction $\phi: F \to H_1$ identical on H_1 and trivial on L. It is sufficient to find a finite index subgroup G_1 in H_1 containing H and not containing g. Then we can set $G = \phi^{-1}(G_1)$.

By Hall's theorem, we see that $H * M$ has finite index in H_1 for some subgroup M. If $H * M$ is a proper subgroup of H_1, we can set $G_1 = H * M$. Now suppose $H * M = H_1$. Consider the retraction $\psi: G_1 \to M$ identical on M and trivial on H. Since $M \neq \{1\}$, we can find a proper normal subgroup M_1 of M of finite index in M which does not contain g. Then we set $G_1 = \psi^{-1}(M_1)$. □

Recently, Wilton [272] has proved that all finitely generated fully residually free groups, i. e., limit groups, are subgroup separable, answering a question of Sela. The next theorem was proved in [143] using Stallings foldings. The proof we give seems to be more direct.

Theorem 3.6.13. *Let H be a finitely generated non-trivial subgroup of a free group F. Then H has finite index in its commensurator.*

Proof. Let $F_1 = \mathrm{comm}_F(H)$. Then F_1 is free and H is a finitely generated subgroup of F_1. As before, there exists a subgroup K of finite index in F_1 such that $K = H * K_1$. No non-trivial element of K_1 can be in the commensurator and hence $K_1 = \{1\}$. Therefore, $K = H$ has finite index in F_1. □

Corollary 3.6.14. *A finitely generated non-trivial subgroup H of a free group F has finite index if and only if $F = \mathrm{comm}_F(H)$.*

In general, if G is any group and $|G : H| < \infty$, then $G = \mathrm{comm}_G(H)$. However, the converse need not be true. As a counterexample, consider the amalgamated free product G of two finite groups G_1, G_2 with amalgamated finite subgroup H with $G_1 \neq H \neq G_2$. Here, $G = G_1 *_H G_2$. Then $\mathrm{comm}_G(H) = G$ but G is infinite, so H has infinite index.

Kapovich and Myasnikov proved Corollary 3.6.14 using a graph-theoretic argument based on Stallings foldings. They then use Corollary 3.6.14 to prove property S for free groups. Abstracting some of these ideas, we define:

Definition 3.6.15. A group G has the commensurator condition if any finitely generated non-trivial subgroup has finite index in its commensurator.

Let G be a group satisfying the commensurator condition. Then a finitely generated non-trivial subgroup H of G has finite index if and only if its commensurator within G is all of G. The following is then easy.

Theorem 3.6.16. *If a group G satisfies the commensurator condition, then it satisfies property S.*

The converse is not true and an infinite finitely generated Abelian group provides a counterexample. In Theorem 3.6.22, we use the commensurator condition that yields an alternative proof of a result of Kapovich that hyperbolic finitely generated fully residually free groups satisfy property S.

Surface groups and property S

Kapovich and Short [144] have proved the following restricted version of property S for torsion-free hyperbolic groups.

Theorem 3.6.17. *Suppose that H is a torsion-free hyperbolic group, and A, B are finitely generated, commensurable, quasi convex subgroups of H. Then $A \cap B$ has finite index in $\langle A, B\rangle$.*

We call a hyperbolic group *QC-free* if any finitely generated subgroup is quasi convex. Free groups and hyperbolic surface groups are *QC*-free. Recall that a surface group G is hyperbolic if $G = \langle a_1, b_1, \ldots, a_g, b_g \mid [a_1, b_1] \cdots [a_g, b_g] = 1\rangle$ with $g \geq 2$ or $G = \langle a_1, \ldots, a_g \mid a_1^2 \cdots a_g^2 = 1\rangle$ with $g \geq 3$. For general finitely generated groups, the *QC*-free property has been called *locally quasi convex*.

Corollary 3.6.18. *Any QC-free hyperbolic group satisfies property S.*

Corollary 3.6.19. *Any hyperbolic surface group satisfies property S.*

This result can be rephrased in the following interesting way.

Theorem 3.6.20. *Let G be a hyperbolic surface group. Suppose that A and B are finitely generated commensurable free subgroups of G. Then the join $\langle A, B\rangle$ is a free group.*

Proof. Suppose that A and B are finitely generated commensurable free subgroups of the hyperbolic surface group G. We recall that subgroups of finite index in G are also surface groups of finite genus while subgroups of infinite index are free. Hence, the join $\langle A, B\rangle$ is either a surface group of finite genus or a free group. From the Greenberg–Kapovich–Short result, the intersection $A \cap B$ has finite index in $\langle A, B\rangle$. The group $A \cap B$ is free. If $\langle A, B\rangle$ is a surface group of finite genus, then it would have a free group of finite index which is impossible. Hence, $\langle A, B\rangle$ must be free. □

We can further extend this.

Corollary 3.6.21. *Suppose that A and B are finitely generated commensurable free subgroups of* $\mathrm{PSL}(2, \mathbb{R})$. *If the join $\langle A, B\rangle$ is discrete, it must also be free.*

The proof of Kapovich and Short uses the boundaries of hyperbolic groups and, of course, quasi convexity. We leave the details as an exercise. This raises the following question: *Is there a purely algebraic or purely function-theoretic proof of the results given in Theorem* 3.6.20 *and Corollary* 3.6.21*?*

Given our proof of property S for free groups using Hall's theorem, this raises the question of whether there is an analogue of Hall's theorem for surface groups. The answer is yes for hyperbolic surface groups of even genus. We obtain the following result. However, it does not allow for a proof of property S to go through.

Theorem 3.6.22. *Let G be a hyperbolic surface group of even genus. Let $A = \{a_1, \ldots, a_n\}$ be a finite set of elements of G. Then there exists a subgroup H of finite index in G such that A is included in a generating system for H.*

Proof. We prove the theorem in the orientable case. The proof for the non-orientable case is analogous. Let K be an orientable surface group of genus $2k$ so that

$$K = \langle x_1, \ldots, x_{2k}, y_1, \ldots, y_{2k} \mid [x_1, x_2] \cdots [x_{2k-1}, x_{2k}] = [y_1, y_2] \cdots [y_{2k-1}, y_{2k}]\rangle.$$

Now consider A to be free on $4k$ generators $x_1, \ldots, x_{2k}, y_1, \ldots, y_{2k}$ and let ϕ be the involution which transposes the x_i and y_i. Let $G = \langle t, A \mid t^{-1}at = \phi(a), a \in A\rangle$. Recall that a finitely generated subgroup H of a group G has the generalized Hall property if whenever $g_0 \notin H$ there is a subgroup H^* of finite index in G such that $g_0 \notin H^*$ and there is a graph decomposition of groups for $H^* = \pi_1(\Gamma, \mathcal{H})$ with $\mathcal{H}(v) = H * K$ for some vertex group $\mathcal{H}(v)$.

From a theorem by Tretkoff [259] the HNN group G given above satisfies the generalized Hall property, where $a = [x_1, x_2] \cdots [x_{2k-1}, x_{2k}]$. Furthermore, if $z_i = t^{-1}x_i t$, then K embeds into G by $x_i \mapsto y_i$ and $y_i \mapsto z_i$.

Let H be a finitely generated subgroup of K considered as being a subgroup of G. Clearly, $t \notin H$. Then from the generalized Hall property, there is a subgroup G_1 of G such that

(a) $G_1 = \pi_1(\Delta, \mathcal{H})$ for some graph of groups $(\Delta, \mathcal{H})$;
(b) For some vertex v of Δ, we have $\mathcal{H}(v) = H * L$; and
(c) $t \notin G_1$.

Now $H \le G_1 \cap K \le K$ and $G_1 \cap K$ has finite index in K. Further, since $\mathcal{H}(v)$ is a free product with H as a factor, a generating system for H is part of a generating system of $\mathcal{H}(v)$. In turn, $\mathcal{H}(v)$ is a vertex group of a graph of groups, so its generating system is part of a generating system for G_1. Therefore, a given generating system for H is part of a generating system for a subgroup of finite index in K. □

Let G be a hyperbolic surface group. A generating system X for G is called a *standard generating system* if X is a generating system over which G splits as a free product with cyclic amalgamation. The problem in applying Theorem 3.6.22 to proving property S for hyperbolic surface groups of even genus is that the generating system that we obtain for A to be part of is not guaranteed to be a standard generating system with A in one of the factors. We pose the question of when we can extend Theorem 3.6.22 so that A is part of a standard generating system and contained in one of the factors expressing G as a cyclically pinched one-relator group. One straightforward condition is for A to actually be part of X.

Theorem 3.6.23. *Let X be a standard generating system for a hyperbolic surface group G. Then given $Y \subset X$, there exists a proper subgroup of finite index H with a decomposition $H = G_1 *_C G_2$ such that Y is part of a generating system for G_1.*

Proof. We prove this theorem only in the orientable case and where G has genus 2. The higher genus case and the non-orientable case are done in an analogous manner. Then let G have the presentation $G = \langle a, b, c, d \mid [a, b] = [c, d]\rangle$. The subsets $\{a, b\}$ and $\{c, d\}$ are already part of standard generating systems for the whole group G. Consider then the subset $\{a, b, c\}$ of the standard generators and we show that it is part of a standard generating system for a subgroup of finite index. Let K be the normal closure of $\langle a, b, c, d^2\rangle$ in G. The quotient is $G/K = \langle d \mid d^2 = 1\rangle$. Hence, K has index 2 in G. Taking the elements 1, d as coset representatives for K and applying the Reidemeister–Schreier rewriting process, we obtain a presentation

$$K = \langle a, b, c, \delta, \alpha, \beta \mid [a, b][c, \delta][\beta, \alpha] = 1\rangle,$$

where $\delta = d^2$, $\alpha = da^{-1}d^{-1}$ and $\beta = db^{-1}d^{-1}$. If we let $G_1 = \langle a, b, c, \delta\rangle$ and $G_2 = \langle \alpha, \beta\rangle$, then $K = G_1 *_C G_2$ with $C = \langle [a, b][c, \delta]\rangle_{G_1} = \langle [\beta, \alpha]^{-1}\rangle_{G_2}$ and the set $\{a, b, c\}$ is part of the generating system for G_1. □

In another direction, Kapovich [142] has proved that all fully residually free groups satisfy property S. This also shows that hyperbolic surface groups satisfy property S (of genus ≥ 4 in the non-orientable case). Recall that a group G is fully residually free if given finitely many non-trivial elements $g_1, \ldots, g_n$ in G, there is a homomorphism $\phi: G \to F$, where F is a free group such that $\phi(g_i) \neq 1$ for all $i = 1, \ldots, n$. Fully residually free groups have played a crucial role in the study of equations and first order formulas over free groups and in the solution of the Tarski problem by Kharlampovich and Myasnikov and independently Sela. A theorem due to Remeslennikov [217] and independently Gaglione and Spellman [108] shows that the finitely generated non-Abelian fully residually free groups coincide with the finitely generated universally free groups, that is, the class of groups having the same universal (equivalently existential) theory as the class of non-Abelian free groups. The structure and properties of fully residually free groups have been extensively studied by Kharlampovich and Myasnikov (see [157] and the references there). It is known that a finitely generated fully residually free group is hyperbolic precisely when all centralizers are cyclic. Wilton [272] has recently proved that all finitely generated fully residually free groups are subgroup separable. Using the concept of Stallings foldings for infinite words, one can give a separate proof of Kapovich's result for hyperbolic finitely generated fully residually free groups; see [18]. We just collect the results here.

Theorem 3.6.24. *A hyperbolic finitely generated fully residually free group satisfies the commensurator condition, that is, any finitely generated non-trivial subgroup is of finite index in its commensurator.*

Corollary 3.6.25. *Any hyperbolic finitely generated fully residually free group satisfies property S.*

As mentioned, Kapovich [142] has proved the whole result, that is, all fully residually free groups satisfy property S.

3.7 Quasi isometries and quasi geodesics

The aim of this paragraph is the proof of Theorem 3.7.15: Let X and Y be two geodesic metric spaces and $F: X \to Y$ be a quasi isometry. If Y is hyperbolic then X is also hyperbolic. This finally proves Theorem 3.3.9 and Corollary 3.3.10. In particular: If Γ_1 is a subgroup of a hyperbolic group Γ_2 of finite index, then Γ_1 is also hyperbolic.

The foundation for the proof of Theorem 3.7.15 is an approximation of certain segments through geodesic segments. In this context the approximation of δ-hyperbolic spaces through trees (see Subsection 3.2.2) plays an important role.

3.7.1 Quasi geodesics, quasi rays and quasi segments

Definition 3.7.1. Let X_0, X be metric spaces, $F: X_0 \to X$ be a function and $\lambda \geq 1$, $c \geq 0$, $L > 0$ be three real numbers. We call F an *isometry* if

$$|F(s) - F(t)| = |s - t| \quad \text{for all } s, t \in X_0$$

(F is not necessarily surjective); we call F a (λ, c, L)-*local quasi isometry* if

$$\frac{1}{\lambda}|s - t| - c \leq |F(s) - F(t)| \leq \lambda|s - t| + c$$

for $|s - t| \leq L$ and we call F a (λ, c)-*quasi isometry* if the above inequalities hold for all $s, t \in X_0$.

Definition 3.7.2. Let X be a metric space and I be an interval in $\mathbb{Z}$ or $\mathbb{R}$.

1. An isometry $g: I \to X$ is called
   - *Geodesic segment* in X if I is bounded.
   - *Ray* or *minimal ray* if I is semi-infinite, that is, I is bounded on one end and unbounded on the other.
   - *Minimal geodesic*, if $I = \mathbb{Z}$ or $I = \mathbb{R}$.
2. A (λ, c)-quasi isometry $f: I \to X$ is called
   - (λ, c)-*quasi segment*, if I is bounded.
   - (λ, c)-*quasi ray*, if I is semi-infinite.
   - (λ, c)-*quasi geodesic*, if $I = \mathbb{Z}$ or $I = \mathbb{R}$ (see Subsection 3.2.2).
3. A (λ, c, L)-local quasi isometry $f: I \to X$ is called
   - (λ, c, L)-*local quasi segment*, if I is unbounded.
   - (λ, c, L)-*local quasi ray*, if I is semi-infinite.
   - (λ, c, L)-*local quasi geodesic*, if $I = \mathbb{Z}$ or $I = \mathbb{R}$.

In the following we characterize the notion of two metric spaces being quasi isometric.

Remark 3.7.3. Metric spaces (X_0, d_0) and (X_1, d_1) are *quasi isometric* if there exist functions

$$X_0 \underset{g}{\overset{f}{\rightleftarrows}} X_1$$

as well as constants $\lambda > 0$ and $c \geq 0$ such that

(a) $d_1(f(x), f(y)) \leq \lambda d_0(x, y) + c$ for all $x, y \in X_0$;
(b) $d_0(g(x), g(y)) \leq \lambda d_1(x, y) + c$ for all $x, y \in X_1$;
(c) $d_0(g \circ f(x), x) \leq c$ for all $x \in X_0$; and
(d) $d_1(f \circ g(x), x) \leq c$ for all $x \in X_1$.

We leave the proof as an exercise.

Theorem 3.7.4. *Metric spaces (X_0, d_0) and (X_1, d_1) are quasi isometric if and only if there exist a quasi isometry $F: X_0 \to X_1$ such that $\sup_{x \in X_1} d_1(x, \operatorname{im}(F)) < \infty$.*

Proof. Suppose (X_0, d_0) and (X_1, d_1) are quasi isometric. Then there are mappings

$$X_0 \underset{g}{\overset{f}{\rightleftarrows}} X_1$$

as well as constants $\bar{\lambda} > 0$ and $\bar{c} \geq 0$, such that the conditions (a)–(d) of Remark 3.7.3 are satisfied. We define $F: X_0 \to X_1$ via $F = f$. According to (a) we have

$$d_1(f(s), f(t)) \leq \bar{\lambda} d_0(s, t) + \bar{c}$$

for all $s, t \in X_0$, hence

$$|F(s) - F(t)| \leq \bar{\lambda}|s - t| + \bar{c}$$

for all $s, t \in X_0$. According to (b) we have

$$d_0(g(f(s)), g(f(t))) \leq \bar{\lambda} d_1(f(s), f(t)) + \bar{c}.$$

Now

$$\begin{aligned} d_0(s,t) &\leq d_0(s, g(f(s))) + d_0(g(f(s)), g(f(t))) + d_0(g(f(t), t)) \\ &\leq \bar{c} + d_0(g(f(s)), g(f(t))) + \bar{c}, \end{aligned}$$

where the last inequality follows from (c), hence $d_0(g(f(s)), g(f(t))) \geq d_0(s,t) - 2\bar{c}$. Thus $d_0(s,t) - 2\bar{c} \leq \bar{\lambda} d_1(f(s), f(t)) + \bar{c}$ and $\frac{1}{\bar{\lambda}} d_0(s,t) - \frac{3c}{\bar{\lambda}} \leq d_1(f(s), f(t))$ or $\frac{1}{\bar{\lambda}}|s-t| - \frac{3\bar{c}}{\bar{\lambda}} \leq |F(s) - F(t)|$ for all $s, t \in X_0$. We set $\lambda = \bar{\lambda}$, $c = \bar{c} + \frac{3\bar{c}}{\bar{\lambda}}$, and hence

$$\frac{1}{\lambda}|s - t| - c \leq |F(s) - F(t)| \leq \lambda|s - t| + c$$

for all $s, t \in X_0$. Furthermore, $\sup_{x \in X_1} |x - \operatorname{im}(F)| < \infty$ because $|x - f(g(x))| \leq \bar{c}$ for all $x \in X_1$ according to (d), that is, $\sup_{x \in X_1} |x - \operatorname{im}(F)| \leq \bar{c}$, because

$$|x - \operatorname{im}(F)| = \inf_{x_0 \in X_0} |x - F(x_0)|.$$

We now suppose that a $(\bar{\lambda}, \bar{c})$-quasi isometry $F: X_0 \to X_1$ with $\sup_{x \in X_1} |x - \operatorname{im}(F)| < \infty$ exists. Let $\bar{\lambda}, \bar{c}$ be the suitable constants with

$$\frac{1}{\bar{\lambda}}|s - t| - \bar{c} \leq |F(s) - F(t)| \leq \bar{\lambda}|s - t| + \bar{c}$$

for all $s, t \in X_0$. Without loss of generality let $\bar{c} \geq \sup_{x \in X_1} |x - \operatorname{im}(F)|$. We would like to construct functions

$$X_0 \underset{g}{\overset{f}{\rightleftarrows}} X_1$$

as well as constants such that (a)–(d) hold. For f we choose F. We construct g. To this end, let $x \in X_1$. As $\sup_{x \in X_1} |x - \mathrm{im}(F)| \le \bar{c}$ there exists a $y_x \in X_0$ with $|F(y_x) - x| \le \bar{c}$. We set $g(x) = y_x$, that is, for all $x \in X_1$ we choose $y_x \in X_0$ with $|F(y_x) - x| \le \bar{c}$ and set $g(x) = y_x$ for this fixed choice of y_x. Because of $|F(s) - F(t)| \le \bar{\lambda}|s - t| + \bar{c}$ we have $d_1(f(s), f(t)) \le \bar{\lambda} d_0(s,t) + \bar{c}$ for all $s, t \in X_0$. Furthermore, we have

$$d_1(f(g(x)), x) = d_1(f(y_x), x) = |F(y_x) - x| \le \bar{c}.$$

Now let $s, t \in X_1$. We look for bounds of $|y_s - y_t|$ in terms of $|s - t|$. We know that

$$\begin{aligned}\frac{1}{\bar{\lambda}}|y_s - y_t| - \bar{c} &\le |F(y_s) - F(y_t)| \\ &\le |F(y_s) - s| + |s - t| + |t - F(y_t)| \\ &\le |s - t| + 2\bar{c},\end{aligned}$$

hence $|y_s - y_t| \le \bar{\lambda}|s - t| + 3\bar{\lambda}\bar{c}$, that is,

$$|g(s) - g(t)| \le \bar{\lambda}|s - t| + 3\bar{\lambda}\bar{c}$$

for all $s, t \in X$. Now let $x \in X_0$. We look for a bound of $d_0(g(f(x)), x)$. Because of $g(f(x)) = y_{f(x)}$ we have $|F(y_{f(x)}) - F(x)| = |f(y_{f(x)}) - f(x)| \le \bar{c}$. Hence $\frac{1}{\bar{\lambda}}|y_{f(x)} - x| - \bar{c} \le \bar{c}$, that is, $|y_{f(x)} - x| \le 2\bar{\lambda}\bar{c}$ or $d_0(g(f(x)), x) \le 2\bar{\lambda}\bar{c}$.

We choose $\lambda = \bar{\lambda}$ as well as $c = \bar{c} + 3\bar{\lambda}\bar{c}$, and this satisfies (a)–(d). □

Remark 3.7.5. The composition of two quasi isometries is again a quasi isometry.

To see this, let $F: X_0 \to X_1$, $G: X_1 \to X_2$ satisfy

$$\begin{aligned}&\frac{1}{\lambda_1}|s - t| - c_1 \le |F(s) - F(t)| \le \lambda_1|s - t| + c_1 \quad \text{for all } s, t \in X_0 \quad \text{and} \\ &\frac{1}{\lambda_2}|s - t| - c_2 \le |G(s) - G(t)| \le \lambda_2|s - t| + c_2 \quad \text{for all } s, t \in X_1.\end{aligned}$$

Then

$$\frac{1}{\lambda_2}|F(s) - F(t)| - c_2 \le |G \circ F(s) - G \circ F(t)| \le \lambda_2|F(s) - F(t)| + c_2,$$

and hence

$$\frac{1}{\lambda_1\lambda_2}|s - t| - \left(c_2 + \frac{c_1}{\lambda_2}\right) \le |G \circ F(s) - G \circ F(t)| \le \lambda_1\lambda_2|s - t| + (c_2 + \lambda_2 c_1).$$

Thus we choose $\lambda := \lambda_1\lambda_2$ and $c := c_2 + \lambda_2 c_1 + \frac{c_1}{\lambda_2}$.

A consequence of this remark is that the image of a quasi segment under a quasi isometry is again a quasi segment; the image of a quasi ray under a quasi isometry is a quasi ray and the image of a quasi geodesic under a quasi isometry is again a quasi geodesic.

Definition 3.7.6. Let X be a metric space. Let Y and Z be non-empty subsets of X.
1. If $H \in (0, \infty) =: \mathbb{R}_+^*$, then we define the *H-neighborhood* of Y in X via
$$V_H(Y) = \{x \in X \mid d(x, Y) \le H\}.$$
2. The *Hausdorff distance* from Y to Z is given by
$$H(Y, Z) := \inf\{H \in (0, \infty) \mid Y \subset V_H(Z) \text{ and } Z \subset V_H(Y)\},$$
if this is meaningful, and if this is not the case, set $H(Y, Z) := \infty$ instead.
3. If A and B are non-empty sets and $f: A \to X$ as well as $g: B \to X$ are mappings, then $H(f, g) := H(f(A), g(B))$ is called the *Hausdorff distance* from f to g.

Remarks 3.7.7. 1. If A, B and C are non-empty subsets of X then we have
$$H(A, B) \le H(A, C) + H(C, B).$$
This is obvious if $H(A, C) = \infty$ or $H(C, B) = \infty$. Now let $H(A, C) < \infty$ and $H(C, B) < \infty$. Let $\varepsilon > 0$ and set $H_\varepsilon := H(A, C) + H(C, B) + \varepsilon$. We show $B \subset V_{H_\varepsilon}(A)$ and $A \subset V_{H_\varepsilon}(B)$, hence $H(A, B) \le H_\varepsilon$. As $\varepsilon > 0$ was chosen arbitrarily, the claim then follows.
We show $B \subset V_{H_\varepsilon}(A)$. Let $b \in B$. Because of $B \subset V_{H(C,B)+\frac{\varepsilon}{2}}(C)$ there exists a $c \in C$ with $|c - b| \le H(C, B) + \frac{\varepsilon}{2}$. Because of $C \subset V_{H(A,C)+\frac{\varepsilon}{2}}(A)$ there exists an $a \in A$ with $|a - c| \le H(A, C) + \frac{\varepsilon}{2}$. It follows that $|a - b| \le |a - c| + |c - b| \le H_\varepsilon$. Hence we have $b \in V_H(A)$. The inclusion $A \subset V_{H_\varepsilon}(B)$ follows analogously.
2. The equation $H(A, B) = 0$ does not imply $A = B$. As an example consider the following: If $A = (0, 1)$, $B = [0, 1]$ and $X = \mathbb{R}$ then $H(A, B) = 0$.

Theorem 3.7.8. *Given three real numbers $\delta \ge 0$, $\lambda \ge 1$ and $c \ge 0$. Then there exists a constant $H = H(\delta, \lambda, c)$ with the following property: If X is a δ-hyperbolic geodesic space, $I = [0, a]$ a bounded interval in either $\mathbb{Z}$ or $\mathbb{R}$, $f: I \to X$ a (λ, c)-quasi segment, $J = [0, |f(a) - f(0)|] \subset \mathbb{R}$ and $g: J \to X$ a geodesic segment with $g(0) = f(0)$ and $g(|f(a) - f(0)|) = f(a)$, then $\operatorname{im}(f) \subset V_H(\operatorname{im}(g))$.*

The proof is based upon some lemmas.

Lemma 3.7.9. *Let F be a metric space and $[x_0, x_n]$ be a geodesic segment in F from x_0 to x_n. Assume there exists a metric tree (T, d'), a mapping $\varphi: F \to T$ and a constant $c' > 0$, such that $\varphi|_{[x_0,x_n]}$ is an isometry and $|u - v| - c' \le |\varphi(u) - \varphi(v)| \le |u - v|$ for all $u, v \in F$. Let $x \in F$ and $y \in [x_0, x_n]$, such that $|x - y| = d(x, [x_0, x_n])$. Let furthermore $z' \in T$ be chosen such that $z' \in \varphi([x_0, x_n])$ and $|z' - \varphi(x)| = d'(\varphi(x), \varphi([x_0, x_n]))$. Let $z \in [x_0, x_n]$ with $\varphi(z) = z'$. Then $|y - z| \le c'$.*

Proof. We have $|x-y| \le |x-z|$ and

$$|x-y| - c' \le |\varphi(x) - \varphi(y)| \le |x-y|,$$
$$|x-z| - c' \le |\varphi(x) - \varphi(z)| \le |x-z|.$$

As the image under φ is contained in a metric tree, we have

$$|\varphi(y) - \varphi(z)| = |\varphi(y) - \varphi(x)| - |\varphi(x) - \varphi(z)|$$

for the chosen x, y, z. Hence

$$|\varphi(y) - \varphi(z)| \le |x-y| - |z-x| + c'.$$

As $\varphi|_{[x_0,x_n]}$ is an isometry, we have $|\varphi(y) - \varphi(z)| = |y-z|$. We furthermore have $|x-z| \ge |x-y|$ because of $|x-y| = d(x, [x_0, x_n])$, thus

$$|x-y| - |x-z| + c' \le c'$$

and further $|y-z| \le c'$. □

The following lemma is a special case of Theorem 3.7.8 for $a \le K$.

Lemma 3.7.10. *Let X be as in Theorem 3.7.8. Let further $K \in (0,\infty)$ and $f: [0,a] \to X$ be a (λ, c)-quasi segment, where $a \le K$. Then there exists an $H = H(\lambda, c, K)$ and a geodesic segment $f': [0, |f(a) - f(0)|] \to X$ from $f(0)$ to $f(a)$, such that* $\mathrm{im}(f) \subset V_H(\mathrm{im}(f'))$.

Proof. Because of $a \le K$ and f being a (λ, c)-quasi segment we have

$$|f(t) - f(0)| \le \lambda|t-0| + c \le \lambda K + c$$

for all $t \in [0,a]$. Hence $\mathrm{im}(f) \subset V_{\lambda K+c}(\{f(0)\})$. If f' is a geodesic segment as above it follows that $f(0) \in \mathrm{im}(f')$, thus $\mathrm{im}(f) \subset V_{\lambda K+c}(\mathrm{im}(f'))$. □

Lemma 3.7.11. *Let $I = [0,a]$ be an interval in $\mathbb{Z}$ or $\mathbb{R}$. Furthermore, let $a > K := \max\{4\lambda c, \lambda c + 2\}$, $\delta \ge 0$, $\lambda \ge 1$ and $c \ge 0$ be as in Theorem 3.7.8. Let $c > 0$. Let $f: [0,a] \to X$ be a (λ, c)-quasi segment. Then there exists a constant $\lambda' = \lambda'(\lambda, c)$, an interval I' in $\mathbb{Z}$ and a $(\lambda', 0)$-quasi segment $f': I' \to X$ from $f(0)$ to $f(a)$, such that* $\mathrm{im}(f) \subset V_H(\mathrm{im}(f'))$, *where*

$$H = \max\{(2\lambda^2+1)c, \lambda^2 c + 2\lambda + c\}.$$

Proof. At first we assume that $[0,a] \subset \mathbb{R}$. As $a > K \ge 4\lambda c$ it follows that $a > 4\lambda c$ and $\frac{a}{2\lambda c} - \frac{a}{4\lambda c} = \frac{a}{4\lambda c} > 1$. Hence there exists an $a' \in \mathbb{Z}$ with $a' > 1$ and $\frac{a}{4\lambda c} < a' < \frac{a}{2\lambda c}$. Let $I' = \{0, 1, \dots, a'\} \subset \mathbb{Z}$. For all $i \in I'$ set $t_i = \frac{i}{a'}a$, hence $t_0 = 0$ and $t_{a'} = a$. We define $f': I' \to X$ via $f'(i) = f(t_i)$ and claim that $\mathrm{im}(f) \subset V_{(2\lambda^2+1)c}(\mathrm{im}(f'))$.

To see this, first observe that $t_i - t_{i-1} = \frac{i}{a'}a - \frac{(i-1)}{a'}a = \frac{a}{a'}$. Because of $\frac{a}{4\lambda c} < a'$ we have $\frac{a}{a'} < 4\lambda c$. It follows that for $t \in [0,a]$ there exists a t_i with $|t - t_i| \le 2\lambda c$. As f is a (λ, c)-quasi isometry it follows that

$$|f(t)-f(t_i)| \le \lambda|t-t_i| + c \le 2\lambda^2 c + c = (2\lambda^2+1)c$$

and hence $\operatorname{im}(f) \subset V_{(2\lambda^2+1)c}(\operatorname{im}(f'))$.

Now let $i,j \in I'$, $i \ne j$. As f is a (λ, c)-quasi segment it follows that

$$\frac{1}{\lambda}|t_i - t_j| - c \le |f(t_i) - f(t_j)| \le \lambda|t_i - t_j| + c.$$

Because of this we have

$$\frac{1}{\lambda}\left|\frac{i}{a'}a - \frac{j}{a'}a\right| - c \le |f(t_i) - f(t_j)| = |f'(i) - f'(j)| \le \lambda\left|\frac{i}{a'}a - \frac{j}{a'}\right| + c$$

and hence

$$\frac{a}{\lambda a'}|i-j| - c \le |f'(i) - f'(j)| \le \frac{a\lambda}{a'}|i-j| + c.$$

As $|i-j| \ge 1$ we have

$$\left(\frac{a}{\lambda a'} - c\right)|i-j| \le \frac{a}{\lambda a'}|i-j| - c \quad \text{and} \quad \frac{a\lambda}{a'}|i-j| + c \quad \le \left(\frac{a\lambda}{a'} + c\right)|i-j|,$$

hence $(\frac{a}{\lambda a'} - c)|i-j| \le |f'(i) - f'(j)| \le (\frac{a\lambda}{a'} + c)|i-j|$. As $\frac{a}{4\lambda c} < a' < \frac{a}{2\lambda c}$ it follows that $2c < \frac{a}{a'\lambda}$, hence $c < \frac{a}{a'\lambda} - c$ and $\frac{a}{a'} < 4\lambda c$, thus $\frac{\lambda a}{a'} + c < 4\lambda^2 c + c$. Set $\lambda' = \max\{\frac{1}{c}, 4\lambda^2 c + c\}$. Then $\lambda' \ge \frac{1}{c}$, hence $c \ge \frac{1}{\lambda'}$, and $\frac{1}{\lambda'} \le \frac{a}{a'\lambda} - c$ (where the inequality follows from $c < \frac{a}{a'\lambda} - c$). Accordingly $\lambda' \ge 4\lambda^2 c + c$ implies that $\frac{\lambda a}{a'} + c \le \lambda'$. With this we have

$$\frac{1}{\lambda'}|i-j| \le |f'(i) - f'(j)| \le \lambda'|i-j|.$$

Of course this also holds for $i = j$, and hence f' is a (λ, c)-quasi segment. This is the case $[0, a] \subset \mathbb{R}$. We now consider the case $\{0, 1, \dots, a\} \subset \mathbb{Z}$. Set $\lambda' = \lambda^2 c + 2\lambda + c$. Let $N \in \mathbb{N}$ such that $\lambda c + 1 \le N < \lambda c + 2$. We know that $a > K \ge \lambda c + 2$, and hence $\frac{a}{N} > 1$. Set $a' = [\frac{a}{N}]$ and $I' = \{0, 1, \dots, a'\} \subset \mathbb{Z}$. We define $f' : I' \to X$ via $f'(i) = f(Ni)$, where $i \in \{0, 1, \dots, a\}$. Set $i' = [\frac{i}{N}]$; then $|i' - \frac{i}{N}| < 1$ and $i' \in I'$. If $i \in \{0, 1, \dots, a\}$, then there exists an $i' \in I'$ with $|i - Ni'| < N$. For $|i - Ni'| < N$ it then follows that

$$\begin{aligned}|f(i) - f'(i')| &= |f(i) - f(Ni')| \\ &\le \lambda|i - Ni'| + c < \lambda N + c \\ &< \lambda(\lambda c + 2) + c = \lambda^2 c + 2\lambda + c \le H,\end{aligned}$$

where the second line follows from f being a (λ, c)-quasi segment and the third line follows from the choice of N. We get $\operatorname{im}(f) \subset V_H(\operatorname{im}(f'))$.

Let $i,j \in I'$ with $i \ne j$. Note that $N \ge \lambda c + 1$ implies that $\frac{N}{\lambda} - c \ge \frac{1}{\lambda}$, and because of $\lambda' = \lambda^2 c + 2\lambda + c > \lambda$ we further have $\frac{1}{\lambda} \ge \frac{1}{\lambda'}$. Because of $N < \lambda c + 2$ we have

$$\lambda N + c < \lambda(\lambda c + 2) + c = \lambda^2 c + 2\lambda + c = \lambda'.$$

As f is a (λ, c)-quasi segment, we have

$$\begin{aligned}\frac{1}{\lambda'}|i-j| &\le \frac{1}{\lambda}|i-j| \le \left(\frac{N}{\lambda} - c\right)|i-j| \\ &= \frac{1}{\lambda}|Ni - Nj| - c|i-j| \le \frac{1}{\lambda}|Ni-Nj| - c \\ &\le \underbrace{|f'(i) - f'(j)|}_{=|f(Ni)-f(Nj)|} \le \lambda|Ni - Nj| + c = \lambda N|i-j| + c \\ &\le (\lambda N + c)|i-j| < \lambda'|i-j|.\end{aligned}$$

Thus f' is a $(\lambda', 0)$-quasi segment. □

Proof of Theorem 3.7.8; *following Thurston.* By Lemmas 3.7.10 and 3.7.11 we can restrict ourselves to the case that $I = \{0, 1, \dots, n\} \subset \mathbb{Z}$ and that $f\colon I \to X$ is a $(\lambda, 0)$-quasi segment. Without loss of generality, let $\delta > 0$. For $i \in I$ let $f(i) = x_i$, and $[x_0, x_n]$ denote a geodesic segment in X. We look for an $H \in (0, \infty)$ with $\{x_i\}_{i\in I} \subset V_H([x_0, x_n])$, where H does only depend on δ and λ ($c = 0$).

We define H in the following way: Choose $N \in \mathbb{Z}$ with $N \ge 1$ and $\log_2(N+2) < \frac{N}{4\lambda\delta} - 2$. Let k denote the integer with $k - 1 < \log_2(N+2) \le k$. Let $c' = 2(k+1)\delta$ and $R = \frac{1}{2}\lambda N + c'$. First observe that $\log_2(N+2) < \frac{N}{4\lambda\delta} - 2$ implies $\frac{1}{\lambda} > \frac{4\delta}{N}(\log_2(N+2) + 2)$ and hence

$$\begin{aligned}\frac{1}{\lambda} - \frac{2c'}{N} &> \frac{4\delta}{N}(\log_2(N+2) + 2) - 2\frac{c'}{N} \\ &> \frac{4\delta}{N}((k-1) + 2) - 2\frac{c'}{N} \\ &= \frac{4\delta}{N}(k+1) - \frac{2}{N}(2(k+1)\delta) = 0.\end{aligned}$$

Now set

$$H = R + \lambda + 2\lambda\left(\frac{1}{\lambda} - 2\frac{c'}{N}\right)^{-1}(R + \lambda + c').$$

As $R < H$ there is nothing to show if $\operatorname{im}(f) \subset V_R([x_0, x_n])$. Because of this let $\operatorname{im}(f) \not\subset V_R([x_0, x_n])$, and set $V_R = V_R([x_0, x_n])$. With this, there exist $u, v \in \{1, \dots, n-1\}$ with $u \le v$, $\{x_u, \dots, x_v\} \subset X \setminus V_R$ and $x_{u-1}, x_{v+1} \in V_R$. Set $I_0 = \{u, \dots, v\} \subset I$. For all $i \in I_0$ let $y_i \in [x_0, x_n]$ such that $|x_i - y_i| = d(x_i, [x_0, x_n])$; compare Figure 3.43.

Let $i, j \in I_0$ with $i \le j$ and $j - i \le N$. Let $F = [x_0, x_n] \cup \{x_i, \dots, x_j\}$, and let x_0 be the basis point of F. Note that as long as $\log_2(N+2) \le k$ we have $N + 2 \le 2^k$. As F can be interpreted as consisting of at most $N + 2$ subrays with basis point x_0, there exists (Theorem 3.2.13) a map $\varphi\colon F \to T$ where T is a real tree with $\varphi(x_0)$ as basis point, such that φ preserves distances from the basis point and such that for all $p, q \in F$ the following inequality holds:

$$|p - q| - c' \le |\varphi(p) - \varphi(q)| \le |p - q|.$$

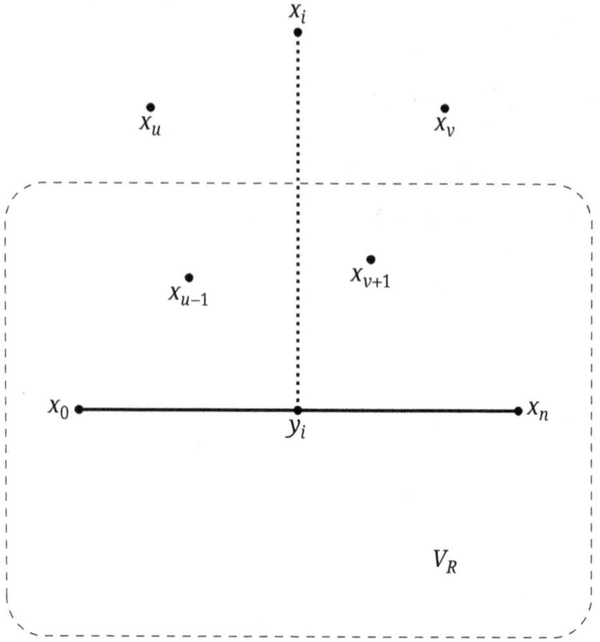

Figure 3.43: Visualization of V_R.

In addition to the statement in Theorem 3.2.13 we remark the following.

Remark 3.7.12. The tree T is constructed such that $\varphi|_{[x_0,x_n]}$ is an isometry. To see this, let $p \in [x_0, x_n]$. We observe that T is a quotient space that contains the quotient subspace $(I_p \dot{\cup} I_{x_n})/\sim$ where $t \sim t'$ if $t = t' \leq (p \mid x_n)_{x_0}$, $t \in I_p$, $t' \in I_{x_n}$. But $p \in [x_0, x_n]$ implies $|x_0 - x_n| = |x_0 - p| + |p - x_n|$, hence

$$(p \mid x_n)_{x_0} = \frac{1}{2}(|x_n - x_0| + |p - x_0| - |x_n - p|) = |p - x_0|.$$

Hence the image of I_p under the quotient map is contained in the image of I_{x_n} under the quotient map. With this $\varphi|_{[x_0,x_n]}$ is an isometry.

This also implies that T is even a metric tree as there can only be finitely many branches. For all $p \in [x_0, x_n]$ we have

$$d(\varphi(x_\nu), \varphi(p)) \geq |x_\nu - p| - c' \geq R - c' = \frac{\lambda N}{2}, \quad \nu = i, j,$$

hence

$$d(\varphi(x_\nu), [\varphi(x_0), \varphi(x_n)]) \geq R - c' = \frac{\lambda N}{2}, \quad \nu = i, j.$$

We further have

$$|\varphi(x_i) - \varphi(x_j)| \leq |x_i - x_j| \leq \lambda|i - j| \leq \lambda N,$$

where the inequalities follow from the property of φ, from f being a $(\lambda, 0)$-quasi segment and the inequality $|i - j| \leq N$, respectively, as well as

$$|\varphi(x_i) - \varphi(x_j)| \leq d(\varphi(x_i), \varphi([x_0, x_n])) + d(\varphi(x_j), \varphi([x_0, x_n])) \geq 2R - 2c' = \lambda N.$$

As $\mathrm{im}(\varphi) \subset T$ and T is a metric tree it follows that, if $p \in \varphi([x_0, x_n])$ with $d(\varphi(x_i), p) = d(\varphi(x_i), \varphi([x_0, x_n]))$ then also $d(\varphi(x_j), p) = d(\varphi(x_j), \varphi([x_0, x_n]))$ (with $x_i, x_j \in F$ as above).

Such a p certainly exists. Suppose this last equation does not hold, then we have the situation depicted in Figure 3.44.

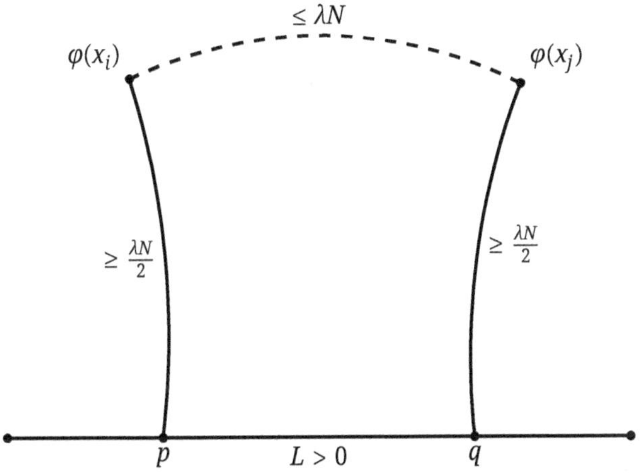

Figure 3.44: Visualization of the contradiction concerning T.

Hence $|\varphi(x_i) - \varphi(x_j)| \geq \frac{\lambda N}{2} + \frac{\lambda N}{2} + L > \lambda N$ (note that T is a tree).

Now let $p \in \varphi([x_0, x_n])$ and $z \in [x_0, x_n]$ with $\varphi(z) = p$ such that

$$d(\varphi(x_i), p) = d(\varphi(x_i), \varphi([x_0, x_n])) \quad \text{and} \quad d(\varphi(x_j), p) = d(\varphi(x_j), \varphi([x_0, x_n])).$$

We can now apply Lemma 3.7.9. Let $y_v \in [x_0, x_n]$, such that—as above—$|x_v - y_v| = d(x_v, [x_0, x_n])$ for $v = i, j$. Then $|y_i - z| \leq c'$ and $|y_j - z| \leq c'$, hence

$$|y_i - y_j| \leq |y_i - z| + |z - y_j| \leq 2c'.$$

In the notation as above we furthermore have

$$|x_u - x_v| \leq |x_u - y_u| + |y_u - y_{u+N}| + |y_{u+N} - y_{u+2N}| + \cdots + |y_v - x_v|,$$

hence $|y_{u+lN} - y_{u+(l+1)N}| \leq 2c'$ and $|y_{u+lN} - y_v| \leq 2c'$, if $v \leq u + (l+1)N$, and finally

$$|x_u - x_v| \leq |x_u - y_u| + |y_v - x_v| + \left(\frac{|u - v|}{N} + 1\right)2c'.$$

We have $|x_u - y_u| \leq |x_u - y_{u-1}|$ where $y_{u-1} \in [x_0, x_n]$ with $|x_{u-1} - y_{u-1}| = d(x_{u-1}, [x_0, x_n])$, hence

$$\begin{aligned} |x_u - y_u| &\leq |x_u - y_{u-1}| \\ &\leq |x_u - x_{u-1}| + |x_{u-1} - y_{u-1}| \\ &\leq \lambda\big|u - (u-1)\big| + R = \lambda + R, \end{aligned}$$

where the last line follows from f being a (λ, c)-quasi segment and from the fact that $x_{u-1} \in V_R$. Analogously we have $|x_v - y_v| \leq R + \lambda$. Altogether

$$|x_u - x_v| \leq 2(R + \lambda) + \left(\frac{|u - v|}{N} + 1\right)2c'.$$

As f is a $(\lambda, 0)$-quasi segment, it follows that $\frac{1}{\lambda}|u - v| \leq |x_u - x_v|$. Hence

$$\frac{1}{\lambda}|u - v| \leq 2(R + \lambda) + \left(\frac{|u - v|}{N} + 1\right)2c',$$

that is, $|u - v|(\frac{1}{\lambda} - \frac{2c'}{N}) \leq 2(R + \lambda + c')$ or $|u - v| \leq 2(\frac{1}{\lambda} - \frac{2c'}{N})^{-1}(R + \lambda + c')$. As f is again a $(\lambda, 0)$-quasi segment and

$$|u - i| \leq |u - v| \quad \text{for } i \in \{u, u+1, \dots, v\}$$

we have

$$|x_n - x_i| \leq \lambda|u - i| \leq 2\lambda\left(\frac{1}{\lambda} - \frac{2c'}{N}\right)^{-1}(R + \lambda + c')$$

for all $i \in \{u, \dots, v\} = I_0$. Because of

$$H - R - \lambda = 2\lambda\left(\frac{1}{\lambda} - \frac{2c'}{N}\right)^{-1}(R + \lambda + c')$$

it follows that $\{x_i\}_{i=u}^{v} \subset V(\{x_n\})$. As $d(x_u, [x_0, x_n]) \leq R + \lambda$ it follows that $\{x_i\}_{i=u}^{v} \subset V_H([x_0, x_n])$. As $\{x_u, \dots, x_v\}$ is an arbitrary connected component of $\mathrm{im}(f) \cap (X \backslash V_R)$, it follows that $\mathrm{im}(f) \subset V_H([x_0, x_n])$. This concludes the proof of Theorem 3.7.8. □

Lemma 3.7.13. *Given real numbers $\lambda \geq 1$, $c \geq 0$ and $H' \geq 0$. Then there exists a constant $H = H(\lambda, c, H')$ with the following property: If X is a geodesic metric space, $I = [p, q]$ and $J = [r, s]$ are two intervals in $\mathbb{Z}$ or $\mathbb{R}$ and given two functions $f: I \to X$ and $g: J \to X$ such that*

(a) *f is a (λ, c)-quasi segment;*
(b) *g is a geodesic segment;*
(c) *$|f(p) - g(r)| \leq H'$ and $|f(q) - g(s)| \leq H'$; and*
(d) *$\mathrm{im}(f) \subset V_{H'}(\mathrm{im}(g))$;*

then $H(f, g) \leq H$.

Proof. We look for an $H = H(\lambda, c, H')$, such that

$$\operatorname{im}(f) \subset V_H(\operatorname{im}(g)) \quad \text{and} \quad \operatorname{im}(g) \subset V_H(\operatorname{im}(f)).$$

We set $H = 2H' + \lambda + c$. We have $H' < H$, that is, (d) implies $\operatorname{im}(f) \subset V_H(\operatorname{im}(g))$. It remains to show that $\operatorname{im}(g) \subset V_H(\operatorname{im}(f))$. Let $p = t_0 < t_1 < \cdots < t_n = q$ be a partition of $[p, q]$, such that $|t_i - t_{i-1}| \leq 1$ for $i = 1, \ldots, n$.

According to (c) and (d) we have $f(t_i) \in V_{H'}(\operatorname{im}(g))$ for $i = 0, \ldots, n$, and there exists a $u_i \in J$ with $|f(t_i) - g(u_i)| \leq H'$ for $i = 0, \ldots, n$. As f is a (λ, c)-quasi segment we have

$$|f(t_{i-1}) - f(t_i)| \leq \lambda |t_{i-1} - t_i| + c \leq \lambda + c.$$

It follows that

$$\begin{aligned} |g(u_{i-1}) - g(u_i)| &\leq |g(u_{i-1}) - f(t_{i-1})| + |f(t_{i-1}) - f(t_i)| + |f(t_i) - g(u_i)| \\ &\leq H' + \lambda + c + H' \\ &= 2H' + \lambda + c. \end{aligned}$$

Note that we may choose $u_0 = r$ and $u_n = s$, and do so. With this we have $[r, s] \subset \bigcup_{i=1}^{n} [u_{i-1}^*, u_i^*]$ with

$$[u_{i-1}^*, u_i^*] = \begin{cases} [u_{i-1}, u_i] & \text{if } u_{i-1} \leq u_i, \\ [u_i, u_{i-1}] & \text{if } u_i < u_{i-1}. \end{cases}$$

If $u \in [r, s]$, then there exists an i with $u \in [u_{i-1}^*, u_i^*]$, hence for $j \in \{i - 1, i\}$ we have

$$\begin{aligned} |u - u_j| &\leq \frac{1}{2} |u_{i-1} - u_i| \leq \frac{1}{2} |g(u_{i-1}) - g(u_i)| \\ &\leq \frac{1}{2}(2H' + \lambda + c) = H' + \frac{1}{2}(\lambda + c), \end{aligned}$$

where we have used the fact that g is a geodesic segment in the second inequality.

As $|g(u) - g(u_j)| = |u - u_j|$ it follows that

$$\begin{aligned} d(g(u), \operatorname{im}(f)) &\leq d(g(u), f(t_j)) \\ &\leq |g(u) - g(u_j)| + |g(u_j) - f(t_j)| \\ &\leq H' + \frac{1}{2}(\lambda + c) + H' \\ &= 2H' + \frac{1}{2}(\lambda + c) < H. \end{aligned}$$

Hence $\operatorname{im}(g) \subset V_H(\operatorname{im}(f))$. □

Theorem 3.7.14. *Given three real numbers $\delta \geq 0$, $\lambda \geq 1$ and $c \geq 0$. Then there exists a constant $H = H(\delta, \lambda, c)$ with the following property: If X is a geodesic δ-hyperbolic metric* *space, $I = [a,b]$ a bounded interval in $\mathbb{Z}$ or $\mathbb{R}$, $f: I \to X$ a (λ, c)-quasi segment, $J \subset \mathbb{R}$ an interval of length $|f(b) - f(a)|$ and $g: J \to X$ a geodesic segment from $f(a)$ to $f(b)$, then we have $H(f, g) \leq H$.*

Proof. This is a direct consequence of Theorem 3.7.8 and Lemma 3.7.13. □

Theorem 3.7.15. *Let X and Y be two geodesic metric spaces and $F: X \to Y$ a quasi isometry. If Y is hyperbolic, then so is X.*

Proof. Let $\delta \geq 0$, $\lambda \geq 1$ and $c \geq 0$ be constants such that
- F is a (λ, c)-quasi isometry; and
- Y is δ-hyperbolic.

Let $H = H(\delta, \lambda, c)$ be the constant from Theorem 3.7.14. Let I be a bounded interval in $\mathbb{Z}$ or $\mathbb{R}$ and $g: I \to X$ be a geodesic segment. As g is an isometry it surely is also a quasi isometry, hence $F \circ g$ is a quasi segment.

In particular, $F \circ g: I \to Y$ is a (λ, c)-quasi segment.

To see this, let $s, t \in I$. Then $\frac{1}{\lambda}|g(s) - g(t)| - c \leq |F(g(s)) - F(g(t))| \leq \lambda|g(s) - g(t)| + c$ as F is a (λ, c)-quasi isometry. As g is an isometry, that is, $|s - t| = |g(s) - g(t)|$, we see that $F \circ g$ is a (λ, c)-quasi segment.

This implies: If $(F \circ g)_0$ is a geodesic segment in Y with the same end points as $F \circ g$, then $H(F \circ g, (F \circ g)_0) \leq H$. Let Δ be a geodesic triangle in X. Then we can consider $F(\Delta)$ as a (λ, c)-quasi geodesic triangle in Y. Let $(F\Delta)_0$ be a geodesic triangle in Y with the same vertices as $F(\Delta)$. According to Theorem 3.2.23 $(F\Delta)_0$ satisfies the Rips condition with the constant 4δ. As now every edge of $F(\Delta)$ is contained in an H-neighborhood of the respective edge of $(F\Delta)_0$ and vice versa, we conclude that $F(\Delta)$ satisfies the Rips condition with the constant $4\delta + 2H$. Now, if x is a vertex of an edge of Δ, there is a point y on one of the other two edges of Δ with $|F(x) - F(y)| \leq 4\delta + 2H$. As F now is a (λ, c)-quasi isometry it follows that

$$\frac{1}{\lambda}|x - y| - c \leq |F(x) - F(y)| \leq 4\delta + 2H.$$

Hence $|x - y| \leq \lambda(4\delta + 2H + c) =: \delta'$. Thus Δ satisfies the Rips condition for δ', hence X is hyperbolic (more precisely $8\delta'$-hyperbolic, compare Theorem 3.2.23). □

Remark 3.7.16. Theorem 3.7.15 is false if X or Y is not a geodesic space.

As an example, consider $Y = [0, \infty) \subset \mathbb{R}$ with the usual metric. Then Y is a tree and hence 0-hyperbolic. Let X be the subspace of $\mathbb{R}^2$ as explained in Figure 3.45 (together with the induced metric).

The space X is homeomorphic to $[0, \infty)$ via the projection on the x-axis but X is not a geodesic space (this also follows from the discussion below together with Theorem 3.7.15). The projection from X onto Y, given by $(x, y) \mapsto x$, is a $(\sqrt{2}, 0)$-quasi isometry.

We now show that X is not hyperbolic: As the equivalence statement in Theorem 3.2.23 requires that the metric space is geodesic we have to show the original def-

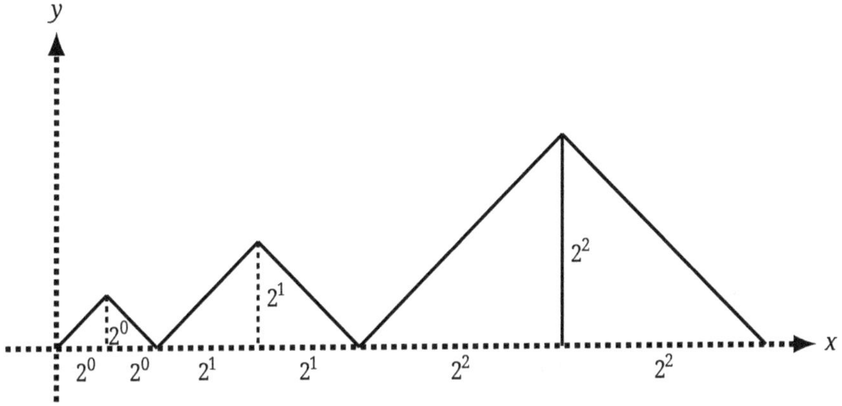

Figure 3.45: Subspace X.

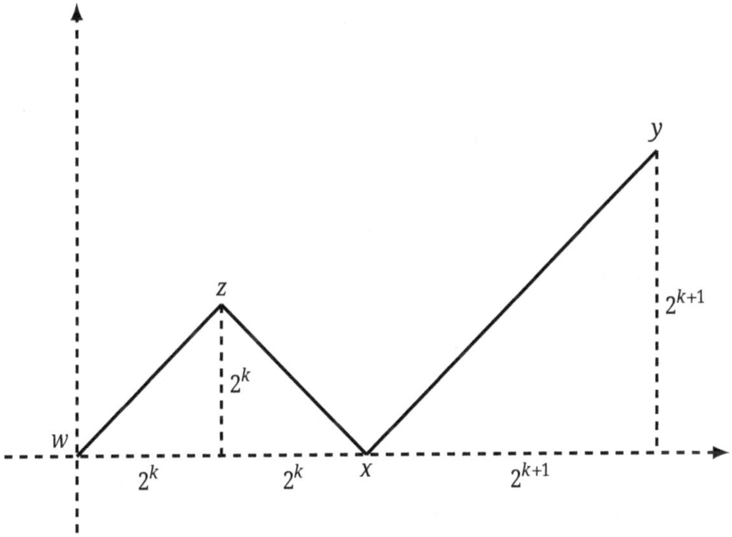

Figure 3.46: X is not hyperbolic.

inition via the Gromov product and use the reformulation for rectangles. We consider the situation in Figure 3.46.

We have $|w-x| = 2^{k+1}$, $|w-y| = 2^{k+1}\sqrt{5}$, $|w-z| = 2^k\sqrt{2}$, $|x-y| = 2^{k+1}\sqrt{2}$, $|x-z| = 2^k\sqrt{2}$, $|y-z| = 2^k\sqrt{10}$. Hence

$$\begin{aligned}
\text{'Top'} + \text{'Bottom'} &= 2^k\sqrt{10} + 2^{k+1}, \\
\text{'Side'} + \text{'Side'} &= 2^k\sqrt{2} + 2^{k+1}\sqrt{2}, \\
\text{'Diagonal'} + \text{'Diagonal'} &= 2^k\sqrt{2} + 2^{k+1}\sqrt{5}.
\end{aligned}$$

We can ask for a $\delta \geq 0$ such that

$$2^k\sqrt{2} + 2^{k+1}\sqrt{5} \leq \max\{2^k\sqrt{10} + 2^{k+1}, 2^k\sqrt{2} + 2^{k+1}\sqrt{2}\} + 2\delta$$

for all $k \in \mathbb{N}$. We have $2^k(\sqrt{2} + 2\sqrt{5}) > 2^k(\sqrt{10} + 2) > 2^k(\sqrt{2} + 2\sqrt{2})$ and $\lim_{k\to\infty}(2^k(\sqrt{2} + 2\sqrt{5}) - 2^k(\sqrt{10} + 2)) = +\infty$, hence such a $\delta \geq 0$ is not possible and X is not hyperbolic.

3.7.2 Independence of the choice of finite generating systems

Corollary 3.7.17. *Let Γ be a group of finite type, S and T be two finite valid generating systems for Γ and $\phi(\Gamma, S)$ and $\phi(\Gamma, T)$ be the Cayley graphs for (Γ, S) and (Γ, T), respectively. Then $\phi(\Gamma, S)$ is hyperbolic if and only if $\phi(\Gamma, T)$ is hyperbolic.*

Proof. The Cayley graph $\phi(\Gamma, S)$ is quasi isometric to (Γ, S)—with respect to the word length metric induced by S—and (Γ, S) is quasi isometric to (Γ, T). Then $\phi(\Gamma, S)$ and $\phi(\Gamma, T)$ are quasi isometric. The claim now follows directly from Theorem 3.7.15. □

Theorem 3.7.18. *Given three real numbers $\delta \geq 0$, $\lambda \geq 1$ and $c \geq 0$. Then there exists a constant $A = A(\delta, \lambda, c)$ with the following properties: If X and Y are two δ-hyperbolic geodesic spaces, $F: X \to Y$ a (λ, c)-quasi isometry and $w, x, y, z \in X$, then we have:*

1.
$$\frac{1}{\lambda}(x \mid y)_w - A \leq (F(x) \mid F(y))_{F(w)} \leq \lambda(x \mid y)_w + A,$$

2.
$$\begin{aligned}\frac{1}{\lambda}\big|(x \mid y)_w - (y \mid z)_w\big| - A &\leq \big|(F(x) \mid F(y))_{F(w)} - (F(y) \mid F(z))_{F(w)}\big| \\ &\leq \lambda\big|(x \mid y)_w - (y \mid z)_w\big| + A.\end{aligned}$$

Proof. 1. Let $H = H(\delta, \lambda, c)$ be the constant from Theorem 3.7.8. Let $[x, y]$ be a geodesic segment in X from x to y and $[F(x), F(y)]$ be one in Y from $F(x)$ to $F(y)$. As X is δ-hyperbolic we see that every geodesic triangle in X is 4δ-thin. Hence, according to Theorem 3.2.23: $(x \mid y)_w \leq d(w, [x, y]) \leq (x \mid y)_w + 4\delta$ and accordingly

$$(F(x) \mid F(y))_{F(w)} \leq d(F(w), [F(x), F(y)]) \leq (F(x) \mid F(y))_{F(w)} + 4\delta.$$

According to Theorem 3.7.15 and Corollary 3.7.17 we have

$$H(F([x, y]), [F(x), F(y)]) \leq H.$$

As F is a (λ, c)-quasi isometry we have

$$\frac{1}{\lambda}|s - t| - c \leq |F(s) - F(t)| \leq \lambda|s - t| + c \quad \text{for all } s, t \in X.$$

We claim

$$\frac{1}{\lambda}d(w,[x,y]) - c \leq d(F(w),F([x,y])) \leq \lambda d(w,[x,y]) + c.$$

To see this, let $\varepsilon > 0$. Then there exists a $z \in [x,y]$ with

$$d(F(w),F(z)) \leq d(F(w),F([x,y])) + \varepsilon,$$

hence

$$\begin{aligned} d(F(w),F([x,y])) &\geq d(F(w),F(z)) - \varepsilon \\ &\geq \frac{1}{\lambda}|w-z| - c - \varepsilon \\ &\geq \frac{1}{\lambda}d(w,[x,y]) - c - \varepsilon. \end{aligned}$$

Hence

$$d(F(w),F([x,y])) \geq \frac{1}{\lambda}d(w,[x,y]) - c - \varepsilon$$

for all $\varepsilon > 0$. Thus we have

$$\frac{1}{\lambda}d(w,[x,y]) - c \leq d(F(w),F([x,y])).$$

Analogously, if $z' \in [x,y]$ with $d(w,z') = d(w,[x,y])$, then

$$\begin{aligned} d(F(w),F([x,y])) &\leq d(F(w),F(z')) \\ &\leq \lambda|w-z'| + c \\ &= d(w,[x,y]) + c. \end{aligned}$$

We claim

$$d(F(w),[F(x),F(y)]) \leq d(F(w),F([x,y])) + H.$$

To see this, let $z \in [x,y]$ with

$$d(F(w),F([x,y])) \geq d(F(w),F(z)) - \varepsilon$$

as above. By definition of H there exists a $z'' \in [F(x),F(y)]$ with $|F(z) - z''| \leq H$. Now

$$\begin{aligned} d(F(w),[F(x),F(y)]) &\leq d(F(w),z'') \\ &\leq d(F(w),F(z)) + d(F(z),z'') \\ &\leq d(F(w),F([x,y])) + \varepsilon + H. \end{aligned}$$

As this holds for all $\varepsilon > 0$ the claim follows.
We now have

$$\begin{aligned}(F(x) \mid F(y))_{F(w)} &\le d(F(w), [F(x), F(y)]) \\ &\le d(F(w), F([x,y])) + H \\ &\le \lambda d(w, [x,y]) + c + H \\ &\le \lambda((x \mid y)_w + 4\delta) + c + H \\ &\le \lambda((x \mid y)_w) + 4\lambda\delta + c + H.\end{aligned}$$

We claim

$$d(F(w), [F(x), F(y)]) \ge d(F(w), F([x,y])) - H.$$

To see this, let $\bar{z} \in [F(x), F(y)]$, such that

$$d(F(w), [F(x), F(y)]) = d(F(w), \bar{z}).$$

By definition of H there exists a $z^* \in [x,y]$ with $|\bar{z} - F(z^*)| \le H$. Hence

$$\begin{aligned}d(F(w), F([x,y])) &\le d(F(w), F(z^*)) \\ &\le |F(w) - \bar{z}| + |\bar{z} - F(z^*)| \\ &\le d(F(w), [F(x), F(y)]) + H.\end{aligned}$$

We now have

$$\begin{aligned}(F(x) \mid F(y))_{F(\omega)} &\ge d(F(w), [F(x), F(y)]) - 4\delta \\ &\ge d(F(w), F([x,y])) - H - 4\delta \\ &\ge \frac{1}{\lambda} d(w, [x,y]) - c - H - 4\delta \\ &\ge \frac{1}{\lambda}(x \mid y)_w - c - H - 4\delta.\end{aligned}$$

Hence $A = 4\lambda\delta + c + H$ is a constant with the desired properties—note that $\lambda \ge 1$.

2. Let $[w,x]$, $[w,y]$, $[w,z]$, $[x,y]$ and $[y,z]$ be geodesic segments in X. Let $\Delta_1 = [w,x] \cup [w,y] \cup [x,y]$, and let $u_1 \in [w,x]$, $u_2 \in [w,y]$, $u_3 \in [x,y]$ be the descriptive triple for Δ_1, that is, $f_{\Delta_1}(u_i)$ is the tripod point in T_Δ. Let $\Delta_2 = [w,y] \cup [w,z] \cup [y,z]$, and let $v_1 \in [w,y]$, $v_2 \in [w,z]$, $v_3 \in [y,z]$ be the descriptive triple for Δ_2; see Figure 3.47. Analogously we have geodesic triangles in Y:

$$\begin{aligned}\Delta_1' &= [F(w), F(x)] \cup [F(w), F(y)] \cup [F(y), F(x)], \\ \Delta_2' &= [F(w), F(y)] \cup [F(w), F(z)] \cup [F(y), F(z)],\end{aligned}$$

and descriptive triples u_1', u_2', u_3' for Δ_1' and v_1', v_2', v_3' for Δ_2', respectively.

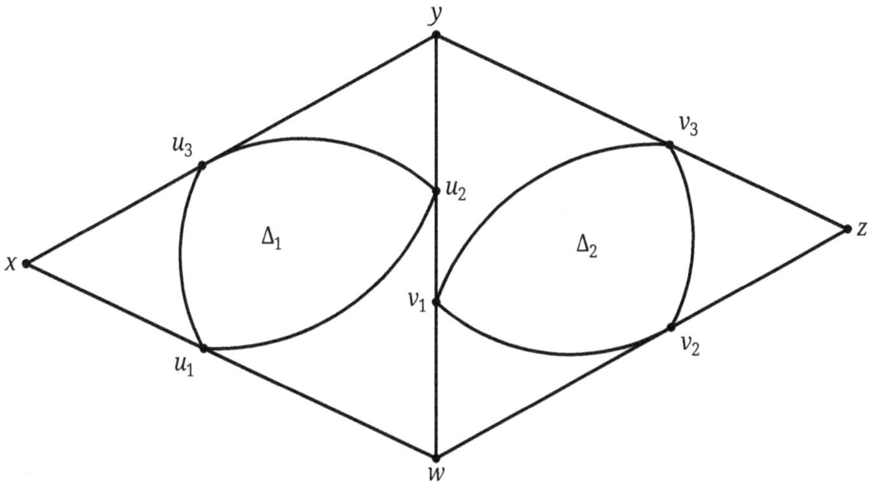

Figure 3.47: Descriptive triples.

We have

$$\begin{aligned}
|w - u_1| &= |w - u_2| = (x \mid y)_w, \\
|w - v_1| &= |w - v_2| = (y \mid z)_w, \\
|F(w) - u_1'| &= |F(w) - u_2'| = (F(x) \mid F(y))_{F(w)}, \quad \text{and} \\
|F(w) - v_1'| &= |F(w) - v_2'| = (F(y) \mid F(z))_{F(w)}.
\end{aligned}$$

According to Theorem 3.7.14 there exist

$$\begin{aligned}
u_1'' &\in [F(w), F(x)], \quad v_1'' \in [F(w), F(y)], \\
u_2'' &\in [F(w), F(y)], \quad v_2'' \in [F(w), F(z)], \\
u_3'' &\in [F(x), F(y)], \quad v_3'' \in [F(y), F(z)]
\end{aligned}$$

with $|F(u_i) - u_i''| \leq H$, $i = 1, 2, 3$ and $|F(v_i) - v_i''| \leq H$, $i = 1, 2, 3$. According to Theorem 3.2.23 the size of Δ_1 as well as the size of Δ_2 is bounded by 4δ, that is, $\operatorname{diam}(\{u_1, u_2, u_3\}) \leq 4\delta$ and $\operatorname{diam}(\{v_1, v_2, v_3\}) \leq 4\delta$. As F is a (λ, c)-quasi isometry it follows that

$$\begin{aligned}
\operatorname{diam}(\{F(u_1), F(u_2), F(u_3)\}) &\leq 4\lambda\delta + c, \\
\operatorname{diam}(\{F(v_1), F(v_2), F(v_3)\}) &\leq 4\lambda\delta + c.
\end{aligned}$$

The triangle inequality then implies that

$$\begin{aligned}
\operatorname{diam}(\{u_1'', u_2'', u_3''\}) &\leq 4\lambda\delta + c + 2H, \\
\operatorname{diam}(\{v_1'', v_2'', v_3''\}) &\leq 4\lambda\delta + c + 2H.
\end{aligned}$$

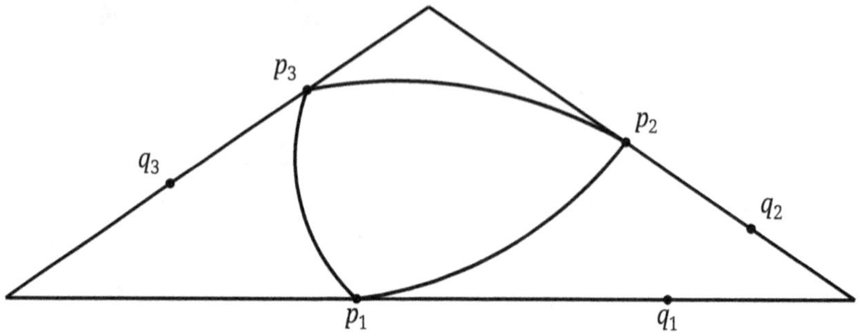

Figure 3.48: Situation of Lemma 3.2.22.

In Lemma 3.2.22 we prove: Given the situation depicted in Figure 3.48 where the diam$(\{q_1, q_2, q_3\}) \leq K$. Then we have $|p_i - q_i| \leq \frac{3}{2}K$ for $i = 1, 2, 3$.
We apply this to our case and get

$$|u_2'' - u_2'| \leq \frac{3}{2}(4\delta\lambda + c + 2H) \quad \text{and}$$
$$|v_1'' - v_1'| \leq \frac{3}{2}(4\delta\lambda + c + 2H).$$

Set $B = \frac{3}{2}(4\delta\lambda + c + 2H) + H$ and $A = 2B + c$. This yields

$$|F(u_2) - u_2'| \leq |F(u_2) - u_2''| + |u_2'' - u_2'| \leq B;$$

and analogously $|F(v_1) - v_1'| \leq B$. It follows that

$$\begin{aligned}
&|(F(x) \mid F(y))_{F(w)} - (F(y) \mid F(z))_{F(w)}| \\
&\quad = ||F(w) - u_2'| - |F(w) - v_1'|| \\
&\quad = |u_2' - v_1'| \\
&\quad \leq |u_2' - F(u_2)| + |F(u_2) - F(v_1)| + |F(v_1) - v_1'| \\
&\quad \leq B + |F(u_2) - F(v_1)| + B \\
&\quad \leq \lambda|u_2 - v_1| + c + 2B \\
&\quad = \lambda|u_2 - v_1| + A \\
&\quad = \lambda||w - u_2| - |w - v_1|| + A \\
&\quad = \lambda|(x \mid y)_w - (y \mid z)_w| + A,
\end{aligned}$$

which shows the desired upper bound. We now prove the lower bound: To this end, note again that $|(x \mid y)_w - (y \mid z)_w| = ||w - u_2| - |w - v_1|| = |u_2 - v_1|$, hence

$$\begin{aligned}\frac{1}{\lambda}|(x \mid y)_w - (y \mid z)_w| - c &= \frac{1}{\lambda}|u_2 - v_1| - c \\ &\le |F(u_2) - F(v_1)| \\ &\le |F(u_2) - u_2'| + |u_2' - v_1'| + |v_1' - F(v_1)| \\ &\le B + |u_2' - v_1'| + B \\ &= |u_2' - v_1'| + 2B \\ &= \big||F(w) - u_2'| - |F(w) - v_1'|\big| + 2B \\ &= |(F(x) \mid F(y))_{F(w)} - (F(y) \mid F(z))_{F(w)}| + 2B.\end{aligned}$$

Finally,

$$\frac{1}{\lambda}|(x \mid y)_w - (y \mid z)_w| - \underbrace{(2B + c)}_{=A} \le |(F(x) \mid F(y))_{F(w)} - (F(y) \mid F(z))_{F(w)}|. \qquad \square$$

Theorem 3.7.14 can be extended analogously for (λ, c, L)-local quasi rays and (λ, c, L)-local quasi geodesics.

Theorem 3.7.19. *Let X be a proper δ-hyperbolic geodesic metric space. Given three real numbers $\delta \ge 0$, $\lambda \ge 1$ and $c \ge 0$. Then there exist constants $H = H(\delta, \lambda, c)$ and $L = L(\delta, \lambda, c)$ with the following properties:*
(a) *If $f: \mathbb{R}_+ \to X$ or $f: \mathbb{N}_0 \to X$ is a (λ, c, L)-local quasi ray, then there exists a minimal ray $g: \mathbb{R}_+ \to X$ with $g(0) = f(0)$ and $H(f, g) \le H$.*
(b) *If $f: \mathbb{R} \to X$ or $f: \mathbb{Z} \to X$ is a (λ, c, L)-local quasi geodesic then there exists a minimal geodesic $g: \mathbb{R} \to X$ with $H(f, g) \le H$.*

For the proof we refer to [114].

3.8 The boundary of a tree

Let X be a locally finite metric tree such that all its edges have length 1. We assume further that every vertex has valency greater or equal to 2 (this guarantees that X is an infinite tree).

3.8.1 Equivalent rays in a metric tree and its boundary

Recall that a ray in X is an isometric embedding of a ray in $\mathbb{R}$ into X. We say that two rays are *equivalent* if their intersection contains a ray, that is, two rays are equivalent if they coincide, compare Figure 3.49.

The equivalence of rays defines an equivalence relation on the set of rays in X.

Definition 3.8.1. The *boundary* of X is the set of equivalence classes (under the equivalence of rays) of rays in X. We denote the boundary of X by ∂X.

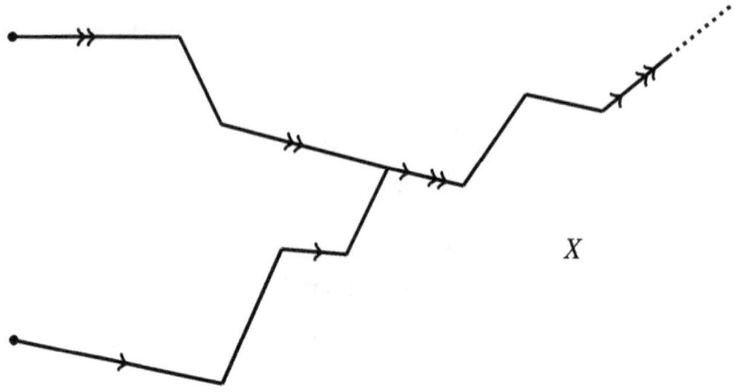

Figure 3.49: Equivalence of rays.

As X is a tree, every equivalence class of rays in X has a unique representative with start point w if we choose w as the basis point in X. Hence every *point* in ∂X can be represented by a uniquely determined ray with start point w.

If $a \in \partial X$, then let $[w, a)$ denote the uniquely determined ray in X with start point w representing a. (We can identify $[w, a)$ with a.)

Definition 3.8.2. For $a, b \in \partial X$ we define the *Gromov product* $(a \mid b)_w = \lim(x \mid y)_w$ where $x \in [w, a)$ and $y \in [w, b)$. The limit is defined in the following way: $(a \mid b)_w = L$ if for all $\varepsilon > 0$ there exists an $N \in \mathbb{R}$ such that $|x - w| > N$ and $|y - w| > N$ implies $|(x \mid y)_w - L| < \varepsilon$. For $\varepsilon > 0$ we define $d_\varepsilon(a, b) = e^{-\varepsilon(a|b)_w}$ for all $a, b \in \partial X$.

The limit can be interpreted as the behavior of $|x - w|$ and $|y - w|$ in ∞. In the following we often write $(a \mid b)$ instead of $(a \mid b)_w$.

Remarks 3.8.3. We study some properties of the above definitions.

1. We have
   (1) $(a \mid b) = +\infty$ and $d_\varepsilon(a, b) = 0$ if and only if $a = b$;
   (2) $(a \mid b) = (b \mid a)$ and $d_\varepsilon(b, a) = d_\varepsilon(a, b)$; and
   (3) $(a \mid c) \geq \min\{(a \mid b), (b \mid c)\}$ as well as $d_\varepsilon(a, c) \leq \max\{d_\varepsilon(a, b), d_\varepsilon(b, c)\}$.

Proof. (1) Let $M \in \mathbb{R}$ and $N \geq M$. If $a = b$ and $|w - x|, |w - y| > N$ then

$$(x \mid y)_w = \min\{|w - x|, |w - y|\} > N \geq M,$$

because x and y lie on the same ray. Hence $a = b$ implies $(a \mid b) = \infty$ and $d_\varepsilon(a, b) = 0$.

Now, let $a \neq b$. As X is a tree there exists an N such that $(x \mid y)$ is equal to a constant for all $x \in [w, a), y \in [w, b)$ with $|w - x| > N$ and $|w - y| > N$ as shown in Figure 3.50.

The distance from w to the tripod point is exactly this constant. Hence $a \neq b$ implies that $(a \mid b)_w$ is this constant, that is, in particular $(a \mid b)_w < \infty$.

(2) These equations are clear.

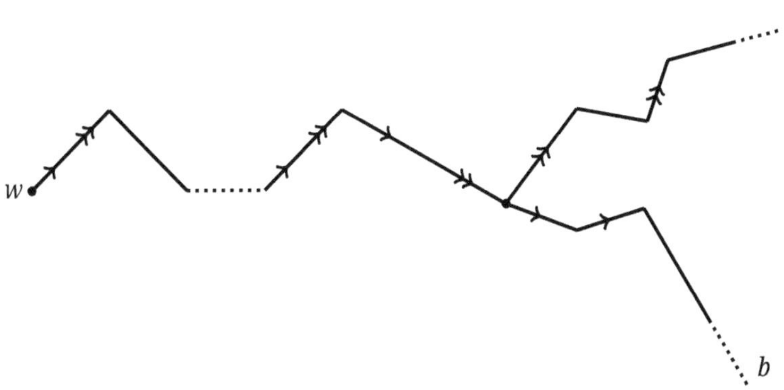

Figure 3.50: Tripod point.

(3) Observe that $(a \mid b)_w$ is the distance from w to the tripod point in $[w, a) \cup [w, b)$. Now (3) follows directly from Definition 3.2.4 and the second remark before Theorem 3.2.8. □

2. The map d_ε is a non-Archimedean metric on ∂X. Set $d_1 = d$, then

$$d_\varepsilon(a, b) = \frac{1}{e^{\varepsilon(a|b)_w}} = \left(\frac{1}{e^{(a|b)_w}}\right)^\varepsilon = d(a, b)^\varepsilon.$$

Hence the topologies induced by the d_ε are equivalent.

3. The values of $d(a, b)$ are of the form e^{-n} with $n \in \mathbb{N}_0 \cup \{\infty\}$. We have $d(a, b) = \frac{1}{e^n}$ if and only if the geodesics $[w, a)$ and $[w, b)$ coincide in the first n edges and do not coincide in all the others. With 2. we can conclude that the topology induced on ∂X is independent of w.
To see this, let w' be another basis point for X with the associated Gromov product $(\,|\,)' \equiv (\,|\,)_{w'}$ and metric $d'_\varepsilon(\,,\,)$ for ∂X. If $|w - w'| = n$ then

$$(a \mid b) - n \leq (a \mid b)' \leq (a \mid b) + n$$

and

$$e^{-\varepsilon n} d_\varepsilon(a, b) \leq d'_\varepsilon(a, b) \leq e^{\varepsilon n} d_\varepsilon(a, b)$$

for all $a, b \in \partial X$.
Figure 3.51 shows a possible case.
We have $(a \mid b) + (a \mid b)' \leq n$. It follows that

$$(a \mid b) - n \leq (a \mid b)' \leq (a \mid b) + n$$

after a discussion of the other possibilities. We further have

$$e^{-\varepsilon n} d_\varepsilon(a, b) \leq d'_\varepsilon(a, b) \leq e^{\varepsilon n} d_\varepsilon(a, b).$$

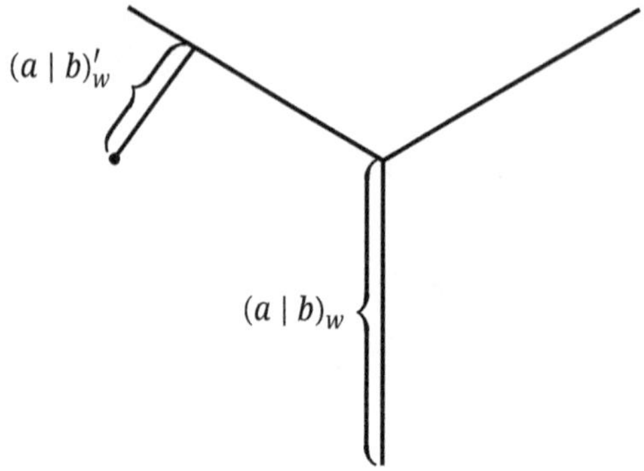

Figure 3.51: Possible case.

We can obtain a slightly better bound with the help of a suitable partition of ∂X. Let $|w - w'| = n$ and $g\colon [0, n] \to X$ be a geodesic from w to w'. As every edge has length 1 we see that $g(0) = w, g(1), \ldots, g(n) = w'$ are the vertices through which the geodesic passes. Set $w_k = g(k)$ and $U_k = \{a \in \partial X \mid [w, a] \cap [w, w'] = [w, w_k]\}$ for $k = 0, 1, \ldots, n$; compare Figure 3.52.

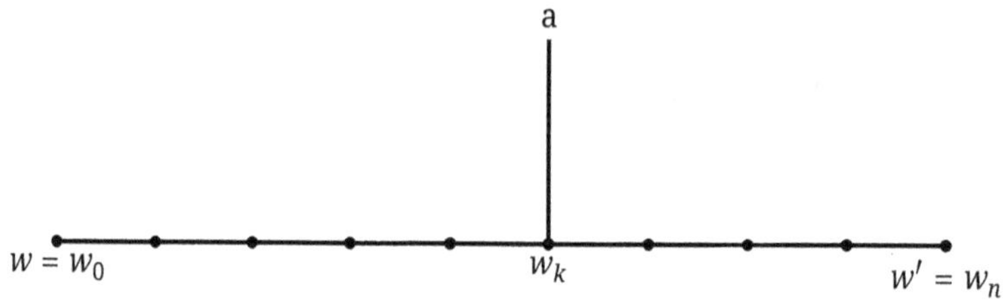

Figure 3.52: Suitable partition.

If $a, b \in U_k$ we have the situation depicted in Figure 3.53.

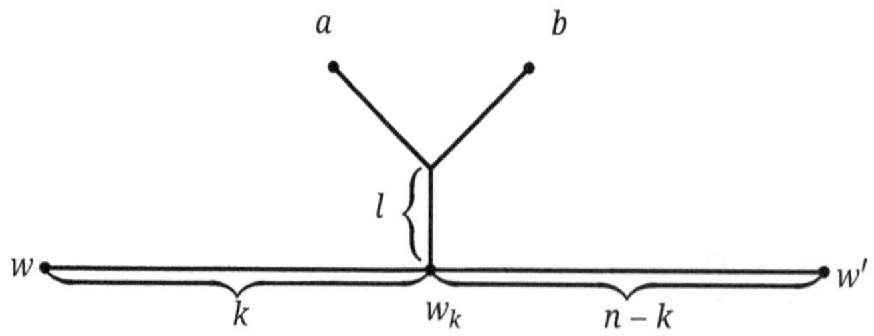

Figure 3.53: Suitable partition with $a, b \in U_k$.

We have $(a \mid b) = l + k$ and

$$(a \mid b)' = l + n - k = (l + k) + n - 2k = (a \mid b) + n - 2k.$$

It follows that $d'_\varepsilon(a, b) = e^{\varepsilon(2k-n)} d_\varepsilon(a, b)$. Note that these equations only hold if a and b are both in U_k.

Theorem 3.8.4. *The boundary ∂X together with the induced topology is compact.*

Proof. As d is a bounded metric on ∂X it suffices to show that ∂X is complete with respect to Cauchy sequences, that is, ∂X is sequentially compact. Now let $(a_n)_{n\in\mathbb{N}}$ be a Cauchy sequence in ∂X. For $k \in \mathbb{N}$ set $\varepsilon_k = \frac{1}{e^k}$. The sequence $(\varepsilon_k)_{k\in\mathbb{N}}$ is a strictly decreasing zero sequence. Let w be the basis point for d. If $a, b \in \partial X$ with $d(a, b) \le \varepsilon_k$ and if we consider $a, b\colon \mathbb{R}_+ \to X$ as geodesic rays with start point w, we have

$$a|_{[0,k]} = b|_{[0,k]}.$$

As $(a_n)_{n\in\mathbb{N}}$ is a Cauchy sequence, there exists an $N_k \in \mathbb{N}$ with

$$d(a_m, a_n) \le \varepsilon_k$$

for all $m, n \ge N_k$.

Without loss of generality let $(N_k)_{k\in\mathbb{N}}$ be strictly increasing. We define a ray $L\colon \mathbb{R}_+ \to X$ via $L|_{[0,k]} = a_{N_k}|_{[0,k]}$ for $k = 1, 2, \ldots$. The sequence $(a_n)_{n\in\mathbb{N}}$ converges to L. □

In the following let X be a metric space such that for all $x \in X$ there exists a decreasing sequence of real numbers ε_k, $k \in \mathbb{N}$, with $\lim_{k\to\infty} \varepsilon_k = 0$ and the ball $B_{\varepsilon_k}(x)$ centered at x with radius ε_k is non-empty for all k.

Note that this hypothesis holds on the boundary of a metric tree as above. Let d be the metric on X and $\varphi\colon X \to X$ be a homeomorphism (not necessarily surjective).

Definition 3.8.5. The *conform dilatation* of φ at $x \in X$ with respect to the distance d is given by

$$H^d_\varphi(x) = \limsup_{\varepsilon\to 0} \frac{\sup\{d(\varphi(x), \varphi(y)) \mid y \in X, d(x,y) = \varepsilon\}}{\inf\{d(\varphi(x), \varphi(y)) \mid y \in X, d(x,y) = \varepsilon\}}.$$

We have $H^d_\varphi(x) \in [1, \infty]$.

3.8.2 Lipschitz and Hölder homeomorphisms

Definition 3.8.6. 1. We call the homeomorphism $\varphi\colon X \to X$

- *Conform* if $H^d_\varphi(x) = 1$ for all $x \in X$.
- *Quasi conform* if $\{H^d_\varphi(x) \mid x \in X\}$ is bounded.
- *K-quasi conform*, if $H^d_\varphi(x) \le K$ for all $x \in X$, where K is a given constant.

2. The homeomorphism $\varphi: X \to X$ is called *Lipschitz* with respect to d, if there exists an $L \geq 1$ with

$$\frac{1}{L} d(x,y) \leq d(\varphi(x), \varphi(y)) \leq L d(x,y)$$

for all $x \in X$.
3. The homeomorphism $\varphi: X \to X$ is called *Hölder* with respect to d if there exist $\alpha > 0$ and $c > 0$ with

$$\frac{1}{c} d(x,y)^{\frac{1}{\alpha}} \leq d(\varphi(x), \varphi(y)) \leq c d(x,y)^{\alpha}$$

for all $x, y \in X$

Remarks 3.8.7. 1. If φ is Lipschitz with respect to d then it is also Hölder with respect to d.
2. If φ is Lipschitz with respect to d then it is also L^2-quasi conform if

$$L \geq 1 \quad \text{and} \quad \frac{1}{L} d(x,y) \leq d(\varphi(x), \varphi(y)) \leq L d(x,y).$$

To see this, let $x \in X$ and $d(x,y) = \varepsilon$ for a $y \in X$. Then

$$\frac{1}{L} \varepsilon \leq d(\varphi(x), \varphi(y)) \leq L\varepsilon,$$

hence

$$\frac{\sup\{d(\varphi(x), \varphi(y)) \mid d(x,y) = \varepsilon\}}{\inf\{d(\varphi(x), \varphi(y)) \mid d(x,y) = \varepsilon\}} \leq \frac{L\varepsilon}{\frac{1}{L}\varepsilon} = L^2.$$

Thus, we have $H_{\varphi}^{d}(x) \leq L^2$ for all $x \in X$.

Remark 3.8.8. As before let X be a metric tree. First let $\gamma: X \to X$ be an isometry (not necessarily surjective). Then γ maps a (geodesic) ray starting at w on a (geodesic) ray starting at $\gamma(w)$. As the rays with start point w correspond uniquely to elements of ∂X and as γ is in particular a homeomorphism, γ induces a homeomorphism of ∂X, which we will also denote by $\gamma: \partial X \to \partial X$. Note that a homeomorphism is not necessarily surjective here.

Theorem 3.8.9. *The induced homeomorphism $\gamma: \partial X \to \partial X$ is conform if $\gamma: X \to X$ is an isometry.*

Proof. We fix $w \in X$. Let $\varepsilon > 0$. Set $w' = \gamma(w)$, and let $n = |w' - w|$. Let $g: [0,n] \to X$ denote the geodesic from w to w'. Set again $w_k = g(k)$ for $k \in \{0, 1, \ldots, n\}$ and

$$U_k = \{a \in \partial X \mid [w,a) \cap [w,w'] = [w,w_k]\}.$$

If $a, b \in U_k$ then $d'_\varepsilon(a,b) = e^{\varepsilon(2k-n)} d_\varepsilon(a,b)$ (this follows from the discussion above) where $d'_\varepsilon(a,b)$ denotes the distance with respect to w'. The U_k partition ∂X in (equivalence) classes. Now let $(a \mid b) := (a \mid b)_w > n$. Geometrically this means that the rays identified with $a, b \in \partial X$ (starting at w) coincide at least in the first $n+1$ edges. Because of $|w - w'| = n$ we have the following: If $a \in U_k$, then also $b \in U_k$. We now compute $H_\gamma^{d_\varepsilon}(a)$. We fix $\bar{\varepsilon}$ small enough such that $d_\varepsilon(a,b) = \bar{\varepsilon}$ implies $(a \mid b) > 2n$ for $n = |w - w'|$. Then the rays associated with a and b (starting at w) coincide at least on the first $2n + 1$ vertices. The images of these rays under γ also coincide, starting at $w' = \gamma(w)$, in at least the first $2n+1$ vertices.

The triangle inequality implies that $\gamma(a)$ and $\gamma(b)$ belong to the same U_k. Now

$$\begin{aligned} d'_\varepsilon(\gamma(a), \gamma(b)) &= e^{\varepsilon(2k-n)} d_\varepsilon(\gamma(a), \gamma(b)) \quad \text{and} \\ d'_\varepsilon(\gamma(a), \gamma(b)) &= d_\varepsilon(a,b) = \bar{\varepsilon}, \\ \text{and hence} \quad d_\varepsilon(\gamma(a), \gamma(b)) &= e^{\varepsilon(n-2k)} \bar{\varepsilon}, \end{aligned}$$

where the second equation follows from γ being an isometry and this holds for all $b \in \partial X$ with $d_\varepsilon(a,b) = \bar{\varepsilon}$. We also have

$$H_\gamma^{d_\varepsilon}(a) = \limsup_{\bar{\varepsilon} \to 0} \frac{e^{\varepsilon(n-2k)} \bar{\varepsilon}}{e^{\varepsilon(n-2k)} \bar{\varepsilon}} = 1,$$

that is, γ is conform at a. □

Theorem 3.8.10. *Let $\gamma: X \to X$ be an isometry (with X a metric tree as above). Then γ induces a Lipschitz homeomorphism $\gamma: \partial X \to \partial X$, that is not necessarily surjective.*

Proof. If $|w - w'| = n$, then

$$e^{-\varepsilon n} d_\varepsilon(a,b) \le d'_\varepsilon(a,b) \le e^{\varepsilon n} d_\varepsilon(a,b)$$

according to the bounds in Remark 3.8.3.3. Hence

$$e^{-\varepsilon n} d_\varepsilon(\gamma(a), \gamma(b)) \le d'_\varepsilon(\gamma(a), \gamma(b)) \le e^{\varepsilon n} d_\varepsilon(\gamma(a), \gamma(b)).$$

Because of $d'_\varepsilon(\gamma(a), \gamma(b)) = d_\varepsilon(a,b)$ we have

$$e^{-\varepsilon n} d_\varepsilon(\gamma(a), \gamma(b)) \le d_\varepsilon(a,b) \le e^{\varepsilon n} d_\varepsilon(\gamma(a), \gamma(b)).$$

Hence

$$e^{-\varepsilon n} d_\varepsilon(a,b) \le d_\varepsilon(\gamma(a), \gamma(b)) \le e^{\varepsilon n} d_\varepsilon(a,b),$$

that is, γ is Lipschitz (the continuity is obvious). □

From now on, let $\gamma: X \to X$ denote a quasi isometry.

Theorem 3.8.11. 1. *Let $f\colon \mathbb{R} \to X$ be a quasi geodesic. Then there exists a uniquely determined geodesic $f_0\colon \mathbb{R} \to X$ with $H(f, f_0) < \infty$.*
2. *Let $f\colon \mathbb{R}_+ \to X$ be a quasi ray. Then there exists a uniquely determined ray $f_0\colon \mathbb{R}_+ \to X$ with $f(0) = f_0(0)$ and $H(f, f_0) < \infty$.*
3. *Let $\gamma\colon X \to X$ be a quasi isometry. Let $f, f'\colon \mathbb{R}_+ \to X$ be two rays that determine the same point in ∂X. Let $(\gamma \circ f)_0$, $(\gamma \circ f')_0$ be the (geodesic) rays connected to $\gamma \circ f$ and $\gamma \circ f'$, respectively, according to 2. Then $(\gamma \circ f)_0$ and $(\gamma \circ f')_0$ also determine the same element of ∂X.*

Proof. Theorem 3.8.11 is an implication of Theorems 3.7.8 and 3.7.19. We just have to show the uniqueness in 1. and 2. The uniqueness in 3. follows because a quasi geodesic ray determines uniquely a point in ∂X. But the uniqueness in 1. and 2. follows from the properties of a tree. If two geodesics or geodesic rays do not have identical ends then their distance (measured in X) is arbitrarily large. Hence, two geodesics or geodesic rays with different ends cannot have a finite Hausdorff distance. □

Theorem 3.8.12. *Let X be a metric space as above. Let $\gamma\colon X \to X$ be a quasi isometry. Then there exists an induced quasi conform Hölder homeomorphism $\gamma\colon \partial X \to \partial X$ that is not necessarily surjective.*

Proof. We proceed in four steps.
1. The quasi isometry $\gamma\colon X \to X$ induces a homeomorphism $\gamma\colon \partial X \to \partial X$ (not necessarily surjective).
   To see this, let $w \in X$ and $\gamma(w) = w'$. If $a \in \partial X$, then we consider again a as a ray starting in w, that is, $a\colon \mathbb{R}_+ \to X$ is a geodesic ray with $a(0) = w$. As γ is a quasi isometry also $\gamma \circ a\colon \mathbb{R}_+ \to X$ is a quasi isometry with $\gamma \circ a(0) = w'$.
   According to Theorem 3.8.11.2 there exists a uniquely determined geodesic ray $(\gamma \circ a)_0\colon \mathbb{R}_+ \to X$ with $(\gamma \circ a)_0(0) = w'$ and $H(\gamma \circ a, (\gamma \circ a)_0) < \infty$ (see Theorem 3.7.14). As $(\gamma \circ a)_0$ is a ray in X it determines an element of ∂X. Hence we can define a function $\gamma\colon \partial X \to \partial X$ via $\gamma(a) = (\gamma \circ a)_0$. Let a and b be distinct elements of ∂X. We would like to show that $(\gamma \circ a)_0 \neq (\gamma \circ b)_0$, that is, $\gamma\colon \partial X \to \partial X$ is injective. It suffices to show that $H(\gamma \circ a, \gamma \circ b) = \infty$ holds. Let γ be a (λ, c)-quasi isometry. Hence

$$\lambda^{-1}|x - y| - c \le |\gamma(x) - \gamma(y)|$$

   for all $x, y \in X$ $(\lambda \ge 1)$. If $a \neq b$, then there exists an $x \in a$, such that $d(x, \text{ray ass. to } b)$ is arbitrarily large and $d(x, \text{ray ass. to } b) = d(x, y)$; compare Figure 3.54.
   It follows that $\lambda^{-1}|x - y| - c$ can be arbitrarily large and hence also $|\gamma(x) - \gamma(y)|$. Thus $H(\gamma \circ a, \gamma \circ b) = \infty$. Hence $\gamma\colon \partial X \to \partial X$ is injective. We remark that in general, γ is not surjective. See Figure 3.55 for an example. Note that γ moves one block horizontally. The map $\gamma\colon X \to X$ is a non-surjective isometry and the image has no quasi ray through w.

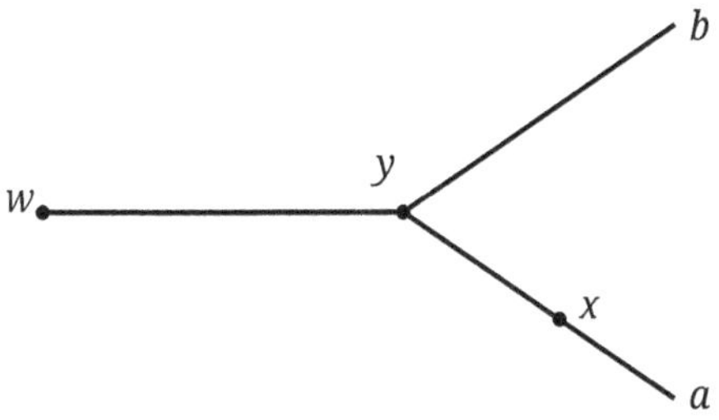

Figure 3.54: Existence of $x \in a$.

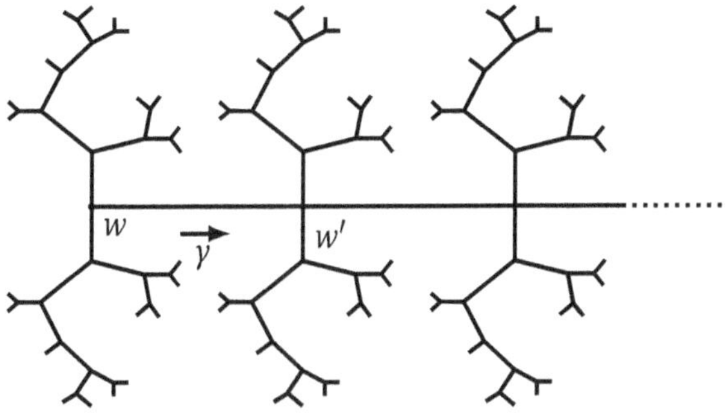

Figure 3.55: Example for non-surjective γ.

2. We now show that $\gamma: \partial X \to \partial X$ is quasi conform and Hölder. To this end, we make some preliminary considerations. Let again $\gamma: X \to X$ be a (λ, c)-quasi isometry. Let $w \in X$, $\gamma(w) = w'$ and $\varepsilon > 0$. Let $H = H(\delta, \lambda, c)$, $\delta = 0$, the constant as in Theorem 3.7.14 or as in Theorem 3.7.19. Hence if $a: \mathbb{R}_+ \to X$ is a geodesic ray, then $H(\gamma \circ a, (\gamma \circ a)_0) \leq H$, where again $(\gamma \circ a)_0$ is the geodesic ray in X with $(\gamma \circ a)_0(0) = \gamma \circ a(0)$ and $H(\gamma \circ a, (\gamma \circ a)_0) < \infty$, that is, $(\gamma \circ a)_0$ is the uniquely determined geodesic ray associated to $\gamma \circ a$ (see Theorem 3.8.11).
Now let a, b be distinct elements in ∂X starting in w such that $(a \mid b)_w = l$. In the following we will show that there is a positive constant $K = K(\delta, \lambda, c)$, $\delta = 0$, such that

$$\frac{1}{\lambda} l - K \leq ((\gamma \circ a)_0 \mid (\gamma \circ b)_0)_{w'} \leq \lambda l + K.$$

We have the situation depicted in Figure 3.56.
Let N be a positive constant with $\frac{1}{\lambda} N - c > 2H$. Along the ray a let r be the point of a with distance $l + N$ from w, that is, with distance N after p along a; see Figure 3.57. We have

$$|\gamma(r) - \gamma(p)| \geq \frac{1}{\lambda} |r - p| - c = \frac{1}{\lambda} N - c > 2H,$$

which follows from γ being a (λ, c)-quasi isometry.
As $|r - s| \geq N$ for all $s \in B$ it follows that $|\underbrace{\gamma(r) - \gamma(b)}_{\text{distance of sets}}| > 2H$ (property of a tree). Let $(\gamma \circ a)_0$ and $(\gamma \circ b)_0$ again be the geodesic rays in X associated to $\gamma \circ a$ and $\gamma \circ b$, respectively. Because of $H(\gamma \circ a, (\gamma \circ a)_0) \leq H$ there exists an $\bar{r} \in (\gamma \circ a)_0$ with $|\gamma(r) - \bar{r}| \leq H$.

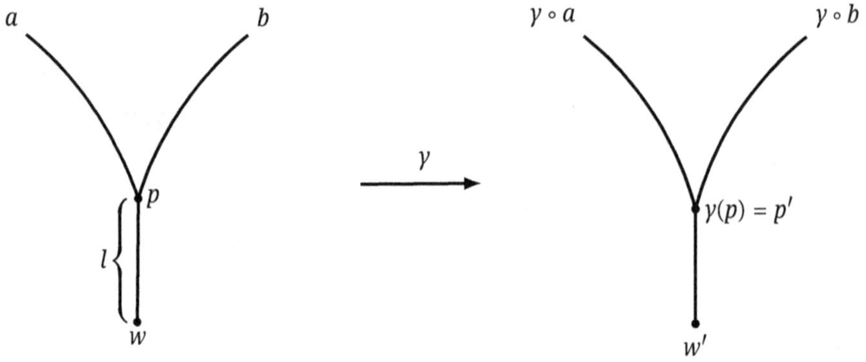

Figure 3.56: Given image.

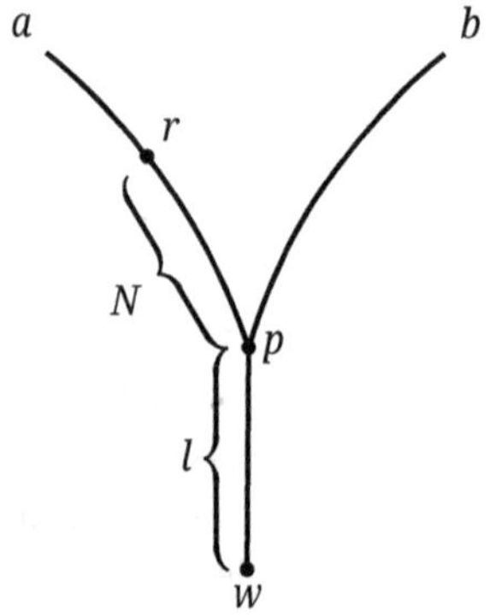

Figure 3.57: Description of *r*.

We claim: $\bar{r} \notin (\gamma \circ b)_0$. To see this, suppose $\bar{r} \in (\gamma \circ b)_0$. Because of $H(\gamma \circ b, (\gamma \circ b)_0) \leq H$ there exists an $s \in \gamma \circ b$ with $|s - \bar{r}| \leq H$, which contradicts the fact that the distance of $\gamma(r)$ and $\gamma(b)$ is strictly larger than $2H$. Hence $\bar{r} \notin (\gamma \circ b)_0$.
This yields $|w' - \bar{r}| > ((\gamma \circ a)_0 \mid (\gamma \circ b)_0)_{w'}$ and we have

$$\begin{aligned} |p' - \bar{r}| &\leq |p' - \gamma(r)| + |\gamma(r) - \bar{r}| \\ &\leq |\gamma(p) - \gamma(r)| + |\gamma(r) - \bar{r}| \\ &\leq \lambda|p - r| + c + H \\ &= \lambda N + c + H. \end{aligned}$$

Because of $p' = \gamma(p) \in \gamma \circ a$ and $p' \in \gamma \circ b$ we see that the distance of p' and $(\gamma \circ a)_0$ as well as the distance p' and $(\gamma \circ b)_0$ is smaller than or equal to H. We have the situation depicted in Figure 3.58.
Here $B = B(p', \alpha)$ is a circle centered at p' with radius $\alpha = \lambda N + c + H$. Because of $\bar{r} \in (\gamma \circ a)_0$ and $\bar{r} \notin (\gamma \circ b)_0$ we conclude that $(\gamma \circ a)_0$ and $(\gamma \circ b)_0$ separate within B. Hence

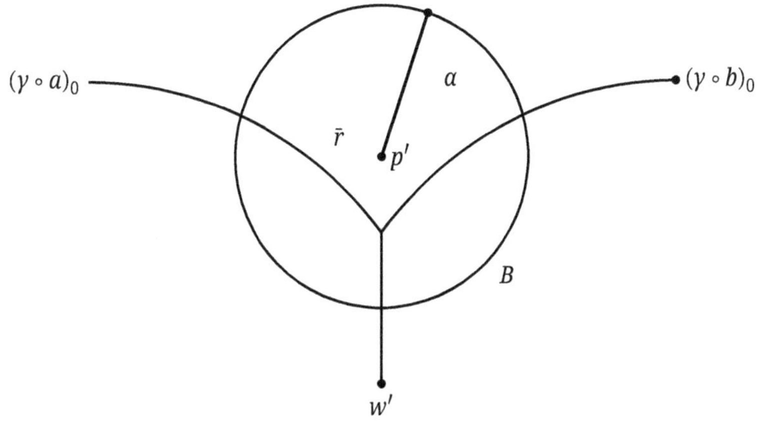

Figure 3.58: Distances for p'.

$$\begin{aligned}|w' - p'| - (\lambda N + c + H) &\leq ((\gamma \circ a)_0 \mid (\gamma \circ b)_0)_{w'} \\ &\leq |w' - p'| + \lambda N + c + H.\end{aligned}$$

Now

$$|w' - p'| = |\gamma(w) - \gamma(p)| \geq \frac{1}{\lambda}|w - p| - c = \frac{1}{\lambda}l - c$$

and

$$|w' - p'| = |\gamma(w) - \gamma(p)| \leq \lambda|w - p| + c = \lambda l + c,$$

thus

$$\frac{1}{\lambda}l - (\lambda N + 2c + H) \leq ((\gamma \circ a)_0 \mid (\gamma \circ b)_0)_{w'} \leq \lambda l + (\lambda N + 2c + H).$$

Hence $K = \lambda N + 2c + H$ has the desired properties.

3. We now show that $\gamma \colon \partial X \to \partial X$ is Hölder. First we have

$$\begin{aligned}((\gamma \circ a)_0 \mid (\gamma \circ b)_0)_{w'} - n &\leq ((\gamma \circ a)_0 \mid (\gamma \circ b)_0)_w \\ &\leq ((\gamma \circ a)_0 \mid (\gamma \circ b)_0)_{w'} + n\end{aligned}$$

if $|w - w'| \leq n$. If we set $\bar{K} = K + n$, the above claim implies

$$\frac{1}{\lambda}(a \mid b) - \bar{K} \leq ((\gamma \circ a)_0 \mid (\gamma \circ b)_0) \leq \lambda(a \mid b) + \bar{K}.$$

This yields

$$e^{-\varepsilon(\lambda(a|b)+\bar{K})} \leq e^{-\varepsilon((\gamma\circ a)_0|(\gamma\circ b)_0)} \leq e^{-\varepsilon(\frac{1}{\lambda}(a|b)-\bar{K})},$$

hence

$$e^{-\varepsilon\bar{K}} d_\varepsilon(a,b)^\lambda \le d_\varepsilon(\gamma(a),\gamma(b)) \le e^{\varepsilon\bar{K}} d_\varepsilon(a,b)^{\frac{1}{\lambda}}.$$

We set $\bar{C} = e^{\varepsilon\bar{K}}$ and $\alpha = \frac{1}{\lambda}$, and then

$$\frac{1}{\bar{C}} d_\varepsilon(a,b)^{\frac{1}{\alpha}} \le d_\varepsilon(\gamma(a),\gamma(b)) \le \bar{C} d_\varepsilon(a,b)^\alpha.$$

Hence $\gamma\colon \partial X \to \partial X$ is Hölder.

4. We now show that $\gamma\colon \partial X \to \partial X$ is quasi conform. To this end, we fix an $a \in \partial X$, and let $b \in \partial X$ with $(a \mid b) = l$. Hence a and b separate at the point $p \in X$; see Figure 3.58. If $d_\varepsilon(a,b') = d_\varepsilon(a,b)$, then a and b' separate at the same point $p \in X$. We must approximate $H_\gamma^{d_\varepsilon}(a)$ and use the inequalities

$$|w'-p'| - (\lambda N + c + H) \le |w'-p'| + \lambda N + c + H$$

and

$$((\gamma \circ a)_0 \mid (\gamma \circ b)_0)_{w'} - n \le ((\gamma \circ a)_0 \mid (\gamma \circ b)_0)_{w'} + n$$

from steps 2 and 3, respectively. Together with the above remark about $d_\varepsilon(a,b') = d_\varepsilon(a,b)$ we get

$$H_\gamma^{d_\varepsilon}(a) \le \frac{e^{-\varepsilon(|w'-p'|-|w'-w|-(\lambda N+c+H))}}{e^{-\varepsilon(|w'-p'|+|w'+w|+(\lambda N+c+H))}}.$$

We set $K^* = |w'-w| + (\lambda N + c + H)$. Then K^* is a constant independent of a and $d_\varepsilon(a,b)$. Hence

$$H_\gamma^{d_\varepsilon}(a) \le e^{-\varepsilon(2K^*)}$$

and this holds for all $a \in \partial X$. Hence $\gamma\colon \partial X \to \partial X$ is quasi conform. Finally we mention that being Hölder yields the continuity of $\gamma\colon \partial X \to \partial X$ and the inverse function on $\gamma(\partial X) \subset \partial X$. □

Remark 3.8.13. We have seen that, in general, quasi isometries $\gamma\colon X \to X$ do not induce surjective maps $\gamma\colon \partial X \to \partial X$ (as above) but only injective ones.

Theorem 3.8.14. *Let $\gamma\colon X \to X$ be a quasi isometry. Then the induced map $\gamma\colon \partial X \to \partial X$ is surjective if and only if $H(\mathrm{id}_X, \gamma) < \infty$, where $H(\mathrm{id}_X, \gamma)$ is the Hausdorff distance from id_X to γ.*

Proof. First let $\gamma\colon \partial X \to \partial X$ be surjective. We observe that $H(\mathrm{id}_X, \gamma) < \infty$ holds if and only if for a given $x \in X$ there exists a $y \in X$ such that $|x-y|$ is bounded above by a fixed constant independent of x.

Let again H be the constant as in Theorem 3.7.19 for (λ, c)-quasi rays in δ-hyperbolic spaces (where $\gamma: X \to X$ is a (λ, c)-quasi isometry and X is δ-hyperbolic). Here $\delta = 0$. We fix $w \in X$ and let $w' := \gamma(w)$. Let $x \in X$. Let b' be a ray in X with $b'(0) = w'$ and $x \in b'$. As $\gamma: \partial X \to \partial X$ is surjective there exists a $b \in \partial X$ with $\gamma(b) = b'$ and $b(0) = w$. According to Theorem 3.7.19 we have $H(\gamma \circ b, b') \leq H$. Hence there exists a $y \in b$ with $|\gamma(y) - x| \leq H$. Hence $H(\mathrm{id}_X, \gamma) \leq H$.

On the other hand, $H(\mathrm{id}_X, \gamma) < \infty$ implies that there is a mapping $\Gamma: X \to X$ such that

$$X \underset{\Gamma}{\overset{\gamma}{\rightleftarrows}} X$$

(with the respective constants) is quasi isometric to itself (compare the beginning of Subsection 3.7.1). The proof showed that $\Gamma: X \to X$ itself is a quasi isometry. The definition of quasi isometric spaces also implies that $H(\gamma \circ \Gamma, \mathrm{id}_X) < \infty$, as well as $H(\Gamma \circ \gamma, \mathrm{id}_X) < \infty$ holds. We already remarked that the composition of quasi isometries is again a quasi isometry. Together with Theorem 3.8.11 this shows that $\gamma \circ \Gamma: \partial X \to \partial X$ and $\Gamma \circ \gamma: \partial X \to \partial X$ both are equal to the identity on X. Hence $\Gamma = \gamma^{-1}$ and γ is surjective as a map $\gamma: \partial X \to \partial X$. □

3.8.3 Applications to groups

Let Γ be the finitely generated free group with basis S^+ (that is, S^+ is a minimal generating system). Recall that the cardinality of elements of such an S^+ is an invariant of Γ, called the rank of Γ. Let S^- be the set of inverses of S^+ and $S := S^+ \cup S^-$. Let $X = \phi(\Gamma, S)$ be the associated Cayley graph (alternatively, X can be considered as the universal covering of a bouquet of $|S^+|$-circles). Let $\gamma: \Gamma \to \Gamma$ be an automorphism. The proof that (Γ, S) and $(\Gamma, \gamma(S))$ are quasi isometric (and hence also $\phi(\Gamma, S)$ and $\phi(\Gamma, \gamma(S))$) implies that γ can be considered as a quasi isometry of $\phi(\Gamma, S)$ in itself with $H(\gamma, \mathrm{id}_{\phi(\Gamma,S)}) < \infty$. Hence, automorphisms of finitely generated free groups define quasi isometries of the associated Cayley graphs.

If $\gamma: \Gamma \to \Gamma$ is an automorphism, then it induces not necessarily an isometry from $\phi(\Gamma, S)$ to $\phi(\Gamma, \gamma(S))$. If γ induces an isometry then it must permute the elements of S. But if $|S^+| \geq 2$ then there exist automorphisms of Γ that do not permute the elements of S.

Example 3.8.15. Let $\Gamma = \langle x, y \mid \rangle$, $\gamma(x) = x$ and $\gamma(y) = yx$. Then γ induces an automorphism of Γ.

Now let X be a metric tree again.

Definition 3.8.16. Let $\gamma: X \to X$ be a surjective isometry of X. Then γ is called *elliptic* if γ has a fixed point in X. If γ does not have a fixed point it is called *hyperbolic*.

Theorem 3.8.17. *Let $\gamma: X \to X$ be a (surjective) hyperbolic isometry of X. Then the induced map $\gamma: \partial X \to \partial X$ has exactly two fixed points a and a'. More precisely, a and a' can be chosen such that they satisfy the following property: For every neighborhood U of a and every neighborhood of U' of a' in ∂X we can choose a $k \in \mathbb{N}$ such that $\gamma^n(\partial X \setminus U') \subset U$ and $\gamma^{-n}(\partial X \setminus U) \subset U'$ for all $n \geq k$.*

In this situation, a is called the *goal* and a' is called the *source* of γ.

Proof. Let $x \in X$. As γ is hyperbolic, we have $\gamma(x) \neq x$. Let $s = [x, \gamma(x)]$ be the uniquely determined geodesic segment in X from x to $\gamma(x)$. We have $\gamma([x, \gamma(x)]) = [\gamma(x), \gamma^2(x)]$, etc. We consider two cases.

1. Let $s \cap \gamma(s) = \{\gamma(x)\}$.
   In this case we have the situation depicted in Figure 3.59.

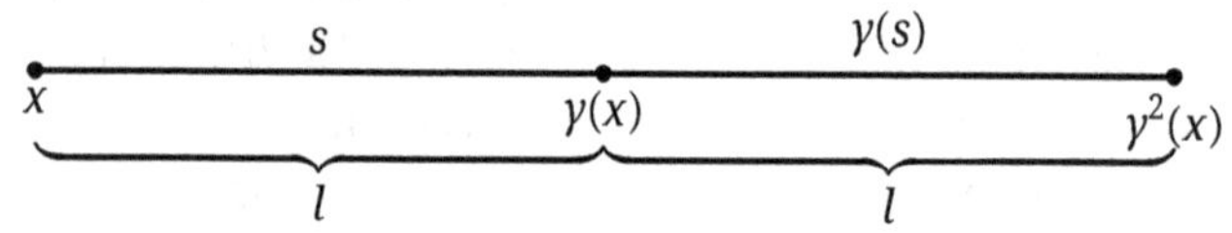

Figure 3.59: Picture for case 1.

We claim $\gamma^2(s) \cap (s \cup \gamma(s)) = \gamma^2(x)$.
To see this, assume $\gamma^2(s) \cap (s \cup \gamma(s))$ contains points other than $\gamma^2(x)$. Then also $\gamma^2(s) \cap \gamma(s)$ would contain points other than $\gamma^2(x)$ and hence $\gamma(s) \cap s = \gamma^{-1}(\gamma^2(s) \cap \gamma(s))$ would contain points other than $\gamma(x)$. Note that l is the length of $\gamma^2(s)$.
Note that the same argument implies for all $n \geq 2$

$$\gamma^n \cap (s \cup \gamma(s) \cup \cdots \cup \gamma^{n-1}(s)) = \gamma^n(x).$$

We claim $\gamma^{-1}(s) \cap (\bigcup_{n \geq 0} \gamma^n(s)) = \{x\}$.
To see this, we can again restrict ourselves to the part $\gamma^{-1}(s) \cap s$ (length argument). If $\gamma^{-1}(s) \cap s$ would contain points other than x, then also $\gamma(\gamma^{-1}(s) \cap s) = s \cap \gamma(s)$ would contain points other than $\gamma(x)$.
Altogether we obtain $\gamma^{-n}(s) \cap (\bigcup_{k > -n} \gamma^k(s)) = \gamma^{-n+1}(x)$ for all $n \in \mathbb{N}$.
We conclude that $\bigcup_{n \in \mathbb{Z}} \gamma^n(s)$ is a geodesic in X that is invariant under γ. We denote a and a' as in Figure 3.60.
(Here, $a, a' \in \partial X$ are fixed by γ.)

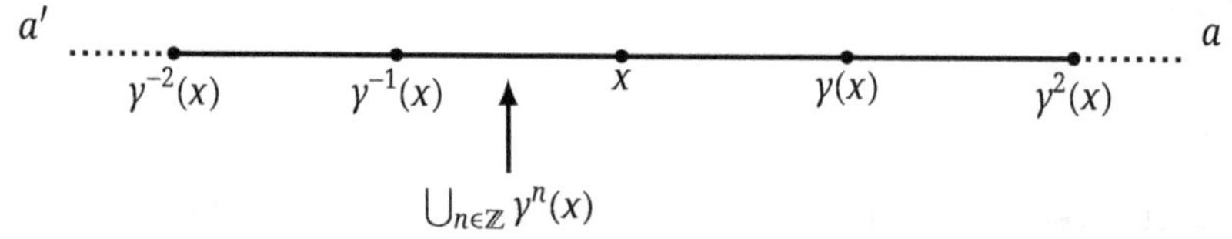

Figure 3.60: Geodesic invariant under γ.

2. Let $\gamma(s) \cap s \neq \{\gamma(x)\}$.
We consider the subcases
(i) $\{x, \gamma(x), \gamma^2(x)\}$ is contained in a geodesic segment in X; and
(ii) $\{x, \gamma(x), \gamma^2(x)\}$ is not contained in a single geodesic segment in X.
We first show that (i) does not occur: As $\gamma|_s$ is an isometric embedding and X is a tree, we must have $\gamma(s) = s$ as sets. But then $\gamma: s \to s$, and γ has a fixed point according to the Brouwer fixed point theorem. This yields a contradiction to γ being hyperbolic. Hence (i) cannot occur.
We now show that (ii) leads us back to case 1: As X is a tree, the set $\{x, \gamma(x), \gamma^2(x)\}$ gives the end points of a tripod in X; compare Figure 3.61.

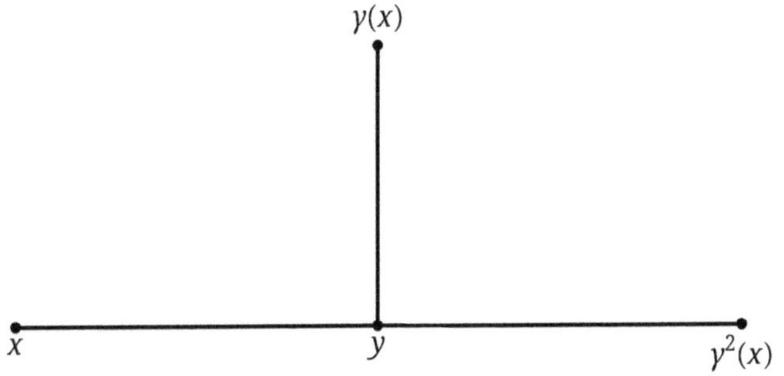

Figure 3.61: Tripod for $\{x, \gamma(x), \gamma^2(x)\}$.

Let y be as above. We have $s = [x,y] \cup [y, \gamma(x)]$ and hence $\gamma(s) = [\gamma(x),y] \cup [y, \gamma^2(x)]$. As l is the length of s and also of $\gamma(s)$ we have the situation depicted in Figure 3.62 and $l = l' + l''$.

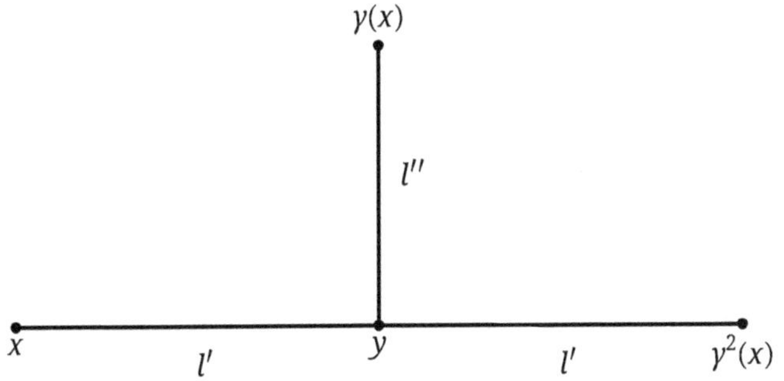

Figure 3.62: Description of l.

We claim: $l' > l''$.

To see this, assume $l'' \geq l'$. If $l'' = l'$, then we necessarily have $\gamma(y) = y$ which yields a contradiction. Hence $l'' > l'$. Now let $g\colon [0, l] \to X$ be the geodesic map with image $s = [x, \gamma(x)]$. We have $\gamma(g([l', l - l'])) = g([l', l - l'])$. Consider Figure 3.63.

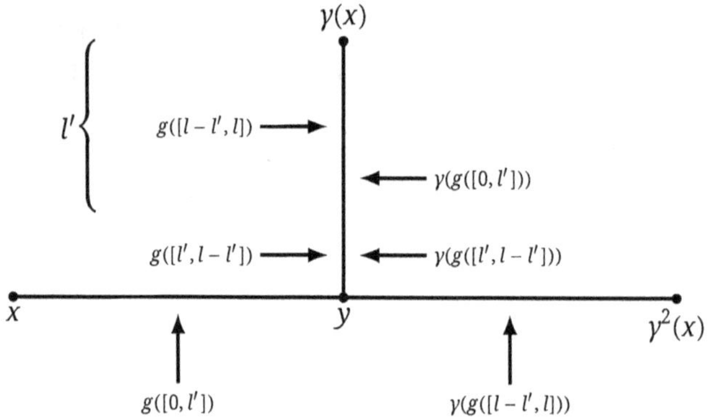

Figure 3.63: Fixed point for γ.

But then γ must have a fixed point again which contradicts the fact that γ is hyperbolic. Hence $l' > l''$.
Let again $g\colon [0, l] \to X$ be as before. We consider $g|_{[l'', l - l'']}$ and Figure 3.64.

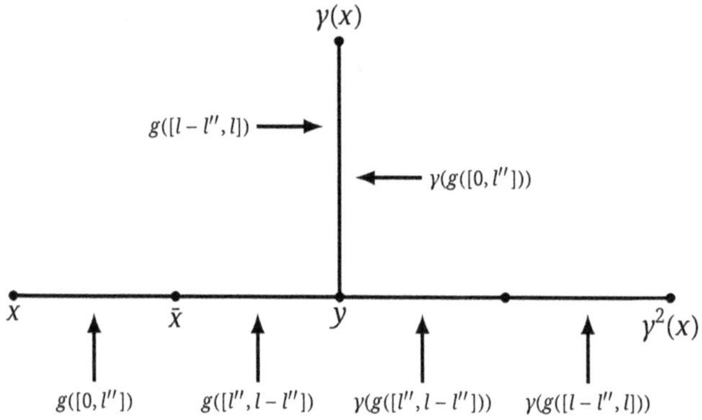

Figure 3.64: Choice of a and a'.

So if $\bar{x} = g(l'')$ then $y = \gamma(\bar{x})$ and $\gamma([\bar{x}, y]) \cap [\bar{x}, y] = \{y\}$, and we are in case 1.

This shows the first part of Theorem 3.8.17. We now consider the supplement. To this end, we have to further discuss the invariant geodesic that we have constructed in case 1. Let a denote this geodesic; see Figure 3.65.

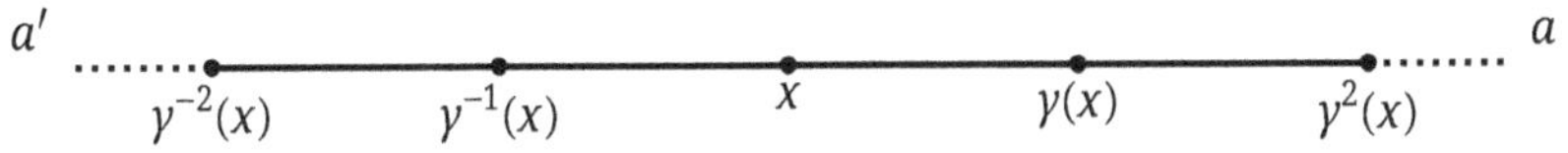

Figure 3.65: Geodesic a.

The isometry γ operates transitively on a from left to right via

$$\gamma([\gamma^n(x), \gamma^{n+1}(x)]) = [\gamma^{n+1}(x), \gamma^{n+2}(x)]$$

for all $n \in \mathbb{Z}$. Let $v_1, \ldots, v_k$ be the vertices of X in $[x, \gamma(x)]$ where v_i occurs before v_{i+1} for $i = 1, \ldots, k-1$ if we go from x to $\gamma(x)$ along $[x, \gamma(x)]$. Now there are subtrees T_{ij}, $j = 1, \ldots, m_i$, of X starting from v_i, for $i = 1, \ldots, k$, as depicted in Figure 3.66.

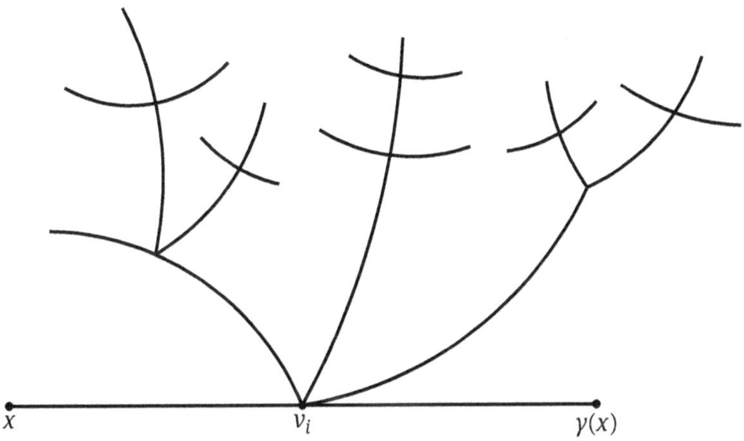

Figure 3.66: Subtrees starting at v_i.

(Recall the assumption on X.)

Now we have $\gamma(v_i) \in [\gamma(x), \gamma^2(x)]$, and $\gamma(T_{ij})$ is a copy of T_{ij} starting in $\gamma(v_i)$. Hence we directly see that there is a K such that $\gamma^n(\partial X \setminus U') \subset U$ and $\gamma^{-n}(\partial X \setminus U) \subset U'$ for all $n \geq K$. □

3.9 The boundary of a hyperbolic space

In the following let $\delta \geq 0$ and let X be a δ-hyperbolic, proper geodesic metric space with basis point $w \in X$. For the Gromov product we write $(x \mid y)$ instead of $(x \mid y)_w$ for $x, y \in X$. As we will need no further specifications, we use the term quasi ray instead of (λ, c)-quasi ray.

Definition 3.9.1. Let $f_1, f_2 \colon \mathbb{R}^+ \to X$ be quasi rays. We call f_1 and f_2 *equivalent* if $H(f_1, f_2) < \infty$.

This defines an equivalence relation on the set of quasi rays to X.

Theorem 3.9.2. *Let $f, g : \mathbb{R}^+ \to X$ be geodesic rays. Then the following are equivalent:*
(i) *We have $H(f, g) < \infty$.*
(ii) *We have $\sup_{t\geq 0} |f(t) - g(t)| < \infty$.*
(iii) *There exists a $t_1 \geq 0$ such that for all $t \geq t_1$ there exists an $s_t \geq 0$ with $|f(t) - g(s_t)| \leq 8\delta$.*
(iv) *There exists a $u \in \mathbb{R}$ and a $t_0 \geq \max\{0, u\}$, such that $|f(t) - g(t-u)| \leq 16\delta$ for all $g \geq t_0$.*

Proof. We prove Theorem 3.9.2 in steps.
- We show that (i) implies (ii). Set $H = H(f, g)$. For every $t \geq 0$ there exists an $s_t \geq 0$ with $|f(t) - g(s_t)| \leq H$. Then

$$\begin{aligned} |s_t - s_0| &= |g(s_t) - g(s_0)| \\ &\leq |g(s_t) - f(t)| + |f(t) - f(0)| + |f(0) - g(s_0)| \\ &\leq H + |f(t) - f(0)| + H \\ &= 2H + t \end{aligned}$$

and

$$\begin{aligned} t &= |f(t) - f(0)| \\ &\leq |f(t) - g(s_t)| + |g(s_t) - g(s_0)| + |g(s_0) - f(0)| \\ &\leq H + |s_t - s_0| + H, \end{aligned}$$

hence $|s_t - s_0| \geq t - 2H$ and $|s_t - s_0| \leq t + 2H$.
Let $s_t - s_0 \geq 0$. Then $s_t - s_0 \leq 2H + t$ and $s_t - s_0 \geq t - 2H$, hence $s_t - t \leq 2H + s_0$ and $s_t - t \geq s_0 - 2H$. Thus $|s_t - t| \leq s_0 + 2H$.
Let $s_t - s_0 < 0$. Then $s_0 - s_t \leq 2H + t$ and $s_0 - s_t \geq t - 2H$, hence $s_0 + 2H \geq s_t + t \geq |s_t - t|$ by the triangle inequality. Hence we always have $|s_t - t| \leq s_0 + 2H$. It follows that

$$|f(t) - g(t)| \leq |f(t) - g(s_t)| + |g(s_t) - g(t)| \leq H + |t - s_t| \leq s_0 + 3H,$$

that is, $\sup_{t\geq 0} |f(t) - g(t)| \leq s_0 + 3H < \infty$.
- We show that (ii) implies (iii): Let $D := \sup_{t\geq 0} |f(t) - g(t)|$ and set $t_1 = D + 4\delta + 1$. Let $t \geq t_1$ and $T \geq t + D + 8\delta + 1$. Let $S \geq 0$ and $|f(T) - g(S)| \leq D$. We consider Figure 3.67 where Δ_1 and Δ_2 are geodesic triangles in X. As X is δ-hyperbolic, Δ_1 satisfies the Rips condition for 4δ. Hence

$$d(f(t), [f(0), g(0)] \cup [g(0), f(T)]) \leq 4\delta.$$

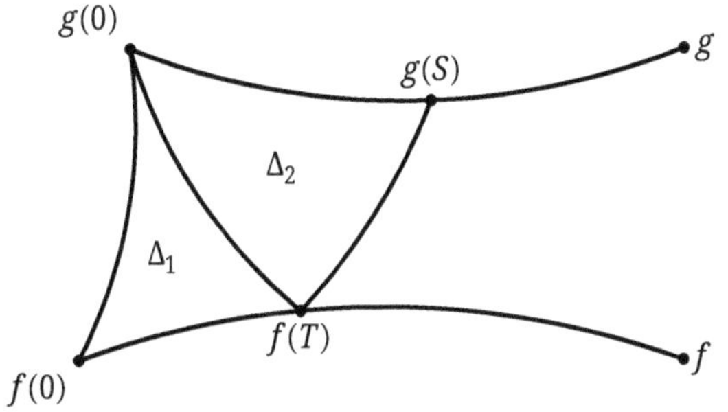

Figure 3.67: Geodesic triangles Δ_1 and Δ_2.

Remark 3.9.3. We have $d(f(t), [g(0), f(T)]) \leq 4\delta$.
To see this, assume that there exists a $y \in [f(0), g(0)]$ with $|f(t) - y| \leq 4\delta$. Then

$$t = |f(0) - f(t)| \overset{\text{tripod map}}{=} |f(0) - y| \leq |f(0) - g(0)| \leq D < t_1.$$

But this contradicts $t \geq t_1$, hence there exists an $x \in [g(0), f(T)]$ with $|f(t) - x| \leq 4\delta$.

Hence we have the situation depicted in Figure 3.68.

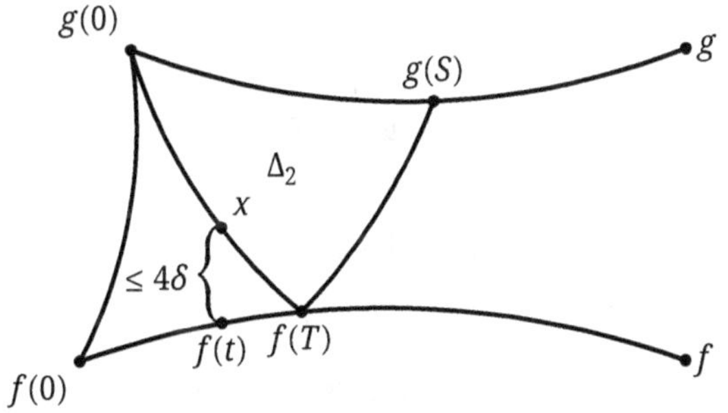

Figure 3.68: Rips condition for 4δ.

Also Δ_2 satisfies the Rips condition for 4δ. Assume there exists a $z \in [f(T), g(S)]$ with $|x - z| \leq 4\delta$. Then

$$\begin{aligned} T - t = |f(T) - f(t)| &\overset{\text{tripod map}}{=} |f(T) - x| \\ &\overset{\text{tripod map}}{=} |f(T) - z| \leq |f(T) - g(S)| \leq D. \end{aligned}$$

Hence $T \leq D + t$. But $T > D + t$, which yields a contradiction. Thus there exists an $s_t \in [0, S]$ with $|x - g(s_t)| \leq 4\delta$. Hence if $t \geq t_1$, then there exists an $s_t \geq 0$ with

$$|f(t) - g(s_t)| \leq |f(t) - x| + |x - g(s_t)| \leq 4\delta + 4\delta = 8\delta$$

(where x is constructed as above with respect to t). Hence (iii) holds.

- We show that (iii) implies (iv). We now know that there exists a $t_1 \geq 0$, such that for all $t \geq t_1$ there exists an $s_t \geq 0$ with $|f(t) - g(s_t)| \leq 8\delta$. In particular, there exists an $s_{t_1} \geq 0$ with $|f(t_1) - g(s_{t_1})| \leq 8\delta$. In the following we just write $s_1 = s_{t_1}$. Let $u := t_1 - s_1$. Hence

$$|f(t_1) - g(t_1 - u)| = |f(t_1) - g(s_1)| \leq 8\delta.$$

As $f\colon \mathbb{R}^+ \to X$ is a geodesic ray we have

$$\lim_{t\to\infty} |f(t) - g(t_1 - u)| = \infty.$$

Because of

$$|f(t) - g(s_t)| \leq 8\delta \quad \text{for all } t \geq t_1,$$

this implies that there exists a $T_1 \geq \max\{t_1, 1\}$ such that $t \geq T_1$ implies $s_t \geq s_1$. Set $t_0 := \max\{T_1, s_1\}$. In particular we have $t_0 \geq t_1$ and $T_1 \geq 1$. If $t \geq t_0$, then

$$\begin{aligned}
|f(t) - g(t-u)| &\leq |f(t) - g(s_t)| + |g(s_t) - g(t-u)| \\
&= |f(t) - g(s_t)| + |s_t - (t-u)| \\
&\leq 8\delta + |s_t - (t-u)| \\
&= |s_t - t - (t_1 - t_1 - u)| + 8\delta \\
&= |s_t - t + t_1 - (t_1 - u)| + 8\delta \\
&= |s_t - t + t_1 - s_1| + 8\delta \\
&= |(s_t - s_1) - (t - t_1)| + 8\delta \\
&= ||g(s_t) - g(s_1)| - |f(t) - f(t_1)|| + 8\delta.
\end{aligned}$$

Because of $|f(t) - g(s_t)| \leq 8\delta$ and $|f(t_1) - g(s_1)| \leq 8\delta$ it follows by the triangle inequality that

$$|f(t) - f(t_1)| - 16\delta \leq |g(s_t) - g(s_1)| \leq |f(t) - f(t_1)| + 16\delta.$$

Hence

$$||g(s_t) - g(s_1)| - |f(t) - f(t_1)|| \leq 16\delta.$$

Thus $t \geq t_0$ implies that

$$|f(t) - g(t-u)| \leq 16\delta + 8\delta = 24\delta.$$

We aim at the stricter estimation

$$|f(t) - g(t-u)| \leq 16\delta.$$

To this end, we need the notion of *convexity* (see Definition 3.2.24).

Definition 3.9.4. Let X be a geodesic metric space. Let $k \geq 0$ and $g\colon [0,k] \to X$ be a geodesic segment. Let $x_0 = g(0)$ and $x_1 = g(k)$. The map $t \mapsto X_t$, $t \in [0,1]$, with $|x_0 - x_t| = t|x_0 - x_1| = tk$ is called the *natural parametrization of* g. If $\delta \geq 0$, then X is called *δ-convex*, if for all naturally parametrized geodesic segments $[x_0, x_1]$ and $[y_0, y_1]$ the inequality

$$|x_t - y_t| \leq (1-t)|x_0 - y_0| + t|x_1 - y_1| + \delta$$

holds for all $t \in [0,1]$.

The following holds: If X is a geodesic metric space, in which all geodesic triangles are δ-thin, then X is 2δ-convex. We have shown this in Theorem 3.2.26.
We will use this for the stricter estimation. As X is δ-hyperbolic every geodesic triangle is 4δ-thin. Hence X is 8δ-convex. Suppose there are a $\varepsilon > 0$ and a $t \geq t_0$ with

$$|f(t) - g(t-u)| \geq 16\delta + \varepsilon.$$

Let $T \geq t$. If $T > t_1$ (we can assume this as $t_1 \leq t_0 \leq t \leq T$) then

$$\begin{aligned} |f(t) - g(t-u)| \leq{}& \frac{t - t_1}{T - t_1}|f(T) - g(T-u)| \\ &+ \left(1 - \frac{t - t_1}{T - t_1}\right)|f(t_1) - g(t_1 - u)| + 8\delta. \end{aligned}$$

This is a consequence of the convexity, applied to the geodesic segments $\bar{t} \mapsto f(\bar{t})$ and $\bar{t} \to g(\bar{t} - u)$ for $\bar{t} \in [t_1, T]$. Let $[x_0, x_1]$ be the natural parametrization of $f([t_1, T])$ and $[y_0, y_1]$ the natural parametrization of $g([t, T] - u)$. We have

$$|x_s - y_s| \leq |1 - s||x_0 - y_0| + s|x_1 - y_1| + 8\delta.$$

Of course we have $x_0 = f(t_1), y_0 = g(t_1 - u), x_1 = f(T)$ and $y_1 = g(T-u)$. Plugging this into the equation for the parametrization yields $f(t) = \frac{x_{t-t_1}}{T-t_1}$ and $g(t-u) = \frac{y_{t-t_1}}{T-t_1}$. We now show the desired stricter estimation. We have

$$|f(t_1) - g(t_1 - u)| \leq 8\delta,$$

hence

$$|f(t_1) - g(t_1 - u)| + \varepsilon \leq 8\delta + \varepsilon.$$

Because of $|f(t) - g(t-u)| \geq 16\delta + \varepsilon$ we have

$$|f(t) - g(t-u)| - 8\delta \geq 8\delta + \varepsilon$$

and hence

$$
\begin{aligned}
\big|f(t_1) - g(t_1 - u)\big| + \varepsilon \le 8\delta + \varepsilon &\le \big|f(t) - g(t-u)\big| - 8\delta \\
&\le \frac{t - t_1}{T - t_1}\big|f(T) - g(T-u)\big| + \left(1 - \frac{t-t_1}{T-t_1}\right)\big|f(t_1) - g(t_1 - u)\big|.
\end{aligned}
$$

Thus

$$\frac{t - t_1}{T - t_1}\big|f(t_1) - g(t_1 - u)\big| + \varepsilon \le \frac{t - t_1}{T - t_1}\big|f(T) - g(T - u)\big|$$

and

$$(t - t_1)\big|f(t_1) - g(t_1 - u)\big| + (T - t_1)\varepsilon \le (t - t_1)\big|f(T) - g(T - u)\big|.$$

As $(T - t_1)\varepsilon \xrightarrow{T\to\infty} \infty$ we have $\lim_{T\to\infty} |f(T) - g(T - u)| = \infty$. But this yields a contradiction to the weaker estimation that we have already obtained. Hence our assumption that there exist an $\varepsilon > 0$ and a $t \ge t_0$ with $|f(t) - g(t - u)| \ge 16\delta + \varepsilon$ is wrong. Hence we have $|f(t) - g(t - u)| \le 16\delta$ for all $t \ge t_0$. This shows (iv).
- We finally remark that it is obvious that (iv) implies (i). This concludes the proof. □

Corollary 3.9.5. *Let X be a δ-hyperbolic geodesic space. Then we have the following.*
1. *Let $f, g\colon \mathbb{R}^+ \to X$ be equivalent rays with $f(0) = g(0)$, then $\sup_{t\ge 0} |f(t) - g(t)| \le 8\delta$.*
2. *Let $f, g\colon \mathbb{R} \to X$ be geodesics with*

$$H(f, g) < \infty \quad \textit{and} \quad H(f(\mathbb{R}^+), g(\mathbb{R}^+)) < \infty.$$

Then there exists a $u \in \mathbb{R}$ with $\sup_{t\in\mathbb{R}} |f(t) - g(t - u)| \le 16\delta$.

Proof. 1. Suppose there exists an $\varepsilon > 0$ and a $t > 0$ with $|f(t) - g(t)| \ge 8\delta + \varepsilon$. For $T \ge t$ we have (see “(iii) implies (iv)” in the proof of Theorem 3.9.2)

$$
\begin{aligned}
\big|f(0) - g(0)\big| + \varepsilon &= \varepsilon \\
&\le \big|f(t) - g(t)\big| - 8\delta \\
&\le \frac{t}{T}\big|f(T) - g(T)\big| + \left(1 - \frac{t}{T}\right)\big|f(0) - g(0)\big| \\
&= \frac{t}{T}\big|f(T) - g(T)\big|.
\end{aligned}
$$

Hence $T\varepsilon \le t|f(T) - g(T)|$ and thus $\lim_{T\to\infty} |f(T) - g(T)| = \infty$ which yields a contradiction to $H(f, g) < \infty$ as $H(f, g) < \infty$ also implies (iv). Hence we have $|f(t) - g(t)| \le 8\delta$ for all $t \in \mathbb{R}^+$.

2. Because of $H(f(\mathbb{R}^+), g(\mathbb{R}^+)) < \infty$ there exist t and s_t in $\mathbb{R}^+$ with $|f(t) - g(s_t)| \le 8\delta$ by Theorem 3.9.2(iii). If we compose f and g with suitable translations of $\mathbb{R}$, we can assume that $|f(0)-g(0)| \le 8\delta$ and $H(f(\mathbb{R}^+), g(\mathbb{R}^+)) < \infty$. The assumptions $H(f,g) < \infty$ and $H(f(\mathbb{R}^+), g(\mathbb{R}^+)) < \infty$ imply that $H(f(\mathbb{R}^-), g(\mathbb{R}^-)) < \infty$. Now suppose there exist $\varepsilon > 0$ and $t > 0$ with $|f(t) - g(t)| \ge 16\delta + \varepsilon$. If $T \ge t$, then we have

$$\begin{aligned}|f(0) - g(0)| + \varepsilon &\le 8\delta + \varepsilon \\ &\le |f(t) - g(t)| - 8\delta \\ &\le \frac{t}{T}|f(T) - g(t)| + \left(1 - \frac{t}{T}\right)|f(0) - g(0)|.\end{aligned}$$

Hence

$$\frac{t}{T}|f(0) - g(0)| + \varepsilon \le \frac{t}{T}|f(T) - g(T)|.$$

Again we have

$$\lim_{T\to\infty} |f(T) - g(T)| = \infty,$$

which contradicts $H(f(\mathbb{R}^+), g(\mathbb{R}^+)) < \infty$. Hence $|f(t) - g(t)| \le 16\delta$ for $t \ge 0$. Changing orientations of f and g and the fact that $H(f(\mathbb{R}^-), g(\mathbb{R}^-)) < \infty$ yield

$$|f(t) - g(t)| \le 16\delta \quad \text{for } t \le 0.$$

Hence $|f(t)-g(t)| \le 16\delta$ for $t \in \mathbb{R}$; after composing with suitable translations. Hence, without such slotting translations in ahead, there exists a $u \in \mathbb{R}$ with $|f(t)-g(t-u)| \le 16\delta$ for all $t \in \mathbb{R}$. □

Now we consider three models for the boundary ∂X of a hyperbolic space X.

First model for ∂X

In this model, often denoted as $\partial_q X$, the boundary of X consists of equivalence classes of quasi rays in X. In this case, if $f\colon \mathbb{R}^+ \to X$ is a quasi ray and a is its equivalence class, we say that f *moves to* a, denoted as $f \to a$ or $f(t) \to a$.

Second model for ∂X

We denote this model by $\partial_{r,w} X$, $w \in X$, or just $\partial_r X$. It consists of the equivalence classes of rays $f\colon \mathbb{R}^+ \to X$ with $f(0) = w$ where w is a basis point. As rays are special cases of quasi rays, the notions of the first model are still applicable. If $f \to a$, we write $[w, a)$ for the image under f. Note that there can be multiple rays starting at w and moving to a, compare Figure 3.69.

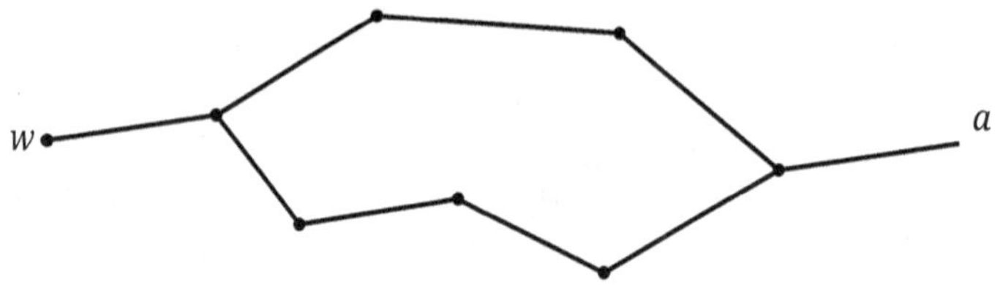

Figure 3.69: Equivalence classes of rays.

Third model for ∂X

Before we give the actual model some preliminary remarks are due.

Definition 3.9.6. The sequence $\{x_n\}_{n\in\mathbb{N}}$ in X moves to ∞, if $\lim_{i,j\to\infty}(x_i \mid x_j)_w = \infty$. If $\{x_n\}_{n\in\mathbb{N}}$ and $\{y_n\}_{n\in\mathbb{N}}$ are sequences in X that move to ∞ we say that they are *equivalent* if $\lim_{i,j\to\infty}(x_i \mid y_j)_w = \infty$, written as $\{x_n\}_{n\in\mathbb{N}} \sim \{y_n\}_{n\in\mathbb{N}}$.

We make two observations.

Remark 3.9.7. The definitions are independent from the choice of the basis point w.

To see this recall Lemma 3.2.16: If $\Delta = [x,y] \cup [y,z] \cup [z,x]$ is a geodesic triangle in the metric space X then we have

1. $(y \mid z)_x \le d(x,[y,z])$; and
2. if Δ is δ-fine, then $d(x,[y,z]) \le (y \mid z)_x + \delta$.

Here X is a geodesic δ-hyperbolic space, in particular X is 4δ-fine. Hence

$$(y \mid z)_x \le d(x,[y,z]) \le (y \mid z)_x + 4\delta \quad \text{for all } x,y,z \in X.$$

Now let $w, w' \in X$ and $\{x_i\}_{i\in\mathbb{N}}$ be a sequence in X with $\lim_{i,j\to\infty}(x_i \mid x_j)_w = \infty$. We have to show that $\lim_{i,j\to\infty}(x_i \mid x_j)_{w'} = \infty$. Let M be a positive real number. We need to find an $N \in \mathbb{N}$ with $(x_i \mid x_j)_{w'} > M$ if $i,j > N$. Let $l = |w - w'|$. Because of $\lim_{i,j\to\infty}(x_i \mid x_j)_w = \infty$ there exists an $N \in \mathbb{N}$ with $(x_i \mid x_j)_w > M + l + 4\delta$ for $i,j > N$. Hence we also have $d(w,[x_i,x_j]) > M + l + 4\delta$ for $i,j > N$. Because of $|w - w'| = l$ and $d(w',[x_i,x_j]) \ge d(w,[x_i,x_j])$ we have $d(w',[x_i,x_j]) > M + 4\delta$ for $i,j > N$. But then $(x_i \mid x_j)_{w'} + 4\delta > M + 4\delta$, and hence $(x_i \mid x_j)_{w'} > M$ for $i,j > N$. Hence $\lim_{i,j\to\infty}(x_i \mid x_j)_{w'} = \infty$. The proof that the equivalence is also independent from the basis point is similar.

Remark 3.9.8. The equivalence is a proper equivalence relation on the set of sequences that move to ∞.

To see this, we first note that the relation is obviously symmetric and reflexive. The transitivity is a consequence of Definition 3.2.4. If X is a δ-hyperbolic space, then

$$(x \mid z)_w \ge \min\{(x \mid y)_w, (y \mid z)_w\} - \delta.$$

Hence, if X is δ-hyperbolic, $\{x_i\} \sim \{y_i\}$ and $\{y_i\} \sim \{z_k\}$, then

$$(x_i \mid z_k)_w \geq \min\{(x_i \mid y_j)_w, (y_j \mid z_k)_w\} - \delta,$$

hence

$$\lim_{i,j\to\infty} (x_i \mid y_j)_w = \infty \quad \text{and} \quad \lim_{j,k\to\infty} (y_j, z_k) = \infty$$

imply that $\lim_{i,k\to\infty}(x_i \mid z_k)_w = \infty$, and hence $\{x_i\} \sim \{z_k\}$.

We now give the third model for ∂X: We denote this model by $\partial_s X$. It consists of the equivalence classes of sequences that move to ∞. If the sequence $\{x_i\}_{i\in \mathrm{N}}$ represents $a \in \partial_s X$, we write $x_i \to a$, that is, x_i moves to a.

Theorem 3.9.9. *There are natural bijections between the three models for ∂X.*

Proof. For a full proof see [114]. We only describe the bijections $\varphi\colon \partial_r X \to \partial_q X$ and $\psi\colon \partial_r X \to \partial_s X$.

1. As every ray $f\colon \mathbb{R}^+ \to X$ is a quasi ray, we can consider φ as an inclusion, that is, φ maps the equivalence class of f into itself. As in both $\partial_r X$ and $\partial_q X$ the definition of equivalence uses the bounded Hausdorff distance, the map φ is injective. The surjectivity follows from the construction in Theorem 3.7.19. For every quasi ray $h\colon \mathbb{R}^+ \to X$ there exists a ray $g\colon \mathbb{R}^+ \to X$ with $g(0) = h(0)$ and $H(h,g) < \infty$.
2. Let $f\colon \mathbb{R}^+ \to X$ be a ray with $f(0) = w$. Set $x_i = f(i)$, $i \in \mathbb{N}$. As f is a ray we have

$$(x_i \mid x_j)_w = \min\{|x_i - w|, |x_j - w|\} = \min\{i,j\};$$

thus

$$\lim_{i,j\to\infty} (x_i \mid x_j)_w = \infty.$$

Hence $\{x_i\}$ moves to ∞. Let $g\colon \mathbb{R}^+ \to X$ be a ray with $g(0) = w$ and $[g] = [f]$ for the equivalence classes. Analogously, set $y_i = g(i)$, $i \in \mathbb{N}$. We would like to show that $\{x_i\} \sim \{y_i\}$. By Corollary 3.9.5 we have $|x_i - y_i| \leq 8\delta$ for all i. Furthermore we have according to Definition 3.2.4

$$(x_i \mid y_j)_w \geq \min\{(x_i \mid y_i)_w, (y_i \mid y_j)_w\} - \delta,$$

and by the definition of the Gromov product (see Definition 3.2.1)

$$(x_i \mid y_i)_w = \frac{1}{2}(|x_i - w| + |y_i - w| - |x_i - y_i|).$$

Because of $|x_i - y_i| \leq 8\delta$ we first have

$$(x_i \mid y_i)_w \geq \frac{1}{2}(|x_i - w| + |y_i - w|) - 4\delta$$

and

$$(x_i \mid y_i) \geq \min\left\{\frac{1}{2}(|x_i - w| + |y_i - w|) - 5\delta, (y_i \mid y_j)_w - \delta\right\}.$$

Because of $\lim_{i\to\infty} |x_i - w| = \infty = \lim_{i\to\infty} |y_i - w|$ and $\lim_{i,j\to\infty}(y_i \mid y_j)_w = \infty$ we have $\lim_{i,j\to\infty}(x_i \mid y_i)_w = \infty$, hence $\{x_i\} \sim \{y_i\}$. Hence the equivalence class map $\psi\colon \partial_r X \to \partial_s X$, $[f] \mapsto [\{x_i\}_{i\in\mathbb{N}}]$ is well defined. The key principle for the proof of the bijectivity of ψ is the Approximation Theorem 3.2.13. □

We now use the notion ∂X for each of the three models for the boundary of X. A consequence of the proof of Theorem 3.9.9 is the following.

Corollary 3.9.10. *Let a and b be different elements of ∂X. Then there is a geodesic $g\colon \mathbb{R} \to X$ with $\lim_{t\to\infty} g(t) = b$ and $\lim_{t\to-\infty} g(t) = a$, that is, $g|_{\mathbb{R}^+} \to b$ and $g|_{\mathbb{R}^-} \to a$.*

Definition 3.9.11. Let $a, b \in \partial X$. We define the Gromov product $(a \mid b) := (a \mid b)_w$, w a basis point, by

$$(a \mid b) := \sup\Big(\lim_{i,j\to\infty} (x_i \mid y_i)\Big),$$

where we evaluate the supremum over all sequences with $x_i \to a$ and $y_j \to b$.

Example 3.9.12. Let X be a metric tree as in Subsection 3.8.1. Let a and b be different elements of ∂X and $(a \mid b)$ the Gromov product as in Subsection 3.8.1. We have $(a \mid b) = (a \mid b)_T$.

Proof. We have the situation depicted in Figure 3.70.

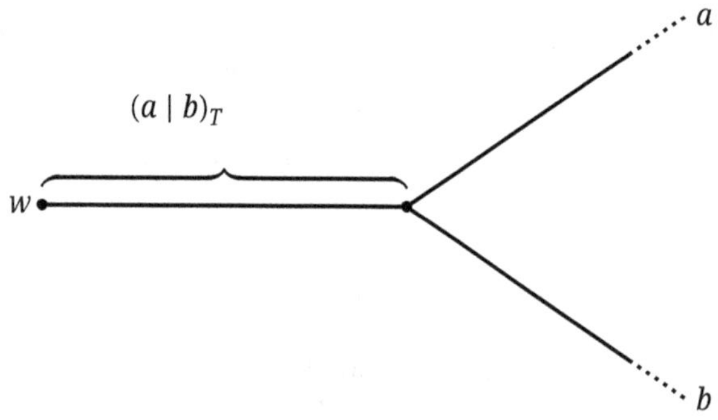

Figure 3.70: Gromov product $(a \mid b)_T$.

Let $x_i \to a$ and $y_i \to b$. Let $\alpha\colon \mathbb{R}^+ \to X$ and $\beta\colon \mathbb{R}^+ \to X$ be rays with $\alpha(0) = w = \beta(0)$ and $\alpha(t) \to a$, $\beta(t) \to b$. Then $\{x_i\} \sim \{\alpha(i)\}$ and $\{y_j\} \sim \{\beta(j)\}$ which follows from the proof that ψ is surjective in the proof of Theorem 3.9.9. Let $M > (a \mid b)_T$. Then there exists an N with $(x_i \mid \alpha(j)) \geq M$ for $i, j \geq N$ and $(y_j \mid \beta(k)) \geq M$ for $j, k \geq N$; compare Figure 3.71.

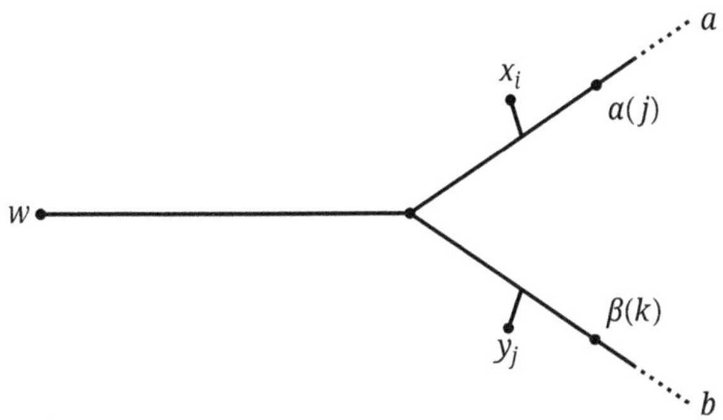

Figure 3.71: Existence of a large *N*.

Hence $(x_i \mid y_j) = (a \mid b)_T$ for all $i, j \geq N$, that is, $\liminf_{i,j\to\infty}(x_i \mid y_j) = (a \mid b)_T$. As this holds for all $x_i \to a$ and $y_j \to b$, we have

$$(a \mid b) = \sup\Big(\liminf_{i,j\to\infty}(x_i \mid y_j)\Big) = (a \mid b)_T. \qquad \square$$

Remark 3.9.13. Let $a, b \in \partial X$ with $a \neq b$, $x_i \to a$, $y_j \to b$. Then we have

$$(a \mid b) - 2\delta \leq \liminf_{i,j\to\infty}(x_i \mid y_j) \leq (a \mid b).$$

This follows from the definition (with respect to X) and the similar considerations in Subsection 3.8.1. We will now construct a topology on ∂X via $(a \mid b)$. For every positive $r \in \mathbb{Q}$ we set

$$V_r = \{(a, b) \in \partial X \times \partial X \mid (a \mid b) \geq r\}.$$

Without loss of generality let $\delta \in \mathbb{Q}$, $\delta \geq 0$, and $s = r + 3\delta$. With the above inequality we can show that $(a, b), (b, c) \in V_s$ implies $(a, c) \in V_r$ as

$$\begin{aligned}(x_i \mid z_k) &\geq \min\{(x_i \mid y_j), (y_j \mid z_k)\} - \delta \\ &\geq \min\{(a \mid b) - 2\delta - \varepsilon, (b \mid c) - 2\delta - \varepsilon\} - \delta,\end{aligned}$$

where we have $\varepsilon > 0$; i, j, k sufficiently large; $x_i \to a$, $y_j \to b$, and $z_k \to c$. Set $V_r(a) = \{b \in \partial X \mid (a, b) \in V_r\}$ for $r \in \mathbb{Q}, r > 0$ and $a \in \partial X$. Then $V_r(a)$ is called a *neighborhood* of a. The $V_r(a)$ where $r \in \mathbb{Q}, r > 0$ and $a \in \partial X$, constitute a basis of a topology on ∂X. This topology is metrizable. This follows from general topology: $\mathfrak{B} = \{V_r \mid r \in \mathbb{Q}, r > 0\}$ is a countable basis for a uniform structure on ∂X with $\Delta = \bigcap_{B\in\mathcal{B}} B$, where $\Delta = \{(a, a) \mid a \in \partial X\}$. By this, the topology on ∂X is metrizable. Together with this metrizable topology ∂X becomes a compact metric space.

For further reference we note this as a theorem.

Theorem 3.9.14. *The boundary ∂X is a compact metric space.*

The explicit description of a metric on ∂X is much more difficult than for metric trees as $\rho_\varepsilon(a,b) = e^{-\varepsilon(a|b)}$ for all $a, b \in \partial X$ in general does not define a metric on ∂X. This is because of

$$(a \mid c) \geq \min\{(a \mid b), (b \mid c)\} - 2\delta \quad \text{for all } a, b, c \in \partial X,$$

that is, $\rho_\varepsilon(a,c) \leq e^{2e\delta} \max\{\rho_\varepsilon(a,b), \rho_\varepsilon(b,c)\}$ for all $a, b, c \in \partial X$, and the coefficient $e^{2e\delta}$ can become large, at least we have $e^{2e\delta} \geq 1$.

But we can proceed in the following way: A *chain connecting a and b (in ∂X)* is a finite sequence $a = a_0, a_1, \ldots, a_n = b$ in ∂X. Let C_{ab} denote the set of all chains connecting a and b. Set

$$\rho_\varepsilon(a_0, a_1, \ldots, a_n) := \sum_{i=1}^{n} \rho_\varepsilon(a_{i-1}, a_i) \quad \text{for } a_0, a_1, \ldots, a_n \in \partial X$$

as well as $d_\varepsilon(a,b) = \inf\{\rho_\varepsilon(\tau) \mid \tau \in C_{ab}\}$. If eventually $e^{2\varepsilon\delta} < \sqrt{2}$, then d_ε defines a metric on ∂X. We can now, analogously to the case of metric trees, consider quasi isometries $\varphi: X \to Y$ (where X, Y are δ-hyperbolic, proper geodesic, metric spaces). This φ again induces a Hölder and quasi conform map $\varphi: \partial X \to \partial Y$, that under certain assumptions is a (not necessarily surjective) homeomorphism. These assumptions are satisfied if X and Y are quasi isometric: For let $\psi: Y \to X$ be a quasi isometry then

$$\sup_{x \in X} |\psi \circ \phi(x) - x| < \infty \quad \text{and} \quad \sup_{x \in Y} |\varphi \circ \psi(y) - y| < \infty.$$

With this we have in particular: Let Γ be a δ-hyperbolic group with S and S' as valid generating systems (symmetric, finite, not containing 1). Let ϕ and ϕ' be the Cayley graphs of S and S', respectively. Then there is a surjective homeomorphism between $\partial\phi$ and $\partial\phi'$. Hence if Γ is a δ-hyperbolic group, then we can speak of the boundary $\partial\Gamma$ of Γ. If Γ' is a subgroup of Γ of finite index, then there exists a surjective homeomorphism between $\partial\Gamma$ and $\partial\Gamma'$.

If $|\Gamma| < \infty$, then of course $\partial\Gamma = \emptyset$. If $|\Gamma : \mathbb{Z}| < \infty$, then $\partial\mathbb{Z}$ only consists of the two points $\{\pm\infty\}$. In general we obtain the following theorem.

Theorem 3.9.15. *Let Γ be a δ-hyperbolic group. Then*
1. $\partial\Gamma = \emptyset$ *if and only if* $|\Gamma| < \infty$;
2. $|\partial\Gamma| = 2$ *if and only if $\mathbb{Z}$ has finite index as a subgroup of Γ; and*
3. $|\partial\Gamma| = \infty$ *if and only if* $|\Gamma| = \infty$ *and Γ does not contain $\mathbb{Z}$ as a subgroup of finite index.*

We leave the details of the proof as an exercise.

Example 3.9.16. Let $\Gamma < \mathrm{PSL}(2,\mathbb{R})$ be a subgroup of the isometry group of the upper half plane $\mathfrak{H} = \{z = x + iy \mid y > 0\}$. Let the action of Γ on $\mathfrak{H}$ be properly discontinuous. Then we have that

1. $\partial\Gamma = \emptyset$ if and only if Γ is finitely cyclic;
2. $|\partial\Gamma| = 2$ if and only if either Γ is infinitely cyclic or $\Gamma = \langle a, b \mid a^2 = b^2 = 1\rangle$, the infinite dihedral group, and $\langle ab\rangle$ is a subgroup of Γ of index 2; and
3. $|\partial\Gamma| = \infty$ if and only if $|\Gamma| = \infty$ and Γ does not contain $\mathbb{Z}$ as a subgroup of finite index. In particular Γ contains two elements of infinite order that do not have a common fixed point.

Interlude on independent elements

If Γ is a hyperbolic group and $g \in \Gamma$ an element of infinite order, then the map $\mathbb{Z} \to \Gamma$, $n \mapsto g^n$, is certainly a quasi isometric embedding.

Now, let Γ be a hyperbolic group, S a valid generating system and $g, h \in \Gamma$ be elements of infinite order. Let $g^n \to a^+ \in \partial\Gamma$, $g^{-n} \to a^- \in \partial\Gamma$, $h^n \to b^+ \in \partial\Gamma$, $h^{-n} \to b^- \in \partial\Gamma$ for $n \in \mathbb{N}$, $n \to \infty$. Recall that $a^+ \neq a^-$ and $b^+ \neq b^-$. From Theorem 3.9.15 we see that $\langle g, h\rangle$ is virtually $\mathbb{Z}$ if $\{a^+, a^-\} \cap \{b^+, b^-\} \neq \emptyset$.

Definition 3.9.17. We call the elements g and h *independent* if

$$\{a^+, a^-\} \cap \{b^+, b^-\} = \emptyset.$$

Theorem 3.9.18. *Let Γ be a δ-hyperbolic group and $g, h \in \Gamma$ be elements of infinite order. If g and h are independent, then there exists an $n \in \mathbb{N}$ such that $\langle g^n, h^n\rangle$ is free of rank 2.*

Proof. As g and h are independent there exists an $r \in \mathbb{R}^+ = \{s \in \mathbb{R} \mid s > 0\}$ such that every geodesic joining point in $\langle g\rangle$ and $\langle h\rangle$ has to pass through the ball $B_r(1)$ with center 1 and radius r. Let $R \in \mathbb{N}$ with $d_S(1, g^n) > 2(r + \delta)$ for all $n \in \mathbb{N}$ with $n > R$.

We define the non-empty sets

$$A = \{x \in \Gamma \mid d_S(x, \langle g\rangle) < d_S(x, \{g^{-R}, \dots, g^R\})\}$$

and

$$B = \{x \in \Gamma \mid d_S(x, \langle h\rangle) < d_S(x, \{h^{-R}, \dots, h^R\})\}.$$

The sets A and B are disjoint. Assume for a contradiction that there exists an $x \in A \cap B$. Let $g_x \in \langle g\rangle$ and $h_x \in \langle h\rangle$ be the closest points to x in $\langle g\rangle$ and $\langle h\rangle$, respectively. Let γ be a geodesic joining g_x and h_x. In particular, γ passes through a point $z \in B_r(1)$. Without loss of generality we may assume that there is a point z' from x to g_x with $d_S(z, z') \le \delta$.

Using the definition of A and g_x we get

$$\begin{aligned} 2(r + \delta) &< d_S(1, g_x) \\ &\le d_S(1, z') + d_S(z', g_x) \\ &\le d_S(1, z') + d_S(z', 1) \end{aligned}$$

$$\leq 2(d_S(1,z) + d_S(z,z'))$$
$$\leq 2(r + \delta),$$

which is a contradiction. Hence $A \cap B = \emptyset$; compare Figure 3.72.

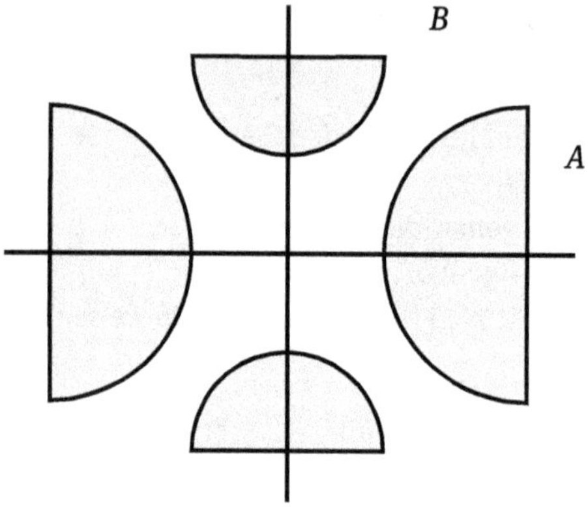

Figure 3.72: Sets A and B.

We apply a ping-pong argument analogously to Sections 1.1 and 1.7. Let $x \in A$. Then $A \cap B = \emptyset$ implies that $x \notin B$. All points in $\langle h \rangle$ that are closest to x lie in $\{h^{-R}, \ldots, h^R\}$. Since d_S is left-invariant, the points in $\langle h \rangle$ closest to $h^{3Rn}x$, $n \in \mathbb{Z}$, $n \neq 0$, will lie in $\{h^{3Rn-R}, \ldots, h^{3Rn+R}\}$ which is disjoint from $\{h^{-R}, \ldots, h^R\}$. Therefore $h^{3Rn}A \subset B$ for all $n \in \mathbb{Z}, n \neq 0$. Analogously $g^{3Rn}B \subset A$ for all $n \in \mathbb{Z}, n \neq 0$. Hence, we may apply the ping-pong argument and get that $\langle g^{3R}, h^{3R} \rangle$ is free of rank 2. □

Corollary 3.9.19. *Let Γ be a torsion-free hyperbolic group. Let $g, h \in \Gamma$. Then either $\langle g, h \rangle$ is cyclic or there exists a large enough $n \in \mathbb{N}$ such that $\langle g^n, h^n \rangle = \langle g^n \rangle * \langle h^n \rangle$ is free of rank 2. Especially, Γ is an RG group.*

3.10 Automatic and parafree groups

In this last section we will give an introduction to the theory of automatic groups. However, our exposition will remain sketchy and we refer to the literature for some of the proofs. Automatic groups arose from the interplay of combinatorial group theory, geometric group theory, geometric tessellations and computer science. More precisely from the implementation of group theoretic algorithms such as Nielsen's cancellation method and the Dehn algorithm.

We use Nielsen's cancellation method in free groups as a running example. Every element of a free group can be represented in a uniquely determined normal form. The free cancellation method is an effective rewrite rule that puts words into their normal form. This principle is well suited for applications in computer science where computer programs can be seen as a manipulation of lists of words in a given alphabet.

A full description of the theory of automatic groups can be found in the book by Cannon, Epstein, Holt, Petersen and Thurston [50] that also includes the definitions, techniques and results for automatic groups we mention here without reference. Within the scope of our group theoretic book we will explain all necessary group theoretic notions. In contrast to that, we will no further explain those notions from computer science like regular language and automata that are involved in the definition of automatic groups. We refer the interested reader to [50].

From a historic point of view, we also have to mention the article [49] by Cannon that can be seen as a first step of the theory.

Now, let X^+ and X^- be two finite, disjoint and non-empty sets of letters with the same cardinality and let $\varphi: X^+ \to X^-$ be a bijection that maps each $x^+ \in X^+$ to exactly one $x^- \in X^-$. Let Γ be the free monoid on $X = X^+ \cup X^-$ (compare [3]).

Let G be a group with generating system X such that $x := x^+$ is inverse to $x^{-1} := x^-$ in G. We assume that X is valid, that is, $1 \notin X$. By construction of X, we see that $x \in X$ implies $x^{-1} \in X$. We consider a map $\mu: \Gamma \to G$ that maps each word $w \in \Gamma$ to the element $\mu(w)$ in G that is represented by w.

A set of *normal forms* for G is a subset T of Γ such that $\mu|_T$ is a bijection. A *hereditary set of normal forms* for G is a set T of normal forms with the following property: If $t \in T$ then also every subword of t is in T. Equivalently, a hereditary set of normal forms is a two-sided Schreier system of representatives for a normal subgroup of a free group $F(X^+)$ with basis X^+. Of course, there exists a hereditary set T of normal forms for G (see Theorem 1.3.18). Hence it is a desirable goal to effectively put words into their normal form.

More formally, a *rewrite rule* is an ordered pair (u, v) of Γ with $\mu(u) = \mu(v)$. An application of the rewrite rule (u, v) is then to replace a subword u in a word w by v in order to obtain a new word w'. Here, the new word w' is not necessarily shorter than w (with respect to the free length).

Let Λ be a set of rewrite rules. Then we call Λ a *rewrite system*. A word w is called Λ-reduced if it is not possible to apply a rewrite rule of Λ to w, that is, no subword of w is the first component of a rewrite rule in Λ. We always assume a set of rewrite rules to be reflexively and transitively closed. For details about arbitrary rewrite systems, see for example [163].

The set Λ is called a *complete rewrite system* if the following holds:
(1) The set of Λ-reduced words is a hereditary set of normal forms.
(2) There are no infinite loops, that is, infinite chains of rewrite rules in Λ, that is, Λ is Noetherian.

Free reduction is a complete rewrite system for a finitely generated free group. If Λ is a complete rewrite system for a finitely generated group G, then the following holds:

(a) Successive application of a rewrite rule of Λ yields exactly one normal form for a word w in Γ.
(b) If $\mu(w_1) = \mu(w_2)$ then applying the rewrite rules of Λ for both w_1 and w_2 yields the same normal form, that is, Λ is *confluent*.
(c) The set of equations $u = v$ with $(u, v) \in \Lambda$ yields a complete system of relations for G.

Hence the following theorem holds immediately.

Theorem 3.10.1. *Let G be a finitely generated group with complete rewrite system Λ. Then G is finitely presented and the set of Λ-reduced words yields a regular language, that is, a language accepted by automata. Furthermore, the word problem is solvable in G.*

From the point of view of computer science the class of finitely generated groups with a complete rewrite system is interesting. Finite groups and finitely free groups are in this class.

As already mentioned above, rewrite rules do not necessarily have to reduce the word length. However, if they do the following theorem is immediate.

Theorem 3.10.2. *Let G be a finitely generated group with a finite, complete rewrite system that reduces (free) lengths. Then this system yields a Dehn algorithm, and thus, G is hyperbolic.*

Remark 3.10.3. The Dehn algorithm solves the word problem for a hyperbolic group. On the other hand, it yields a specific rewrite system for a hyperbolic group. This rewrite system is in general not complete. In the above theorem we assumed that the rewrite system was complete. The hyperbolic groups that we indeed get from Theorem 3.10.2 are those that contain a free subgroup of finite index (compare [184]).

We now aim at the definition of an automatic group. To this end, let G be again a finitely generated group with a finite and valid generating system X. Let Γ and μ be as above. A *synchronous automatic structure* for G relative to X is a regular language $L \subset \Gamma$ with $\mu(L) = G$ together with a finite synchronous two-tape automaton M that accepts the set of pairs (u, v) of elements in L with $\mu(u) = \mu(vx)$ for an $x \in X \cup \{1\}$. Note that X is valid. The pair (L, M) is then called an *automatic structure* for G.

If in addition there exists a finite automaton M' that accepts the set of pair in $\{(u, v) \mid u, v \in L, \mu(u) = \mu(xv) \text{ for an } x \in X \cup \{1\}\}$ then the triple (L, M, M') is called a *biautomatic structure* for G.

If M is asynchronous instead, then (L, M) is called an *asynchronous automatic structure* for G.

Definition 3.10.4. A finitely generated group G
1. Is called *automatic* if it admits an automatic structure;
2. Is called *biautomatic* if it admits a biautomatic structure; and
3. Is called *asynchronous automatic* if it admits an asynchronous automatic structure.

Observe that the property of G being automatic does not depend on choice of the generating set X. Of course, biautomatic groups are automatic. However, it is not known whether automatic groups are also biautomatic. The class of asynchronous automatic groups is larger than the class of automatic groups.

Theorem 3.10.5. 1. *Automatic groups are finitely presented.*
2. *If G is automatic then G has a solvable word problem.*
3. *If G is biautomatic then G has a solvable conjugacy problem.*

See [50] for a proof. More strongly, we can state that the word problem for automatic groups (and hence also for biautomatic groups) can be solved in quadratic time. This is due to the fact that words w, w' can be put into canonical form in quadratic time and then it only remains to check if their canonical forms represent the same element; also compare [50].

For automatic groups we have $A(w) \le f(|w|)$ for the combinatorial area where f is a quadratic polynomial. Hence we have the following theorem.

Theorem 3.10.6. *An automatic group admits a quadratic isoperimetric inequality.*

Before we consider examples of automatic groups we will give an equivalent characterization that adds a geometric perspective and is known as the *fellow traveler property*; compare [45].

Remark 3.10.7. Let G and X be as above and consider the Cayley graph Y of (G, X). For $x, y \in X$ let $d(x, y)$ denote the distance between x, y in Y. Then the group G is automatic with respect to an automatic structure (L, M) if and only if there exists a constant $C \in \mathbb{N}$ such that, for all words $u, v \in M$ which differ by at most one generator, the distance between the respective prefixes of u and v is bounded by C.

We would like to point out that the interplay between automatic groups, hyperbolic groups and the constant C of the fellow traveler property is an important issue in research. We end our detour of automatic groups with a list of examples.

Examples 3.10.8. 1. Finite groups are biautomatic (to see this, take the regular language to be the set of all words in the finite group).
2. Finitely generated free groups are biautomatic.
3. Finitely generated Abelian groups, in particular $\mathbb{Z}^k$, $k \in \mathbb{N}$, are automatic.
4. Hyperbolic groups are biautomatic: If G is a hyperbolic group with finite valid generating system X and Δ a Dehn algorithm for (G, X) then the Δ-reduced words yield a regular language. Moreover, the set of geodesics in the Cayley graph $\psi(G, X)$ with start point 1 (considered as group element) yields a regular language.
5. The group $G = A *_C B$ is automatic, if
   (a) A and B are finitely generated Abelian groups; or
   (b) A and B are hyperbolic and C is cyclic.
6. Groups of F-type are automatic by Corollary 3.5.28.

Finally we mention that the Baumslag–Solitar groups $BS(m, n)$ constitute an important family of non-automatic groups.

In Subsection 1.6.3 we considered the isomorphism problem for the groups $G_{i,j}$. These groups are parafree. In general, a group G is called *parafree* if its quotients by the terms of its lower central series are the same as those of free groups. The two-generator subgroups of parafree groups are free (see [16]). If G is a parafree group then every finitely generated normal subgroup has finite index (see [36]).

There is an interesting survey article by G. Baumslag on parafree groups [17] where especially the following aspects are discussed. Parafree groups arise naturally in homotopy theory (see [33]). Questions of automaticity can be linked to the study of parafree groups (see [19]). Various attempts to prove that all finitely generated parafree groups have finitely generated second homology groups and more generally that they are finitely related have been unsuccessful (see [120]). The diversity of presentations that give rise to such groups mark them as a fascinating generalization of free groups. As an example of this diversity consider the two examples of one-relator parafree groups $G = \langle a, b, c \mid a^2b^3c^5 = 1\rangle$ and $H = \langle a, b, c \mid a = [a, b][c, b]\rangle$.

They are not isomorphic because H has exactly one Nielsen equivalence class of minimal generating systems and G has exactly two, and neither of them is free. Both groups are automatic, in fact they are hyperbolic (see Subsection 3.5.4). But this remark underlines in principle the difficulty of the isomorphism problem for parafree groups, especially for one-relator parafree groups in the class of one-relator groups. One of the problems here is to reconcile conditions imposed on the lower central series of a group with the geodesics in its Cayley graph (with respect to a finite generating system).

We finish with some questions on parafree groups which mirror some aspects of our book.

Questions. 1. Are parafree groups hyperbolic or at least automatic?
2. Are parafree groups linear?
3. Is the isomorphism problem for one-relator parafree groups solvable in the class of one-relator groups?
4. Are parafree groups residually finite?
5. Suppose that G is a finitely generated, residually finite group with the same finite images as a free group. Is G parafree?
6. Let G be a (finitely generated) parafree group. Does G only have finitely many Nielsen equivalence classes of minimal generating system?

3.11 Stallings foldings and subgroups of free groups

As we have seen, the Cayley graph provides a very useful geometric way of studying finitely generated groups. For free groups F, the group can be identified with the fundamental group of this graph and hence from covering space theory each subgroup corresponds to a covering map of another graph to the original graph. Using this viewpoint Stallings [244] introduced the concept of foldings of graphs now called *Stallings foldings*. These have been used in many different applications, and especially in the study of subgroups of free groups. In this section we introduce the fundamental idea. Kapovich and Myasnikov [143] recast Stallings foldings in a non-combinatorial framework, and applied them systematically to the study of subgroups of free groups. In this section we introduce the fundamental idea.

Let F be a free group with a free basis X. Then $S = X \cup X^{-1}$ is a valid generating system for F. The basic idea in the use of Stallings foldings in the study of the subgroup structure of free groups is to associate to each subgroup H of F a directed graph $\Gamma(H)$, whose edges are labeled by the element of S, and which carries all the essential information about the subgroup H. Further, if H is finitely generated then $\Gamma(H)$ is finite and easy to construct. A finite labeled directed graph can be identified with a finite state automaton and thus for each finitely generated subgroup H the constructed $\Gamma(H)$ is a finite state automaton which accepts precisely the reduced words in S which belong to H.

This automaton approach provides an extremely useful method to handle computational and algorithmic problems in free groups. Topologically a free group on X can be identified as the fundamental group of a wedge of circles of cardinality $|X|$. The graph automaton $\Gamma(H)$ represents the topological core of the covering space relative to H, of this wedge of circles. Algebraically $\Gamma(H)$ can be viewed as the essential part of the relative coset Cayley graph of F/H with respect to X, also known as the *Schreier graph* F/H with respect to X. It is a graph where the vertices are the right cosets Hg for $g \in F$ and the edges are of the form (Hg, Hgx_i), $x_i \in S$. The free group F acts on its standard Cayley graph which is a regular tree. There is a unique H-invariant subtree $T(H)$ that is minimal among H-invariant subtrees containing the identity. The graph automaton $\Gamma(H)$ can be identified with the quotient graph $T(H)/H$. Now, let $F = F(X)$ be a free group with free basis $X = \{x_1, x_2, \dots, x_m\}$ and let H be a subgroup of F generated by elements $h_1, h_2, \dots, h_n \in F$. We show how to construct a finite automaton $\Gamma(H)$ which accepts precisely the reduced words in $S = X \cup X^{-1}$ which belong to H.

Before we do this in detail we need to describe finite state automata in terms of directed graphs, but we keep especially in mind that we consider groups.

Let $S = X \cup X^{-1}$ be a finite subset of a group G and Γ be a directed S-labeled graph with vertex set $V(\Gamma)$, that is, the edges of Γ are labeled by S-elements. Γ has the *one way reading property*, denoted *OR*, if for each vertex $v \in V(\Gamma)$ and each label $s \in S$, there exists at most one edge in Γ starting at v with label s. The tuple (Γ, s, v_0, Y, μ) defines a finite state automaton $\mathcal{A}$ where Γ is a directed S-graph with property *OR*, $v_0 \in V(\Gamma)$ is a distinguished vertex, $Y \subset V(\Gamma)$ and μ is a *transition function* (defined below). The automaton $\mathcal{A}$ has the set of states $Q = V(\Gamma) \cup \bigoplus$, where $\bigoplus$ is the failure state, the alphabet S, the starting state $v_0 \in V(\Gamma)$, the accepting states Y and the transition function $\mu : Q \times S \to Q$,

$$\mu(v, s) = \begin{cases} w, & \text{if } v \in V(\Gamma) \text{ and there is an edge } v \text{ to } w \text{ with label } s, \\ \bigoplus, & \text{otherwise.} \end{cases}$$

Let S^* denote the set of all words in the alphabet S. Then a subset $L \subset S^*$ is *regular* if there is a finite state automaton which has L as its accepting language.

If L is regular then the set of words I that consists of reduced words from L is also regular.

Example 3.11.1. Let Γ be the following directed labeled S-graph in Figure 3.73 where $S = \{a, a^{-1}\}$.

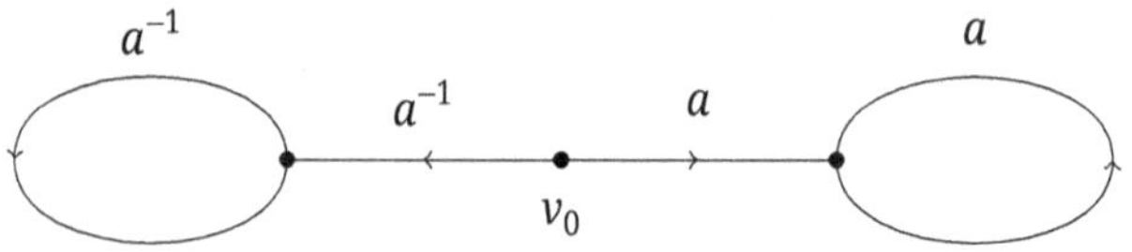

Figure 3.73: Directed S-graph.

Let $Y = V(\Gamma)$. Then Γ corresponds to a finite automaton $\mathcal{A}$ which accepts all words of the form a^n, $n \in \mathbb{Z}$. Hence, the language $L(\mathcal{A})$ is precisely the infinite cyclic groups on a. In this case we say that $\mathcal{A}$ accepts the infinite cyclic group generated by a.

Note that for any group G with valid generating system S the Cayley graph $\Gamma(G, S)$ satisfies the one way reading property. If we chose $v_0 = \{1\}$ and $Y = \{1\}$ then the Cayley graph will lead to an automaton $\mathcal{A}$ which is infinite if G is infinite and which accepts a word w in the alphabet S if and only if $w = 1$ in G. To start the construction of $\Gamma(H)$ we begin with a finite non-oriented labeled graph Γ with labels from $S = X \cup X^{-1}$, a distinguished vertex 1 and n simple loops $p_1, p_2, \ldots, p_n$ with labels $h_1, h_2, \ldots, h_n$. In a labeled graph, if x is a label on an edge e then x^{-1} is the label of the inverse edge e^{-1}, defined by $e^{-1}(t) = e(1-t)$, $t \in [0, 1]$. In the graph Γ we place an orientation by making the positive edges those with labels $x_1, x_2, \ldots, x_m$.

As an example consider the free group F on two generators x, y, and let H be the subgroup generated by the words

$$h_1 = xyx^{-1}y^{-1}x \quad \text{and} \quad h_2 = xy^{-1}.$$

The corresponding graph Γ is pictured in Figure 3.74.

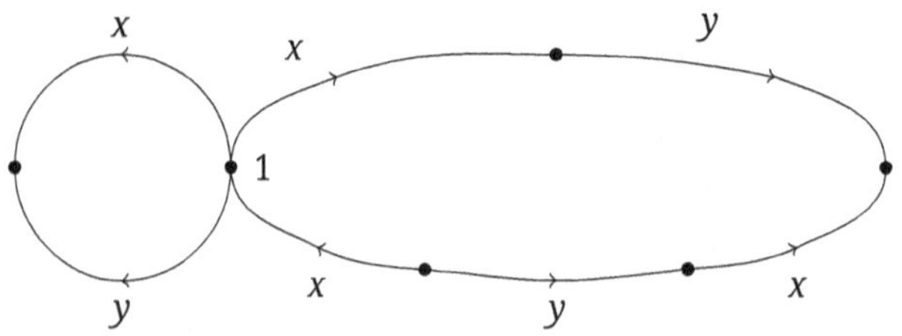

Figure 3.74: Graph Γ.

Notice that every loop p in Γ at the vertex 1 is a product of finitely many simple loops $p_1^{\pm 1}, p_2^{\pm 1}, \dots, p_n^{\pm 1}$. It follows that the label of p is a finite product of the generators $h_1^{\pm 1}, h_2^{\pm 1}, \dots, h_n^{\pm 1}$ and hence p defines an element of H. Conversely if a word $w \in F$ is in H there must be a loop at 1 with label w. Hence for any subgroup $H = \langle h_1, h_2, \dots, h_n \rangle$ the graph Γ has the following property:

Property L: An element $f \in F$ belongs to H if and only if there exists a loop in Γ at the vertex 1 with label w where the word w defines $f \in F$.

Notice that in general the graph Γ does not satisfy the one way reading property *OR* and therefore does not give rise immediately to an automaton. The problem, as illustrated by the example above, is that there may be two different edges starting at the vertex 1 with the same label. To correct this problem we *fold* the graph.

Suppose that in a labeled graph Γ there are two different edges starting at the same vertex v and labeled by the same label x. We transform the graph Γ into the graph Γ' by identifying the two edges and their terminal vertices. This is called a *folding* at v. If there are no such vertices then we say that the graph is *completely folded*. It is clear that a finite graph can be folded in a finite number of steps to a completely folded graph. What is crucial is the following which is straightforward.

Lemma 3.11.2. *If a graph satisfies Property L and Γ' is a folding of Γ, then Γ' also satisfies Property L.*

Now let Γ be a graph constructed from the subgroup H. The finite automaton that goes with H is $\Gamma(H)$, the completely folded graph from Γ.

Now $\Gamma(H)$ satisfies both the one way reading property and Property L. Hence it forms an automaton. We put $v_0 = 1$ and $Y = \{1\}$. Then it follows that $\Gamma(H)$ is a finite automaton which accepts H, that is, $\Gamma(H)$ accepts a reduced word in $S = X \cup X^{-1}$ if and only if this word defines an element of H. Further, this construction is algorithmically effective and hence we have the following theorem.

Theorem 3.11.3. *Let $H = \langle h_1, h_2, \dots, h_n \rangle$ be a finitely generated subgroup of a free group $F = F(x_1, x_2, \dots, x_m)$. Then one can effectively construct a finite automaton $\Gamma(H)$ which accepts H.*

There are many classical results on free groups that can be proved more easily using this automaton (see [143] for many of them). We mention several further items here.

There are at most

$$cm(|h_1| + |h_2| + \cdots + |h_n|)^2$$

steps necessary to construct $\Gamma(H)$ where c is a constant. This provides a bound for the solution of the membership problem (or generalized word problem) in a free group. Hence we have in addition to Theorem 1.2.14 the following.

Corollary 3.11.4. *Let F be a free group of finite rank. Then the membership problem in F is decidable in quadratic time complexity. That is, given words $w, h_1, h_2, \ldots, h_n \in F$, one can effectively decide whether or not $w \in \langle h_1, h_2, \ldots, h_n \rangle$ and it takes at most $C(|h_1| + |h_2| + \cdots + |h_n|)^2$ steps, where C is a constant which depends linearly on the rank of F.*

This method also provides another proof of the Nielsen Theorem 1.2.9 for finitely generated subgroups of free groups of finite rank.

The graph automaton $\Gamma(H)$ also provides a method to determine if a subgroup H is of finite index. In particular, we say that an S-labeled graph Γ is regular if for any vertex $v \in \Gamma$ and any label $x \in S$ there exists an edge in Γ starting at v and with label x.

Theorem 3.11.5. *A finitely generated subgroup H of F has finite index if and only if $\Gamma(H)$ is a regular S-labeled graph.*

Proof. Suppose $\Gamma(H)$ is not a regular S-labeled graph. Then there exists a vertex $v \in \Gamma(H)$ and a label $x \in S$ such that there is no edge in $\Gamma(H)$ starting at v and labeled by x. Let p be a reduced path from 1 to v in $\Gamma(H)$ and let g be the label of the path p. Choose a letter $y \in S$ not equal to x or x^{-1} and define a word w as follows:

$$w = \begin{cases} gx, & \text{if } g \text{ does not begin with } x^{-1}, \\ gxy, & \text{otherwise.} \end{cases}$$

Clearly w is a cyclically reduced word and hence any power w^k is reduced as written. Further, $w \notin H$ since we cannot read $fx \in \Gamma(H)$ where $f \in F$ is defined by the word w. It follows that

$$hw^k = hw^\ell \Rightarrow w^{k-\ell} \in H \Rightarrow k = \ell.$$

Therefore in this case H must be of finite index in F.

Conversely, suppose that $\Gamma(H)$ is a regular S-labeled graph. Let T be a maximal subtree of $\Gamma(H)$. For any vertex $v \in \Gamma(H)$ denote by $\gamma(v)$ the path from 1 to v inside T. If w is a reduced word in S then we can read off w inside $\Gamma(H)$ starting at 1.

Let v_w be the vertex where we finish reading w inside $\Gamma(H)$. Then $w\gamma(v_w)^{-1}$ is the label of a loop in $\Gamma(H)$ at 1 and hence $w\gamma(v_w)^{-1} \in H$. Therefore, $Hw = H\gamma(v_w)$, and $\gamma(v_w)$ is a representative of w modulo H. Put

$$R_\Gamma = \{\gamma(v) \mid v \in \Gamma(H)\},$$

then R_Γ is a transversal of H in F. Since $\Gamma(H)$ is a finite graph, it follows that $|R_\Gamma|$ is finite, and H is of finite index in F. □

We close this section by briefly explaining the relationship between $\Gamma(H)$ and the Cayley graph $\Gamma(F, S)$ of F with respect to S.

Let G be a group, S be a valid generating system of G and $\Gamma = \Gamma(G, S)$ be the Cayley graph of G with respect to S. The group G acts on $\Gamma(G, S)$ by left multiplication $g : v \mapsto gv$, where here v is a vertex in Γ and $g \in G$. This map is an isomorphism of Γ.

Let H be a subgroup of G. Then H also acts on Γ. Denote by Γ_H the quotient graph under this action $\Gamma_H = \Gamma/H$. The vertices of Γ_H are the orbits of the action of H on Γ, and hence they are just the right cosets of H in G:

$$Hf = \{g \in G \mid g = hf \text{ for some } h \in H\}.$$

There is an edge with label $x \in S$ from Hf into Hg if and only if $Hfx = Hg$. This turns Γ_H into a S-labeled graph. Notice that every loop in Γ_H at the vertex H defines an element of H. Moreover, Γ_H has the one way reading property. Hence, if we put $v_0 \in H$, $Y = \{H\}$ then Γ_H gives rise to an automaton that accepts H. If G is a free group then $\Gamma(H)$ is embeddable into Γ_H and any loop of H in Γ_H is a product of loops in $\Gamma(H)$.

Bibliography

[1] M. Aab and G. Rosenberger. Subgroup separable free products with cyclic amalgamations. *Results Math.*, 28:185–194, 1995.

[2] M. T. Abu Osman and G. Rosenberger. Embedding property of surface groups. *Bull. Malays. Math. Soc. II Ser.*, 3:21–27, 1980.

[3] P. Ackermann, V. große Rebel, and G. Rosenberger. *Algebraische Strukturen und universelle Algebren für Informatiker*. Shaker, 2004.

[4] P. Ackermann, G. Rosenberger, and M. Näätänen. Fuchsian groups with signature $(0; 2, 2, 2, q)$. In *Recent Advances in Group Theory and Low-Dimensional Topology*, 1–9. Heldermann Verlag, 2003.

[5] S. I. Adyan. The unsolvability of certain algorithmic problems in the theory of groups. *Tr. Mosk. Mat. Obŝ.*, 2:231–298, 1957.

[6] R. B. J. T. Allenby. Conjugacy separability of a class of one-relator groups. *Proc. Am. Math. Soc.*, 116:621–628, 1992.

[7] J. Alonso, T. Brady, D. Cooper, V. Ferlini, M. Lustig, M. Mihalik, M. Shapiro, and H. Short. Notes on word hyperbolic groups. In *Group Theory from a Geometrical Viewpoint*, 3–63, 1991.

[8] G. Arzhantseva, A. Minasyan, and D. Osin. The SQ-universality and residual properties of relatively hyperbolic groups. *J. Algebra*, 315:165–177, 2007.

[9] H. Bass. Finitely generated subgroups of $GL(2, \mathbb{C})$. In *The Smith Conjecture*. J. Morgan, Wiley, New York, 1984.

[10] B. Baumslag. Residually free groups. *Proc. Lond. Math. Soc.*, 17:402–418, 1967.

[11] B. Baumslag, F. Levin, and G. Rosenberger. A cyclically pinched product of free groups which is not residually free. *Math. Z.*, 21:533–534, 1993.

[12] B. Baumslag and S. Pride. Groups with two more generators than relators. *J. Lond. Math. Soc.*, 17:425–426, 1978.

[13] G. Baumslag. On the residual finiteness of generalized free products of nilpotent groups. *Trans. Am. Math. Soc.*, 106:193–209, 1963.

[14] G. Baumslag. On generalized free products. *Math. Z.*, 78:423–438, 1962.

[15] G. Baumslag. Residual nilpotence and relations in free groups. *J. Algebra*, 2:271–285, 1965.

[16] G. Baumslag. Groups with the same lower central series as a relatively-free group II. *Trans. Am. Math. Soc.*, 142:507–538, 1969.

[17] G. Baumslag. Parafree groups. *Prog. Math.*, 248:1–14, 2005.

[18] G. Baumslag, O. Bogopolski, B. Fine, A. Gaglione, G. Rosenberger, and D. Sepllmann. On some finiteness properties in infinite groups. *Algebra Colloq.*, 15:1–22, 2008.

[19] G. Baumslag and S. Cleary. Parafree one-relator groups. *J. Group Theory*, 7:191–201, 2006.

[20] G. Baumslag, B. Fine, M. Kreuzer, and G. Rosenberger. *A Course in Mathematical Cryptography*. De Gruyter, 2015.

[21] G. Baumslag, B. Fine, and G. Rosenberger. One-relator groups: An overview. In *Groups St. Andrews 2017 in Birmingham*, 119–157. London Math. Soc. Lecture Note Series #445, 2019.

[22] G. Baumslag, J. W. Morgan, and P. Shalen. Generalized triangle groups. *Math. Proc. Camb. Philos. Soc.*, 102:25–31, 1987.

[23] G. Baumslag and P. Shalen. Groups whose three generator subgroups are free. *Bull. Aust. Math. Soc.*, 40:163–174, 1989.

[24] C. Baer. *Klassifikation arithmetischer Fuchsscher Gruppen der Signatur* $(0; e_1, e_2, e_3, e_4)$. Dissertation, Dortmund, 2001.

[25] G. Beardon. *The Geometry of Discrete Groups*. Springer, 1983.

[26] V. V. Benyash-Krivets. Decomposing finitely generated groups into free products with amalgamation. *Sb. Math.*, 192:163–186, 2001.

[27] M. Bestvina and M. Feighn. A combination theorem for negatively curved groups. *J. Differ. Geom.*, 35:85–101, 1992.

https://doi.org/10.1515/9783111340043-004

[28] M. Bestvina and M. Handel. Train tracks and automorphisms of free groups. *Ann. Math.*, 135:1–51, 1992.

[29] O. Bogopolski. A surface analogue of a theorem of Magnus. *Contemp. Math.*, 352:55–89, 2005.

[30] M. Boileau, D. J. Collins, and H. Zieschang. Genus 2 decompositions of small Seifert manifolds. *Ann. Inst. Fourier*, 4:1005–1024, 1991.

[31] A. Borel. Commensurability classes and volumes of hyperbolic three-manifolds. *Ann. Sc. Norm. Pisa*, 8:1–33, 1981.

[32] A. Borel and Harish-Chandra Mehrota. Arithmetic subgroups of algebraic groups. *Ann. Math. II Ser.*, 75:485–535, 1962.

[33] A. K. Bousfield and D. M. Kan. *Homotopy Limits, Completion and Localization*. Springer, 1972.

[34] M. Bridson and J. Howie. Normalisers in limit groups. *Math. Ann.*, 337:385–394, 2007.

[35] M. R. Bridson. *Geodesics and Curvature in Metric Simplicial Complexes*. Cornell University, 1991.

[36] M. R. Bridson and A. W. Reid. Nilpotent completions of groups, Grothendieck pairs, and four problems of Baumslag. *Int. Math. Res. Not.*, 2015:2111–2140, 2015.

[37] K. Brown. *Cohomology of Groups*. Springer, 1982.

[38] A. M. Brunner. A group with an infinite number of Nielsen inequivalent one-relator presentations. *J. Algebra*, 42:81–84, 1976.

[39] A. M. Brunner, R. G. Burns, and D. Solitar. The subgroup separability of free products of two groups with cyclic amalgamation. *Contemp. Math.*, 33:90–115, 1984.

[40] I. Bumagin. Examples of amalgamated products of groups. *Isr. J. Math.*, 124:279–284, 2001.

[41] I. Bumagin, O. Kharlampovich, and A. Myasnikow. The isomorphism problem for finitely generated fully residually free groups. *J. Pure Appl. Algebra*, 208:961–977, 2007.

[42] G. Burde. Zur Theorie der Zöpfe. *Math. Ann.*, 151:101–107, 1963.

[43] J. O. Button. Acylindrical hyperbolicity, non simplicity and SQ-universality of groups splitting over $\mathbb{Z}$. *J. Group Theory*, 20:371–383, 2017.

[44] J. O. Button and R. P. Kropholler. Nonhyperbolic free-by-cyclic and one-relator groups. *N.Y. J. Math.*, 22:755–774, 2016.

[45] C. M. Campbell, E. F. Robertson, N. Ruskuc, and R. M. Thomas. Automatic semigroups. *Theor. Comput. Sci.*, 250:36–391, 2001.

[46] T. Camps, M. Dörfer, and G. Rosenberger. A recurrence relation for the number of free subgroups in free products of cyclic groups. In *Ascpects of Infinite Groups*, 54–74. Algebra and Discrete Mathematics #1. World Scientific Publ., 2008.

[47] T. Camps, V. große Rebel, and G. Rosenberger. *Einführung in die Kombinatorische und die Geometrische Gruppentheorie*. Heldermann Verlag, 2008.

[48] T. Camps, S. Kühling, and G. Rosenberger. *Einführung in die mengentheoretische und die algebraische Topologie*. Heldermann, Lemgo, 2006.

[49] J. W. Cannon. The combinatorial structure of cocompact discrete hyperbolic groups. *Geom. Dedic.*, 16:755–774, 1984.

[50] J. W. Cannon, D. B. A. Epstein, D. F. Holt, M. S. Paterson, and W. P. Thursten. *Word Processing in Groups*. Jones and Bartlett, 1972.

[51] C. Carstensen-Opitz, B. Fine, A. Moldenhauer, and G. Rosenberger. *Abstract Algebra*, 2nd edition. De Gruyter, 2019.

[52] M. Casals-Ruiz, I. Kazachkov, and A. Zakharov. Commensurability of Baumslag–Solitar groups, 2019. arXiv:1910.02117v1.

[53] I. Chiswell. Euler characteristics of groups. *Math. Z.*, 147:1–11, 1976.

[54] L. Ciobanu, B. Fine, and G. Rosenberger. The surface group conjecture: Cyclically pinched and conjugacy pinched one-relator groups. *Results Math.*, 64:1175–1184, 2013.

[55] L. Ciobanu, B. Fine, and G. Rosenberger. Groups generalizing a theorem of Benjamin Baumslag. *Commun. Algebra*, 677:656–667, 2016.

[56] D. Collins and H. Zieschang. Rescuing the Whitehead method for free products I. *Math. Z.*, 185:487–504, 1984.
[57] D. Collins and H. Zieschang. Rescuing the Whitehead method for free products II. *Math. Z.*, 186:335–361, 1984.
[58] D. Collins and F. Perraud. Cohomology and finite subgroups of small cancellation quotients of free products. *Math. Proc. Camb. Philos. Soc.*, 37:243–259, 1984.
[59] D. Collins and H. Zieschang. On the Nielsen method in free products with amalgamated subgroups. *Math. Z.*, 197:97–118, 1987.
[60] D. Collins and H. Zieschang. A presentation for the stabilizer of an element in a free product. *J. Algebra*, 106(106):53–77, 1987.
[61] D. Collins and H. Zieschang. Combinatorial group theory and fundamental groups. In *Algebra VII*. Springer, 1993.
[62] D. J. Collins. Representations of the amalgamated free product of two infinite cycles. *Math. Ann.*, 237:99–120, 1978.
[63] D. J. Collins and F. Levin. Automorphisms and hopficity in certain Baumslag–Solitar groups. *Arch. Math.*, 40:385–400, 1987.
[64] L. Comerford, C. Edmunds, and G. Rosenberger. Commutators as powers in free products. *Proc. Am. Math. Soc.*, 122:47–52, 1994.
[65] D. A. Cox. *Primes of the Form* $x^2 + ny^2$. J. Wiley & Sons, 2013.
[66] H. S. M. Coxeter and W. O. J. Moser. *Generators and Relations for Discrete Groups*. Springer, 1972.
[67] P. Csörgö, B. Fine, and G. Rosenberger. On certain equations in free groups. *Acta Sci. Math.*, 68:895–905, 2002.
[68] M. Culler and P. Shalen. Varieties of group representations for discrete groups and splittings of three manifolds. *Ann. Math.*, 117:109–147, 1983.
[69] F. Dahmani and V. Guirardel. The isomorphism problem for all hyperbolic groups. *Geom. Funct. Anal.*, 21:223–300, 2011.
[70] M. Dörfer and G. Rosenberger. Zeta functions of finitely generated nilpotent groups. In A. C. Kim, D. L. Johnson, editors, *Groups Korea '94*, 35–46. Walter de Gruyter, 1995.
[71] S. Dowdall and S. J. Taylor. Rank and Nielsen equivalence in hyperbolic extensions. *Int. J. Algebra Comput.*, 29:615–625, 2019.
[72] C. Drutu and M. Sapir. Non-linear residually finite groups. arXiv:math/0405470v3, 2004.
[73] C. C. Edmunds. A short combinational proof of the Vaught conjecture. *Can. Math. Bull.*, 18:607–608, 1983.
[74] A.-K. Engel, B. Fine and G. Rosenberger. Test elements, generic elements and almost primitivity in free products. *Algebra Colloq.*, 23:263–280, 2016.
[75] B. Farb and L. Mosher. A rigidity theorem for the solvable Baumslag–Solitar groups. *Invent. Math.*, 131:419–451, 1998.
[76] B. Fine. A note on the two-square theorem. *Can. Math. Bull.*, 20:93–94, 1977.
[77] B. Fine, A. Gaglione, A. Moldenhauer, G. Rosenberger, and D. Spellman. *Geometry and Discrete Mathematics*. De Gruyter, 2018.
[78] B. Fine, A. Gaglione, A. Myasnikov, G. Rosenberger, and D. Spellman. *The Elementary Theory of Groups*. De Gruyter, 2014.
[79] B. Fine, A. Gaglione, G. Rosenberger, and D. Spellman. On CT and CSA groups and related ideas. *J. Group Theory*, 19:923–940, 2016.
[80] B. Fine, A. Gaglione, G. Rosenberger, and D. Spellman. Commutative transitivity and the CSA property. In P. Baginski, B. Fine and A. Gaglione, editors, *Infinite Group Theory*, 95–117, 2018.
[81] B. Fine, J. Howie, and G. Rosenberger. One-relator quotients and free products of cyclics. *Proc. Am. Math. Soc.*, 102:1–6, 1988.
[82] B. Fine, O. G. Kharlampovich, A. G. Myasnikov, V. N. Remeslennikov, and G. Rosenberger. On the surface group conjecture. *Scientia, Ser. A, Math. Sci.*, 15:1–15, 2007.

[83] B. Fine, M. Kreuzer, and G. Rosenberger. Faithful real representations of cyclically pinched one-relator groups. *Int. J. Group Theory*, 3:1–8, 2014.

[84] B. Fine, A. Moldenhauer, and G. Rosenberger. Faithful real presentations of groups of F-type. *Int. J. Group Theory*, 9:143–155, 2020.

[85] B. Fine, A. Myasnikov, V. große Rebel, and G. Rosenberger. A classification of conjugately separated Abelian, commutative transitive, and restricted Gromov one-relator groups. *Results Math.*, 50:183–193, 2007.

[86] B. Fine, A. Myasnikov, V. große Rebel, and G. Rosenberger. Erratum to: A classification of conjugately separated Abelian, commutative transitive, and restricted Gromov one-relator groups. *Results Math.*, 61:421–442, 2012.

[87] B. Fine, A. Rosenberger, and G. Rosenberger. A note on Lyndon properties in one-relator groups. *Results Math.*, 59:239–250, 2011.

[88] B. Fine, A. Rosenberger, and G. Rosenberger. Quadratic properties in group amalgams. *J. Group Theory*, 657–671, 2011.

[89] B. Fine and G. Rosenberger. Conjugacy separability and related questions. *Contemp. Math.*, 109:11–18, 1990.

[90] B. Fine and G. Rosenberger. Generalizing algebraic properties of Fuchsian groups. In Groups St. Andrews 1989, 124–147. London Math. Soc. Lecture Notes Ser. #159, 1991.

[91] B. Fine and G. Rosenberger. On restricted Gromov groups. *Commun. Algebra*, 20:2171–2182, 1992.

[92] B. Fine and G. Rosenberger. Classification of all generating pairs of two generator fuchsian groups. In *Groups St. Andrews 1993*, 205–232. London Math. Soc. Lecture Note Ser. #211, 1995.

[93] B. Fine and G. Rosenberger. *Algebraic Generalization of Discrete Groups*. Marcel Dekker, 1999.

[94] B. Fine and G. Rosenberger. Surface groups with Baumslag doubles. *Proc. Edinb. Math. Soc.*, 54:91–97, 2011.

[95] B. Fine and G. Rosenberger. Faithful representations of limit groups II. *Groups Complex. Cryptol.*, 5:91–96, 2013.

[96] B. Fine and G. Rosenberger. Groups which admit essentially faithful representations. *N.Z. J. Math.*, 25:1–7, 1996.

[97] B. Fine, G. Rosenberger, D. Spellman, and M. Stille. Test words, generic elements and almost primitivity. *Pac. J. Math.*, 190:277–297, 1999.

[98] B. Fine, G. Rosenberger, and M. Stille. Nielsen transformations and applications: A survey. In *Groups – Korea 94*, 69–105, 1995.

[99] B. Fine, G. Rosenberger, and M. Stille. The isomorphism problem for a class of para-free groups. *Proc. Edinb. Math. Soc.*, 40:541–549, 1997.

[100] B. Fine, G. Rosenberger, and R. Weidmann. Two generator subgroups of free products with commuting subgroups. *J. Pure Appl. Algebra*, 172:193–204, 2001.

[101] B. Fine, F. Röhl, and G. Rosenberger. On HNN groups whose three generator subgroups are free. In *Infinite Groups and Group Rings*, 13–37. World Scientific, 1993.

[102] B. Fine, F. Röhl, and G. Rosenberger. A Freiheitssatz for certain one-relator amalgamated products. In *Combinatorial Geometric Group Theory*, 73–86. Lecture Notes Ser. #204. London Math. Soc., 1995.

[103] B. Fine and M. Tretkoff. On the SQ-universality of HNN-groups. *Proc. Am. Math. Soc.*, 73:283–290, 1979.

[104] J. R. Ford. *Automorphic Functions*. Chelsea, New York, 1951.

[105] W. Fulton. *Algebraic Topology*. Springer, 1995.

[106] G. Gardam, D. Kielak, and A. D. Logan. Private communication.

[107] D. Gaborian, G. Levitt, and F. Paulin. Pseudographs of isometries of $\mathbb{R}$ and Rips's theorem on free actions on $\mathbb{R}$-trees. *Isr. J. Math.*, 87:403–428, 1994.

[108] A. Gaglione and D. Spellman. Even more model theory of free groups. In *Infinite Groups and Group Rings*, 37–40, 1993.

[109] A. Gaglione and D. Spellman. Some model theory of free groups and free algebras. *Houst. J. Math.*, 21:225–245, 1993.

[110] A. Gaglione and D. Spellman. More model theory of free groups. *Houst. J. Math.*, 21:225–245, 1995.

[111] F. W. Gehring, C. Maclachlan, G. J. Martin, and A. W. Reid. Arithmeticity, discretness and volume. *Trans. Am. Math. Soc.*, 349:3611–3643, 1997.

[112] S. M. Gersten. On fixpoints of automorphism of finitely generated free groups. *Bull. Am. Math. Soc.*, 8:451–454, 1983.

[113] S. M. Gersten. On Whitehead's algorithm. *Bull. Am. Math. Soc.*, 10:281–284, 1984.

[114] E. Ghys and P. de la Harpe. *Sur les groupes hyperbolique d'après Mikhael Gromov*. Birkhäuser, 1990.

[115] D. Gildenhuys, O. Kharampovich, and A. Myasnikov. CSA groups and separated free construction. *Bull. Aust. Math. Soc.*, 52:63–84, 1995.

[116] G. Gratzer. *Universal Algebra*. Van Nostrand, Princeton, 1968.

[117] L. Greenberg. *Commensurable Groups of Moebius Transformations*. Princeton University Press, 1974.

[118] M. Gromov. Hyperbolic groups. In S. M. Gersten, editor, *Essays in Group Theory*, 75–263. Mathematical Sciences Research Institute Publications #8, 1987.

[119] F. Grunewald, D. Segal, and G. Smith. Subgroups of finite index in nilpotent groups. *Invent. Math.*, 9:185–223, 1988.

[120] M. Gutierrez. Homology and completions of groups. *J. Algebra*, 51:354–366, 1978.

[121] M. Hall. Subgroups of finite index in free groups. *Can. J. Math.*, 1:187–190, 1949.

[122] P. R. Halmos. *Naive Set Theory*. Springer-Verlag, 1974.

[123] H. Helling. Bestimmung der Kommensurabilitätsklasse der Hilbertsche Modulgruppe. *Math. Z.*, 92:269–280, 1966.

[124] H. Helling. Diskrete Untergruppen von $SL(2,\mathbb{R})$. *Invent. Math.*, 17:217–229, 1972.

[125] H. Helling. On the commensurability class of the rational modular group. *J. Lond. Math. Soc.*, 2:67–72, 1970.

[126] J. Hempel. 3-*Manifolds*. Princeton Univ. Press, 1976.

[127] A. Hennekemper, S. Kühling, and G. Rosenberger. *Einführung in die Funktionentheorie*. Shaker, 2003.

[128] P. J. Higgens and R. C. Lyndon. Equivalence of elements under automorphisms of a free group. *J. Lond. Math. Soc. (2)*, 8:254–258, 1974.

[129] G. Higman. A finitely related group with an isomorphic factor group. *J. Lond. Math. Soc.*, 26:59–61, 1951.

[130] P. Hill and S. J. Pride. Commutators, generators and conjugacy equations in groups. *Arch. Math.*, 44:1–14, 1985.

[131] J. Howie. The quotient of a free product of groups by a single high-powered relator I. *Proc. Lond. Math. Soc.*, 59:507–540, 1988.

[132] J. Howie. The quotient of a free product of groups by a single high-powered relator II. *Proc. Lond. Math. Soc.*, 61:33–92, 1990.

[133] J. Howie. Some results on one-relator surface groups. *Bol. Soc. Mat. Mexicana*, 10:255–262, 2004.

[134] J. Howie. Generalized triangle groups of types $(3,q;2)$. *Algebra Discrete Math.*, 15:1–18, 2013.

[135] J. Howie and A. Duncan. One-relator products with high powered relators. In *Proc. of the Geometric Group Theory Symposium*, Univ. of Sussex, 1990.

[136] A. Hulpke, T. Kuusalo, M. Näätänen, and G. Rosenberger. On orbifold coverings of genus 2 surfaces. *Scientia, Ser. A, Math. Sci.*, 11:45–55, 2005.

[137] A. Jaikin-Zapirain and M. Linton. On the coherence of one-relator groups and their group algebras. arXiv:2303.05976v3, 2023.

[138] W. H. Jaco and P. B. Shalen. *Seifert Fibered Spaces in* 3-*Manifolds*. Memoirs AMS #192, 1979.

[139] T. Jørgensen. A note on subgroups of $SL(2,\mathbb{C})$. *Q. J. Math. Oxf. Ser. (2)*, 18:209–212, 1977.

[140] A. Juhász and G. Rosenberger. On the combinatorial curvature of groups of F-type and other one-relator products of cyclics. *Contemp. Math.*, 169:373–384, 1994.

[141] R. N. Kalia and G. Rosenberger. Über Untergruppen ebener diskontinuierlicher Gruppen. *Contemp. Math.*, 33:308–327, 1984.

[142] I. Kapovich. Subgroup properties of fully residually free groups. *Trans. Am. Math. Soc.*, 354:335–362, 2001.

[143] I. Kapovich and A. Myasnikov. Stallings foldings and subgroups of free groups. *J. Algebra*, 248:608–668, 2002.

[144] I. Kapovich and H. Short. Greenberg's theorem for quasiconvex subgroups of word hyperbolic groups. *Can. J. Math.*, 48:1124–1144, 1996.

[145] I. Kapovich and R. Weidmann. Nielsen equivalence in small cancellation groups. arXiv:1011.5862v3, 2012.

[146] I. Kapovich and R. Weidmann. On the structure of two-generators hyperbolic groups. *Math. Z.*, 231:783–801, 1999.

[147] I. Kapovich and R. Weidmann. Two generator groups acting on trees. *Arch. Math.*, 73:172–181, 1999.

[148] I. Kapovich and R. Weidmann. Nielsen methods and groups acting on hyperbolic spaces. *Geom. Dedic.*, 98:95–121, 2003.

[149] I. Kapovich and R. Weidmann. Acylindrical assessibility for groups acting on trees. *Math. Z.*, 249:773–782, 2005.

[150] I. Kapovich and R. Weidmann. Nielsen equivalence in a class of random groups. *J. Topol.*, 9:502–534, 2016.

[151] A. Karrass and D. Solitar. The subgroups of a free product of two groups with amalgamated subgroup. *Trans. Am. Math. Soc.*, 150:227–255, 1970.

[152] S. P. Kerckhoff. The Nielsen realization problem. *Ann. Math.*, 117:235–265, 1983.

[153] G. Kern-Isberner and G. Rosenberger. Einige Bemerkungen über Untergruppen der $PSL(2, \mathbb{C})$. *Results Math.*, 6:40–47, 1983.

[154] G. Kern-Isberner and G. Rosenberger. A note on numbers of the form $x^2 + Ny^2$. *Arch. Math.*, 43:148–156, 1984.

[155] G. Kern-Isberner and G. Rosenberger. Normalteiler vom Geschlecht eins in freien Produkten endlicher zyklischer Gruppen. *Results Math.*, 11:272–288, 1987.

[156] O. Kharlampovich and A. Myasnikov. Hyperbolic groups and free constructions. *Trans. Am. Math. Soc.*, 350:571–613, 1998.

[157] O. Kharlampovich and A. Myasnikov. Algebraic geometry over free groups: lifting solutions into generic points. In *Groups, Languages, Algorithms*, 213–318. Contemp. Math. #378, 2005.

[158] O. Kharlampovich and A. Myasnikov. Effective JSJ decompositions. *Contemp. Math.*, 378:87–211, 2005.

[159] O. Kharlampovich and A. Myasnikov. Affine varieties over a free group II. Systems of trianggular quadi-quadratic form and a description of residually free groups. *J. Algebra*, 200:517–569, 1998.

[160] G. Kim. Cyclic subgroup separability of HNN extensions. *Bull. Korean Math. Soc.*, 30:285–293, 1993.

[161] S. H. Kim and S. Oum. Hyperbolic surface subgroups of one-ended doubles of free groups. *J. Topol.*, 7:927–947, 2014.

[162] J. Konieczny, G. Rosenberger, and J. Wolny. Tame almost primitive elements. *Results Math.*, 38:116–129, 2000.

[163] M. Kreuzer and S. Kühling. *Logik für Informatiker*. Pearson Studium, 2006.

[164] M. Kreuzer and G. Rosenberger. Growth in Hecke groups. In *Riemann and Klein Surfaces, Automorphisms, Symmetries and Moduli Spaces*, 261–280. AMS Contemporary Mathematics #629, 2014.

[165] M. Kreuzer and G. Rosenberger. On the numbers of the form $x^2 + 11y^2$. In *Elementary Theory of Groups and Groups Rings and Related Topics*. de Guyter Verlag, 2020.

[166] R. S. Kulkarni. Geometry of free products. *Math. Z.*, 193:613–624, 1986.

[167] F. Levin. Factor groups of the modular group. *J. Lond. Math. Soc.*, 43:195–203, 1968.

[168] F. Levin and G. Rosenberger. On power-commutative and commutation-transitive groups. In *Proceedings of Groups – St. Andrews 1985*, 249–253. London Math. Soc. Lect. Note #121, 1986.

[169] D. D. Long and G. A. Niblo. Subgroup separability and 3-manifold groups. *Math. Z.*, 207:209–215, 1991.
[170] K. I. Lossov. SQ-universality of free products with amalgamated finite subgroups. *Sib. Math. J.*, 27:890–899, 1986.
[171] L. Louder. Nielsen equivalence of generating sets for closed surface groups. *Preprint*, 2010.
[172] L. Louder and H. Wilton. Negative immensions and one-relator groups. *Preprint*.
[173] L. Louder and H. Wilton. One-relator groups with torsion are coherent. arXiv:1805.11976v3, 2020.
[174] M. Lustig. Nielsen equivalence and simple homotopy type. *Proc. Lond. Math. Soc.*, 62:537–562, 1991.
[175] M. Lustig and Y. Moriah. Nielsen equivalence in Fuchsian groups. *Algebraic Geom. Topol.*, 22:189–226, 2022.
[176] R. C. Lyndon. Cohomology theory of groups with a single defining relation. *Ann. Math.*, 52:650–665, 1950.
[177] R. C. Lyndon and P. E. Schupp. *Combinatorial Group Theory*. Springer, 1977.
[178] I. G. Lysenok. On some algorithmic properties of hyperbolic groups. *Math. USSR, Izv.*, 35:145–163, 1990.
[179] M. Maclachlan and A. W. Reid. *The Arithmetic of Hyperbolic 3-Manifolds*. Spinger-Verlag, 2003.
[180] C. Maclachlan and G. Rosenberger. Two-generator arithmetic Fuchsian groups. *Math. Proc. Camb. Philos. Soc.*, 93:383–391, 1983.
[181] C. Maclachlan and G. Rosenberger. Two-generator arithmetic Fuchsian groups II. *Math. Proc. Camb. Philos. Soc.*, 111:7–24, 1992.
[182] C. Maclachlan and G. Rosenberger. Commensurability classes of two-generator Fuchsian groups. In *Discrete Groups and Geometry*, 171–189. LMS Lecture Note Ser. #173, 1992.
[183] C. Maclachlan and G. Rosenberger. Arithmetic Fuchsian groups of signature $(0; m_1, m_2, m_3, m_4)$. *Scientia, Ser. A, Math. Sci.*, 24:1–24, 2013.
[184] K. Madlener and F. Otto. About the descriptive power of certain classes of string-writing systems. *Theor. Comput. Sci.*, 67:143–172, 1989.
[185] W. Magnus, A. Karras, and D. Solitar. *Combinatorial Group Theory*. Wiley, New York, 1966.
[186] A. Majeed. *Two generators subgroups of* $SL(2, \mathbb{C})$. Thesis, Ottawa, 1974.
[187] A. Malcev. On isomorphic matrix representations of infinite groups. *Mat. Sb.*, 8(50):405–422, 1940.
[188] A. Mann. *How Groups Grow*. London Mathematical Society, Lecture Note Series #395, 2011.
[189] J. McCool. A presentation for the automorphism group of a free group of finite rank. *J. Lond. Math. Soc.*, 8:259–266, 1974.
[190] N. S. Mendelsohn and R. Ree. Free subgroups of groups with a single defining relation. *Arch. Math.*, XIX:577–580, 1968.
[191] S. Meskin. Non residually finite one-relator groups. *Trans. Am. Math. Soc.*, 164:105–114, 1972.
[192] C. F. Miller. *On Group-Theoretic Decision Problems and Their Classification*. Princeton Univ. Press, 1971.
[193] M. R. R. Moghaddam, G. Rosenberger and M. A. Rostamyari. Commutative transitivity property to more classes of groups and Lie algebras. In *Elementary Theory of Groups and Group Rings and Related Topics*, 225–231. Proc. in Math. de Gruyter, 2020.
[194] E. E. Moise. *Geometric Topology in Dimensions 2 and 3*. Springer, 1977.
[195] A. Moldenhauer, G. Rosenberger, and K. Rosenthal. On the Tits alternative for a class of finitely presented groups with a special focus on symbolic computation. *Contemp. Math.*, 677:145–169, 2016.
[196] J. R. Munkres. *Elements of Algebraic Topology*. Addison-Wesley, 1984.
[197] T. Nakanishi, M. Näätänen, and G. Rosenberger. Arithmetic Fuchsian groups of signature $(0; e_1, e_2, e_3, e_4)$ with $2 \geq e_1 \geq e_2 \geq e_3, e_4 = \infty$. *Contemp. Math.*, 240:269–277, 1999.
[198] H. Neumann. *Varieties of Groups*. Springer, 1967.
[199] P. M. Neumann. The SQ-universality of some finitely presented groups. *J. Aust. Math. Soc.*, 11:1–6, 1973.
[200] M. Neumann-Brosig. A note on the hyperbolicity of strict pride groups. *Contemp. Math.*, 582:181–185, 2012.

[201] M. Neumann-Brosig and G. Rosenberger. A note on the homology of hyperbolic groups. *Groups Complex. Cryptol.*, 2:203–212, 2010.

[202] B. B. Newman. Some results on one-relator groups. *Bull. Am. Math. Soc.*, 74:568–571, 1968.

[203] M. Neumann-Brosig and G. Rosenberger. *Preprint*.

[204] G. A. Niblo. Fuchsian groups are strongly subgroup separable. *Preprint*.

[205] G. A. Niblo. H. N. N. extensions of a free group $\mathbb{Z}$ which are subgroup separable. *Proc. Lond. Math. Soc.*, 61:18–32, 1990.

[206] J. Nielsen. Die Automorphismen der allgemeinen unendlichen Gruppen in zwei Erzeugenden. *Math. Ann.*, 78:385–397, 1918.

[207] J. Nielsen. Die Isomorphismengruppen der freien Gruppen. *Math. Ann.*, 91:169–209, 1924.

[208] J. O'Neill and E. C. Turner. Test elements and the retract theorem in hyperbolic groups. *N.Y. J. Math.*, 6:107–117, 2000.

[209] K. Ohshika. *Discrete Groups*. AMS, 2002.

[210] N. Peczynski and W. Reiwer. On cancellations in HNN-groups. *Math. Z.*, 158:79–86, 1978.

[211] N. Peczynski, G. Rosenberger, and H. Zieschang. Über Erzeugende ebener diskontinuierlicher Gruppen. *Invent. Math.*, 29:161–180, 1977.

[212] H. Petersson. Über die Konstruktion zykloidaler Kongruenzgruppen in der rationalen Modulgruppe. *J. Reine Angew. Math.*, 250:182–212, 1971.

[213] W. Plesken and G. Rosenberger. Simultanious conjugation in quaternion algebras. *Results in Math.*, 25:120–124, 1994.

[214] S. J. Pride. The isomorphism problem for two generator one-relator groups with torsion is solvable. *Trans. Am. Math. Soc.*, 227:109–139, 1977.

[215] S. J. Pride. The two generator subgroups of one-relator groups with torsion. *Trans. Am. Math. Soc.*, 234:483–496, 1977.

[216] M. O. Rabin. Recursive unsolvability of group theoretical problems. *Ann. Math.*, 67:172–194, 1979.

[217] V. N. Remeslennikov. ∃-Free groups. *Sib. Math. J.*, 30:998–1001, 1989.

[218] D. J. S. Robinson. *A Course in the Theory of Groups*. Graduate Texts in Mathematics #80, 1982.

[219] G. Rosenberger. *Zum Rang- und Isomorphieproblem für freie Produkte mit Amalgam*. Universität Hamburg, 1974.

[220] G. Rosenberger. Applications of Nielsen's reduction method in the solution of combinatorial problems in group theory. In *Homological Group Theory*, 339–358. London Math. Soc. Lecture Notes #36, 1979.

[221] G. Rosenberger. Tschebyscheff-Polynome, Nicht-Kongruenzuntergruppen der Modulgruppe und Fibonacci-Zahlen. *Math. Ann.*, 246:193–204, 1979.

[222] G. Rosenberger. Gleichungen in freien Produkten mit Amalgam. *Math. Z.*, 137:1–12, 1980. Correction *Math. Z.*, 178:579, 1981.

[223] G. Rosenberger. Bemerkungen zu einer Arbeit von R. C. Lyndon. *Arch. Math.*, 40:200–207, 1983.

[224] G. Rosenberger. Some remarks on a paper of A. F. Beardon and P. L. Waterman about strongly discrete subgroups of $SL_2(\mathbb{C})$. *J. Lond. Math. Soc.*, 27:39–42, 1983.

[225] G. Rosenberger. On one-relator groups that are free products of two free groups with cyclic amalgamation. In *Groups St. Andrews (1981)*, 328–344. LMS Lecture Note Ser. #71, 1982.

[226] G. Rosenberger. Über Darstellungen von Elementen und Untergruppen in freien Produkten. In *Proc. Groups – Korea 1983*, 142–160. Springer Lecture Notes in Math. #1098, 1984.

[227] G. Rosenberger. Minimal generating systems for plane discontinuous groups and an equation in free groups. In *Groups – Korea 1988*, 170–186. Lecture Notes in Math. #1398. Springer, 1989.

[228] G. Rosenberger. Linear representations of cyclically pinched one-relator groups. *Sib. Math. J.*, 32:203–206, 1991.

[229] G. Rosenberger. On subgroups of free products of cyclics. *Contemp. Math.*, 131:315–324, 1992.

[230] G. Rosenberger. The isomorphism problem for cyclically pinched one-relator groups. *J. Pure Appl. Algebra*, 95:75–86, 1994.

[231] G. Rosenberger and S. L. Sasse. Residual properties of HNN-extensions with cyclic associated subgroups. *Algebra Colloq.*, 3:91–96, 1996.
[232] G. S. Sacerdote and P. E. Schupp. SQ-universality in HNN-groups and one-relator groups. *J. Lond. Math. Soc.*, 7:733–740, 1974.
[233] M.-P. Schützenberger. Sur l'équation $a^{2+n} = b^{2+m}c^{2+p}$ dans un group libre. *C. R. Acad. Sci. Paris, Ser. I*, 248:2435–2436, 1959.
[234] G. P. Scott. Finitely generated 3-manifold groups are finitely presented. *J. Lond. Math. Soc.*, 6:437–440, 1973.
[235] P. Scott. Subgroups of surface groups are almost geometric. *J. Lond. Math. Soc.*, 17:555–565, 1978.
[236] Z. Sela. The isomorphism problem for hyperbolic groups I. *Ann. Math.*, 141:217–283, 1995.
[237] Z. Sela. Diophantic geometry: Makanin–Razborov diagrams. *Publ. Inst. Hautes études Sci.*, 93:31–106, 2001.
[238] A. Selberg. On discontinuous groups in higher-dimensional symmetric spaces. In *Int. Colloq. Function Theory*, Tata Institute, Bombay, 1960.
[239] P. Shalen. Linear representations of certain amalgamated products. *J. Pure Appl. Algebra*, 15:187–197, 1979.
[240] A. Shenitzer. Decomposition of a group with single defining relator into a free product. *Proc. Am. Math. Soc.*, 6:273–279, 1955.
[241] D. Singerman. Finitely maximal Fuchsian groups. *J. Lond. Math. Soc.*, 6:29–39, 1972.
[242] I. Snopce and S. Tamushevski. Test elements in pro-p groups with applications in discrete groups. *Isr. J. Math.*, 219:783–816, 2017.
[243] I. Snopce and S. Tamushevski. Almost primitive and generic elements in free pro-p groups. *Preprint.*
[244] J. Stallings. Topology of finite graphs. *Invent. Math.*, 71:551–565, 1983.
[245] J. Stallings. Problems about free quotients of groups. In *Geometric Group Theory*, 165–182. Ohio State University Math. Res. Inst. Publ. #3, 1995.
[246] I. A. Stewart. *Obtaining Nielsen reduced sets in free groups*. Technical Report Series No. 293, 1989.
[247] R. Stöcker and H. Zieschang. *Algebraische Topologie*, 2nd edition. B. G. Teubner, 1994.
[248] R. Stöhr. Groups with one more generator than relators. *Math. Z.*, 182:45–47, 1983.
[249] J. Sunaga. Some arithmetic Fuchsian groups with signature $(0; e_1, e_2, e_3, e_4)$. *Tokyo J. Math.*, 20:435–451, 1997.
[250] J. Sunaga. Some arithmetic Fuchsian groups with signature $(0; e_1, e_2, e_3, e_4)$ II. *Saitama Math. J.*, 15:15–46, 1997.
[251] K. Takeuchi. A characterization of arithmetic Fuchsian groups. *J. Math. Soc. Jpn.*, 27:600–612, 1975.
[252] K. Takeuchi. Commensurability classes of arithmetic triangle groups. *J. Fac. Sci., Univ. Tokyo, Sect. 1A, Math.*, 24:201–212, 1977.
[253] K. Takeuchi. Arithmetic triangle groups. *J. Math. Soc. Jpn.*, 29:91–106, 1977.
[254] K. Takeuchi. Arithmetic Fuchsian groups of signature $(1; e)$. *J. Math. Soc. Jpn.*, 35:381–407, 1983.
[255] K. Takeuchi. Subgroups of the modular group with signatur $(0; e_1, e_2, e_3, e_4)$. *Saitama Math. J.*, 14:55–78, 1996.
[256] C. Y. Tang. Some results on one-relator quotients of free products. *Contemp. Math.*, 109:165–177, 1990.
[257] T. Tatuzawa. On a theorem of Siegel. *Jpn. J. Math.*, 24:163–178, 1951.
[258] J. Tits. Free groups in linear groups. *J. Algebra*, 20:250–270, 1972.
[259] M. Tretkoff. Covering spaces, subgroup separability and the generalized M. Hall property. *Contemp. Math.*, 109:179–191, 1990.
[260] B. L. van der Waerden. Free products of groups. *Am. J. Math.*, 70:527–528, 1948.
[261] B. von Querenburg. *Mengentheoretische Topologie*, 2001.
[262] R. Weidmann. *Über den Rang von amalgamierten Produkten und NEC-Gruppen*. Dissertation, Ruhr-Universität, Bochum, 1997.
[263] R. Weidmann. On the rank of amalgamated products and product knot groups. *Math. Ann.*, 312:761–771, 1998.

[264] R. Weidmann. A Grushko theorem for 1-acylindrical splittings. *J. Reine Angew. Math.*, 540:77–86, 2001.
[265] R. Weidmann. The Nielsen method for groups acting on trees. *Proc. Lond. Math. Soc.*, 85:93–118, 2002.
[266] R. Weidmann. A rank formula for amalgamated products with finite amalgam. *Preprint*, 2005.
[267] R. Weidmann. Generating tuples of free products. *Bull. Lond. Math. Soc.*, 39:393–403, 2007.
[268] R. Weidmann. A rank formula for acylindrical splittings. *Ann. Fac. Sci. Toulouse*, XXIV:1057–1078, 2015.
[269] P. J. Weinberger. Exponents of the class group of complex quadratic fields. *Acta Arith.*, 22:117–124, 1973.
[270] J. H. C. Whitehead. On certain sets of elements in a free group. *Proc. Lond. Math. Soc.*, 41:48–56, 1936.
[271] J. H. C. Whitehead. On equivalent sets of elements in a free group. *Ann. Math.*, 37:782–800, 1936.
[272] H. Wilton. Elementarily free groups are subgroup separable. *Proc. Lond. Math. Soc.*, 95:473–496, 2007.
[273] H. Wilton. Hall's theorem for limit groups. *Geom. Funct. Anal.*, 18:271–303, 2008.
[274] H. Wilton. One ended subgroups of graphs of free groups with cyclic edge groups. *Geom. Topol.*, 16:665–683, 2008.
[275] D. Wise. The structure of groups with a quasiconvex hierarchy. *AIMS Math.*, 16:44–55, 2009.
[276] K. Wohlfahrt. An extension of Klein's level concept. *Ill. J. Math.*, 8:529–535, 1964.
[277] A. Zastrow. Construction of an infinitely generated group that is not a free product of surface and Abelian groups but which acts free on an $\mathbb{R}$-tree. *Proc. R. Soc. Edinb.*, 128:433–445, 1998.
[278] H. Zieschang. Über die Nielsensche Kürzungsmethode in freien Produkten mit Amalgam. *Invent. Math.*, 10:4–37, 1970.
[279] H. Zieschang. Generators of the free product with amalgamation of two infinite cyclic groups. *Math. Ann.*, 227:195–221, 1977.
[280] H. Zieschang. *Finite Groups of Mapping Classes of Surfaces*. Lecture Notes in Math. #875. Springer, 1981.
[281] H. Zieschang, E. Vogt, and H.-D. Coldewey. *Surfaces and Planar Discontinous Groups*. Lecture Notes in Math. #835. Springer, 1980.

Index

https://doi.org/10.1515/9783111340043-005

www.ingramcontent.com/pod-product-compliance
Lightning Source LLC
Chambersburg PA
CBHW080648220326
41598CB00033B/5142